Sustainable Smart Cities 2.0

Priyanka Mishra • Ghanshyam Singh

Sustainable Smart Cities 2.0

Enabling Research Toward SDG 11

Springer

Priyanka Mishra
Auckland Park Kingsway Campus
University of Johannesburg
Johannesburg, South Africa

Ghanshyam Singh
Auckland Park Kingsway Campus
University of Johannesburg
Johannesburg, South Africa

ISBN 978-3-032-01101-5 ISBN 978-3-032-01102-2 (eBook)
https://doi.org/10.1007/978-3-032-01102-2

This Springer imprint is published by the registered company Springer Nature Switzerland AG
The registered company address is: Gewerbestrasse 11, 6330 Cham, Switzerland

Competing Interests The authors have no competing interests to declare that are relevant to the content of this manuscript.

Preface

The urbanization of the world is increasing, and by 2030 more than 60% of the world's community will be living in cities. The cities and metropolitan areas generate a substantial amount of global GDP not only as epicentres of economic activity, but at the same time, they also take a large amount of world resources and discharge an oversized volume of carbon. Cities also account for more than 60% of global resource consumption and 70% of greenhouse emissions. Rapid urbanization presents one of many challenges as the world becomes more and more urban, particularly its slum residents. Air pollution is getting worse, infrastructure and services are overburdened, and uncontrolled growth of cities continues. These challenges also threaten the quality of life for residents. Cities, despite these issues, offer a platform for developing innovative solutions. To fully realize the potential of urban environments, governments, private companies, and public entities need to use transformative technologies. At a global level, the smart city movement utilizes technology to drive innovation and support community development, ultimately increasing quality of life. One of its key focus areas is sustainability. Modern technology plays a key role in community development and growth. Social progress can happen through innovation, entrepreneurship, and starting new businesses. By managing processes well, it can also cut public spending on services and provide efficient, high-quality services. Smart cities and sustainable urbanization comprise precisely those strategies that focus on more efficient and more livable living environments. To address some of these challenges, various strategies have been presented, including the use and integration of advanced technologies into cities. Sustainable smart city development will improve people's lives by raising urban living standards through innovative technology. It will minimize environmental impact and use resources efficiently. An integrated approach in urban living covers governance, transportation services, energy management, healthcare delivery, and manufacturing processes. Various smart city initiatives are being done all over the world to promote sustainability. Cities are smartening up because of technologies advancing, benefitting different sectors such as government, transport, energy management, healthcare, production, and education. Smart cities combine intelligent government, transportation, energy, healthcare, factories, tech, security, farming, and learning to challenge urban issues broadly. Digital tools are used by smart governments to increase effectiveness, transparency, and citizen participation. Mobility-as-a-Service, or MaaS, combines modes of travel to reduce traffic and carbon emissions in urban areas. A sustainable and resilient energy supply is reachable through renewable energy sources and microgrids. Artificial intelligence and telemedicine have redesigned healthcare delivery, especially aiding underprivileged groups. Industry 5.0 technologies allow sustainable manufacturing by cutting down waste and energy usage. IoT and edge computing give city planners real-time data processing and connectivity. Net-zero energy constructions and eco-friendly infrastructure contribute to forming sustainable urban environments. Vertical farming and detailed agriculture optimize food production, enhancing sustainability and effectiveness. Digital technologies transform education into a lifelong learning space accessible to all. In order to promote inclusive and sustainable development, smart cities create a balance between ethical and sociological concerns and technology breakthroughs. Data privacy, cybersecurity, and the digital divide are examples of issues that continue to exist in spite of advancements. Addressing these issues necessitates ongoing research to guide future urban sustainability innovations. A sustainable smart city enhances efficiency, enriches the quality of living, and reduces environmental effects through advanced technology. However, in this book, we have tried to cover most representative achievements with emphasis on thoughts, philosophy, and methodology for advanced sustainable smart cities.

This book is dedicated to a discussion about the future developments, needs, and sustainability in smart cities, and its significant roles in advancing the governance, transportation, energy, healthcare, factories, technologies, securities, agriculture, and education sector. The book also discusses about the impact of enabling technologies on smart cities and how they enhance citizens' lives, information sharing with the public, and coming up with a finer attribute of citizen welfare and government services. This book provides an overview of advanced technologies and sustainable city development, which is vital to the development of smart cities. It focuses on how smart city technologies can mitigate issues like these, resulting in a more sustainable and livable urban environment. Smart cities must also contribute to the UN's Sustainable Development Goals (SDGs). From smart grids to intelligent traffic management systems, this book explores the impact of Internet of Things solutions on urban sustainability. By improving decision-making processes and optimizing resource allocation, artificial intelligence and machine learning are also transforming urban systems. Several applications of artificial intelligence and machine learning in urban settings, including the use of predictive infrastructure maintenance and the provision of personalized public services, are discussed in detail. The potential of blockchain technology to improve urban management efficiency and trust is highlighted. The revolutionary potential of 6G-IoT for urban environments is also covered in detail. Telemedicine, health monitoring, and data-driven healthcare solutions are discussed for city residents. Further, this book also explores a wide range of ITS revolutions, including autonomous vehicles, traffic management systems, and Mobility-as-a-Service. It also discusses how digital learning environments can promote lifelong learning and skill development in smart cities. Furthermore, it addresses the future of energy in smart cities, highlighting modernizations that will drive the energy revolution. Eventually, academicians/researchers will profit because it will cover all aspects of advanced sustainable smart cities in one book. These unique features make the book a comprehensive guide for anyone involved in urban planning, policy-making, and technology integration, ensuring it serves as an essential resource for developing sustainable and advanced smart cities. Eventually, academicians/researchers will profit because it will cover all aspects of sustainable smart cities in one book.

- Provides an overview and discussion of comprehensive vision and global insights of advanced sustainable smart cities, characteristics, requirements, and distinct architectural design of sustainable smart cities.
- Provides an extensive overview on cutting-edge technological integration, like artificial intelligence, Internet of Things, blockchain, and 6G-IoT, that can be put into practical use in order to assist cities prosper.
- Explores possible solutions for key urban challenges in areas such as renewal energy integration, digital education, smart transportation, and smart healthcare.
- Explores the role of Sustainable Development Goals and Society 5.0 with their significant contributions with social and ethical considerations in a variety of domains to the development of sustainable smart cities.
- Explores the significance of sustainability and ethical considerations for sustainable urbanization, green infrastructure, and net-zero energy buildings.
- Addresses recent advancements along with relevant prospects on opportunities and challenges with potential research direction in the development of advanced sustainable smart cities.

The topics covered in the book are concerned with sustainable smart cities and their development for better life of inhabitants and industries. In this way it will attract various stages of readers as given below:

1. This book will assist researchers working in academia and industry in conceptualizing the development of future sustainable smart cities.
2. It encourages researchers in academia and industry to learn more in this evolving field of sustainable smart cities.
3. This book brings together a comprehensive collection of significant research works on establishing sustainable smart cities from the perspectives of Internet of Things and artificial intelligence.
4. It discusses about the integration of enabling technologies for the development of sustainable smart cities based on next-generation communication systems.

This book is intended for senior undergraduate and graduate students working in electronics/electrical and communications engineering; however, it can also be used as a reference book for engineering professionals and scientists working in academia and industry. In this book, we have tried to cover most representative achievements, with emphasis on thoughts, philosophy, and methodology for sustainable smart city accomplishment. Indeed, the existing knowledge is undoubtedly important. However, in this book, we have provided our thoughts and methodology which helps in creating new knowledge for the future sustainable smart cities.

This book comprises 12 chapters. In Chap. 1, the concept of sustainable smart city development is explored, focusing on integrating and enhancing strategies to reconstruct urban environments for technological advancements. Key approaches include digital democracy, smart governance, Mobility-as-a-Service (MaaS), renewable energy and microgrids, healthcare innovations, Industry 5.0, sustainable manufacturing, IoT, edge computing, Net-ZEBs, green infrastructure, precision agriculture, vertical farming, and education. Society 5.0 is emphasized as a foundation for human-centred sustainability, merging cyberspace with physical space to address economic and social challenges, ethical, social, equitable, private, and diverse principles are highlighted as critical for this transition. Adopting these strategies and technologies enables sustainable development that prioritizes both people and the environment.

In Chap. 2, the relationship between smart city development and the United Nations Sustainable Development Goals (SDGs) is explored as a foundation for sustainable urban growth. This emphasizes how technological advancements enhance resource utilization, comfort, and efficiency in urban environments. Wireless technologies, IoT, artificial intelligence, and big data play a crucial role in energy, waste, and, transport management while supporting SDGs such as affordable clean energy, sustainable cities, and climate action. Smart city strategies align with global objectives to address poverty, resources inequality, and environmental degradation. These initiatives aim to promote inclusivity, improve climate resilience, and optimize resource use. Aligning smart city solutions with SDGs is essential for achieving sustainable economic, social, and environmental outcomes. Ongoing research and innovation remain necessary to overcome existing challenges and offers full potential of smart cities in meeting the 17 SDGs.

In Chap. 3, the integration of smart cities with IoT technologies is examined, highlighting innovative approaches that optimize urban living. The implementation of next generation IoT networks and 5G introduces ultra-low latency, essential for real-time applications. IoT applications in urban mobility, resource management, air and water quality monitoring, smart waste management, and energy grids contribute to creating efficient and sustainable city systems. Ensuring secure transactions requires privacy-preserving methods supported by technologies such as blockchain and artificial intelligence. However, interoperability and scalability remain significant challenges, necessitating proper protocols and edge computing solutions for large-scale deployment. Key consideration includes policy frameworks, security measures, and device integration for a coherent IoT ecosystem. Future directions involve combining IoT with augmented reality, advancing healthcare solutions, autonomous IoT devices, and self-governing networks, offering opportunities for improved urban living while presenting challenges such as compatibility and security threats.

In Chap. 4, the role of artificial intelligence in smart cities is explored as a step toward sustainable urban development through advanced analytics and decision-support systems. Machine learning and deep learning techniques are applied for predictive maintenance, traffic modelling, and resource optimization. AI-based systems enhance traffic flow by forecasting congestion patterns and improve environmental management through sensor-based air and water quality monitoring. Explainable AI is emphasized to ensure transparency and reliability in decision-making. Despite these advancements, challenges remain in addressing scalability and handling the complexity of urban data, highlighting the need for research to achieve intelligent and sustainable urban growth.

In Chap. 5, the potential of blockchain technology in developing sustainable smart cities and addressing urban challenges is examined. This explains how blockchain enhances trust, transparency, and efficiency through distributed ledger technology, cryptography, and consensus mechanisms. It differentiates public, private, and consortium blockchains with examples of their applications in areas such as governance, waste management, transportation, and energy trading. Blockchain is presented as a cost-effective solution for improving city operations and citizen engagement. The discussion also highlights research challenges, including scalability, interoperability, and regulatory frameworks, which are essential for ethical and secure implementation. Consequently, blockchain integration supports operational efficiency, sustainability, and quality in smart cities.

In Chap. 6, the role of 6G technology in advancing smart city systems is explored, highlighting its ability to create sustainable, efficient, and responsive urban systems. Key features such as higher frequency bands, larger bandwidth, and massive device connectivity take 6G essential for IoT applications. It supports real-time traffic management, pollution monitoring, intelligent buildings, telemedicine, and energy-efficient solutions. Integration with artificial intelligence and machine learning enhances decision-making, while blockchain ensures secure and transparent data exchange. Edge computing enables low-latency processing for real-time applications. Despite these advancements, challenges remain in spectrum sharing, interoperability, privacy, and standardization. Addressing these issues is crucial for resilient, smart, and sustainable urban environments.

In Chap. 7, the shift from conventional healthcare to modern approaches driven by the Internet of Medical Things (IoMT) is explored. IoMT incorporates wearable, implantable, and remote monitoring devices that capture real-time data for improved medical services. Key applications include telemedicine, AI-based diagnosis, and personalized medicine supported by big data for customized treatment plans. Mobile health solutions and wearables technologies enhance patient engagement and self-management. IoMT also aids in the planning and management of healthcare resources for efficient service delivery. Despite these advancements, challenges remain in data protection, device interoperability, and system integration. Addressing these issues is essential for realizing Healthcare 5.0 in smart cities.

In Chap. 8, the significance of smart mobility in addressing challenges related to urban expansion, environmental sustainability, and technological advancement is highlighted. Initiated through the application of ICT in transportation, smart mobility integrates intelligent transportation systems (ITS), real-time data processing, and wireless communication technologies. Key components include connected vehicles, EV infrastructures, and traffic regulation systems, aimed at reducing emissions and improving efficiency. Sustainable approaches also involve pedestrian-friendly designs to complement advanced transport solutions. Despite progress, challenges such as scalability, interoperability, and data privacy persist. Further research is essential to enhance smart mobility technologies and ensure accessibility for urban residents.

In Chap. 9, the role of smart education in transforming learning through immersive technologies and metaverse integration is emphasized. Virtual classrooms enable real-time interaction and personalized distance learning, while 5G and 6G networks enhance inclusive access to educational resources. Advanced tools such as gamification, augmented reality, and virtual reality improve engagement and support diverse learning needs. Open Educational Resources and digital libraries provide cost-effective knowledge access, reducing educational disparities. Higher education institutions utilize these innovations to promote equality and inclusivity. Future research must address challenges related to digital skills, policy frameworks, and technological advancements to ensure effective and sustainable education systems within smart cities.

In Chap. 10, the focus is on smart energy systems as a critical component of sustainable urban development. Rapid urbanization increases energy demand, requiring innovative solutions for efficiency and resilience. Smart Energy 4.0/5.0 introduces technologies such as smart grids, smart metering, energy storage, and the Internet of Energy supported by computational intelligence. Real-time monitoring, predictive maintenance, and demand bidding optimize energy usage. Integration of renewal energy sources contributes to decarbonization, while Digital Twins enhance planning and management. Future advancements involve improved energy storage materials, flexible grids designs, and robust data frameworks. These technologies ensure supply and sustainability for modern smart cities.

In Chap. 11, smart governance emerges as a critical element of modern city management. Traditional governance models offer limited citizen interaction, while smart governance applies ICT for data-driven decisions and responsive policies. This approach enhances service delivery, improves quality of life, supports environmental sustainability, and drives economic growth through innovation and strategic partnerships. Core components include advanced technology, strong infrastructure, effective data analytics, and robust governance framework. Despite the advantages, challenges such as data privacy, cybersecurity, and inclusiveness remain significant. Addressing these issues through innovative models is essential for developing efficient, transparent, and sustainable governance systems in smart cities.

Finally, a cybersecurity approach in sustainable smart cities is presented in Chap. 12 to address vulnerabilities in digitally connected urban systems. Rapid integration of IoT and artificial intelligence increases risks such as data breaches, Denial of Service (DoS) attacks, ransomware, and compromised IoT connections. Blockchain and machine learning play a key role in securing infrastructures by ensuring tamper-proof data records and transparent operations. A sustainable cybersecurity framework introduces zero-trust models, continuous threat monitoring, and secure IoT deployment for safeguarding citizen information. Strong security strategies enhance public trust, reduce operational disruptions, and protect against cyber-physical threats. Future research focuses on AI-driven security, scalable IoT protection, and privacy-preserving analytics.

In summary, the book provides a unified view of the state-of-the-art sustainable smart city, which should be accessible to the readers with basic knowledge emerging technology and their role to accomplish sustainable smart city. The readership may find the rich set of references in each of the chapters very useful. The authors have performed a good job by providing a concise summary of all the chapters at the preface of the book. I would strongly recommend the book to senior undergraduate/graduate students, researchers, and engineers working or intending to work in the area of sustainable smart city and emerging technologies required to establish sustainable smart city. Although numerous journal/conference publications, tutorials, and books on the sustainable smart city have been published in the last few years, the vast majority of them focus on the various components of the smart city. However, this book distinguishes itself from the existing prosperous literature on sustainable smart city and its enabling technologies.

The authors are greatly indebted to numerous colleagues for the valuable suggestions during the entire period of manuscript preparation. Dr. Priyanka Mishra is thankful to her mentor Prof. (Dr.) Ghanshyam Singh for providing an opportunity and motivation to explore the potential and work in the field of Sustainable Smart City. She also extends her deepest gratitude to her father, Mr. Hira Mani Mishra, her mother, Mrs. Sita Mishra, whose unwavering blessings and countless sacrifices have been a constant source of strength and encouragement. Their immense patience and unconditional support have played an invaluable role throughout the writing of this manuscript. Words can never truly capture the depth of her appreciation for all they have done. We would also like to thank publishers at Springer, in particular Charles B. Glaser, Brian P. Halm, and Pradheepa Vijay for their helpful guidance and encouragement during the creation of this book. The authors would not justify their work without showing the gratitude to their family members who have always been the source of strength to tirelessly work to accomplish the assignment. The last but not the least, I thank the ultimate source of energy of every particle in the universe, the Almighty, for giving enough energy and strength to complete the work. All praise and gratitude belong to him.

Johannesburg, South Africa Priyanka Mishra

Johannesburg, South Africa Ghanshyam Singh

Contents

List of Figures

List of Tables

About the Authors

Priyanka Mishra is currently pursuing a Senior Postdoctoral Research Fellowship with the Centre for Smart Information and Communication Systems in the Department of Electrical and Electronic Engineering Sciences, Auckland Park Kingsway Campus, University of Johannesburg, South Africa. She has more than 9 years of teaching and research experience. She has authored more than 50 publications in various journals, conferences, books, and book chapters. She is a reviewer for several reputed journals and conferences and has supervised various master's thesis. Her areas of interest include next-generation communication systems (5G/6G), applications of 5G/6G in sustainable smart cities, industry 4.0/5.0, healthcare 4.0, intelligent transport systems (ITS), energy management (IoE), unmanned aerial vehicles, wireless communication, signal processing, and NOMA.

Ghanshyam Singh (Member, IEEE) received his Ph.D. degree in electronics engineering from the Indian Institute of Technology (Banaras Hindu University) Varanasi, Varanasi, India, in 2000. Currently, he is a Full Professor and Director of the Centre for Smart Information and Communication Systems, Department of Electrical and Electronic Engineering Science, Auckland Park Kingsway Campus, University of Johannesburg, South Africa. He was associated with the Central Electronics Engineering Research Institute, Pilani, and the Institute for Plasma Research, Gandhinagar, India, where he was a Research Scientist. He also worked as an Assistant Professor in the

Electronics and Communication Engineering Department, Nirma University of Science and Technology, Ahmedabad, India. He was a Visiting Researcher at Seoul National University, Seoul, South Korea. He also served as a Professor in the Department of Electronics and Communication Engineering, Jaypee University of Information Technology, Waknaghat, Solan, India. He has more than 23 years of teaching and research experience. His research and teaching interests include millimeter/THz wave technologies and their applications in communication and imaging; next-generation communication systems (5G/6G), cognitive radio, and NOMA; resource allocation and management; interference management; and applications of 5G/6G in sustainable smart city, industry 4.0/5.0, healthcare 4.0, intelligent transport systems, energy management (IoE), and digital farming. He has authored or co-authored more than 290 scientific papers in refereed journals and international conferences, as well as several books and book chapters published by Springer, Wiley, IET, CRC, and Academic Press. He has supervised various M.Tech. and Ph.D. theses. He has also served as an editor, associate editor, and reviewer for several reputed journals and conferences.

Chapter 1
Introduction: Sustainable Advances in Urban Environment

Contents

P. Mishra, G. Singh, *Sustainable Smart Cities 2.0*,
https://doi.org/10.1007/978-3-032-01102-2_1

1.1 Introduction

Urbanization is a key phenomenon of the current stage of human development within the context of the anthropocene, which defines the formation of megalopolises and the main issues of sustainable development and quality of life. In this regard, the global population rate is expected to reach two-thirds of the population living in urban areas by 2050—the pursuit of sustainable urbanization is more significant than ever before [1]. Leading this change are smart cities, which are a multiplicity of urban systems and structures that integrate the usage of digital technology, data analytics, and infrastructure in an effort to allow sustainable use of resources. Smart cities are essentially a new means of managing and developing cities by using Information and Communication Technology (ICT) to promote the functionality of a city with respect to economic success while at the same time taking into account the effects that their actions have on the environment and the people living in that city. This evolution stems from the desire to effectively overcome problems pertaining to pollution, traffic jams, energy utilization, and waste disposal within cities. Smart cities will realize better, more effective, and more adaptive urban systems through an integrated approach. In the context of smart cities, it is important to note that sustainability is a core value in the achievement of smart cities [2]. Thus, the use of renewable energy sources, an efficient transportation system, and smart grid infrastructure simultaneously decreases the environmental pressure on a city and enhances energy security. Smart cities are very good at how they use and preserve resources and at minimizing waste and emissions to create a more sustainable future. For example, smart grids manage energy supply and demand and distribute heat and electricity in a precise manner, reducing energy waste and dependence on fossil-based resources. Furthermore, smart cities capture the idea of the efficient use of technology in the advancement of mobility. Another example is Intelligent Transport Systems (ITS), which help to improve the flow of traffic, thereby enhancing the efficiency status of a city in as much as it reduces congestion and pollution. Apart from enhancing the convenience of residents in transportation, they also help achieve the vision of lowering the overall emissions of greenhouse gases. The core elements in smart cities that are applicable to waste management are advanced technologies in recycling and waste disposal, which contribute largely towards waste management in smart cities. The intelligent waste collection path enhances the recycling rate while reducing the direct cost to the surroundings of cities within the waste management system. With the help of data analysis, it becomes possible to predict the generation of waste and adjust the schedule of its collection, which minimizes the negative impact of waste on the environment in large cities. Smart cities also embrace sustainable planning and development. This is because enhanced models and geographic information systems assist planners in developing efficient as well as energy-sustainable cities. It helps in enhancing green spaces and efficient housing and improving infrastructure that will enable the integration of living facilities and conservation of the environment when there is physical expansion. Social sustainability also plays the same key role in the vision of smart cities [3]. As such, smart city advertisements are intended to bring about success with equal access and improve the living standards of all users. The advent of smart healthcare, smart governance, e-governance, and other digital education forums enlightens citizens and encourages them to be involved in the general governance system of cities. A variety of components are essential for creating sustainable smart cities, including healthcare, education, governance, infrastructure, security, and transportation, as shown in Fig. 1.1. Therefore, smart cities utilize the concept of advocacy of technology and data science to comprehensively solve some of the main problems that result from urbanization. These cities are yet to offer the potential of establishing long-lasting and forthcoming green cities for the people in a way that they meet the ever-changing human expectations as well as secure the future of the Earth. In the future, the government, private sectors, and citizens will have a bigger role to play in integrating smart technology into making smart cities and contributing towards a sustainable future.

Fig. 1.1 Smart city development for a sustainable future

1.1.1 Sustainable Smart City Development: Vision and Objectives

Sustainable smart city development refers to the changing process of urbanization that embraces ICT and sustainability as the major enhancements in urban management and governance [4]. The idea of creating smart cities represents innovative and sustainable solutions to the most complex challenges of urban development, climate change, and scarcity of resources, which will define life in the future. This vision originates from the concept of sustainable economic development for cities, not only building the economy of the cities but also providing social equity, opportunities, and quality of life for the public. Therefore, it is vital to highlight that the objectives of developing a sustainable smart city are numerous and interconnected, as the many challenges faced by modern cities require integrated solutions. Firstly, these goals are associated with the use of digital devices and data analytics for innovative improvements in urban structures and resources, as well as in municipal services. In addition, smart cities seek to increase society's cohesion in as much as to ensure equal technological growth for all citizens. This includes training citizens in the use of digital devices, making communication accessible to everyone, and offering interactive, internet-enabled means of addressing, monitoring, and participating. Consequently, cities can ensure that citizens play an active role in decision making. Another goal is economic, as smart cities aim to promote economic growth, attract investment, and generate employment. Cities should thus support innovation so that startups and other technology-based businesses become important focal points. In other words, the idea of a smart city is not only to enhance existing technologies continuously but also to build urban environments that are stronger and better.

1.1.2 Sustainable Urbanization and Smart Cities: Strategies and Goals

Sustainable urbanization and the development of smart cities, one of the emerging trends of the new millennium, are an innovative perspective in the management of modern cities based on the practical principles of reasonable economic growth, social equity, and environmental integration [5]. Africa is already experiencing a shift to urbanization due to migration and population increases within a particular area, making smart and sustainable city development a necessity. Smart cities, therefore, do incorporate the latest technologies, data, and infrastructure that enhance the use of scarce resources and the provision of effective and efficient services to their population. This vision of urbanization is not only about how developing countries can address the pressures felt in the course of the rapid growth of the urban population or how inadequate measures have been taken to counterbalance these pressures, but also, more prominently, about effective and efficient opportunities that stem from pressures felt in the course of such growth. The concept of smart cities corresponds to ideas of applying ICT to urban structures in order to monitor and, if necessary, manage the flow of utilities such as electricity, water, and traffic in cities. Technological advancements assist in making the physical environment more intelligent to deal with issues of carbon footprint, efficient ways of waste disposal, and quality air. Also, smart cities ensure citizen-centred initiatives and smart governance with the effective participation of citizens in decision-making processes. This strategy makes certain that urban development goals are in tune with the community's vision and firmly supports the sense of fairness and engagement of society. Sustainable urbanization within a smart city is attained through the provision of smart grids, ITS, and green buildings, which reduce negative environmental impacts while enhancing efficiency in urban systems. These goals include the promotion of economic growth and investment, innovation, economic diversification, building the ability of cities to adapt to challenges, including climate change, and making cities friendly to all inhabitants, providing equal access to basic services and equal opportunities. Thus, in terms of the intelligent and sustainable cities of the future, the synergy of the technological, regulatory, and public variables will be the key to achieving the true goal of creating sustainable smart cities that could provide the inhabitants with effective, efficient, and human living environments. Such sustainability and intelligence of the cities need to be incorporated for the growth demands of the future and the next generation.

1.2 Current Research Trends of Smart Cities

This study investigates the trends, contributions, and impact of research in smart cities from 2015 to 2025 through a bibliometric analysis of Scopus data [6]. The Bibliometrix R package, along with Origin software, was used to analyse and depict the data. This approach ensures that researchers are presented with the most current research developments in the field, enabling them to stay informed on emerging trends and advancements. In March 2025, a bibliometric analysis was conducted utilizing the Scopus core collection database to thoroughly investigate the concept and understanding of smart cities. Ultimately, 15,351 articles were exported and saved in BibTeX format; they were included and examined in the bibliometric study. The primary information, syntax, and specific criteria used in screening for this study are shown in Table 1.1.

1.2.1 Annual Scientific Production and Top 10 Most Published Journals/Cited Papers

Figure 1.2 shows the annual publication trends from 2015 to 2025, highlighting the growth in research output. The early years show a steady increase in publications, rising from 2015 (262 docs) to 2024 (2325 docs), indicating a growing interest in the field. In 2025 (597 docs), the number continues to rise and may increase further until the end of the year. Figure 1.3 presents the highest number of articles in the smart city research domain published in "IEEE Access" (806 docs), "Sustainability" (718 docs), "IEEE Internet of Things Journal" (453 docs), and "Sustainable Cities and Society" (423 docs), which indicates a higher output of articles compared to the

Table 1.1 Summary of data characteristics and main information of the bibliometric study

Description	Specific Standard Requirements/Results
Research database	Scopus
Search query	("smart") AND ("City")
Time span	2015–2025
Document type	Articles, conferences, book chapters, etc.
Language	English
Data export format	BibTeX
Documents	15,351
Annual growth rate %	8.58
Document average age	3.86
Average citations per doc	24.88
Keywords plus (ID)	43,531
Author's keywords (DE)	33,769
Authors	34,757
Authors of single-authored docs	976
Single-authored docs	1101
Co-authors per doc	4.02
International co-authorships %	35.26

other journals. In addition, some other journals, such as "Sensors" (383 docs), "Sensors (Switzerland)" (318 docs), "Smart Cities" (302 docs), "Applied Sciences" (275 docs), "Energies" (259 docs), and "Future Generation Computer Systems" (197 docs), are in this domain. Table 1.2 displays the top ten most cited papers from the study, which present both the ranking and authors with their published works. This section presents information about authors, article titles, and their corresponding publications in journals, with documentation for publication years and citation frequency. The ranking focuses on smart city articles in the engineering field to demonstrate their total citation (TC) counts, annual citation rates, and normalized total citation (NTC) data that collectively signify their impact in the field.

1.2.2 Sankey Diagram (Three-Field Plot and Thematic Map)

A Sankey diagram illustrates the three-field plot thematic progression to visualize the flow and distribution in scientific research, as depicted in Fig. 1.4. In this case, keywords, abstracts, and titles are added, acting as indicators to highlight the most influential research papers on the subject. These keywords indicate vital concepts and themes that stand out in academic discussions of the smart city domain.

Figure 1.5 categorizes various smart city themes based on their development and relevant degrees. In the top right quadrant (motor themes), which represents themes that are highly developed and highly relevant, are terms like "internet of things", "network security", "automation", "deep learning", "learning systems", and

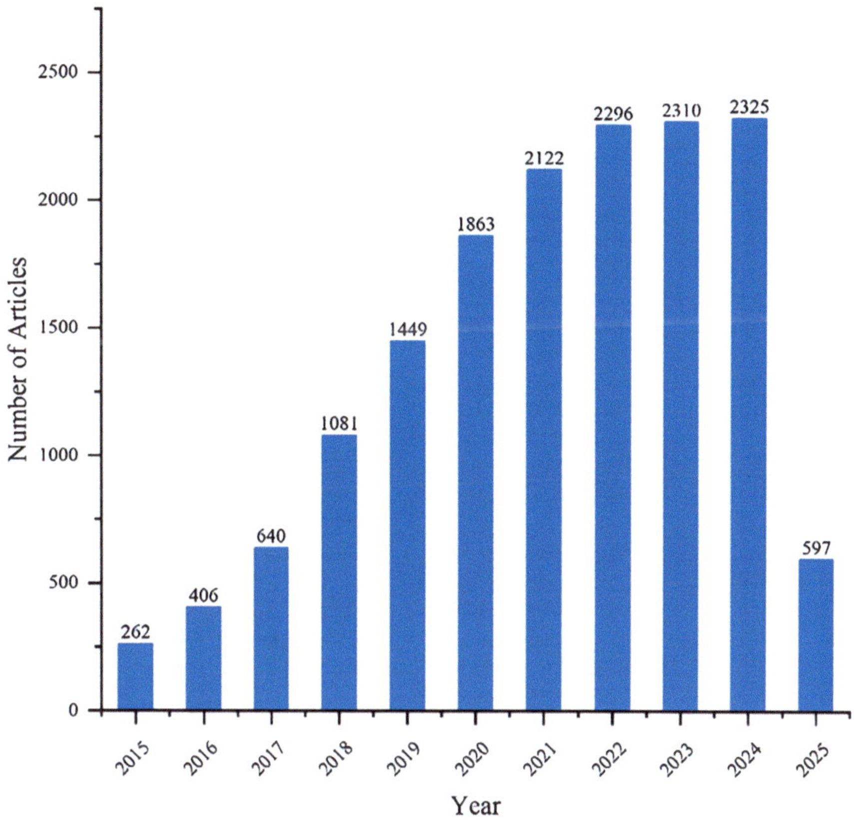

Fig. 1.2 Yearly trend of publication numbers in smart cities

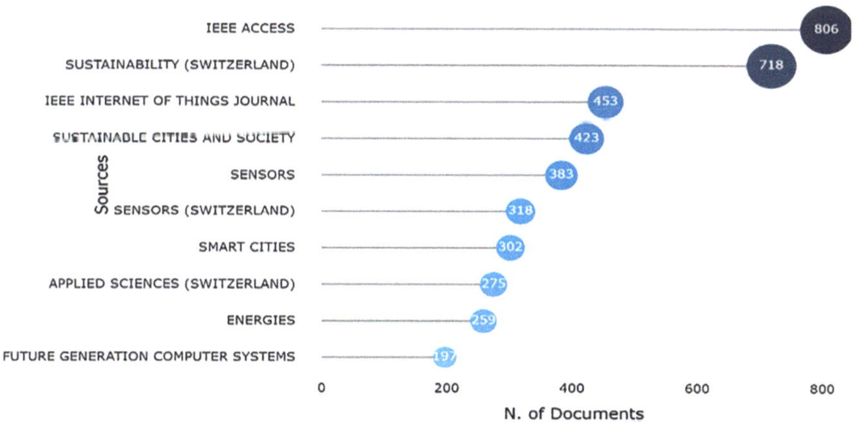

Fig. 1.3 Most relevant sources with the number of articles on smart cities

Table 1.2 Most globally cited documents in smart cities

Rank	Author's name	Paper Title	Journal's Name	TC	TC/ Year	NTC
1	W. Shi et al. (2016) [7]	Edge Computing: Vision and Challenges	IEEE Internet of Things Journal	6033	603.30	88.03
2	J. Lin et al. (2017) [8]	A Survey on Internet of Things: Architecture, Enabling Technologies, Security and Privacy, and Applications	IEEE Internet of Things Journal	2214	246.00	40.47
3	M.A. Khan et al. (2018) [9]	IoT Security: Review, Blockchain Solutions, and Open Challenges	Future Generation Computer Systems	1942	242.75	38.98
4	A. Botta et al. (2016) [10]	Integration of Cloud Computing and Internet of Things: A survey	Future Generation Computer Systems	1907	190.70	27.83
5	U. Raza et al. (2017) [11]	Low Power Wide Area Networks: An Overview	IEEE Communications Surveys & Tutorials	1560	173.33	28.52
6	A. Fuller et al. (2020) [12]	Digital Twin: Enabling Technologies, Challenges and Open Research	IEEE Access	1432	238.67	40.10
7	M. Centenaro et al. (2016) [13]	Long-Range Communications in Unlicensed Bands: The Rising Stars in the IoT and Smart City Scenarios	IEEE Wireless Communications	931	93.10	13.58
8	C. Moreno et al. (2021) [14]	Introducing the "15-Minute City": Sustainability, Resilience and Place Identity in Future Post-Pandemic Cities	Smart Cities	864	172.80	32.03
9	I. A.T. Hashem et al. (2016) [15]	The Role of Big Data in Smart City	International Journal of Information Management	854	85.40	12.46
10	A. G. Gad (2022) [16]	Particle Swarm Optimization Algorithm and Its Applications: A Systematic Review	Archives of Computational Methods in Engineering	829	207.25	44.47

"forecasting". In the left quadrant (niche themes), represents, "smart power grids", "electric power transmission networks", and "smart grids". This indicates a strong focus on these areas, suggesting they are central to smart city research and practice. This quadrant reflects the areas where significant advances and investments are made, highlighting their critical role in smart cities. In contrast, the bottom left quadrant, which represents themes that are either emerging or declining and have

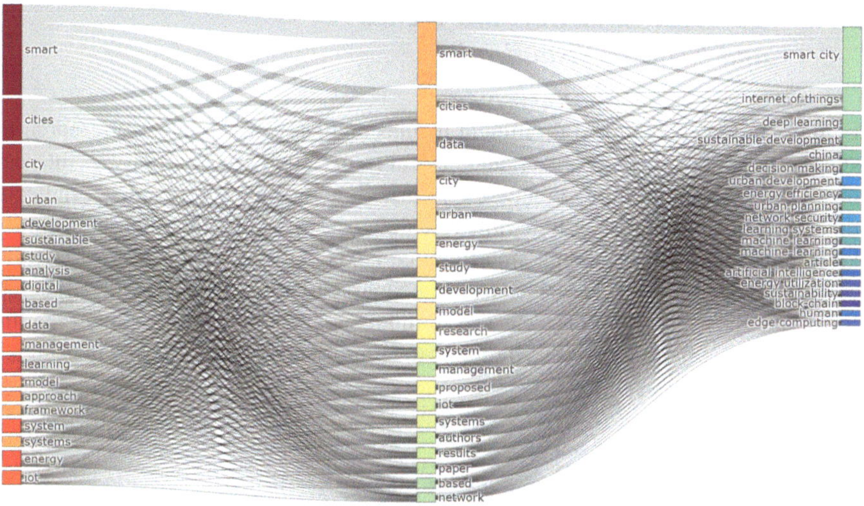

Fig. 1.4 Three Field Plot (thematic development) using Sankey diagram in smart cities

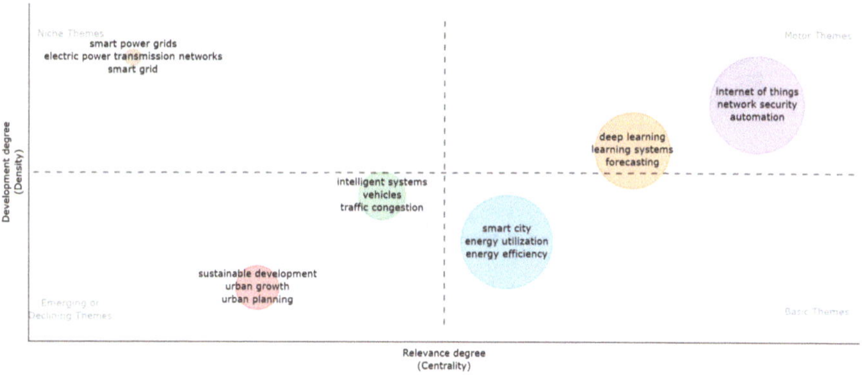

Fig. 1.5 Thematic map in smart city research

lower relevance, includes terms like "intelligent systems", "vehicles", "traffic congestion", "sustainable development", "urban growth", and "urban planning". These themes might be considered foundational but must be emphasized more in smart cities. This visual categorization provides valuable insights for educators, curriculum developers, and researchers, helping them understand which areas are gaining traction and which might need more attention or innovation.

1.2.3 Co-occurrence Keyword Trends and Network Visualization

Figure 1.6 depicts the most prominent themes and keywords related to research in smart cities. The most significant and central terms, such as "smart city", "internet of things", "deep learning", "network security", and "energy efficiency", indicate the core focus of the research on how researchers grasp and refine their conceptual knowledge, particularly within smart city contexts. An illustration in Fig. 1.7 displays word occurrence frequencies from the Smart City Research Network. Research focuses mainly on "smart city", as this term appears 5744 times throughout the literature. Research on smart cities indicates the importance of the "Internet of Things"

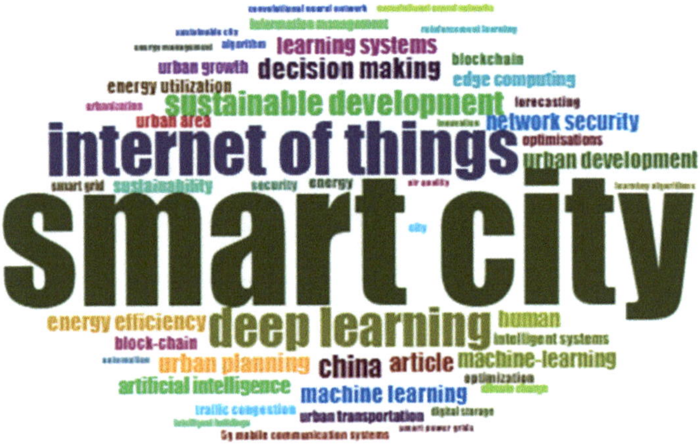

Fig. 1.6 Co-occurrence keyword trends in smart city domain

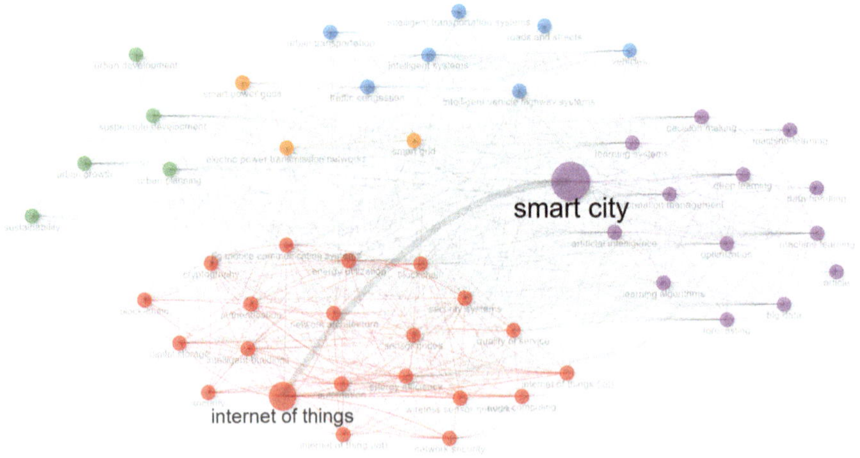

Fig. 1.7 Co-occurrence network of words in smart city research domain

based on its 3162-word occurrences. Machine Learning technology through "Deep Learning" (1233) and "Network Security" (852) combinations shows that artificial intelligence, together with cybersecurity, occupies crucial importance in emerging technology fields.

The sustainability challenges in energy consumption are shown through terms like "Energy Efficiency" (845) along with "Energy Utilization" (840). The documents place emphasis on data-driven planning through combined mentions of "Decision Making" (813) and "Sustainable Development" (739). Additionally, "Learning Systems" (660) and "Intelligent Systems" (649) demonstrate the role of automation and smart technologies in urban environments. The data show that smart city research revolves primarily around several major subjects.

1.2.4 Countries' Scientific Production

The scientific publication frequencies of various countries around the world are represented in Fig. 1.8. Scientific contributions from China stand at 8321, which demonstrates its powerful role in research development. The growing scientific power of India displays itself through its 4522 scientific publications, which rank second after China. The research hub of the USA produces 2655 scientific papers among its total contributions. Numerous publications emerge from Saudi Arabia and Spain, together with Italy, given their research output comprises 1785, 1420, and 1531 papers, respectively. Academic and industrial research sectors of the UK and South Korea produce approximately 1400 and 1387 scholarly works, respectively. Australia and Canada have smaller publication numbers, but their contributions amount to 965 and 732 research items, respectively. The measured

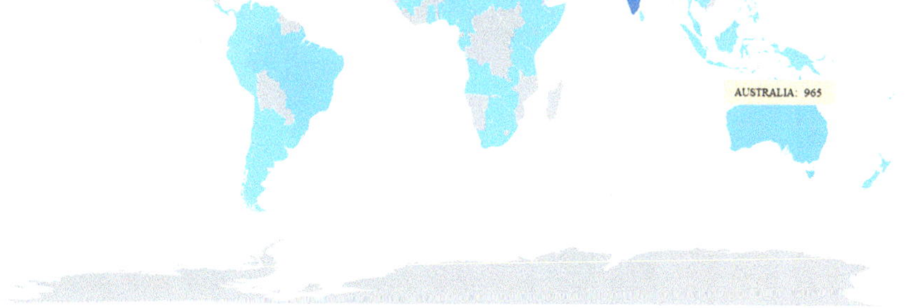

Fig. 1.8 Top 10 countries' scientific production of various countries based on frequency

data indicate that Asian countries significantly influence the global scientific research output.

1.2.5 Mapping Smart City Research to Sustainable Development Goals (SDGs)

SDGs, their targets (bubble series), and the percentage (bar graph) of corresponding keywords detected in the text (calculated as the ratio of keywords in one goal to the total number of keywords detected) are depicted in Fig. 1.9. For SDG mapping, the study utilized the SDG Mapper from the "Knowledge Base for the Sustainable Development Goals", provided by the European Commission [17]. This tool uses bubble charts to depict SDG interconnections and relevance among bibliometric records. The size of each bubble correlates with the proportion of associated keywords found, which is calculated as the number of goal-specific keywords divided by the total keyword count. As such, SDG 11 is marked by 84 keyword occurrences. This is followed by SDG 7, with 43 occurrences, and SDG 12, with 22 occurrences, underlining their relative prominence in the research database. Within SDG 11, there are several key subtargets that the research addresses: Subtarget 11.3 is shown as a distinct bubble within the larger SDG 11 circle, indicating focused research on

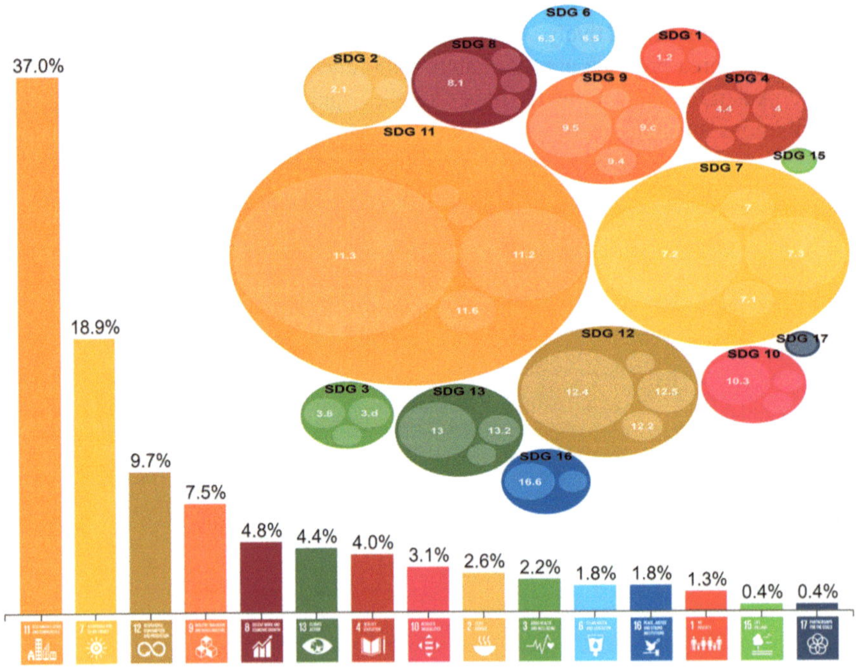

Fig. 1.9 Mapping smart city research to relevant SDGs and targets

smart cities. Subtarget 11.2, which aims to provide access to safe, affordable, accessible, and sustainable transport systems for all, is another bubble within the SDG 11 area. This highlights efforts in research to improve public transport in a smart city. Subtarget 11.6 focuses on reducing the adverse per capita environmental impact of cities through smart city solutions. Figure 1.9 shows that the size of the SDG 7 bubble implies that advancements in smart cities are being focused on affordable and clean energy that prioritizes sustainability. Similarly, SDG 12 (Responsible Consumption and Production) is also closely linked to SDG 11 and suggests that research should explore how to achieve the environmentally sound management of chemicals and all waste in order to minimize their adverse impact on humans and to achieve the sustainable management and efficient use of natural resources within the context of smart cities. The detailed descriptions of other relevant SDGs are depicted in Table 1.3.

Table 1.3 SDGs ranking, targets, and relevance in the smart city research domain

SDGs (Rank)	Description	Targets	No. of Occurrence		Relevance
SDG11	Sustainable cities and communities	11.3	59	84	37.0%
		11.2	19		
		11.6	4		
		11.7	1		
		11.5	1		
SDG 7	Affordable and clean energy	7.2	25	43	18.94%
		7.3	12		
		7.1	3		
		7	3		
SDG 12	Responsible consumption and production	12.4	15	22	9.7%
		12.5	4		
		12.2	2		
		12	1		
SDG 9	Industry, innovation, and infrastructure	9.5	8	17	7.5%
		9.c	5		
		9.4	2		
		9.3	1		
		9.1	1		
SDG 8	Decent work and economic growth	8.1	8	11	4.8%
		8.2	1		
		8.3	1		
		8.5	1		
SDG 13	Climate action	13	7	10	4.4%
		13.2	2		
		13.1	1		

(continued)

Table 1.3 (continued)

SDGs (Rank)	Description	Targets	No. of Occurrence		Relevance
SDG 4	Quality education	4	3	9	4.0%
		4.4	3		
		4.a	1		
		4.3	1		
		4.5	1		
SDG 10	Reduced inequalities	10.3	5	7	3.1%
		10.2	1		
		10.1	1		
SDG 2	Zero hunger	2.1	5	6	2.6%
		2.5	1		
SDG 3	Good health and Well-being	3.8	2	5	2.2%
		3.d	2		
		3.4	1		
SDG 6	Clean water and sanitation	6.5	2	4	1.8%
		6.3	2		
SDG 16	Peace, justice, and strong institutions	16.6	3	4	1.8%
		16.7	1		
SDG 1	No poverty	1.2	2	3	1.3%
		1.4	1		
SDG 15	Life on land	15.3	1	1	0.4%
SDG 17	Partnerships for the goals	17.17	1	1	0.4%

1.3 Important Features of Urban Development

With an increase in the expansion of urban areas, smart cities have become the future vision for developing urban areas. These cities assist in improving the globe and making them a better place to reside in by utilizing technology, data, and innovation in making the cities better. The principles of smart cities are numerous since they encompass nearly all aspects of human life within cities, including infrastructure, transportation, governance, and people's participation. The following aspects aim to focus on the important features of urban development of particular interest to smart cities: integration, inclusiveness, and sustainability.

1.3.1 Sustainability and Environmental Responsibility

Sustainability is one of the most significant aspects of the development of smart cities [18]. The cities are seen to be one of the biggest contributors to the deterioration of the environment, and this is where smart cities step in to reduce the effects. This involves the use of renewable energy, increased energy efficiency, and encouraging the adoption of green buildings. In consideration of minimizing the impacts on the environment, smart cities ensure that there is low carbon dioxide emission through utilizing smart technologies like smart grids and smart transport systems. Also, smart waste management is employed through monitoring of waste through Internet of Things (IoT) sensors to determine the level of waste in bins to determine the right time to collect waste. Gardening is also included in urban land use to increase air quality and the availability of food produce in the urban centres. Investing in sustainability in smart cities also benefits the environment and the residents in the cities due to the concern for the infrastructure of the cities.

1.3.2 Data-Driven Decision Making

The foundation of the smartness of a smart city is the capacity to support real-time and crucial information. The collection and analysis of data play an important role in the assessment of urban dynamics and key issues like traffic, pollution, and security [19]. Smart cities consist of a number of interconnected computers that gather data about different aspects of municipal life, such as traffic and energy consumption. This is then fed through intricate algorithms and artificial intelligence (AI) to determine patterns, requirements for the future, and the improvements to city services. For example, public mobility is enhanced through access to information on routes and scheduling of public transport, reducing citizens' waiting time. Smart cities use data to make better decisions that will positively impact the lives of citizens.

1.3.3 Connectivity and Digital Infrastructure

Connectivity is a prominent attribute when it comes to the emergence of smart cities. Reliable and fast internet connection and developed digital infrastructure provide the basis for the technologies used in the smart city [20]. It includes smart meters, smart streetlights, self-driving cars, and digital services for citizens. The respective investments in 5G networks and broadband ensure that all individuals gain the fundamental right of access to internet services, hence addressing the issue of digital inequality. Smart cities also pay much attention to the security measures in relation to cyber threats to networks and systems. Thus, through the focus on connectivity, smart cities enable technology to enhance and develop the quality of living in urban areas.

1.3.4 Inclusive and Equitable Development

Equality is one of the main themes for the smart city concept, and smart cities must adopt it. Modernized cities should have equal opportunities for all people in the context of contemporary achievements based on smart technologies [21]. This includes making housing that is inexpensive, food, water, medical services, and making the internet available to the public domain. Alternatively, smart cities also address the aspect of citizenship, which engages people in planning and decision-making processes. It means that the goals of such city development would be made from the beginning and involve the participation of all the members of the community. Smart cities are effective in providing the basic needs of their citizens, from basic needs to economic well-being, and this articulates how smart city designs and planning goals ensure that the entire society is included and all the citizens are treated equally.

1.3.5 Efficient Transportation and Mobility

Transportation has always been considered a key aspect of urban growth, as well as the development of smart cities. It involves the use of buses, bicycle lanes, and friendly walkways, to discourage personal car use to reduce traffic congestion. Smart cities feature advanced traffic control systems, real-time updates on the status of transport through social media, and self-driving cars [22]. Other ride-sharing services include micromobility services such as e-scooters and bike-sharing services, which allow for flexible means of transport. Smart cities promote a smart transportation system that increases access and reduces environmental effects.

1.3.6 Resilience and Adaptability

Citizens' resilience is one of the primary concepts used in the construction of smart cities, especially when it comes to dealing with climate change and other disasters. Smart cities are thus sustainable infrastructure and systems that can easily cope with a disaster and recover quickly. This includes sensors and other infrastructures that can sense changes, for example, in the environment, like flood alarms and smart electricity networks that can identify deficits and still deliver power to other areas [23]. Smart cities are also about planning for the worst and having effective early warning systems and management of disasters in urban areas. In this way, smart cities adequately prepare for these shocks, and the residents of the cities can withstand the impacts of the shocks.

1.3.7 *Smart Governance and Civic Engagement*

The advancement of smart city strategies depends on good governance. E-governance is the process of employing technology for improving the quality of services offered and the degree of transparency and accountability in the work of the government [24]. This comprises electronic government or governance, which aims at providing information on various services to the citizens, feedback, and other related activities. Another component of the smart cities is the open data policies, which imply the availability of government information for the public and the promotion of innovation. Through smart governance, smart cities come with a sense of a new government being formed that is more efficient in its operations than traditional government.

1.3.8 *Economic Growth and Innovation*

A smart city is expected to be an essential centre of innovation and an engine of the economy. By giving an investor-friendly environment and encouraging research, innovation, entrepreneurship, and startups, smart cities attract human capital [25]. This is done by making accessible resources such as incubators, accelerators, and innovation labs, and also developing the partnership between the public and private parties. Smart cities also participate in training and educating residents' skills suitable for the job market and employment opportunities of the future. Economic Redevelopment and Innovation (ERI) as a smart city strategy outlines the need to succeed in creating vibrant economic success in the city's environment.

1.3.9 *Quality of Life and Well-Being*

A smart city is the qualitative improvement of the life of the inhabitants of a particular city. It involves the production of spaces that are habitable, sustainable, and full of energy in the form of human spirit within the urban environment [25]. The other dimensions of a smart city involve a focus on public health, which includes technologies aimed at enhancing air quality monitoring, encouraging movement and physical exercise, and providing health care. Social facilities are also provided through parks, landmarks, and museums that contribute to the quality of living standards of the societies. Hence, smart city systems, needs to put emphasis on the quality of individuals' lives and aesthetics to provide conditions under which life can be upgraded essentially.

1.4 Driving and Guiding Principles for Developing Sustainable Smart Cities

Smart cities comprise numerous sustainability components, including smart government, infrastructure, buildings, transportation, energy, healthcare, industry, technology, agriculture, and education. These dimensions address a range of issues such as participation in services and governance, innovation, entrepreneurship, healthcare, education, ICT infrastructure, sustainability, transport systems, environmental protection, and sustainable resource management. To ensure inclusivity, a smart city must align with the six interconnected principles. These principles serve as guiding guidelines for decision making during the identification, planning, and implementation of smart initiatives and technologies [26]. Smart initiatives and technologies should be shaped by the following guidelines: designed for everyone and everywhere, enabling technologies, localization, community and city needs, partnerships, collaboration, innovation, robustness, safety, and sustainability. Figure 1.10 illustrates how the principles shown in the following sections are used to plan and implement smart cities.

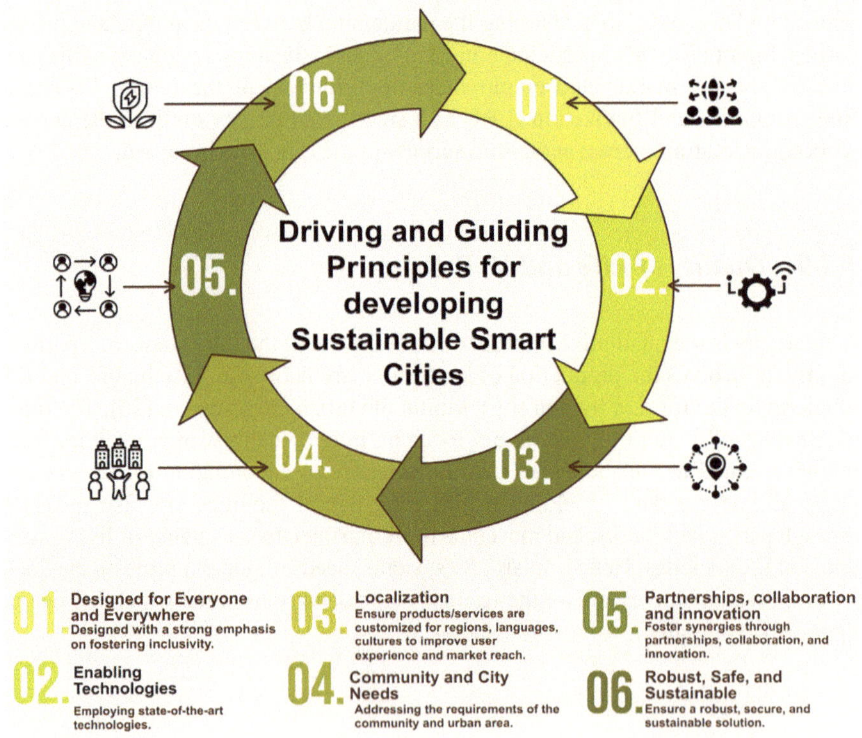

Fig. 1.10 The essential concepts and guidelines for developing sustainable smart cities

1.4.1 Designed for Everyone and Everywhere

Enhanced urban well-being, which incorporates advanced technology and initiatives, is at the core of a smart city [27]. Such initiatives should not prioritize certain segments of society or cities at the expense of others. The benefits of a smart city should be distributed equally among residents. Financial resources and technological expertise should not be barriers to accessing smart initiatives. The most effective way to align these initiatives with smart city principles is to innovatively plan and execute them. Smart initiatives should ensure accessibility and deliver benefits to society, regardless of their financial situation.

1.4.2 Enabling Technologies

Information and communication technologies (ICTs) enable and enhance sustainable smart cities. Emerging technologies enable cities and societies to provide services with greater effectiveness and efficiency. [28]. The selection of technologies should be based on the local and available resources. An industry or system that self-optimizes, self-configures, self-diagnoses, and is transparent is known as a cyber-physical system (CPS). It is also called Industry 4.0 or the Fourth Industrial Revolution. These developments rely on IoT, VR, AR, near-field communication, advanced materials, rapid connectivity (e.g., 5G), 4IR, and other technologies. Several technologies and concepts are associated with the Internet of Things (IoT), including human enhancement technologies, artificial intelligence, smart grids, bots, drones, satellite-enabled services, facial recognition, and autonomous vehicles [29].

1.4.3 Localization

When planning smart city technologies and initiatives, it is important to consider the local context. Cities and communities are also affected by poverty, inequality, and unemployment [30].

1.4.4 Community and City Needs

The active participation of the community in smart city initiatives can significantly contribute to their success, as well as to their identification, development, and implementation. The participatory process involves people in the decision-making process from the beginning. In addition to participating in planning and design, they should also assist in executing and managing the city [31]. A city is generally understood to be composed of all relevant stakeholders, and other parties involved. Community contributions also enhance an initiative when adopted as part of community engagement [32]. With sharing information, exchanging data and information, and assisting with the monitoring and assessment processes, individuals could potentially make the initiative successful.

1.4.5 Partnerships, Collaboration, and Innovation

Developing smart cities requires public and private initiatives, innovation, partnerships, and collaborations [33]. To achieve the same goals and values, it is critical to identify partners who exhibit the same objectives and values. All partners must have a clear understanding of their roles. Smart city agendas are sometimes driven by the private sector rather than by the government. It is imperative for local governments to guarantee that smart city efforts are directed by appropriate policies, strategies, plans, and frameworks related to urban development and planning [34].

1.4.6 Robust, Safe, and Sustainable

Building an inclusive, robust, safe, and sustainable city is vital. As a result, sustainability, resilience, and safety concepts must be included in smart city initiatives. Through innovative interventions and technologies, sustainable development addresses the human, social, economic, and environmental dimensions. The sustainability of cities must be enhanced using multifaceted approaches [35]. Smart technology has the potential to improve community safety. By increasing the effectiveness and efficiency of limited resources, law enforcement authorities, including the police, can both prevent and detect criminal activity. Technologies such as e-policing, AI, face recognition, data, predictive policing, and community safety apps improves community security and safety [36].

1.5 Performance Measures for Sustainable Smart Cities

Information and communication technology and other technologies are used in smart sustainable cities to improve the standard of living, increase the effectiveness and competitiveness of urban operations and services, and address the cultural, social, economic, and environmental demands of present and future generations [37–39], as illustrated in Fig. 1.11.

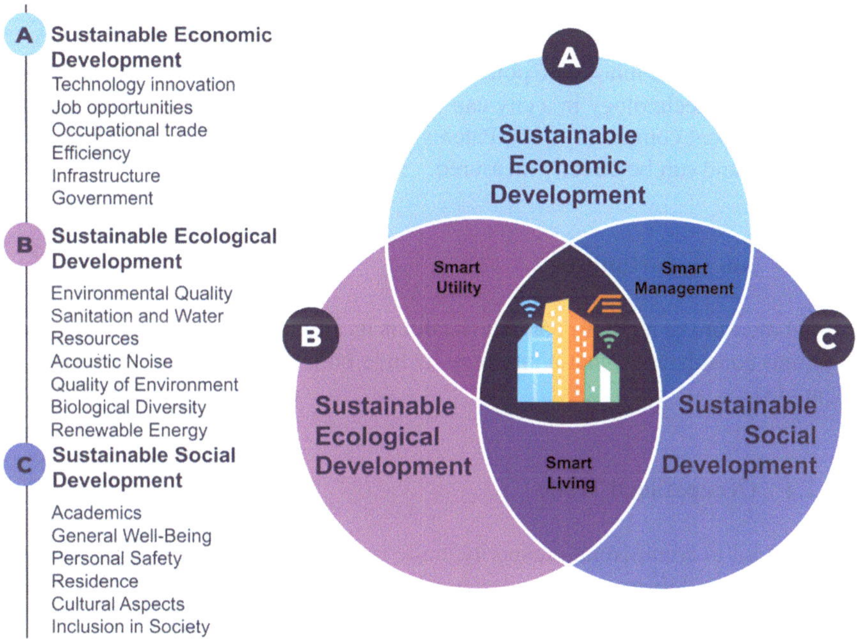

Fig. 1.11 Sustainable smart city concept

1.5.1 Sustainable Economic Development

To assess the possibility of an eco-friendly smart city economy [40], the following fields should be considered: infrastructure for information technology, technology innovation, job opportunities, occupational trade, efficiency, infrastructure, and governance. Sustainable smart cities should study the local economies.

1.5.1.1 Infrastructure for Information Technology

ICT infrastructure serves as a framework to provide additional ICT solutions, making cities intelligent and sustainable. Information platforms and services, along with terminals, networks, and access, are also considered part of the ICT infrastructure [41]. ICT infrastructure measurements require the consideration of a wide range of terminals (computers, mobile phones, tablets, etc.), the internet, backbone networks, cloud computing platforms, wireless/fixed broadband, and data centres.

1.5.1.2 Technology Innovation

An evaluation of the innovative potential of a city must consider various viewpoints. An innovative technology in a city can quickly adapt to changes and is at the forefront of its local community [42]. Patents are a good indicator of R&D investment and output and can be directly measured.

1.5.1.3 Job Opportunities

A vital measure of a city's economic health is its employment rate. It is essential to take into consideration that a city may include both formal and informal employment [43].

1.5.1.4 Occupational Trade

With regard to commercial prosperity, trade is the foundation for success. A country's exports and imports are used as a measure of trade. Furthermore, trade in sustainable smart cities is heavily reliant on e-commerce [44].

1.5.1.5 Efficiency

Productivity assessments must rely on information and media to achieve sustainable smart cities. Business leadership, products, and process innovation were included in this category. Smart cities have succeeded because of their ability to use data effectively [45]. It also relies on how inventive its goods and services are and how successfully they draw in customers and suppliers. As a result, the city is more productive as a result of all these factors.

1.5.1.6 Infrastructure

A city should improve its water supply, electric power, infrastructure for healthcare, transport, road infrastructure, buildings, and urban planning, among others [46]. City officials need to deal with all of these categories to ensure that city runs smoothly and its citizens have a high quality of life. A more efficient use of resources will be possible with an enhanced electricity and water supply. Moreover, increased effectiveness and equitable city environments are created by improving health, transportation, roads, and building infrastructure.

1.5.1.7 Governance

The public sector is responsible for providing most of the government's services. Public sector businesses are usually run by the government, social services like welfare and unemployment settlements, as well as facilities like public transport, schools, and hospitals. Many government services are provided to the public through taxpayer money. A sustainable smart city should prioritize ICT to improve its service efficiency. The use of ICT in services in the public sector makes them more efficient, cost-effective, and user-friendly. [47]. Better quality of life for the people and more sustainable service delivery might lead to better outcomes. Furthermore, ICT improves accountability in the public sector by reducing bureaucracy and increasing transparency.

1.5.2 Sustainable Ecological Development

The following six categories are utilised for measuring environmental sustainability in sustainable smart cities: environmental quality, sanitation and water resources, acoustic noise, quality of environment, biological diversity, and renewable energy.

1.5.2.1 Environmental Quality

Air quality is one of the most important factors affecting the quality of life in many cities. Residents of cities are concerned about air pollution, which needs to be carefully observed, and the data made public [48]. There is also the matter of CO_2-e emissions, where "-e" represents equivalents and CO_2 is the result of converting all other greenhouse gases into CO_2.

1.5.2.2 Sanitation and Water Resources

Environmental protection should consider water distribution, water resources, water conservation, drainage, sanitation, and wastewater treatment [49].

1.5.2.3 Acoustic Noise

Noise exposure in the city was considered in this category. The quality of the environment in a city is significantly affected by the noise levels [50]. The effects of noise exposure include hearing loss, stress, and sleep disturbance.

1.5.2.4 Quality of Environment

There are quantitative and qualitative methods for assessing environmental quality in green spaces, solid waste, and electromagnetic fields. The computation involves two metrics: solid waste generated and the proportion of waste that is reused [51]. It is possible to measure a green space by taking into account its size and the quantity of vegetation it has. It is possible to assess the quality and availability.

1.5.2.5 Biological Diversity

Biodiversity measurement is a challenging process with several issues. The main problem in characterising biodiversity is its complexity. The number of species in a given area can be exceptionally challenging to evaluate since some species may be tremendously tiny and tough to find. It is difficult to assess biodiversity across diverse areas since it may consist of different species. The natural environment and the inherent species are two examples of elements that are included in a smart city plan [52].

1.5.2.6 Renewable Energy

Cities consume energy in a variety of ways, such as through the usage of electricity, renewable energy sources, and energy-saving home practices [53]. All electrical appliances in a city are considered to be part of the electricity usage. The use of energy generated from renewable sources, including solar and wind power, supports sustainable energy consumption. An effort to minimise household energy use, such as installing energy-efficient equipment or improving lighting, is known as an energy conservation measure.

1.5.3 Sustainable Social Development

The performance of these six domains should be considered in a sustainable smart city: academics, general well-being, personal safety, residence, cultural aspects, and inclusion in society.

1.5.3.1 Academics

Education and training play a critical role in developing human resources and stimulating innovation [54]. The knowledge as well as the skills needed for creativity and productivity are provided by education and training. Today's workforce needs crucial reasoning, interaction, addressing issues, and other vital abilities for a successful profession. The assessment of educational developments is approached from multiple perspectives, including increased funding for education, the use of ICT as a tool, enhancement of student abilities, and adult literacy.

1.5.3.2 General Well-being

It is often acknowledged that medical and healthcare services are essential parts of health services. With the aim of improving and preserving individual health, these services are crucial for the diagnosis, prevention, and treatment of conditions, including diabetes and cancer [55]. Health administration (disease control, epidemic prevention, resource distribution, and investment), health service organizations (hospitals, pharmacies, healthcare centres, and health insurance providers), and health status indicators (life expectancy, morbidity, and mortality rates) are some of the perspectives that must be considered when assessing a city's health.

1.5.3.3 Personal Safety

Security and safety for citizens have always been essential civil services since the beginning of civil government. Safety plans deal with how to handle accidents and natural disasters, while security concerns are primarily about man-made dangers like crime and terrorism. ICTs are essential for improving safety and security because they facilitate large-scale data collection, analysis, and communication [56]. These technologies are critical for dealing with threats (both natural and man-made), in an effective manner. ICTs also make it easier for the public and different government agencies to collaborate better, which is essential for disaster preparedness, response, and recovery.

1.5.3.4 Residence

The average living space and expenditures are among the factors that contribute to sustainable smart cities. Accordingly, it is also one of the Sustainable Development Goals (SDGs) of the UN that poverty will be eradicated by the year 2030, and a very important aspect of this is reducing slum communities located within cities [57]. Smart cities can give residents access to safe and affordable properties in addition to essential amenities like clean water and sanitary facilities. Furthermore, they can assist in creating jobs and generating income. This will contribute to the reduction of poverty and the creation of a better living environment for residents.

1.5.3.5 Cultural Aspects

In sustainable smart city assessments, libraries, theatres, museums, galleries, and other cultural/knowledge infrastructures are considered. Cities greatly benefit from these places because they make knowledge, culture, and history accessible [58]. A sense of community and civic engagement can also be developed through education, recreation, and social interactions.

1.5.3.6 Inclusion in Society

Equity in income and consumption, accessibility to infrastructure and services, transparency and public involvement, and governance are all important components of an equitable and inclusive smart city. The government and public services have a significant impact on social development [59]. Efficiency and transparency should be hallmarks of modern government. A constant state of turbulence, may undermine a country's stability and progress. Moreover, based on the social harmony and administrative efficiency expected to be achieved through the use of ICT in this category.

1.6 Importance of Enabling Technologies in Urban Development

Today's cities are faced with multiple issues such as traffic jams, pollution, resource management, and utilization of infrastructure. While cities continue to expand steadily, there is a higher demand to create intelligent, sustainable solutions to address this issue. The incorporation of enabling technologies in cities is key to sustainable development since they offer practical solutions for the improvement of the standard of living and resource efficiency, as well as decreasing adverse effects on the environment. A smart city is an urban form that applies information technology and telecommunications technologies in an advanced manner to enhance the standard of services provided to citizens. Intelligent technologies such as IoT, AI, Big Data and Analytics, and blockchain technology form the foundations of smart cities. These technologies enhance data collection, management, and analysis that improve the decision-making process and add a variety of solutions in the urban environment. The significance of enabling technologies is discussed as follows and illustrated in Fig. 1.12.

1.6.1 Enhancing Urban Mobility

Among all the problems that occur in the cities, one of the most significant ones is congestion. Enabling technologies provide a set of systems for improving transport in and around urban areas. However, on the other hand, smart traffic systems utilize the IoT to gather data from traffic signals and traffic cameras to manage traffic congestion. These two have a tremendous effect on decreasing the congestion level and

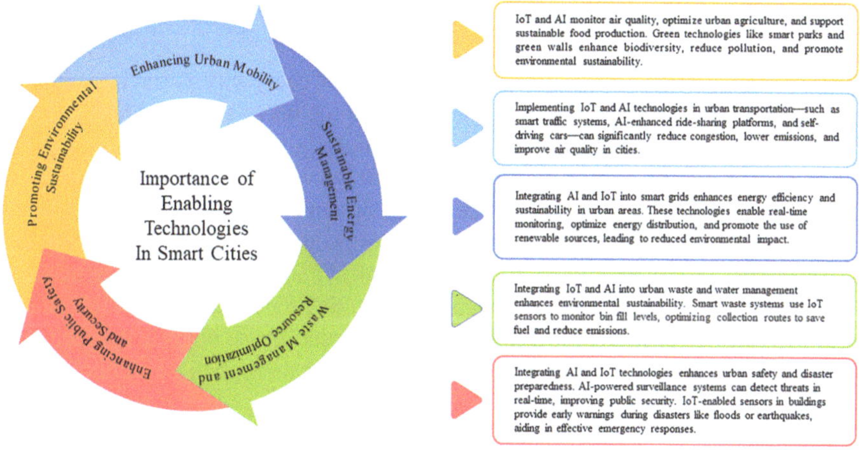

Fig. 1.12 The importance of enabling technologies in smart cities

avoiding long-distance travel. Furthermore, the use of AI in ride-sharing platforms helps in identifying drivers and passengers in the shortest time, hence reducing the number of cars on the road, hence lowering emissions. Such systems involve IoT and AI smart technologies to give real-time updates about the buses, view traffic maps, and change the routing system, hence making the public transit systems smarter and more efficient. Self-driving cars, which are controlled by AI and ML algorithms, are predicted to change the way traffic in cities is managed, and there will also be a decrease in the number of accidents. The technologies play a crucial role in decreasing transport's carbon footprint and enhancing the quality of air in urban regions [60].

1.6.2 Sustainable Energy Management

Energy consumption, especially in urban areas, is another reason that has been identified to cause environmental degradation. Generally, enablers are essential for the support of sustainable energy management [61]. With the use of IoT and AI, smart grids help the distribution and consumption of electricity to be effective. It enabled real-time monitoring and control of energy usage as well as a responsive energy supply, while promotng conservation. Solar and wind energy are available sources that can be incorporated into smart grids to deliver cleaner, less polluted energy than that of fossil energy sources. Smart meters and the smart energy management system equip consumers with the real-time information to influence energy conservation. Another application of AI is to control, monitor, and predict the distribution of energy or predict the needs for maintenance in supply chains, making them more efficient and sustainable.

1.6.3 Waste Management and Resource Optimization

Waste management is critical in making sure that every city is environmentally sustainable [25]. Progressive approaches to managing waste and using resources are progressively implemented with advanced technology. Waste disposal may use IoT sensors placed on waste bins to identify bins' fill levels to guide schedules for waste collection vehicles, hence saving fuel. The programs that are implemented by AI algorithms can assess the amount of waste produced and, thus monitor waste and recycling in city centres. Also, with the help of IoT and AI, smart water management can be built, which must be capable of detecting the leakage and water quality continuously and ensuring the efficient usage of water. It also helps in the recirculation of water and thus the rational utilization of water through these technologies.

1.6.4 Enhancing Public Safety and Security

The safety of the people is always a major concern in most urban areas. There are various enablers that improve security and emergency measures. Security cameras and surveillance systems connected with the internet and then combined with AI technology can help in threat identification as well as prevent such acts in real-time, contributing to enhancing the safety of the public. [29]. Advanced drones, along with robotic systems, are useful for search and rescue missions, as they are effective in responding to an emergency. Smart city facilities in buildings have sensors and connected devices that help in providing warnings during calamities such as floods or earthquakes. These methods of disaster management can prevent loss of lives and damage to properties, hence supporting the development of stronger cities.

1.6.5 Promoting Environmental Sustainability

Enabling technologies are helpful in the sustenance of the environment in the urban environment. The system is aware of the air quality, using IoT and AI to monitor pollution levels and the source of emissions as well. This means that the data can be used for policy formulation as well as engaging in activities that will seek to achieve a better state of the air. IoT and AI-integrated urban agriculture make use of resources in the best manner and also contribute to the reduction of food insecurity [29]. Irrigation systems and precision agriculture technologies also help in the balanced use of water and nutrients, as well as the reduction of adverse effects on the environment and the promotion of sustainable production of foods. Green systems, such as smart parks and green walls, are incorporated into urban environment to improve air quality, mitigate the urban heat island effect, and enhance biodiversity. Some technologies that support green structures help in the maintenance and sustainability of these green areas for the benefit of the ecology.

Enabling technologies play the crucial role in creating sustainable change in urban spaces, which actually turns the cities smart, efficient, and resilient. These technologies solve the major challenges of urbanization, including better mobility of people and goods, proper management of energy and waste, public safety, and ideas of environmentally sustainable cities. Working together with governments, companies, and residents of societies or communities, smart cities will be in a position to apply these enabling technologies in order to develop livable societies for the present population as well as those of the next generations.

1.7 Technology Enablers for Sustainable Smart Cities

Smart cities utilize a diverse range of technological solutions to establish interconnected networks of devices and entities, thereby reshaping urban lifestyles. Figure 1.13 emphasizes the critical importance of key enabling technologies in the emerging vision of smart city initiatives. Smart cities utilize modern technologies to autonomously manage various urban entities, facilitated by the growth of smart devices and IoT networks. As cities integrate ICT and IoT technologies, new applications and services emerge to meet evolving citizen needs, promoting global awareness of smart city concepts [62, 63]. In the pursuit of sustainable smart cities, it is essential to establish deep connectivity among the city's components and layers through advanced data management technologies, such as IoT, Big Data, and Cloud Computing. A few recent advances in communication technology are discussed in this section.

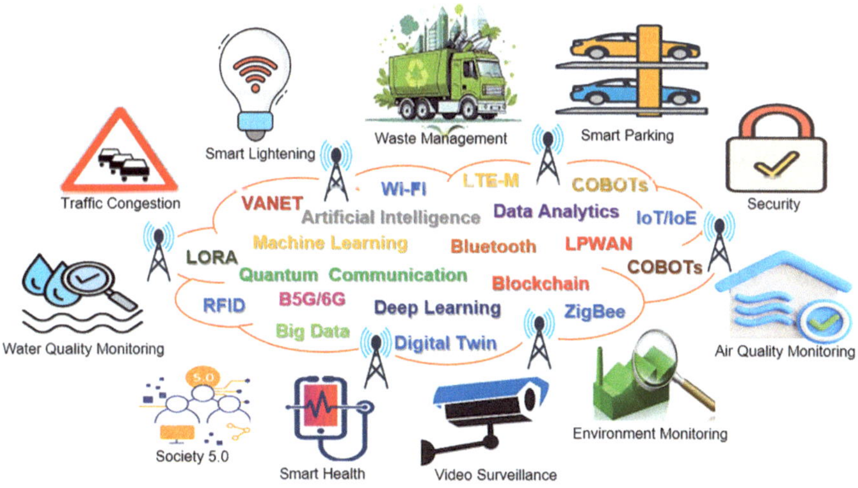

Fig. 1.13 Enabling technologies used in several sustainable smart city areas

1.7.1 Sixth Generation (6G)

As smart cities continue to advance, 5G networks offer enhanced flexibility and faster information-sharing capabilities. The envisioned 6G network must possess intelligence capable of autonomously learning service characteristics and adapting to changes accordingly [64]. The architecture and functionalities should be designed for easy updates in response to dynamic shifts. Expected to deliver three times the data rate of 5G, with reduction in latency and vastly improved reliability, 6G will revolutionize connectivity [65]. Furthermore, the Internet of Everything, the Internet of Nano-Things, and the Internet of Bodies will be reinforced by 6G, making it easier for robots and drones connected to IoT devices to connect. Numerous services, including digital sensing, telemedicine, smart traffic management, environmental monitoring and control, and high-definition video transmission, are provided by these networked systems. By leveraging a combination of radio and fibre optic technologies, the 6G Internet will provide fast connection speeds unaffected by distance, delivering Internet access at unprecedented rates.

1.7.2 Quantum Communication

Quantum computers have gained substantial attention in recent years because of their ability to solve complex problems more efficiently than classical computers. This transformative technology holds promise for various applications, including smart cities. With quantum computing, smart cities can make real-time decisions and analyse data at unprecedented speeds. The greatest potential lies in application involving massive data processing, where absolute accuracy is not crucial. Such situation will be abundant in future smart cities. Quantum computing can be employed to analyse vast amounts of data, optimize traffic routes, and reduce travel times for all vehicles [66]. Similarly, optimizing electrical grids is another compelling use case. Quantum computing holds immense potential for revolutionizing smart city development, enabling real-time decision making and optimization across various domains such as transportation and energy management.

1.7.3 The Internet of Everything (IoE) and Internet of Things (IoT)

Through smart computing, embedded systems, and sensor networks, IoT devices enable simultaneous communication among various devices, empowering users to interact with more complex automated systems [67, 68]. Automation systems also heavily rely on the architecture made possible by IPv6, flexible IoT design frameworks, and energy-efficient management systems. Particularly, a large number of application-focused studies have examined the many applications of IoT [69]. In modern cities, the IoT finds a multitude of applications, encompassing cyber vehicles, digital cities, information cities, smart cities, and connected cities [70].

1.7.4 Artificial Intelligence

As technology advances, we are more capable of solving complex issues, and a greater number of domains are integrating AI into their everyday operations. There is a high probability that this integration will continue to grow, resulting in even greater potential for innovative solutions as AI technology advances [71]. This is done by providing faster and more accurate diagnoses and monitoring of patient health to predict hospitalization. AI enhances entertainment through video games and pattern analyses. The use of AI makes cities more sustainable, safer, and efficient. The implementation of renewable energy is also facilitated by AI. In the travel and transportation sectors, AI helps with reservations and route optimization [72]. Car companies are developing virtual assistants to enhance customer service. Farmers can predict crop yields, identify diseases, and manage irrigation more efficiently by using AI-driven robots, sensors, and predictive analyses. Farming equipment is also automated, which improves efficiency. AI simplifies machine-to-machine communication, data processing, and analysis. Employing ML and deep learning (DL) technologies enables predictive and preventative decision-making models. Moreover, these technologies facilitate a comprehensive understanding of both intra- and intersystem dynamics [73, 74].

1.7.5 Big Data and Data Analytics

The development of sustainable smart cities depends on the utilization of big data, enabling cities to extract valuable perspectives from vast and diverse sources of information [75]. Effective data management enhances public services, reduces operational costs, and optimizes resource utilization [76]. However, integrating such extensive datasets poses significant challenges, not only from a technical standpoint but also in addressing socio-technical considerations such as privacy and power dynamics. With the convergence of ICTs and big data analytics, smart cities can make better decisions in real time, maximizing efficiency and effectiveness. Intelligence, connectivity, and information sharing are fundamental to the development and management of smart cities [77]. By analysing urban situations in real time, big data enables precise insights and supports decision-making processes, promotes sustainability, and offers new services to citizens. Despite its potential, big data presents challenges, particularly in terms of processing and storage.

1.7.6 Blockchain

Blockchain technology offers a seamless means of interconnecting IoT devices, thereby facilitating distributed data management across networks. Each device can securely access and exchange data with other devices through a decentralized ledger stored on the blockchain. Autonomous communication enables IoT devices to interact independently, ensuring secure, efficient, and cost-effective data management without human intervention. It has become the most widely used technology for enabling initiatives related to smart cities as it guarantees reliability, accuracy, and data security [78]. The blockchain technology in smart cities advances the security and flexibility of the exchange of information, providing possibilities for applications in the city.

1.8 Smart City Components with Advanced Technology

Smart city means an intelligent utilization of information technologies in order to design and develop effective city growth and functionality for living. They have core components from intelligent transportation systems that employ the use of sensors plus analytics in the traffic patterns with a view to avoiding traffic congestion. Smart grids refer to the effective control of the electrical grid, where the distribution of electricity depends on renewable energy sources. The implementation of IoT creates an effective monitoring process of utilities to ensure the proper functioning of water, waste, and air quality services. The exceptional communication networks ensure the availability of effective means of communication that can be used by citizens to access various urban services. Data analysis and the utilization of AI in decision making help in solving the urban problems and also enhance public security. Various smart city components with advanced technology are discussed as follows and illustrated in Fig. 1.14.

1.8.1 Digital Democracy and Smart Governance

Democracy 2.0 and Smart Governance are important elements for making the landscape smart and turning cities into smart cities by making use of advanced technology for citizen involvement, managing municipal services, and advancing sustainable development. The three key elements of an intelligent city involve IoT sensors, big data, and AI, helps in collecting and analysing real-time data. This is

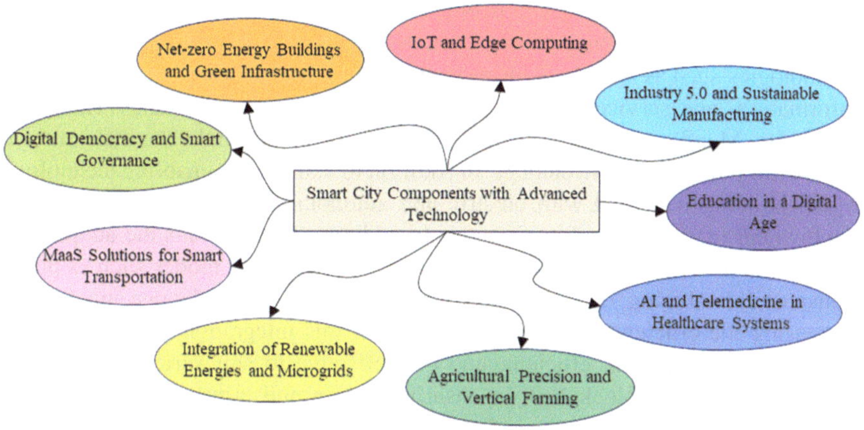

Fig. 1.14 Smart city components with advanced technology

helpful for city officials in two ways: for budgetary purposes and to improve city planning. Digital democracy in smart cities also focuses on enhancing the levels of engagement from the citizens through digital means. E-governance tools facilitate the citizens' opportunity and involvement in decision making, expression, and information acquisition in an effective and efficient manner. With the help of mobile applications and web-based portals, people are allowed to lodge complaints, submit service requests, and attend online meetings. Besides promoting an increase in civic participation, it also promotes the aspect of people's ownership and responsibility towards the fulfilment of good governance. Smart governance also includes the use of smart infrastructures like ITS and smart buildings to minimize the negative effects on the environment and enhance the living standards of the people. Transport networks and the use of intelligent traffic lights, in combination with self-driving cars, are included in smart transportation systems, which help to reduce traffic jams and pollution. Moreover, smart grids have evolved to enhance the distribution of electricity in the network and reduce wastage, and increase the acceptance level of renewable energy sources. Furthermore, information security helps to protect the big data produced and guarantees users' confidentiality and integrity in interaction. Hence, as smart cities advance, information democracy and smart administration will remain important in the creation of strong, progressive, and resourceful, as well as sustainable urban environments globally.

1.8.2 MaaS Solutions for Smart Transportation

MaaS is the way that the future smart city is redefining mobility to improve transportation systems, sustainability, and accessibility through the use of new technologies [79]. The essence of this process is the unification of the transportation means in one application that will allow people to choose the most suitable type of transport. Through integration with big data, analytics, IoT, as well as other related technologies such as AI, MaaS offers customized travel advisory, real-time shuttle information, and service logistics that would minimize traffic jams. Thus, in a smart city, MaaS can be seen as an essential element, as it links the different systems within the urban environment. In addition to being a key enabling technology, MaaS is being deployed with electric and autonomous vehicles, as well as the deployment of standard safety features. In addition, MaaS means the adoption of different modes of transport, which involves promoting key modes such as public transport, ride-sharing, bike-sharing, and micro-mobility, all under one app. Besides increasing the efficiency in utilizing resources, this integration also brings forth ideas that are practical, efficient, and cost-effective to suit various commuters. There is the use of blockchain in ensuring secure and transparent transactions, which enhances people's and service providers' trust and confidence. It increases the quality of people's lives and, at the same time, advances city-wide goals and priorities for the widespread development of intelligent cities. As technological advancements become more culturally integrated, the relationship between MaaS and smart city elements will become more helpful and transformative for the urban landscape [80].

1.8.3 Integration of Renewable Energies and Microgrids

Renewable energy systems and microgrids are the key components in the generation of smart cities since they utilize smart technologies to design and build environment-friendly, efficient, and reliable smart cities [81]. Renewable energy sources like solar electricity, wind electricity, and bioelectricity, with microgrids, are deregulated systems of energy that are independent of the main power grid to supply commercial and residential areas with alternative power. It contributes to energy security when adopted, offers flexibility in the sources of energy that are used, and equally promotes sustainability as it reduces the use of fossil sources of energy, hence cutting down on carbon emissions. Microgrids can self-organize and control their operation with the help of remote control systems, energy storage devices, and real-time data, and deliver energy supply on demand and manage supply and demand variability and intermittency of renewable energy resources while providing a localized supply independently from the main grid if necessary [82]. Monitoring of energy flows is possible through IoT technology, and the AI predictions and ML algorithms predict and manage the consumption of energy and distribution. Like smart lighting, energy-efficient buildings, and electric vehicle charging stations, are allowed to benefit from this integration in that renewable-powered microgrids supply energy to these smart city components. The combination of renewable energy, microgrids, and the use of technology is the smart city approach in view of improving the livability of cities. It not only satisfies the need for energy in growing populations of cities but also helps lead cities into a green, smart, and sustainable future.

1.8.4 Artificial Intelligence and Telemedicine in Healthcare

The utilization of AI, especially in combination with telemedicine, as part of the healthcare system, is one of the essential key trends in the formation of smart cities. The implementation of AI in the diagnosis of illnesses and in the formulation of individualistic treatments, as well as the rational use of health resources, is remarkable in the healthcare industry. The usage of big data from electronic health records, images, and genomic data by using ML algorithms to determine the predictors of a certain survival, such as the ability to detect cancer at an early stage. This ability is vital in preventive health care, hence easing the load on the healthcare facilities of the smart city. Additionally, the current advanced technologies in AI have chatbots and virtual health assistants that offers advice, timed prescriptions, and mental health, and thus increase patients' involvement and compliance [83]. While telemedicine takes healthcare a step forward by allowing healthcare consultations, monitoring, and remote management of patients with chronic diseases. In smart cities, through the use of electronic technology in diagnosing the signs, the patient can reach the relevant specialist and obtain medical assistance irrespective of geographical location. This is especially useful for the elderly or patients who have difficulties in travelling in search of health facilities. Telemedicine decreases the number of face-to-face contacts, hence limiting the spread of diseases within the facilities and the crowding of patients.

1.8.5 Industry 5.0 and Sustainable Manufacturing

Industry 5.0 is the latest industry model in manufacturing, which provides more emphasis on digitalized technologies along with human improvement and sustainability [84]. This iteration aims to tackle global problems by incorporating efficiency promoted by Industry 4.0, innovating through automation, data exchange, and smart technologies. For the development of smart cities, there is a new direction of Industry 5.0 that helps in redesigning the urban environment to enhance a sustainable and resourceful environment where civilization resides. Various technologies like AI, IoT, and blockchain make the components of smart cities more efficient, save waste, and enable real-time data sharing to make the city smart and to adapt to the challenges. Industry 5.0 involves the elastic element for green people, and smart production for sustainable manufacturing, where organizations opt for renewable energy, circular economy, and green technologies. These practices are useful to smart city components through the utilization of sustainable materials and energy in constructions, efficient methods of managing waste, and other optimal approaches.

Smart city technologies that include predictive analysis and IoT sensors ensure efficiency in the utilization of resources as well as control of environmental conditions that recommend proper growth and development of the city, as well as the inhabitants' welfare. Moreover, Industry 5.0 promotes the integration of man and machines with improved performance regarding creativity, design, and problem-solving. Self-driving cars, ITS, and the smart grid are examples in which technology can be utilized to develop better sustainable methods of transportation. Industry 5.0 and sustainable manufacturing together are the future of smart cities [85]. All these initiatives complement the functionality of the components, however, at the same time increase the sustainability level and provide an opportunity to make cities more livable, productive, and environmentally sustainable.

1.8.6 Internet of Things and Edge Computing

The efficient use of IoT and edge computing in the components of smart cities is evolving sustainable smart cities through the implementation of innovative technologies that enable urban development in terms of productivity, environmental conservation, and human well-being [86]. The principles of this transformation are based on the essence of IoT devices that work as data input points where data from the traffic signals, physical facilities, and environmental monitors is accumulated. They allow for the monitoring and control of the infrastructure of cities so as to improve the decision-making processes. By that, edge computing gets the data close to the source to manage it in a way that reduces latency, maintains minimal bandwidth consumption, and ensures that actionable information is processed as quickly as possible. For such a reason, local data processing capability will be necessary when the application must be capable of responding instantly, as in the case of self-driving vehicle control, emergency response services, and safety measures. Smart city systems include smart transport systems, energy management grids, and smart waste management systems, and most of them integrate IoT and edge computing.

1.8.7 Net-Zero Energy Buildings and Green Infrastructure

Energy-positive infrastructures, sustainable residential and commercial buildings, and green infrastructure are part of smart cities as they embrace the use of technology to address the sustainability [87]. Net-zero energy buildings are infrastructures that use energy to the extent that it is produced by active renewable resources such as solar energy, wind energy, or geothermal energy. These structures frequently use energy-saving solutions that include smart thermostats, LED lights, and enhanced insulation, to reduce energy requirements. These buildings use data analytics and IoT devices to track energy usage and efficiency at the same time. Green roofs, together with rain gardens and permeable pavements, are considered green infrastructure elements that contribute to the improvement of the ecosystem of urban areas. Green infrastructure is applicable in ensuring increased stability of the urban areas through integration with other smart city elements, including water management and transportation systems [88]. These systems are integrated through sensors and automatic control systems, which allow the efficient use of resources within a building and also minimizing environmental changes. Another feature of intelligent cities is that advanced technology acts as support in the introduction of net-zero energy buildings and green infrastructure. For example, smart grids are efficient in distributing green energy, and with the help of AI-based platforms, it is efficient to plan urban infrastructure. Also, using blockchain technology in the performance and recording of energy transactions will increase trust and collaboration among the stakeholders. These innovations not only contribute to the improvement of the sustainability of big cities but also lead to an increased quality of life for the citizens of such cities. The integration of technology-based net-zero energy buildings, green infrastructures, and smart city features is ensured in supporting sustainable city development.

1.8.8 Agricultural Precision and Vertical Farming

Agricultural precision and vertical farming are some of the significant areas that will define a smart city through the use of smart technologies in the farming sector, with special focus on food production that is efficient, sustainable, and environmentally friendly [89]. Precision farming in agriculture involves data analytics, IoT, and AI in the management of resources that are light, such as water, fertilizers, and pesticides, so that while the yield will increase, the negative impact on the environment will be reduced. This is done by incorporating information from the sensors that provide the real-time conditions of the soil and prevailing weather conditions and the health of crops in the fields. This accomplishes the reduction of wastage, increased efficiency in production, and promotion of the right food usage essential for large populations living within these urban cities. Vertical farming illustrates innovation through redesigning the technique of agriculture in the centralized areas that are urban. It helps minimize the widespread losses of land use since food is produced near consumers in urban areas, hence cutting down on the costs of transport as well as emissions. Technological products like LED lighting and hydro-nutrient and aqua-nutrient techniques allow farming all year round without regard to the outdoor weather. It has high productivity while using comparatively less water and no pesticides and therefore is identified as a sustainable measure towards smart cities. The use of these developed agricultural practices in smart cities also complements other advanced technologies.

1.8.9 Education in a Digital Age

Employing education in a smart city provides one of the components of smart cities that work under an advanced model of technology integration for learning purposes [90]. Educational institutions provide the perfect environment for connecting with the growing technologies of smart urbanization to offer unique and advanced learning experiences for students. There is a need to make reference to smart city components; for example, ubiquitous internet connection, IoT devices, and data analytical platforms in the redesigning of the frameworks governing education. In a smart city, connectivity facilitates access to huge content on the internet, which the students use to learn at their own pace and style. Smart gadgets in the framework of IoT can also be used in the learning process by applying virtual and augmented reality, which makes it possible to study using real-life devices and situations. Data analytics, the other facet of smart cities, can further change the educational practices by improving the understanding of the students' achievements and activity levels. The data allows the educators to make proactive decisions regarding course content and to identify learning deficiencies. This not only makes the education system more effective and teachers oriented towards the needs of their students but also provides an effective way of resource allocation in regard to students most in need. Also, the use of AI and ML in the expansion of smart cities empowers academic institutions to achieve administrative decision making and to spare time for teaching [91]. It also supports the development of environments that could dynamically adapt to students and change the content and related assessments according to the student's pace, need, and/or performance. In other words, education in a digital age with the support of smart city technology is about drawing a certain environment or a set of conditions in which technological and informational solutions contribute to the improvement of the educational process. It not only makes learners capable of handling an era that is already characterized by an advanced technological environment but also shapes their ability to continue learning, which is crucial in the society of the future.

1.9 Society 5.0 and Smart City Sustainability

Society 5.0 can be defined as a system that encompasses smart city sustainability in its development process in urban surroundings. It is described as a society in which the two environments, the physical environment and the cyber environment, are integrated to improve the well-being of people, with consideration of the various challenges that come with the advancement of technology [2]. First of all, Society 5.0 is aimed at creating smart cities, which is a concept of the urban area using ICT, data analysis, and IoT to enhance people's well-being, businesses/economic productivity, and environmental conservation. Smart cities represent the concept of the new approach towards the sustainable development of cities based on the utilization of real-time data and magnificent systems regarding the reliable usage of resources, transport, energy, and waste disposal systems. It makes it easier for cities to take appropriate measures in handling the challenges regarding the environment and minimizing the impact on the environment by reducing carbon emissions. One of the essential progresses is the emergence and implementation of the smart grids and the renewable energy plants. This is so due to the fact that they optimize the flow of energy demand and supply, minimize the dependency on hydrocarbons, and encourage the utilization of renewable energy. Moreover, the use of smart solutions in transportation through electrically charged vehicles as well as self-driven traffic systems reduces emissions and improves the quality of the air. Both Society 5.0 and smart city sustainability are to develop sustainable and resilient urban environments that build a symbiosis between technology, environment, and society [92]. Educating communities in the values of innovation and sustainability, they demonstrate how future cities should be created that focus on the people's quality of life as well as on a responsible attitude towards the environment. This advancement will help meet some of the current issues posed by urbanization and global climate change, thereby shaping cities of the future for sustainable and prosperous living.

1.9.1 Ethical and Social Implications

With the further advancement of technologies in the coming years, advances in Society 5.0 and smart city technology will revolutionize aspects of human lives. The concept of Society 5.0 has been developed in Japan and is based on the idea of the integrated human society, which is based on the development of technologies and human needs. Smart city concepts are the current attempt at the sustainability of cities with the help of technologies. These innovations are resources for an array of social, political, and ethical issues that enables better social utility and growth.

1.9.1.1 Data Privacy and Security

It is therefore quite dominant to address a critical issue of data privacy and protection in the transition to Society 5.0 and smart city designs. These activities are mainly based on data collection and analysis of the programs. Of the billions of connected IoT devices, including surveillance cameras and even biometric systems, the amount of data that is being generated cannot be overemphasized. Although such data helps to increase the efficiency of city organizations and enhance the level of residents' comfort.

Ethical Implications: It is a potential violation of the privacy rights of the people by acquiring a lot of personal details without their knowledge. The dangers include cases of invasion of privacy, one's identity being stolen, being discriminated against, or being exploited in many other ways.

Social Implications: A recommendation is made that the majority of the implementation of these technologies should be founded on public confidence. For this reason, citizens may refuse to employ new technologies if they ever feel their privacy is at risk. Also, since stigmatized groups are often targeted for surveillance, the advancement in technology can increase social injustices.

1.9.1.2 Digital Divide

Society 5.0 and smart cities are such wonders that can only exist if everyone is adequately connected to digital technology. However, the truth is that it is still present today, with a clear division between developed and developing countries. This divide is defined by the technology gap, lack of necessary knowledge in using technology, and utilization of digital tools to improve one's standard of living.

Ethical Implications: Technology brings social and economic inequality in regard to access to the technologies and impacts of their use. Some people will be left stranded without using the digital tools for chances to be able to engage in any sector of the digital economy or even access some vital services.

Social Implications: Bridging the digital divide means that government and private enterprise need to have specific policies and spend cash to put in, as well as technology within a society. This encompasses the hope to broaden internet connection and digitization, and guarantee that the use of innovations in society is not a preserve of a certain class or standard.

1.9.1.3 Surveillance and Autonomy

The use of high-level security systems in employing smart cities increases the invention of autonomy and freedom issues. The public and private areas currently have extensive surveillance, which will result in the formation of a surveillance society that infringes the freedom of persons.

Ethical Implications: Privacy and security remain one of the most sensitive issues that should be taken into consideration. Having too many countries will inevitably result in a compromise of personal freedom and human rights. There is a possibility of developing a society where citizens will be under constant surveillance and, therefore, feel threatened into self-censorship.

Social Implications: Despite the acceptance of surveillance measures, an individual must understand the cultural and social differences that exist. In certain situations, an increase in security can be seen as positive; in others, it is perceived as a restriction of rights. Public debate and the active involvement of the public in decision making are necessary to address such problems and guarantee the proper use of surveillance technologies.

1.9.1.4 Environmental Sustainability

There are Intelligent Cities which have some principles dealing with pollution, waste disposal, or energy usage. Society 5.0 also has communication with the aspects of technology and nature management.

Ethical Implications: This is a simple but significant point that states that technology may be used for sustainable purposes, but on the other hand, it may be the cause of the problem. The making and discarding of electronic gadgets lead to e-waste, and the power demand for digital structures undermines environmental savings.

Social Implications: A number of issues arise when designing sustainable technological solutions for urban areas, which involve solving for environmental sustainability and ecological footprint while implementing technological solutions for urban development. Policies should thus support the utilization of renewable energy, and an appropriate consumption pattern and ensure that new technologies do not harm the environment.

1.9.1.5 Economic Disruption and Employment

It is desirable to explore and analyse how the growth of Society 5.0 and Smart Cities is likely to affect present occupations and industries. These kinds of changes may cause job loss since there are certain tasks that can be handled by equipment and robotics.

Ethical Implications: As much as technology is considered a beneficial factor in society, there is an ethical question of how much it can change the unemployment and economic status. Technological advancement should be achieved responsibly, meaning that the workers should not be provided inadequate society, which can only be availed through technology.

Social Implications: It is necessary to invest in the re-skilling and up-skilling of the workforce to deal with economic disruption. Education systems should change a bit so as to prepare people for the right skills needed in the emerging economy. Moreover, there is a need to increase the buffers that are social support mechanisms that are provided to the people who are affected by these technologies.

1.9.1.6 Governance and Accountability

The possibility of Society 5.0 and the smart cities approach can only be realizable by the integration of proper governance and accountability. The complexity of these initiatives necessitates transparent decision-making processes and robust regulatory frameworks.

Ethical Implications: The breakthrough of these technologies in the hands of a few technology companies or government bodies may result in ethical issues. The case is that there is a potential to develop 'technocratic' structures of decision making that are almost immune to interference from the public.

Social Implications: It is important to be accountable while implementing digital technologies in different organizations so that people can have confidence in the end product. In this case, with the consciousness achieved by citizens on participation in decision making on technological developments that affect them, this means that technological solutions that are being implemented meet the standards and values of society.

It can be stated that there are great opportunities associated with the transformation to Society 5.0 and smart cities in terms of improving the quality of life in an urbanized society and dealing with various problems. Such a change must be done under the key principles of ethics, human rights, democracy, and economic justice. It is, therefore, imperative to discuss issues like data and privacy, the digital divide, surveillance, the environment, and economic impact, as well as the impact of governance, to make these initiatives as productive as possible. When engaging in this technological context, it is necessary to make sure that people's rights are respected and that the technology processes are conducted in a fair manner. In this way, it is possible to build a new smart and sustainable urban environment. It is a shift towards not only the technological transformations of Society 5.0 and smart cities but also the development of the socio-economy envisaged by citizens themselves.

1.10 SWOT Analysis for Sustainable Smart City

A sustainable smart city may address its limitations and reduce potential threats by exploiting its opportunities and strengths, enabling for a more prosperous and sustainable urban future. The SWOT analysis is illustrated in Fig. 1.15.

1.10.1 Strengths

1.10.1.1 Advanced Technology Integration (S1)

Advanced technologies like IoT, AI, and big data are used by sustainable smart cities to improve urban management. This fusion of technology results in improved efficacy in the usage of energy, resource allocation, and traffic management. Smart grids reduces waste by instantly modifying the amount of electricity that is delivered and needed. Predictive infrastructure services are enabled by AI-driven analytics, which reduces the frequency of significant failures. The connected systems

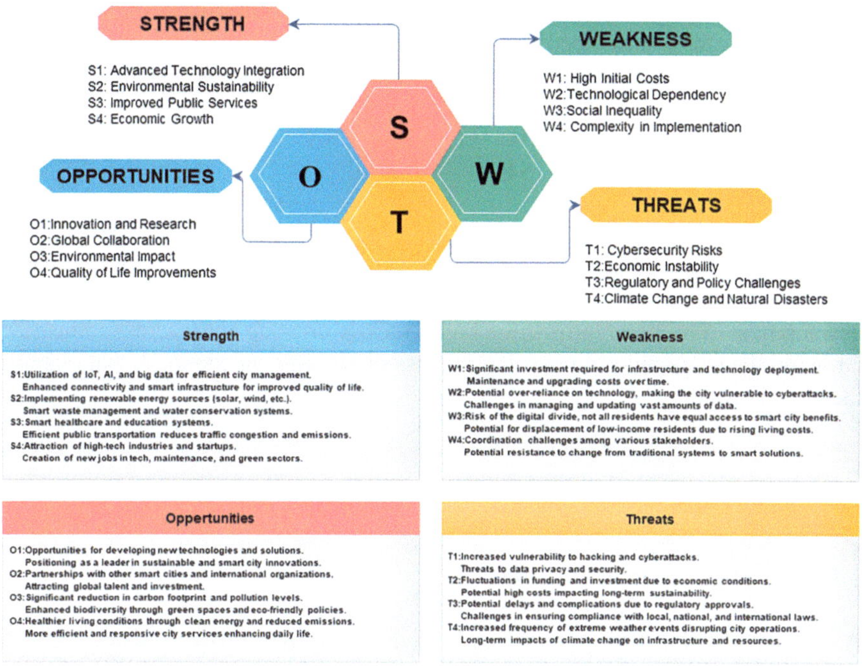

Fig. 1.15 SWOT analysis for sustainable smart cities

improve the performance of the city, making it more convenient for citizens to live within.

1.10.1.2 Environmental Sustainability (S2)

Cities focus on renewable energy sources, incorporating solar electricity, wind, and geothermal as a means to reduce carbon emissions. Energy-efficient technologies employed in smart buildings minimize the amount of energy used. Waste management is improved by automated recycling and sorting systems, which considerably reduce the quantity of landfill area needed. Smart irrigation and leak detection mechanisms are two ways to conserve water. By executing green projects, cities contribute to global efforts to reduce climate change and protect the environment for future generations.

1.10.1.3 Improved Public Services (S3)

The capacity to remotely monitor patients through smart healthcare technologies minimizes the need for hospital visits and accelerates emergency response times. Digital classrooms and e-learning help educational organizations by making learning more accessible to all. Public transportation is improved using real-time data, which leads to enhanced reliability and faster services. The city is now more beneficial to both residents and visitors due to these developments in public services.

1.10.1.4 Economic Growth (S4)

Developing a sustainable smart city attracts high-tech growth companies and businesses, which stimulates entrepreneurship and innovation. Investors and scientists from all over the world move to the city because it has a centre for research and development. The hasty growth of businesses is creating jobs across industries. Growing the economy promotes the process of building an economy that changes the global economy.

1.10.2 Weaknesses

1.10.2.1 High Initial Costs (W1)

A substantial amount of expenses must be allocated to infrastructure and technology for the development of smart cities. This involves the expenditure on emerging sensors, implementing smart grids, and improving existing systems. These anticipated expenses can be excessive for several towns, particularly with limited funding. The costs of retaining, enhancing, and implementing safety measures can ultimately limit expenses, making it challenging to implement the plans without stable funding.

1.10.2.2 Technological Dependency (W2)

Cities that rely heavily on technology could be more susceptible to cyberattacks and other technical challenges. For example, a power grid intrusion producing a disruption in the distribution of electricity might lead to extensive panic. Immense hindrances may arise when it comes to the security of management and data. An excessive dependence on technology may also lead to the loss of conventional knowledge and abilities, which makes it more challenging for the city to manage regularly in a situation of a technical malfunction.

1.10.2.3 Social Inequality (W3)

Citizens may not have awareness or understanding of the technology required to take advantage of smart city services, and social inequality may be escalated by the technological gap. Therefore, wealthy communities can get more benefits than less rich ones. As property prices and living costs rise, the city's growing interest may lead to the removal of low-income citizens and the subsequent growth of the socio-economic gap.

1.10.2.4 Complexity in Implementation (W4)

The implementation of a smart city is coordinated by stakeholders, government organizations, and businesses. This complexity could cause higher costs and inefficiencies. Resistance to the shift from conventional to smart systems is another big challenge that may come. Securing the collaboration and alignment of all investors is essential; however, it can be challenging and hinder the progress and financial achievement of smart city initiatives.

1.10.3 Opportunities

1.10.3.1 Innovation and Research (O1)

Sustainable smart cities draw researchers and professionals' interest in developing new ideas because they offer an ideal atmosphere for innovation. The effective testing of evolving technologies is possible by research labs and related programmes, which gain rapid deployment and prototyping. Developments in fields like sustainable agriculture, renewable energy, and urban mobility evolve via academic collaboration and research. Because of steady technological progress, which promotes economic growth, the city is considered the standard in sustainability.

1.10.3.2 Global Collaboration (O2)

Cities may interchange useful data, resources, and time management by creating partnerships with other smart cities and globalized entities. These collaborations enhance the city's skills to think resourcefully and solve issues more promptly. Involving itself in international networks also draws funding and investment worldwide, further driving the city's prosperity. With these collaborations, the city can progress, potentially leading to its sustainability targets, joint ventures, targeting funding opportunities, and research initiatives. Developing networks with other smart cities and globalized entities enables cities to share best practices, expertise, and resources.

1.10.3.3 Environmental Impact (O3)

The environmental impact of smart cities may be significantly reduced with efficient energy usage and emission control. Smart grids, green buildings, and electric public transportation all aid in minimizing pollution. Enhancing urban green spaces and sustainable approaches improves ecological stability and biodiversity. Smart cities play a major role in global initiatives to alleviate change in climatic conditions and stimulate environmental sustainability.

1.10.3.4 Quality of Life Improvements (O4)

Regular advancement in education and healthcare is balanced by a lowering in traffic congestion and pollution induced by smart technology integration. Smart homes have mechanized infrastructure that offers energy efficiency. A group that has better emergency services and public safety is more reliable and secure. These updates will enhance the city's services, capacity, and liability to encourage the comfort of its people.

1.10.4 Threats

1.10.4.1 Cybersecurity Risks (T1)

Smart cities are a significant aim for cybercriminals since they rely more on integrated technologies. By limiting vital services like transportation, healthcare, power, and water, cyberattacks have the ability to damage the system. Ensuring robust cybersecurity defence requires monitoring, investment, and updates. Data violations have the possibility to compromise the effectiveness of smart city systems by initiating privacy breaches and a loss of trust among residents.

1.10.4.2 Economic Instability (T2)

Financing deviations and challenges could risk the sustainability of smart city initiatives. If the economy declines, high and ongoing expenses may become unaffordable. Dependence on third-party funds involves further threats in the event that financiers minimize the investments. To reduce these risks and maintain the rate of smart city growth, an appropriate and diverse financial system is vital.

1.10.4.3 Regulatory and Policy Challenges (T3)

Managing a complex framework of rules and regulations might cause to smart city project implementation to be significantly delayed. Resisting agendas or bureaucratic challenges at various levels of government might hinder progress. Moreover, guaranteeing that a range of national, regional, and worldwide regulations are complied with requires a considerable investment of resources. To simplify processes and reduce obstacles, it is crucial to communicate proactively with lawmakers and regulatory bodies.

1.10.4.4 Climate Change and Natural Disasters (T4)

Even if it is intended to be robust, natural disasters and climate change have an impact on smart cities. More often, extreme weather conditions, like floods, storms, and heat waves, impede services and infrastructure. Energy systems, quality of life, and water supply may be affected by climate change. Strong disaster response schemes and flexible infrastructure are vital for reducing these threats and protecting the city's resilience.

1.11 Case Studies on Sustainable Urban Development: Big Data, Social IoT, City Models, and Citizen Involvement

A systematic discussion of sustainable urban development strategies is provided in the four case studies. Each case study includes various aspects of the development of smart cities, like incorporating the concepts of smart, circular, and green cities; employing big data to manage cities; and utilizing the Social Internet of Things (SIoT) to promote urban mobility. The analysis of several city models demonstrates how cities enhance sustainability with the adoption of policies. Big data applications are essential to attaining Sustainable Development Goal 11 in a phased manner. The SIoT case study illustrates how to use IoT and AI to improve traffic flow and alleviate congestion. The impact of participation in smart city initiatives is highlighted by a study of Quayside, Toronto, community involvement. The key aspects of the case studies mentioned are illustrated in Table 1.4.

1.11.1 Case Study 1: Sustainable Development in City Models

This case study explores how evolving city models like smart, circular, and green cities are crucial to accomplishing sustainable urban development [93]. The study explores how these models promote sustainability by using content analysis and

Table 1.4 The key aspects of the case studies

Aspect	Case Study 1	Case Study 2	Case Study 3	Case Study 4
Introduction	Analysis of smart, circular, and green city models	Utilization of big data in smart cities	Integration of SIoT for urban mobility	Evaluation of citizen participation in the quayside, Toronto
Primary aim	Understand the contribution to sustainable development	Explore the effective use of big data to achieve SDG 11	Evaluate the SIoT architecture for managing urban mobility	Assess the quality of citizen engagement and develop new heuristics
Methodology	Content analysis and concept mapping using Leximancer	Analysis of data applications across three development phases	Implementation of the SIoT system with AI and ML techniques	Document analysis of citizen engagement initiatives
Phases/areas of focus	Identified 24 research areas for circular cities	Initial, intermediate, and advanced phases	Real-time data collection and analysis	Different initiatives with varying quality of participation
Key tools/ frameworks	Conceptual model incorporating the DMAIC tool	Framework for data orchestration strategies	AI, ML, sensors for real-time traffic data	Cardullo and Kitchin's scaffold of smart citizen participation
Case study application	Maribor, Slovenia	City development	Cagliari, Italy	Quayside, Toronto
Challenges identified	Data collection, measurement inconsistencies	Data availability, interoperability, and privacy issues	Traffic flow, congestion, and data security	Variation in engagement quality, need for nuanced heuristic
Results	Validated conceptual model, identified challenges	Developed a framework for phased data orchestration	Significant improvement in traffic flow and reduction in congestion	Varied engagement quality, identified gaps in existing heuristics
Conclusion	Importance of evolving city models for sustainability	Importance of a phased approach to data orchestration	Effectiveness of SIoT in managing urban mobility	Need for a new heuristic to assess citizen engagement complexity

concept mapping to evaluate research paper outcomes. The main aim was to understand and construct a conceptual model for circular cities. The research validated 24 pivotal key areas for the growth of circular cities through Leximancer software. Based on the content analysis, city models are emphasizing sustainable development and increasingly combining useful solutions. The DMAIC tool was combined into the established conceptual model, providing a systematized approach for addressing challenges. The model's testing on Maribor discovered issues with data collection and erratic measurements that made it difficult to track the shift to circularity. The method assists cities in heading towards sustainability despite these hindrances.

The study highlights the consequences of robust frameworks for data collection and monitoring, including the dynamic integration of city models. The outcomes show how imperative it is to merge green, circular, and smart city solutions for incorporating strategies for urban sustainability. Urban planners and executives gain from the intellectual model's practical advice; however, further investigation is essential to improve its usage. The study generates substantial support for our understanding of how city models aid in sustainable urban development.

1.11.2 Case Study 2: Big Data for Smart City Development

The case study studies the key role that big data plays in the growth of sustainable smart cities with the aim of attaining Sustainable Development Goal 11, to make cities comprehensive, safe, robust, and sustainable [94]. The study differentiates three distinct stages of smart city evolution: beginning, moderate, and advanced. A distinguished set of data orchestrating approaches is needed at each stage. During the first stage, the key focus is on building a robust governance framework and data infrastructure. The intermediate stage emphasizes interoperability and data scalability, while the advanced stage deals with more complicated issues such as security, data privacy, and real-time processing. A comprehensive framework is generated to guide data orchestration throughout these stages as part of the study's approach. The observation demonstrates that customised data orchestration techniques are crucial at each stage of the development. The first stage deals with data silos and limited access, requiring stakeholder collaboration and core capabilities. The advanced phase demands AI, ML, and strong data security measures due to its enhanced real-time processing and data orchestration. The research concludes that a phased method to data orchestration is required for the development of sustainable smart cities. By handling specific data-related issues at each stage, cities use big data to boost urban management and attain sustainability goals, which will eventually advance the quality of life for their residents. The suggested framework provides useful insights to researchers and practitioners in the big data and smart city domains.

1.11.3 Case Study 3: Social IoT for Urban Mobility

This case study investigates the development of an SIoT system in Cagliari, Italy, within the confines of the infrastructure of the smart city prototype, with the goal of enhancing urban mobility [95]. The SIoT system utilizes real-time data collection from both public and private vehicles, as well as pedestrians, to improve traffic flow and minimize congestion. The strategies comprised putting sensors and devices in cars and on pathways for pedestrians to collect traffic data that an AI system processed. The technology utilizes AI and ML to analyse traffic patterns and provide optimal routes and management strategies. Metrics involving traffic flow and congestion levels were compared between the pre-installation traffic data and the execution of the SIoT system. Additionally, a scheme was implemented that offers citizens and local government deputies real-time traffic updates. Based on the results, daily traffic congestion was decreased by up to 35%, and traffic flow was considerably improved. The elements of the SIoT system include reliable data exchange and increased defences against threats. This impactful application shows how integrating IoT with AI and ML can deal with urban mobility challenges in a dynamic way. The increased urban mobility in the study's outcome serves as verification of how well it manages traffic and mitigates congestion. This accomplishment creates a model for other communities to enhance urban transportation using cutting-edge technologies, creating urban systems that are more efficient, sustainable, and interconnected. The impact highlights the transformational effect of SIoT solutions on smart city advancement for a better living environment in cities.

1.11.4 Case Study 4: Citizen Participation in Smart Cities

This case study employs Cardullo and Kitchin's Scaffold of Smart Public Participation as a heuristic framework to analyse public participation in Quayside, Toronto's first smart city neighbourhood [96]. Its main aim is to assess the ability of citizen engagement schemes and determine how well the latest heuristics confine the complexity of this type of participation. By analysing 17 reports and advertising related to public engagement, the study shows a variance in the level of participation within schemes. Certain schemes exhibit significant participation, whereas others currently have restricted possibilities for public feedback. This case study emphasizes the need for more complex frameworks that deal with the challenges of citizen participation in smart city advancement. These case studies illustrate frameworks, challenges, and methodologies to provide a thorough understanding of sustainable urban development. This thorough analysis employs inclusive, effective, and sustainable cities by aiding academics and professionals to understand strategies that are efficient and areas that still require improvement.

1.12 Organization of the Book

Chapter 1: In this chapter, we discussed the idea of sustainable smart city development that helps to integrate and enhance the process of restructuring the urban environments to be smart to face the challenges due to advancements in technology. Some strategies, like digital democracy and smart governance, MaaS, renewable energies and microgrids, healthcare, Industry 5.0 and sustainable manufacturing, IoT and edge computing, Net-ZEBs and green infrastructures, agricultural precision and vertical farming, and education, are highlighted in detail for the further development of the cities and making it them more livable for the citizens.

Furthermore, Society 5.0 is essential to the fundamentals of smart city sustainability regarding a human-oriented society where economic growth and social issues are postulated to be fixed via the confluence of cyberspace and the physical space. However, it is very essential that the transition principles be ethical-social, equal, private, and diverse. In this way, if smart cities adopt these ideas and technologies, they can find a way of elaborating a sustainable development that respects both people and the environment.

Chapter 2: Smart city development and the United Nations (UN) Sustainable Development Goals (SDGs) show a perfect relationship to serve as a key to sustainable city development. Smart cities deal with the utilization of technological improvements to increase the comfort and effective use of the available resources for the benefit of the city. In this chapter, we discussed the awareness of the incorporation of smart cities in enhancing the achievement of the goals set by the UN-SDGs, besides pointing out major research directions. Technology is also an important factor that needs to be embraced in order to achieve the set goals within smart city solutions related to energy, waste, and transport. Wireless technologies and big data allow the collection of reliable and up-to-date data, improving the efficiency of city management. Also, technological management in smart cities largely deals with the application of IoT, AI, and Big Data for better quality of life of civil structures and systems with the alignment of SDGs such as affordable and clean energy, sustainable cities and communities, and climate change. They offer a clear approach to meeting global challenges such as poverty, unfair distribution of resources, and environmental destruction. Particularly in relation to these goals, the approaches that are adopted by smart cities can be further defined. The goal of sustainable developments in smart cities includes inclusion for all in the urban environment, improvement of climate change, and better utilization of resources. By adopting smart city initiatives that are aligned with the SDG goals, it is possible for cities to achieve sustainable economic, social, and environmental development. Conclusively, the combination of smart cities and SDGs has the potential to envision sustainable living environments of the future. However, more research and development efforts are essential in eradicating the barriers that confront the smart city to unlock the optimum potential of attaining the 17 SDGs.

Chapter 3: In this chapter, we discussed the integration of smart cities with IoT technologies. They offer new approaches that optimize urban life in several ways.

The emergence of new-generation IoT networks and the implementation of 5G enhance the promise of low latency, which is crucial for applications that employ real-time operation. The application of IoT in urban mobility, resource management, remote tracking systems for air and water quality, smart waste management, and energy grids help to make systems of cities more efficient and sustainable. To execute secure transactions, other methods such as privacy-preserving data are needed, and that requires technologies like blockchain and artificial intelligence. Interoperability as well as scalability remain major issues with IoT systems, and therefore, proper protocols and edge computing should be established to address large-scale implementations. Policy, security, and integration are crucial factors for a coherent IoT environment in terms of devices. Future directions of IoT technology include the combination of the IoT with augmented reality and health, autonomous IoT devices, and self-governing networks. These developments provide a sense of possibility for better living in urban environments but also create new research opportunities, like compatibility issues and security threats.

Chapter 4: In this chapter, the use of smart technologies that help implement advanced artificial intelligence in smart cities is discussed, which is one of the significant steps towards creating sustainable urban development through better analytical activities and decision-support systems. Machine learning and deep learning techniques are being introduced in data analysis to aid the processes in urban development. It involves models that are intelligent and used for making predictions for maintenance and keeping track of useful data. Applications of AI have been used in traffic management domains, especially in the management of traffic and signals. By application of ML in traffic modelling, AI enables the identified urban areas to reduce congestion by foreseeing traffic flow patterns and increasing traffic flow. Finally, smart cities are applying AI for the supervision and control of environmental resources. Sensors for collecting data on the quality of air and water are also adopted and integrated with AI to employ deep learning in the prevention and remediation of unhealthy urban living conditions. The intelligent cities are also growing forward with enhanced Explainable AI, making it safe to use artificial intelligence results in making decisions. This is especially essential, for cities that are keen on achieving sustainable development, meeting the needs of the residents. However, it is important to focus on some of the open research issues and opportunities of the field. It is vital to work on these issues in the further development of smart and wise cities, which are important for intelligent growth.

Chapter 5: This chapter aims to analyse the capability of blockchain technology in developing sustainable smart cities and tackling the challenges of urban development and improving the operation of the cities. Blockchain technology applications have presented an opportunity to enhance the management of the various components that are within the urban environment. It highlights how blockchain is potentially influencing the future development of such cities by addressing issues related to sustainable development and city management. In applying blockchain to smart cities, the technology improves the level of trust and openness necessary for public acceptance and joint activities in developing new approaches to urban life. The more details on the categories of understanding in explaining what blockchain is by using distributed ledger technology, cryptography, and consensus mechanisms, which are the basic building blocks of any blockchain system. A classification between public, private, and consortium blockchains is done with examples that outline how it is suitable for a certain application in an urban environment. It also explores how government security and transparency, management of wastes, transport options, as well as energy trading and management, can benefit from blockchain. Blockchain technology can be seen as an efficient and cost-effective solution when it comes to improving the work in cities, as well as the engagement of the citizens. The existing and potential research issues and possible future work, including scalability, compatibility of the different types of blockchain, and concerns about the right regulation that brings about the right ethical and secure use. Through these challenges, it would be possible to get better and advanced future designs of smart city infrastructures. However, the use of blockchain technology is one of the efficient tools for the sustainable development of cities. The integration of blockchain technology into smart city infrastructures further milestones of the city in terms of operational efficiency, sustainability, and quality, and this can propel humanity to a sustainable urban existence.

Chapter 6: The benefits of 6G technology with smart city systems are making city ecosystems sustainable, efficient, and responsive in terms of connectivity, data speed, and reduced latency. Some of the characteristics of sixth-generation wireless networks are higher frequency bands, larger bandwidth, and the capability to support a large number of devices, which makes it a key technology for IoT in cities. Smart traffic control and ITSs are offered by 6G, which leads to efficient traffic control and minimization of traffic rates in cities. Further, intelligent building automation systems and telemedicine systems as a part of smart solutions improve the quality and standard of life in the urban environment. Machine learning and artificial intelligence add to the elements from emerging technologies to enrich 6G-IoT in terms of analysis and decision making. Blockchain guarantees security and openness in exchanges of data, which is very important in smart city systems. Edge computing improves the processing of raw data on a real-time basis, hence minimizing the delay and transmission of data. In the education and training system, especially in the application of augmented reality (AR) and virtual reality (VR), enhancement by the 6G technology plays a crucial role in the city. However, there are still some open research issues to explore, including spectrum sharing, interconnection and compatibility and privacy issues. Further research should be aimed at working on standardization issues, security requirements, and possible options for the use of 6G infrastructure in the context of sustainable energy. The foundation by which smart cities will be realized through 6G-IoT is that it enable residents' to live in more comfortable environments, enhance the resilience of such environments, and make the world smarter and more sustainable.

Chapter 7: There is a growing transformation from the conventional system of healthcare towards modern approach anticipated by smart technologies, specifically as an Internet of Medical Things (IoMT). It extends through devices such as wearable devices, implantable devices, remote monitoring devices, and other smart devices that are connected through architectures that enable the capturing of real-time data. In this chapter, we discussed the importance of integrating smart technology in smart cities for a sustainable and advanced form of healthcare referred to as healthcare 5.0. Mobile health applications and wearable health technologies also enhance the support of patients since they enable patient engagement and self-management of health. In addition, IoMT provides solutions for planning and management of the medical care facilities and resources to render more efficient services. However, several research issues are still open. More research work is needed in the areas of data protection, handling, and management of different types of devices and systems, and the need to overcome the interoperability challenges. Mitigating these challenges will therefore enable the achievement of a higher standard of healthcare 5.0 with the desired goal of achieving a sustainable and efficient healthcare system for smart cities.

Chapter 8: It has become evident that smart mobility is among the most crucial needs in smart city development to deal with issues of city expansions, environment, and technology. Smart mobility was initiated through the implementation of ICT in the transport system as a way of improving the mode of transport. The concept of smart mobility becomes more relevant to support such development of the urban environment, as traffic systems and adverse impact on cities expand simultaneously. Modern developments in transportation systems, i.e. ITS, which uses real-time information processing and analysis in relation to transportation. These advances are built on wireless technologies such as vehicular communication systems, cellular mobile communications, short-range static communications, etc. Integrated intelligent smart mobility needs the ITS to support it, the organization of smart and connected automobiles, and advanced wireless systems. All these systems combined make for a more effective and efficient system of transport in the urban environment. Sustainable personal transport aims at moving away from the current high CO_2 emissions, including through electric vehicles (EVs) and auxiliary means of transport, as well as supporting sustainable urban mobility policies. The organization of EVs infrastructure and charging systems, as well as the traffic regulation offered by ITS, are examples of smart mobility in practice. In the aspect of transport and urban design, pedestrian- and bicycle-friendly measures constitute another significant model of sustainable mobility. However, there are three main challenges that are still open and are relevant: the first challenge is scalability, the second is interoperability, and the third and most important is data privacy. Further research work should be conducted to improve these technologies and develop new solutions to support the mobility of people in urban areas, as well as make smart mobility technologies accessible to all the inhabitants of such cities.

Chapter 9: Education through smart education, due to the applicability of immersive technologies, defines the shift in the system of education. In this chapter, we discussed the integration of the metaverse in the education environment that modifies the previously distinct options and becomes an effective tool for engaging learning. As the current educational forms are in constant development, different types of coursework and virtual classrooms appear as sustainable solutions. Virtual classrooms increase the interactivity to some extent as they allow real-time communication and cooperation, which is a great asset in providing a personal approach to distance learning. There is a need to enhance the digital skills development of the urban workforce for the provision of the emerging economic environment. Smart technologies, such as 5G/6G, help in providing education inclusively in smart cities. It has been established that these technologies improve enrolment and access to educational material, which helps meet the diversity of learners' requirements. These advancements are used by higher education institutions to give marginalized people an equal chance to access education. Modern approaches like gamification and exposure to learning through virtual and augmented reality define the options of learning. These tools also enhance learning as they engage more knowledge delivery modes and increase the students' interest in learning. Open Educational Resources (OER) and digital libraries should be embraced by ensuring education is afforded to as many people as possible. They allow people to gain knowledge of a wide range of topics for free of cost and also contribute to the elimination of the gap in education. This is because the research challenges and future directions in smart education offer suggestions on proper strategies for addressing the needs emerging in smart education. It is important that further studies are carried out to develop more advanced technological techniques in the provision of learning. To overcome these issues, citizens involved in the practice of education, technological advances, and policy-making should come together and put in place structural solutions for learning experience designs as well as policies that promote the use of ICTs in education.

Chapter 10: Smart cities are a new generation in urban evolution, as smart energy systems have been an integral constituent in coping with the demands of energy in these urban areas. With increased rates of expansion in cities, there is an increased demand for energy in these areas, and therefore, new methods of using energy must be developed. Without smart energy systems, renewable energy generation will not be included, and their supporting technologies, such as Smart Grids, Smart Metering and Energy Storage, will also not be integrated into the process. This is the case for Smart Energy 4.0/5.0, where the Internet of Energy and computational intelligence are at the forefront of the coming revolution. This chapter focuses on the significance of smart energy technologies as a means of achieving energy efficiencies, resilience, and sustainability of urban cities. They include real-time monitoring and control of energy usage, predictive maintenance, and unique demand bidding useful for the successful functioning of modern cities. The adoption of renewable energy sources also adds to the transition to a decarbonized economy, conforming to the transformation of the energy sector worldwide. Smart cities also employ such concepts as Digital Twins for better planning and management. The application of these technologies not only promotes energy saving but also allows the development of sustainable urban environments. Some of the possible advancements include the use of new materials in energy storage, improving the flexibility of the smart grids, and designing effective frameworks for utilization of data in energy management. With the changing energy scenario, Smart Energy Systems have significant processes towards the development of smart cities with sustainable as well as reliable energy supply systems.

Chapter 11: The changes to city management and people's interactions in the habitation require the new smart governance, which applies ICT to manage the cities. The current forms of governance include bureaucracies with very less interface with citizens, while smart government is all about openness and responsiveness. In this chapter, we discussed the concept of smart governance, which is the implementation of ICT as the environment for a city where most decisions and policies concerning development are based on data and information. These principles ensure that certain service delivery in the public sector is effective, the quality of life is enhanced, policies and measures for environmental conservation are promoted. Innovation of financial and other capital resources to promote our economic growth, and the achievement of a safe society is enhanced. Smart government is therefore made up of adequate and strong physical infrastructure, high-quality and advanced technology, adequate and effective data and analytics, solid governance policies, and appropriate strategic partnerships. These elements make it possible for cities to make good use of information in decision making and reduce the chances of taking reactive measures in managing the cities. Smart government is thus highly beneficial in smart cities, as it has several advantages. Environmental policies are implemented with quantifiable approaches towards resource supply, which leads to the generation of low ecological impacts. In addition, the process of innovation as well as economic development is enhanced by the establishment of a favourable business environment, support for small businesses, and technology development. However, there are still some open issues and challenges to be discussed in this regard. Challenges including data privacy and protection, cybersecurity, as well as the inclusiveness of smart solutions in governance, are dominant to the proper implementation of smart solutions. Subsequent studies should aim at novel approaches to the challenges that need to be addressed in order to develop adequate smart governance models for effective support of urban populations' needs.

Chapter 12: The fast-changing urban infrastructure transition into smart cities creates new ways of city management through digital technology integration along with IoT and AI. A massive cybersecurity vulnerability exists for critical urban infrastructure because smart systems connect to digital networks for efficient resource management and data-based decision systems. A comprehensive analysis of smart city sustainability and cybersecurity connections through core vulnerability analysis and proposed robust defensive measures for new security threats is highlighted in this chapter. The study examines several prominent attack paths, which include data breaches, denial-of-service (DoS) attacks, ransomware incidents, and problems within IoT connections. Technologies like blockchain and ML are developed to secure smart city frameworks. Such systems built using blockchain create unchangeable, tamperproof transaction records to increase transparency and trust between parties, specifically when handling urban data management tasks. The sustainable cybersecurity framework specifically designed for smart cities serves as the basis for developing resilient urban development. The security framework presents methods for protected IoT device installation, along with constant threat surveillance, together with a zero-trust infrastructure topology and systems to safeguard citizen information. The advantage of robust cybersecurity measures for sustainable smart cities is that they create better public trust levels and minimize operational disruptions, and boost resistance to cyber-physical attacks. The existing research challenges that require attention are secure AI decision systems, scalable IoT security protocols, and privacy-protected public service analytics.

1.13 Summary

The vision and objectives for sustainable smart city development depend on the effectiveness of the innovative concept and planning to deliver intelligent methods that enhance the organizational efficiency of a living complex and a sustainable city. These cities are on the cutting edge of the multi-faceted issue of sustainable cities development through the use of technology to improve the lives of the inhabitants. The considerations concerning the creation of smart cities are based on IoT, AI, and edge computing that help enhance efficiency in numerous contexts. These are the enablers to the creation of the digital democracy and smart governance, which include citizen involvement and participation in governance. Furthermore, there is an integration of MaaS and smart transportation systems that are taking the form of innovation in shared, integrated, accessible, fair, and smart mobility solutions that will help in avoiding traffic congestion and pollution. The usage of renewable energy and microgrids is the most proper example of shifting towards green energy sources to make the cities more and more energy-reliant and sustainable. In healthcare, both artificial intelligence and telemedicine offer the best way to reach out for services, especially in rural areas, to achieve the objectives of the healthcare sector. Furthermore, Industry 5.0, is a new way of manufacturing that involves the introduction of smart technologies along with creativity in order to produce smart products with a focus on sustainability. The use of Society 5.0 as a human-centred society has a similar concept to smart cities, as it attempts to find a balance between technology and people's quality of life. It is important that technological advancements introduce ideas of a sustainable smart city in order to inform the need to be technologically advanced as well as consistent with ecological capacities.

References

1. A. Valavanidis, A. Valavanidis, Half of the Global Population Lives in MegaCities. Megacities are hotspots for slums, GHG emissions, toxic air pollution, climate change, deteriorating water and soil quality. [Online] (2024). Available: https://www.researchgate.net/publication/379655450
2. P. Mishra, P. Thakur, G. Singh, Sustainable smart city to society 5.0: State-of-the-art and research challenges. SAIEE Afr. Res. J. **113**(4), 152–164 (2022). https://doi.org/10.23919/SAIEE.2022.9945865
3. E. Eizenberg, Y. Jabareen, Social sustainability: A new conceptual framework. Sustainability **9**(1), 68 (2017). https://doi.org/10.3390/su9010068

4. B. Ben Hichou, H. Bachiri, M. Mouhti, Smart cities' key indicators, in *Smart cities for inclusive innovation*, (The Institution of Engineering and Technology, Stevenage, UK, 2024), pp. 19–26. https://doi.org/10.1049/PBBE005E_ch2
5. Z. Yemets, M. Smoliarchuk, L. Demchuk, I. Koreneva, V. Omelyanenko, The ecosystem of modern cities in the context of sustainable development, in *Big data in finance: transforming the financial landscape. Studies in big data*, vol. 164, (Springer, Cham, 2025), pp. 477–484. https://doi.org/10.1007/978-3-031-75095-3_37
6. N. Donthu, S. Kumar, D. Mukherjee, N. Pandey, W.M. Lim, How to conduct a bibliometric analysis: An overview and guidelines. J. Bus. Res. **133**, 285–296 (2021). https://doi.org/10.1016/j.jbusres.2021.04.070
7. W. Shi, J. Cao, Q. Zhang, Y. Li, L. Xu, Edge computing: Vision and challenges. IEEE Internet Things J. **3**(5), 637–646 (2016). https://doi.org/10.1109/JIOT.2016.2579198
8. J. Lin, W. Yu, N. Zhang, X. Yang, H. Zhang, W. Zhao, A survey on internet of things: Architecture, enabling technologies, security and privacy, and applications. IEEE Internet Things J. **4**(5), 1125–1142 (2017). https://doi.org/10.1109/JIOT.2017.2683200
9. M.A. Khan, K. Salah, IoT security: Review, blockchain solutions, and open challenges. Futur. Gener. Comput. Syst. **82**, 395–411 (2018). https://doi.org/10.1016/j.future.2017.11.022
10. A. Botta, W. de Donato, V. Persico, A. Pescapé, Integration of cloud computing and internet of things: A survey. Futur. Gener. Comput. Syst. **56**, 684–700 (2016). https://doi.org/10.1016/j.future.2015.09.021
11. U. Raza, P. Kulkarni, M. Sooriyabandara, Low power wide area networks: An overview. IEEE Communications Surveys and Tutorials **19**(2), 855–873 (2017). https://doi.org/10.1109/COMST.2017.2652320
12. A. Fuller, Z. Fan, C. Day, C. Barlow, Digital twin: Enabling technologies, challenges and open research. IEEE Access **8**, 108952–108971 (2020). https://doi.org/10.1109/ACCESS.2020.2998358
13. M. Centenaro, L. Vangelista, A. Zanella, M. Zorzi, Long-range communications in unlicensed bands: The rising stars in the IoT and smart city scenarios. IEEE Wirel. Commun. **23**(5), 60–67 (2016). https://doi.org/10.1109/MWC.2016.7721743
14. C. Moreno, Z. Allam, D. Chabaud, C. Gall, F. Pratlong, Introducing the '15-Minute City': Sustainability, resilience and place identity in future post-pandemic cities. Smart Cities **4**(1), 93–111 (2021). https://doi.org/10.3390/smartcities4010006
15. I.A.T. Hashem et al., The role of big data in smart city. Int. J. Inf. Manag. **36**(5), 748–758 (2016). https://doi.org/10.1016/j.ijinfomgt.2016.05.002
16. A.G. Gad, Particle swarm optimization algorithm and its applications: A systematic review. Archives of Computational Methods in Engineering **29**(5), 2531–2561 (2022). https://doi.org/10.1007/s11831-021-09694-4
17. S. Borchardt, G.B. Vignola, D. Buscaglia, M. Maroni, *Mapping EU policies with the 2030 agenda and SDGs-fostering policy coherence through text-based SDG mapping*. https://doi.org/10.2760/110687
18. P. Mishra, G. Singh, *Sustainable smart cities* (Springer, Cham, 2023). https://doi.org/10.1007/978-3-031-33354-5
19. S.E. Bibri, Data-driven smart sustainable cities of the future: Urban computing and intelligence for strategic, short-term, and joined-up planning. Comput. Urban Sci. **1**(1), 8 (2021). https://doi.org/10.1007/s43762-021-00008-9
20. I. Yaqoob, I.A.T. Hashem, Y. Mehmood, A. Gani, S. Mokhtar, S. Guizani, Enabling communication technologies for smart cities. IEEE Commun. Mag. **55**(1), 112–120 (2017). https://doi.org/10.1109/MCOM.2017.1600232CM
21. S. Flynn, *Equality in the city: imaginaries of the smart future* (Intellect, 2022)
22. P. Mishra, G. Singh, Internet of vehicles for sustainable smart cities, in *Sustainable smart cities*, (Springer, Cham, 2023), pp. 175–193. https://doi.org/10.1007/978-3-031-33354-5_8

23. O. Arshi, S. Mondal, Advancements in sensors and actuators technologies for smart cities: A comprehensive review. Smart Construction and Sustainable Cities **1**(1), 18 (2023). https://doi.org/10.1007/s44268-023-00022-2
24. G.V. Pereira, P. Parycek, E. Falco, R. Kleinhans, Smart governance in the context of smart cities: A literature review. Information Polity **23**(2), 143–162 (2018). https://doi.org/10.3233/IP-170067
25. P. Mishra, G. Singh, Introduction: importance of sustainable smart city, in *Sustainable smart cities*, (Springer, Cham, 2023), pp. 1–37. https://doi.org/10.1007/978-3-031-33354-5_1
26. A. Abu-Rayash, I. Dincer, Development of integrated sustainability performance indicators for better management of smart cities. Sustain. Cities Soc. **67**, 102704 (2021). https://doi.org/10.1016/J.SCS.2020.102704
27. S. Rani, P. Bhambri, A. Kataria, A. Khang, Smart city ecosystem, in *AI-centric smart city ecosystems*, (CRC Press, Boca Raton, 2022), pp. 1–20. https://doi.org/10.1201/9781003252542-1
28. P. Mishra, P. Thakur, G. Singh, Enabling technologies for IoT based smart city, in *2021 Sixth International Conference on Image Information Processing (ICIIP)*, (IEEE, 2021), pp. 99–104. https://doi.org/10.1109/ICIIP53038.2021.9702673
29. P. Mishra, G. Singh, Enabling technologies for sustainable smart city, in *Sustainable smart cities*, (Springer, Cham, 2023), pp. 59–73. https://doi.org/10.1007/978-3-031-33354-5_3
30. T. Chen, J. Ramon Gil-Garcia, M. Gasco-Hernandez, Understanding social sustainability for smart cities: The importance of inclusion, equity, and citizen participation as both inputs and long-term outcomes. Journal of Smart Cities and Society **1**(2), 135–148 (2022). https://doi.org/10.3233/SCS-210123
31. A.M. Levenda, N. Keough, M. Rock, B. Miller, Rethinking public participation in the smart city. Canadian Geographies/Géographies canadiennes **64**(3), 344–358 (2020). https://doi.org/10.1111/cag.12601
32. B. Anthony, The role of community engagement in urban innovation towards the co-creation of smart sustainable cities. J. Knowl. Econ. **15**, 1592 (2023). https://doi.org/10.1007/s13132-023-01176-1
33. D.E. Mills, I. Izadgoshasb, S.G. Pudney, Smart city collaboration: A review and an agenda for establishing sustainable collaboration. Sustainability **13**(16), 9189 (2021). https://doi.org/10.3390/su13169189
34. H. Jiang, S. Geertman, P. Witte, Smart urban governance: an alternative to technocratic ‘smartness’. GeoJournal **87**(3), 1639–1655 (2022). https://doi.org/10.1007/s10708-020-10326-w
35. B. Giles-Corti, M. Lowe, J. Arundel, Achieving the SDGs: Evaluating indicators to be used to benchmark and monitor progress towards creating healthy and sustainable cities. Health Policy (New York) **124**(6), 581–590 (2020). https://doi.org/10.1016/J.HEALTHPOL.2019.03.001
36. P. Sarzaeim, Q.H. Mahmoud, A. Azim, G. Bauer, I. Bowles, A systematic review of using machine learning and natural language processing in smart policing. Computers **12**(12), 255 (2023). https://doi.org/10.3390/computers12120255
37. A. Quijano et al., Towards sustainable and smart cities: replicable and KPI-driven evaluation framework. Buildings **12**(2) (2022). https://doi.org/10.3390/buildings12020233
38. M. Hara, T. Nagao, S. Hannoe, J. Nakamura, New key performance indicators for a smart sustainable city. Sustainability (Switzerland) **8**(3) (2016). https://doi.org/10.3390/su8030206
39. Z. Sang, K. Li, *ITU-T standardisation activities on smart sustainable cities* (Wiley, 2019). https://doi.org/10.1049/iet-smc.2019.0023
40. B. Li, Effective energy utilization through economic development for sustainable management in smart cities. Energy Rep. **8**, 4975–4987 (2022). https://doi.org/10.1016/J.EGYR.2022.02.303
41. A.H. Fakhimi, A.H. Khani, J.M. Sardroud, Smart-city infrastructure components, in *Solving Urban Infrastructure Problems Using Smart City Technologies: Handbook on Planning, Design, Development, and Regulation*, (Elsevier, 2021), pp. 17–54. https://doi.org/10.1016/B978-0-12-816816-5.00002-4

42. A. Omri, Technological innovation and sustainable development : Does the stage of development matter? Environ. Impact Assess. Rev. **83**, 106398 (2020). https://doi.org/10.1016/J.EIAR.2020.106398
43. M.J. Nikki Han, M.J. Kim, A critical review of the smart city in relation to citizen adoption towards sustainable smart living. Habitat Int. **108**, 102312 (2021). https://doi.org/10.1016/J.HABITATINT.2021.102312
44. M. Viu-Roig, E.J. Alvarez-Palau, The impact of E-commerce-related last-mile logistics on cities: A systematic literature review. Sustainability **12**(16), 6492 (2020). https://doi.org/10.3390/su12166492
45. M. Talebkhah, A. Sali, M. Marjani, M. Gordan, S.J. Hashim, F.Z. Rokhani, IoT and big data applications in smart cities: Recent advances, challenges, and critical issues. IEEE Access **9**, 55465–55484 (2021). https://doi.org/10.1109/ACCESS.2021.3070905
46. A. Kasznar et al., Multiple dimensions of smart cities' infrastructure: A review. Buildings **11**(2), 73 (2021). https://doi.org/10.3390/buildings11020073
47. I. Nastjuk, S. Trang, E.I. Papageorgiou, Smart cities and smart governance models for future cities. Electron. Mark. **32**(4), 1917–1924 (2022). https://doi.org/10.1007/s12525-022-00609-0
48. M. Bakirci, Smart city air quality management through leveraging drones for precision monitoring. Sustain. Cities Soc. **106**, 105390 (2024). https://doi.org/10.1016/J.SCS.2024.105390
49. A. Rao, J.S. Laura, G. Dhania, Water conservation for environmental sustainability. Water, the Environment, and the Sustainable Development Goals, 85–106 (Jan. 2024). https://doi.org/10.1016/B978-0-443-15354-9.00012-8
50. K. Samuel, S. Yakubu, O. Durowoju, Noise in the city: a socio-spatial analysis of the actual and perceived noise levels in a medium-sized urban center. Coğrafya Dergisi J. Geography **0**(47), 1–13 (2024). https://doi.org/10.26650/JGEOG2023-1189396
51. D.I. Hussain, D.A. Elomri, D.L. Kerbache, D.A. El Omri, Smart city solutions: Comparative analysis of waste management models in IoT-enabled environments using multiagent simulation. Sustain. Cities Soc. **103**, 105247 (2024). https://doi.org/10.1016/J.SCS.2024.105247
52. R. Chen, J. Carruthers-Jones, S. Carver, J. Wu, Constructing urban ecological corridors to reflect local species diversity and conservation objectives. Sci. Total Environ. **907**, 167987 (2024). https://doi.org/10.1016/J.SCITOTENV.2023.167987
53. L.Y. Rock, F.P. Tajudeen, Y.W. Chung, Usage and impact of the internet-of-things-based smart home technology: A quality-of-life perspective. Univ. Access Inf. Soc. **23**(1), 345–364 (2024). https://doi.org/10.1007/s10209-022-00937-0
54. D. Scala, Á.I. Aguilar Cuesta, M.Á. Rodríguez-Domenech, M. del Cañizares Ruiz, Bibliometric study on the conceptualisation of smart city and education. Smart Cities **7**(1), 597–614 (2024). https://doi.org/10.3390/smartcities7010024
55. K.A. Bin Ahmad, H. Khujamatov, N. Akhmedov, M.Y. Bajuri, M.N. Ahmad, A. Ahmadian, Emerging trends and evolutions for smart city healthcare systems. Sustain. Cities Soc. **80**, 103695 (2022). https://doi.org/10.1016/J.SCS.2022.103695
56. M.H. Panahi Rizi, S.A. Hosseini Seno, A systematic review of technologies and solutions to improve security and privacy protection of citizens in the smart city. Internet of Things **20**, 100584 (2022). https://doi.org/10.1016/J.IOT.2022.100584
57. R. Sietchiping, J. Omwamba, The future of urban policy in Africa. Localization of SDG 11 and the new urban agenda, in *Local governance in the new urban agenda. local and urban governance*, (Springer, Cham, 2020), pp. 275–291. https://doi.org/10.1007/978-3-030-47135-4_14
58. F. Zhao, O.I. Fashola, T.I. Olarewaju, I. Onwumere, Smart city research: A holistic and state-of-the-art literature review. Cities **119**, 103406 (2021). https://doi.org/10.1016/J.CITIES.2021.103406
59. V. Lapuente, S. Van de Walle, The effects of new public management on the quality of public services. Governance **33**(3), 461–475 (2020). https://doi.org/10.1111/gove.12502
60. X. Wang, X. Dong, Z. Zhang, Y. Wang, Transportation carbon reduction technologies: A review of fundamentals, application, and performance. Journal of Traffic and Transportation

Engineering (English Edition) **11**(6), 1340–1377 (2024). https://doi.org/10.1016/J.JTTE.2024.11.001
61. M. Golušin, S. Dodić, S. Popov, Sustainable Energy Management. Elsevier (2013). https://doi.org/10.1016/C2011-0-05403-0
62. M.A. Ahad, R. Biswas, Request-based, secured and energy-efficient (RBSEE) architecture for handling IoT big data. J. Inf. Sci. **45**(2), 227–238 (2019). https://doi.org/10.1177/0165551518787699
63. M.A. Ahad, S. Paiva, G. Tripathi, N. Feroz, Enabling technologies and sustainable smart cities. Sustain. Cities Soc. **61**, 102301 (2020). https://doi.org/10.1016/J.SCS.2020.102301
64. P. Mishra, G. Singh, 6G-IoT framework for sustainable smart city: vision and challenges, in *Sustainable smart cities*, (Springer, Cham, 2023), pp. 97–117. https://doi.org/10.1007/978-3-031-33354-5_5
65. L. Mohjazi, B. Selim, M. Tatipamula, M.A. Imran, The journey toward 6G: A digital and societal revolution in the making. IEEE Internet of Things Magazine **7**(2), 119–128 (2024). https://doi.org/10.1109/IOTM.001.2300119
66. A. Taneja, S. Rani, Quantum-enabled intelligent resource control for reliable communication support in internet-of-vehicles. IEEE Trans. Consum. Electron. **70**, 1–1 (2024). https://doi.org/10.1109/TCE.2024.3376701
67. A.H. Alavi, P. Jiao, W.G. Buttlar, N. Lajnef, Internet of things-enabled smart cities: State-of-the-art and future trends. Measurement **129**, 589–606 (2018). https://doi.org/10.1016/J.MEASUREMENT.2018.07.067
68. L.U. Khan, W. Saad, Z. Han, E. Hossain, C.S. Hong, Federated learning for internet of things: Recent advances, taxonomy, and open challenges. IEEE Communications Surveys and Tutorials **23**(3), 1759–1799 (2021). https://doi.org/10.1109/COMST.2021.3090430
69. H. Salehi, R. Burgueño, Emerging artificial intelligence methods in structural engineering. Eng. Struct. **171**, 170–189 (2018). https://doi.org/10.1016/J.ENGSTRUCT.2018.05.084
70. H. Samih, Smart cities and internet of things. J. Inf. Technol. Case Appl. Res. **21**(1), 3–12 (2019). https://doi.org/10.1080/15228053.2019.1587572
71. H. Nguyen, D. Nawara, R. Kashef, Connecting the indispensable roles of IoT and artificial intelligence in smart cities: A survey. Journal of Information and Intelligence (2024). https://doi.org/10.1016/J.JIIXD.2024.01.003
72. T. Alam, Blockchain cities: The futuristic cities driven by Blockchain, big data and internet of things. GeoJournal **87**(6), 5383–5412 (2022). https://doi.org/10.1007/s10708-021-10508-0
73. Z. Allam, Z.A. Dhunny, On big data, artificial intelligence and smart cities. Cities **89**, 80–91 (2019). https://doi.org/10.1016/J.CITIES.2019.01.032
74. N. Komninos, C. Kakderi, A. Panori, P. Tsarchopoulos, Smart City planning from an evolutionary perspective. J. Urban Technol. **26**(2), 3–20 (2019). https://doi.org/10.1080/10630732.2018.1485368
75. R. Matheus, M. Janssen, D. Maheshwari, Data science empowering the public: Data-driven dashboards for transparent and accountable decision-making in smart cities. Gov. Inf. Q. **37**(3), 101284 (2020). https://doi.org/10.1016/J.GIQ.2018.01.006
76. M. Batty, Big data and the City. Built Environ. **42**(3), 321–337 (2016). https://doi.org/10.2148/benv.42.3.321
77. R. Khatoun, S. Zeadally, Smart cities. Commun. ACM **59**(8), 46–57 (2016). https://doi.org/10.1145/2858789
78. S. Underwood, Blockchain beyond bitcoin. Commun. ACM **59**(11), 15–17 (2016). https://doi.org/10.1145/2994581
79. C.O. Cruz, J.M. Sarmento, 'Mobility as a service' platforms: A critical path towards increasing the sustainability of transportation systems. Sustainability **12**(16), 6368 (2020). https://doi.org/10.3390/su12166368
80. I. Lopez-Carreiro, A. Monzon, E. Lopez, MaaS implications in the Smart City: A multi-stakeholder approach. Sustainability **15**(14), 10832 (2023). https://doi.org/10.3390/su151410832

81. M.S. Sami et al., Energy Management of Microgrids for smart cities: A review. Energies (Basel) **14**(18), 5976 (2021). https://doi.org/10.3390/en14185976
82. A. Sanchez-Miralles, C. Calvillo, F. Martín, J. Villar, Use of renewable energy systems in smart cities, in *Use, operation and maintenance of renewable energy systems. Green energy and technology*, (Springer, Cham, 2014), pp. 341–370. https://doi.org/10.1007/978-3-319-03224-5_10
83. J.G. Fernandes, Artificial intelligence in telemedicine, in *Artificial intelligence in medicine*, (Springer, 2022), pp. 1219–1227. https://doi.org/10.1007/978-3-030-64573-1_93
84. G.B. Narkhede, B.N. Pasi, N. Rajhans, A. Kulkarni, Industry 5.0 and sustainable manufacturing: A systematic literature review. Benchmarking **32**, 608 (2024). https://doi.org/10.1108/BIJ-03-2023-0196
85. A. Adel, N.H.S. Alani, *Human-centric collaboration and industry 5.0 framework in smart cities and communities: fostering sustainable development goals 3, 4, 9, and 11 in society 5.0* (Multidisciplinary Digital Publishing Institute (MDPI), 2024). https://doi.org/10.3390/smartcities7040068
86. Y. Liu, C. Yang, L. Jiang, S. Xie, Y. Zhang, Intelligent edge computing for IoT-based energy management in smart cities. IEEE Netw. **33**(2), 111–117 (2019). https://doi.org/10.1109/MNET.2019.1800254
87. Y. Jiao, H. Kang, H. Sun, An intelligent landscaping framework for net-zero energy smart cities: A green infrastructure approach. Sustain Energy Technol Assess **64**, 103665 (2024). https://doi.org/10.1016/J.SETA.2024.103665
88. M. Artmann, M. Kohler, G. Meinel, J. Gan, I.C. Ioja, How smart growth and green infrastructure can mutually support each other — A conceptual framework for compact and green cities. Ecol. Indic. **96**, 10–22 (2019). https://doi.org/10.1016/J.ECOLIND.2017.07.001
89. S. Oh, C. Lu, Vertical farming - smart urban agriculture for enhancing resilience and sustainability in food security. J. Hortic. Sci. Biotechnol. **98**(2), 133–140 (2023). https://doi.org/10.1080/14620316.2022.2141666
90. A. Sharifi, M. Amirzadeh, A.R. Khavarian-Garmsir, The metaverse as a future form of smart cities: A systematic literature review of co-benefits and trade-offs for sustainable development goals. Cities **161**, 105879 (2025). https://doi.org/10.1016/J.CITIES.2025.105879
91. I.H.Y. Yim, J. Su, Artificial intelligence (AI) learning tools in K-12 education: A scoping review. J. Comput. Educ. (2024). https://doi.org/10.1007/s40692-023-00304-9
92. P.K. Priya, R. Sivaranjani, M. Sathyamoorthy, R.K. Dhanaraj, Efficient network and communication Technologies for Smart and Sustainable Society 5.0, in *Networked Sensing Systems*, (Wiley, 2025), pp. 63–100. https://doi.org/10.1002/9781394310890.ch3
93. K. Brglez, M. Perc, R.K. Lukman, A conceptual model for a circular city: A case study of Maribor, Slovenia. Clean Techn. Environ. Policy **26**(1), 45–65 (2024). https://doi.org/10.1007/s10098-023-02579-z
94. D. Zhang, L.G. Pee, S.L. Pan, L. Cui, Big data analytics, resource orchestration, and digital sustainability: A case study of smart city development. Gov. Inf. Q. **39**(1), 101626 (2022). https://doi.org/10.1016/j.giq.2021.101626
95. M. Anedda, M. Fadda, R. Girau, G. Pau, D. Giusto, A social smart city for public and private mobility: A real case study. Comput. Netw. **220**, 109464 (2023). https://doi.org/10.1016/j.comnet.2022.109464
96. W. Chantry, 'Built from the internet up': Assessing citizen participation in smart city planning through the case study of Quayside, Toronto. GeoJournal **88**(2), 1619–1637 (2023). https://doi.org/10.1007/s10708-022-10688-3

Chapter 2
Smart Cities and Sustainable Development Goals

Contents

P. Mishra, G. Singh, *Sustainable Smart Cities 2.0*,
https://doi.org/10.1007/978-3-032-01102-2_2

2.1 Introduction

Smart cities are defined as innovative urban developments that employ social technologies and the effective use of data and information. With the increase in population density, especially in urban centres, where over 50% of the global population lives, issues related to sustainability, resource management, and the delivery of public services are compounded. In this respect, smart cities have been established as a fundamental approach to solving the conflict between the growth of cities and sustainable development. This advancement in the use of technology in urban planning and management meets the goals of the United Nations (UN) Sustainable Development Goals (SDGs 2030), which laid down the UN's global roadmap to address some focal challenges facing humanity [1]. Smart cities use the Internet of Things (IoT), Artificial Intelligence (AI), big data, and other advanced techniques to enhance the functions in urban areas and the built environment. This is covered by smart highways that reduce traffic jams and minimize emissions/noise pollution and by smart electricity infrastructures that improve the energy distribution. That is how smart cities seek to optimize urban processes and, enhance citizen's quality of life by providing with more information and services. Cities can use this technological infrastructure in a sustainable way and use and control resources and simultaneously maximize the liveability and sustainability of cities [2]. For example, SDG 11 aims at making cities and human settlements available, safe, inclusive, productive, and sustainable. Smart cities help to achieve these goals of providing better mobility and security solutions by installing intelligent surveillance systems and the inclusive provision of services such as health and education. Also, smart cities have positive impact on ensuring the achievement of SDG 7, which focuses on affordable, reliable, sustainable, and modern energy for all. The smart city is capable of utilizing energy and also of regulating its usage effectively, reducing wastage of energy and carbon footprint with the help of smart grids and renewable energy systems. Some of these strategies have a positive impact on environmental sustainability, together with adding to energy stability and cost to the urban population. Waste management is another area in smart cities where efficient collection, recycling, and reduction of waste will reduce the usage of landfills and pollution. Therefore, implementing information technologies in waste management contributes to the achievement of the Sustainable Development Goal, in particular, SDG 12, which deals with responsible consumption and production. Significant contribution to SDG 13 in the area of climate action is another area where smart cities can make a contribution. Specifically, through the use of real-time data and big data analytics, cities can prepare and plan for climate problems, including climate disasters like storms, floods, and heat waves. Incorporation of green buildings and urban green spaces also complement the initiative towards making the city more resilient to climate change. The impacts are not only beneficial to the global climate targets but also protect the dense urban communities from the consequences of climate change. Besides the environmental connection, smart cities also adopt development and innovation in the economy and business, corresponding to SDG 9, is known as 'Industry, Innovation and Infrastructure'. Smart cities have the potential to facilitate the growth and development of new technologies and opportunities for employment by creating a favourable environment for entrepreneurship. Technological advances, which include the incorporation of digital technology in urban areas, mean that new industries such as clean energy, smart mobility, and digital services are sustainable [3]. In addition, smart cities are also helpful in education and improving digital literacy to contribute towards attaining the Sustainable Development Goal for quality education and lifelong learning—SDG 4. Since the technology is continually growing, its facilitation of education in cities offers residents knowledge in the digital economy. Besides, it upgrades the potential for every single person and contributes toward better quality and depends on people in cities, which in turn makes them more innovative and competitive. This is especially the case with health and well-being, which are the most explicit targets of SDG 3. Hence, by adopting telemedicine, digital record systems, and smart health centres, the city can meet this objective and achieve better health for the population [4]. Therefore, utilizing air quality sensors, noise, and any environmental factors can assist in making policies for better living and avoiding diseases. Another component of smart cities is focusing on equality and reducing the inequality level, as stated in SDG 10. Smart city ideas can benefit the disadvantaged populace by influencing the provision of cheap houses, better transport systems, and connectivity. Thus, a smart city is a new concept of cities where all ideas are combined to make the city smart through the use of technology, sustainability, and social inclusiveness [5]. Through considering the key principles of the Sustainable Development Goals, it is possible to state that the promotion of the concept of smart cities can successfully handle the complex issues of urbanization in favour of the Sustainable Development Goals. Generally, smart city projects require partnerships between the state and various government entities, the private sector, citizens, and other stakeholders. This also calls for sound governance infrastructures, data protection measures, non-discrimination measures, and policies on how the gains from smart city technologies could be availed to everyone. Having seen the increase in the growth of the urban areas, developing a smart city is a way to ensure that cities are sustainable and provide for the current and future generations. Through innovation, effective partnerships, and references to the Sustainable Development Goals, cities have employed for changing the world [6].

2.2 Current Research Trends of SDGs in Smart Cities

This study investigates the trends, contributions, and impact of research in smart cities with Sustainable Development Goals from 2015 to 2025 through a bibliometric analysis of Scopus data [7]. The Bibliometrix R package, along with Origin software, was utilized to analyse and depict the data. This approach ensures that researchers are presented with the most current research developments in the field, enabling them to remain updated on emerging trends and advancements. In March 2025, a bibliometric analysis was conducted utilizing the Scopus core collection database to thoroughly investigate the concept and understanding of smart cities. Ultimately, 1534 articles were exported and saved in BibTex format before being included and examined in the bibliometric study. The primary information, syntax, and specific criteria used in screening for this study are shown in Table 2.1.

2.2.1 Annual Scientific Production and Top 10 Most Published Journals/Cited Papers

Figure 2.1 shows the annual publication trends from 2015 to 2025, highlighting the growth in research output. The early years show a steady increase in publications, rising from 2015 (15 docs) to 2021 (193 docs), indicating a growing interest in the field. There is a slight decrease in the number of documents in 2022 (187), followed by an increase in 2023 (247) and 2024 (347). In 2025 (597), the number continues to rise and may continue increasing until the end of the year. Figure 2.2 presents the highest number of articles in Sustainable Development Goals in the smart city research domain published in "Sustainability (Switzerland)" (109 docs), "IOP Conference Series: Earth and Environmental Science" (35 docs), "Energies" (32 docs), and "Sustainable Cities and Society" (32 docs), which indicates a higher output of articles compared to the other journals. In addition, some other journals, such as the "Lecture Notes in Networks and Systems" (27 docs), "Cities" (22 docs), "ACM International Conference Proceeding Series" (20 docs), "E3S Web of Conferences" (20 docs), "Lecture Notes in Civil Engineering" (18 docs), and "Urban Book Series" (18 docs), are in this domain.

Table 2.1 Summary of data characteristics and main information of the bibliometric study

Description	Specific Standard Requirements/Results
Research database	Scopus
Search query	(“smart”) AND (“cities”) AND (“sustainable”) AND (“development”) AND (“goals”)
Time span	2015–2025
Document type	Articles, conference papers, book chapters, etc.
Language	English
Data export format	BibTeX
Documents	1534
Annual growth rate %	19.35
Document average age	3.39
Average citations per doc	18.54
Keywords plus (ID)	5988
Author’s keywords (DE)	4322
Authors	4708
Authors of single-authored docs	213
Single-authored docs	262
Co-authors per doc	3.5
International co-authorships %	29.66

Table 2.2 displays the top ten most cited papers from the study, which presents both the ranking and authors with their published works. This section presents information about authors, article titles, and their corresponding publications in journals with documentation for publication years and citation frequency. The ranking focuses on engineering field smart cities articles to demonstrate their total citation (TC) counts and annual citation rates and normalized total citation (NTC) data, which collectively signify their impact in the field.

2.2.2 *Sankey Diagram (Three-Field Plot and Thematic Map)*

In Fig. 2.3, the Sankey diagram illustrates the three-field plot thematic progression to visualize the flow and the distribution in scientific research. In this case, keywords, abstracts, and titles are added, acting as indicators to highlight the most influential research papers on the subject. These keywords indicate vital concepts and themes that stand out in academic discussions of the smart city domain. Figure 2.4 categorizes various smart city themes based on their Sustainable Development Goals and relevant degrees. In the top right quadrant (motor themes), which represents themes that are highly developed and highly relevant. The terms like “sustainability”, “urban development”, and “urbanization” are found in the middle of the top right and left quadrants, which is an intermediate theme suggesting bridging highly developed and strong focus in this field, whereas terms like “economics”, “environmental technology”, and “economic and social effects” are

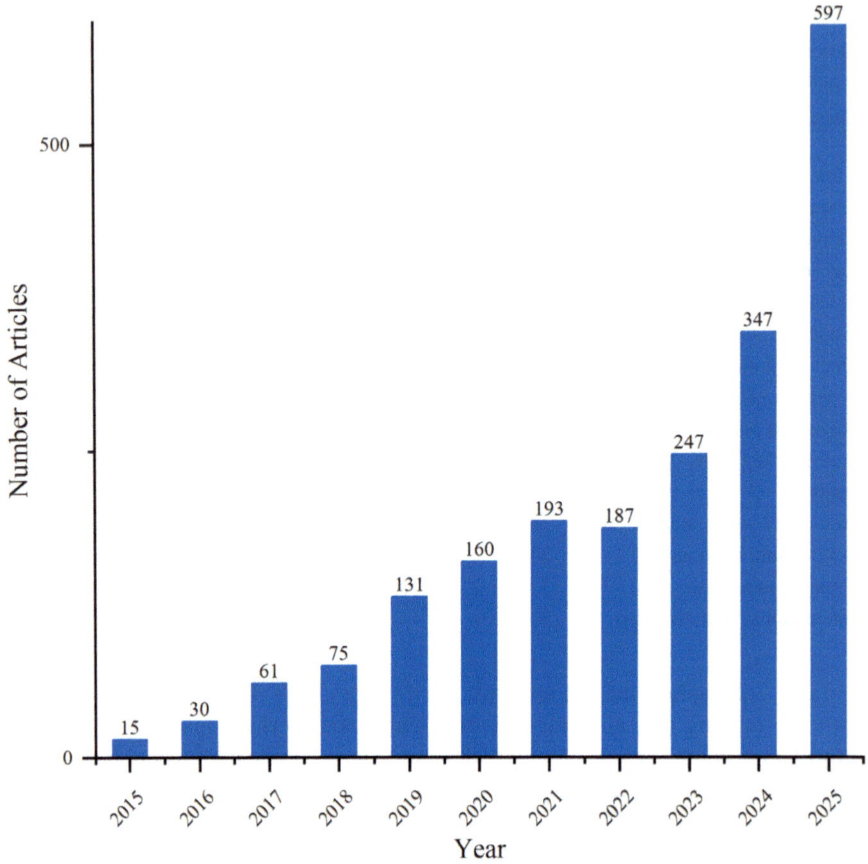

Fig. 2.1 Yearly trend of publication numbers in smart cities with SDGs

Fig. 2.2 Most relevant sources with the number of articles

Table 2.2 Most globally cited documents in the smart city research domain with SDGs

Rank	Author's name	Paper Title	Journal's Name	TC	TC/ Year	NTC
1	Bibri et al. (2017) [8]	Smart sustainable cities of the future: An extensive interdisciplinary literature review	Sustainable Cities and Society	1054	117.11	15.26
2	Ahvenniemi et al. (2017) [9]	What are the differences between sustainable and smart cities?	Cities	1001	111.22	14.50
3	Moreno et al. (2021) [10]	Introducing the "15-Minute City": Sustainability, resilience, and place identity in future post-pandemic cities	Smart Cities	889	177.80	38.33
4	Allam et al. (2019) [11]	On big data, artificial intelligence, and smart cities	Cities	770	110.00	20.07
5	Nizetic et al. (2020) [12]	Internet of Things (IoT): Opportunities, issues, and challenges towards a smart and sustainable future	Journal of Cleaner Production	679	113.17	27.06
6	Ismagilova et al. (2019) [13]	Smart cities: Advances in research—An information systems perspective	International Journal of Information Management	652	93.14	16.99
7	Bibri (2018) [14]	The IoT for smart sustainable cities of the future: An analytical framework for sensor-based big data applications for environmental sustainability	Sustainable Cities and Society	579	72.38	16.16
8	Yigitcanlar et al. (2019) [15]	Can cities become smart without being sustainable? A systematic review of the literature	Sustainable Cities and Society	504	72.00	13.14
9	Fatimah et al. (2020) [16]	Industry 4.0-based sustainable circular economy approach for smart waste management system to achieve sustainable development goals: A case study of Indonesia	Journal of Cleaner Production	466	77.67	18.57
10	Kaika (2017) [17]	'Don't call me resilient again!': The new urban agenda as immunology … or … what happens when communities refuse to be vaccinated with 'smart cities' and indicators?	Environment & Urbanization	380	42.22	5.50

found in bridging the top right and bottom right of the theme, suggesting the basic and more developed research in this field. In the left quadrant (niche themes), there are representations of "energy efficiency", "energy urbanization", and "energy", along with "urban transportation", "intelligent systems", and "traffic congestion". This indicates a strong focus on these areas, suggesting they are central to smart city research and Sustainable Development Goals practice. This quadrant reflects the

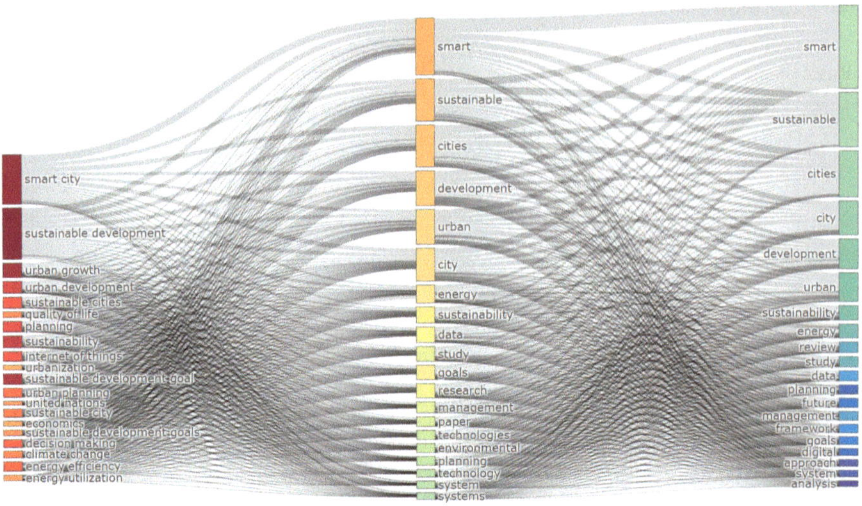

Fig. 2.3 Three-field plot (thematic development) using a Sankey diagram in smart cities with SDGs

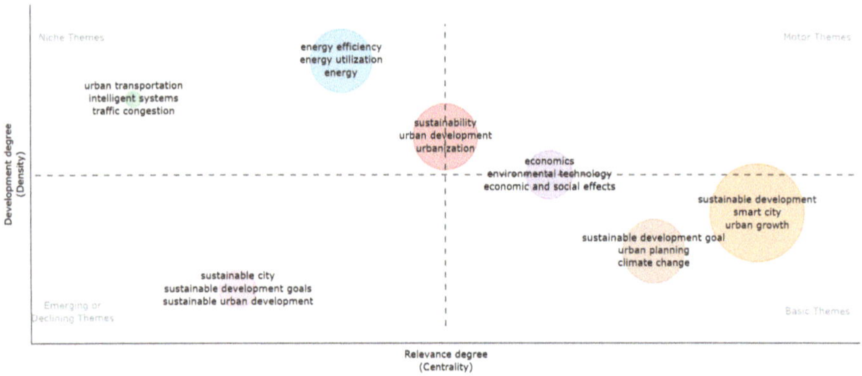

Fig. 2.4 Thematic map in smart city research with SDGs

areas where significant advances and investments are made, highlighting their critical role in the Sustainable Development Goals in a smart city. In contrast, the bottom left quadrants, which represent themes that are either emerging or declining and have lower relevance, include terms like "sustainable city", "sustainable development goal", and "sustainable urban development". Additionally, the terms like "sustainable development goals", "urban planning", "climate change", "sustainable development", "smart city", and "urban growth" form the foundational information for further analysis. These themes might be considered foundational but must be emphasized more in smart cities. This visual categorization provides valuable insights for educators, curriculum developers, and researchers, helping them understand which areas are gaining traction and which might need more attention or innovation.

2.2.3 Co-occurrence Keyword Trends and Network Visualization

Figure 2.5 depicts the most prominent themes and keywords related to research in "Smart City" with "Sustainable Development Goals". The most significant and most central terms, such as “sustainable development”, “smart city”, “urban growth”, “sustainable development goal”, and “sustainability”, indicate the core focus of the research on how researchers grasp and alter their conceptual knowledge, particularly within smart city contexts.

An illustration in Fig. 2.6 displays word occurrence frequencies from the Sustainable Development Goals in the smart city research network. Research focuses mainly on “sustainable development”, as this term appears 614 times throughout the literature. Research on Sustainable Development Goals in smart cities indicates the importance of “smart city” based on its 579 word occurrences. The “urban growth” (142) and “sustainable development goal” (132) combinations show the growth and sustainability in cities. The sustainability challenges in

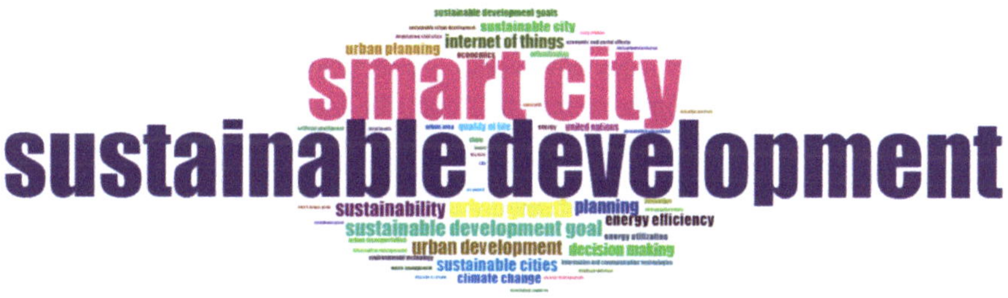

Fig. 2.5 Co-occurrence keyword trends in the smart city domain and Sustainable Development Goals

Fig. 2.6 Co-occurrence network of words in the smart city research domain with SDGs

development are shown through terms like "sustainability" (124) along with "urban development" (120). The documents place emphasis on data-driven planning through combined mentions of "planning" (113) and "internet of things" (107). Additionally, "sustainable cities" (105) and "decision making" (99) show the important role of sustainability and smart technologies in urban environments. The data shows that Sustainable Development Goals and smart city research revolve primarily around several major subjects.

2.2.4 Countries' Scientific Production

The scientific publication frequencies of various countries in the world are represented in Fig. 2.7. Scientific contributions from India stand at 425, which demonstrates their powerful role in research development. The growing scientific power of China displays itself through its 278 scientific publications, which rank second after India. The research hub of Italy produces 193 scientific papers among its total contributions. Numerous publications emerge from the USA and Spain, together with the UK, given their research output comprises 176, 132, and 130 papers, respectively. Academic and industrial research sectors of Brazil and Norway produce approximately 118 and 105 scholarly works, respectively. Portugal and Malaysia have smaller publication numbers, but their contributions amount to totals of 100 and 97 research items, respectively.

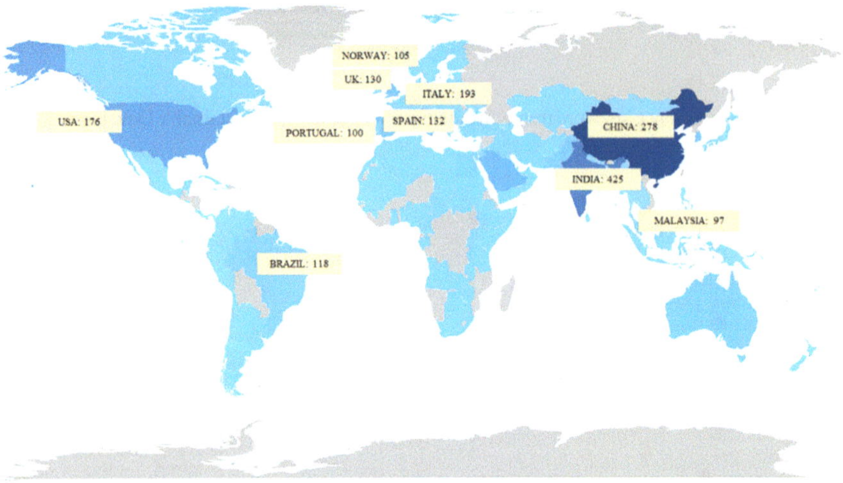

Fig. 2.7 Top 10 countries' scientific production of various countries based on frequency

2.2.5 *Mapping Smart City Research to Sustainable Development Goals (SDGs)*

Sustainable Development Goals, their targets (bubble series), and the percentage (bar graph) of corresponding keywords detected in the text (calculated as the ratio of keywords in one goal to the total number of keywords detected) are depicted in Fig. 2.8. For SDG mapping, the study utilized the SDG Mapper tool from the "Knowledge Base for the Sustainable Development Goals", provided by the European Commission [18]. This tool uses bubble charts to depict SDG interconnections and relevance among bibliometric records. The size of each bubble correlates with the proportion of associated keywords found, which is calculated as the number of goal-specific keywords divided by the total keyword count. As such, SDG 11 is marked by 81 keyword occurrences. This is followed by SDG 7, with 45 occurrences, and SDG 8, with 42 occurrences, underlining their relative prominence in the research database. Within SDG 11, there are several key subtargets that the research addresses. Subtarget 11.3 is shown as a distinct bubble within the larger

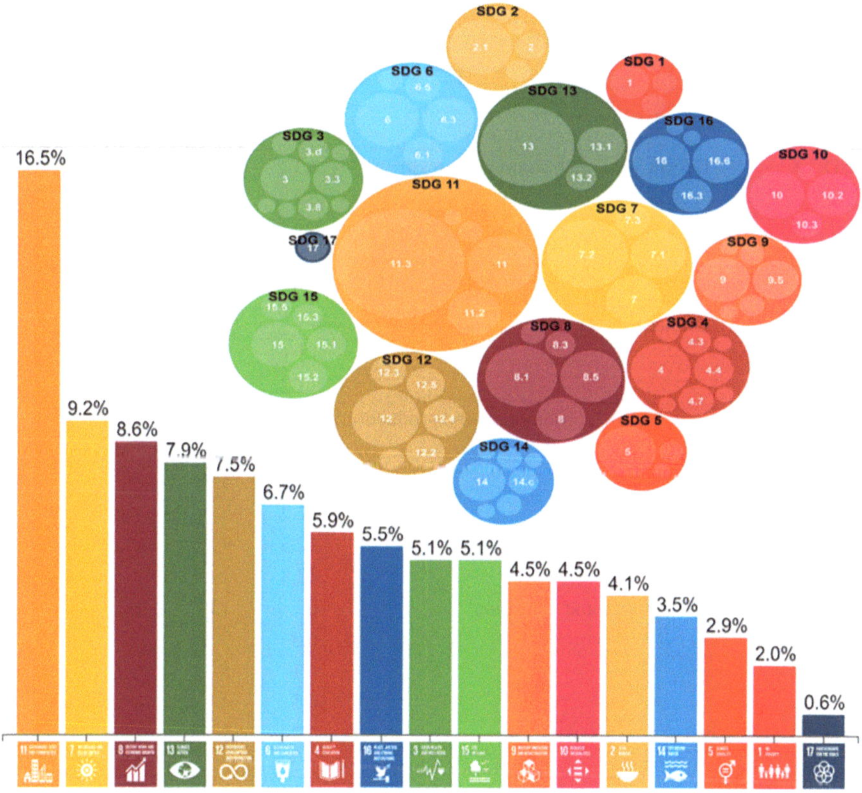

Fig. 2.8 Mapping smart city research to relevant SDGs and targets

SDG 11 circle, indicating focused research on smart cities. Subtarget 11, which aims at sustainable cities and communities, is another bubble within the SDG 11 area. This highlights the efforts in research to improve the cities. Subtarget 11.2 focuses on providing safe, accessible, and sustainable smart transportation. Figure 2.8 shows that the size of the SDG 7 bubble implies that advancements in smart cities are being focused on affordable and clean energy that prioritizes sustainability. Similarly, SDG 8 (Decent Work and Economic Growth) is also closely linked to SDG 11 and suggests that research should explore how to achieve economic growth, productive employment, and decent work for all women and men. The detailed descriptions of other relevant SDGs are depicted in Table 2.3.

Table 2.3 SDGs ranking, targets, and relevance in the smart city research domain with SDGs

SDGs (Rank)	Description	Targets	No. of Occurrence		Relevance
SDG 11	Sustainable cities and communities	11.3	54	81	16.5%
		11	15		
		11.2	9		
		11.7	2		
		11.5	1		
SDG 7	Affordable and clean energy	7.2	21	45	9.2%
		7.1	10		
		7	10		
		7.3	4		
SDG 8	Decent work and economic growth	8.1	17	42	8.6%
		8.5	12		
		8	8		
		8.3	3		
		8.2	2		
SDG 13	Climate action	13	29	39	7.9%
		13.1	7		
		13.2	3		
SDG 12	Responsible consumption and production	12	15	37	7.5%
		12.4	6		
		12.5	5		
		12.2	5		
		12.3	4		
		12.8	2		
SDG 6	Clean water and sanitation	6	13	33	6.7%
		6.3	9		
		6.1	5		
		6.5	4		
		6.2	2		

(continued)

Table 2.3 (continued)

SDGs (Rank)	Description	Targets	No. of Occurrence		Relevance
SDG 4	Quality education	4	12	29	5.9%
		4.4	5		
		4.7	4		
		4.3	3		
		4.5	2		
		4.1	1		
		4.a	1		
		4.b	1		
SDG 16	Peace, justice, and strong institutions	16	11	27	5.5%
		16.6	9		
		16.3	5		
		16.4	1		
		16.5	1		
SDG 3	Good health and well-being	3	8	25	5.1%
		3.3	5		
		3.8	3		
		3.d	3		
		3.2	2		
		3.1	1		
		3.4	1		
		3.5	1		
		3.7	1		
SDG 15	Life on land	15	9	25	5.1%
		15.1	5		
		15.2	5		
		15.3	3		
		15.5	3		
SDG 9	Industry, innovation, and infrastructure	9	9	22	4.5%
		9.5	7		
		9.2	2		
		9.1	2		
		9.3	1		
		9.b	1		
SDG 10	Reduced inequalities	10	10	22	4.5%
		10.2	8		
		10.3	3		
		10.4	1		
SDG 2	Zero hunger	2.1	13	20	4.1%
		2	3		
		2.4	2		
		2.3	1		
		2.2	1		

(continued)

Table 2.3 (continued)

SDGs (Rank)	Description	Targets	No. of Occurrence		Relevance
SDG 14	Life below water	14	7	17	3.5%
		14.c	3		
		14.1	2		
		14.4	2		
		14.2	1		
		14.5	1		
		14.6	1		
SDG 5	Gender equality	5	11	14	2.9%
		5.c	2		
		5.6	1		
SDG 1	No poverty	1	5	10	2.0%
		1.2	2		
		1.5	2		
		1.1	1		
SDG 17	Partnerships for the Goals	17	3	3	0.6%

2.3 United Nations Sustainable Development Goals

In September 2015, the United Nations General Assembly adopted the 2030 Agenda for Sustainable Development and drew up the 17 Sustainable Development Goals. The purpose of these goals is to tackle the wide variety of global challenges, from poverty eradication, health improvement, and environmental protection to economic empowerment. SDG has a specified background and objective, and each also depends on the other, highlighting the importance of a comprehensive view of global development. The SDGs still seek to facilitate equal opportunity and environmental justice and promote partnerships among states while attaining sustainable development. The 17 SDGs span from clean water and sanitation to gender equality, reducing inequalities, and others. An implementation opportunity lies in addressing, in part, economic, social, and environmental sustainability and a more equitable and sustainable world. The approach to the SDGs is supported by the human rights approach, which is the commitment to realizing the rights of every individual and community. This fulfils a comprehensive framework upon which the complex issues and opportunities facing development on a global scale can be addressed and processes taken towards a sustainable and equitable world. Table 2.4 is a list of the SDGs showing the nature of the goals in their brief description of what they seek to achieve, one related target, and the year the goal is set to be accomplished by, with all goals set to be achieved by 2030. These SDGs are integrated into achieving a common goal of eradicating inequality, preserving the planet, and ensuring prosperity for all through addressing these connected global issues.

Table 2.4 Sustainable Development Goals with description, target, and strengths

SDGs	Description	Target (by 2030)	Strengths
1	End poverty in all its forms everywhere	Eradicate extreme poverty for all people everywhere	It aims to eliminate poverty from the lives of every individual, hence enhancing his/her standard and resource endowment
2	Ensure access by all people, and especially children, to safe, nutritious, and sufficient food all year round and at an affordable price	Eradicate malnutrition, both undernutrition and obesity, and increase agricultural productivity by doubling	The sponsoring organization aims at eradicating hunger, ensuring food security, reducing starvation, promoting proper nutrition for everyone, and sustainable means of farming
3	Access affordable medicines and promote the overall health of the global population of all ages	Integrate commitment to achieve a global maternal mortality rate below 70 per 100,00 live births	It focuses on achieving healthy lives and promoting the well-being of all individuals, irrespective of age
4	Provide quality education through equity to all so that effective and efficient learning can be offered to all	Every girl and boy should complete primary and secondary education for free	There is alignment with Goal 4, which is inclusive and equitable quality education for sustainable development and equal access to quality lifelong learning
5	Ensure women's economic independence and reproductive rights	End female genital mutilation and other harmful practices	In seeking to promote the rights for gender equality, empowerment for women and girls, as well as social, economic, and political
6	Ensure the availability and sustainable management of water and sanitation for all	Achieve universal and equitable access to safe and affordable drinking water	By making water resources accessible and sustainable for all, with a focus on sanitation, a healthy life is attained as well, and disparities are reduced
7	Ensure access to affordable, reliable, sustainable, and modern energy for all	Increase the share of renewable energy in the global energy mix	Focuses on the policy intervention to increase the affordable, reliable, sustainable, and modern assets for the people to support sustainable development and to reduce emissions
8	Sustain per capita economic growth; forecast sustainable consumption and production patterns; and advance the social, economic, and labour dimensions of sustainable development	Full and productive employment and decent work for all women and men	Ensures intended, long-term economic development, as well as the expansion of employment and workplaces for all individuals

(continued)

Table 2.4 (continued)

SDGs	Description	Target (by 2030)	Strengths
9	Ensure the availability of affordable, reliable, sustainable, and modern energy	To build other means of quality, reliable, sustainable, and resilient infrastructure for everyone	The overall intention is to strengthen architectural measures, support the organization of industrialization that is sustainable and inclusive, and stimulate innovations
10	To increase equality between the population and within countries	Lessen population and territorial disparities in terms of outcomes	Intends to remove disparities both between as well as within countries, espousing for fairness and equity
11	Make cities and human settlements inclusive, safe, resilient, and sustainable	Provide universal access to safe, inclusive, and accessible green and public spaces	It is dedicated to promoting sustainable and resilient human settlements with a specific focus on inclusive and safe cities and living
12	Ensure sustainable consumption and production patterns	Halve per capita global food waste at the retail and consumer levels	Tries to promote sustainable consumption and production to minimize waste and consumption of the planet's resources
13	Integrate coping and mitigation measures for climate change into comprehensive and urgent plans	Enhance the countries' abilities to reduce vulnerability and the catastrophic effects of climate-sensitive events	Examples of concrete goals are to fight climate change and its consequences in the context of the construction of sustainability measures and the reduction of disaster risk
14	Proportion of marine populations utilizing resources from ancillary sources, proportion of marine populations using resources related to unsustainable fisheries and aquaculture	Reduce the incidence and negatively impacted area of marine pollution of all kinds, reducing the incidence and extent of marine pollution	Aim at the conservation and sustainable use of the ocean, seas, and their resources for development
15	Ensure conservation of the ecosystems, restoration of the forests, and combating desertification and termination and reversal of the decline in the land's biological productivity of the land	Repair of the depleted soil, caused by the depletion of the forest cover	Implementation of measures to conserve, restore, and sustainably use terrestrial ecosystems, sustainable forest management, and controlling desertification

(continued)

Table 2.4 (continued)

SDGs	Description	Target (by 2030)	Strengths
16	Support peace, justice, and institutions for development; guarantee equal justice to everyone; and establish effective, accountable, and inclusive institutions starting at the national level	Substantially prevent and curb all forms of bribery of officials, smuggling of small arms and light weapons, as well as recovery of stolen assets	It has two more merits, which include supporting local cultures of non-violence and of sustaining peaceful societies, ensuring that everyone has appropriate access to justice, and contributing to state- and national-level accountable institutions
17	Mobilize all forms of financing for development for the implementation of policies financed by partner countries	Mobilize international trade as a factor promoting economic growth	Aims at enhancing the implementation and prolonging the partnership for development in a sustainable manner so that it is mutual

2.3.1 Goal 1: No Poverty

Reducing poverty in all its forms and dimensions is the goal set out by the Sustainable Development Goals in Goal 1 to be reached by 2030. Poverty has many definitions, which have evolved over time, from which three are considered primary: absolute, relative, and multidimensional. Absolute and relative poverty measures concern an individual or a household's ability to have access to the basic life attributes like shelter, education, food, and so on, using income and expenditure as basic indicators. By the international standard, a person is considered poor if a citizen lives on less than US$1 or US$2 a day in purchasing power parity. However, this definition has recently been challenged as arbitrary and is now being accepted as the term multidimensional poverty, which takes into cognizance multiple welfare impoverishment indicators altogether [19]. The 17 SDGs from the UN's 2030 agenda for Sustainable Development aim to balance economic progress and protection of the environment and to overcome the gap between industrialized and developing countries. The SDGs create an opportunity for advancing equal opportunity, employing economic empowerment, and promoting sustainable development worldwide [20].

2.3.2 Goal 2: Zero Hunger

The second goal of the Sustainable Development Goals is to end hunger, achieve food security, improve nutrition, and promote sustainable agriculture. To eradicate hunger, targets and indicators should be tied to the four pillars of food security: availability, access, utilization, and stability [21]. By 2030, the SDGs also aim to provide all people, including vulnerable populations and children, with access to safe, nutritious, and sufficient food and to reduce premature mortality from non-communicable diseases, as well as reduce food waste [22]. Rapidly diminishing public investment coupled with low agricultural productivity and gender disparities alongside socio-cultural factors represent further challenges to the nutrition of women and children in low- and middle-income countries (LMICs). Achieving Goal 2 is closely correlated with tracking progress and challenges through the performance of food systems, which is important to achieving all SDG goals, especially in areas of undernutrition and growing risks of non-communicable diseases. The challenge of flexible indicators and a systematic approach is emphasized, prompting monitoring of performance to be used as a way to inform priority actions and channel the scarce resources effectively. For effective approaches to global hunger and food insecurity, it is imperative to strengthen collaborations across disciplines, increase the research around food systems, and provide the appropriate funding, financial investments, and coordinated actions from many of the stakeholders.

2.3.3 Goal 3: Good Health and Well-being

In the Sustainable Development Goals, Goal 3 promotes the assurance of good health and well-being. Included in this aim are various targets ranging from reducing maternal and child mortality to eradicating diseases, such as AIDS, tuberculosis, malaria, and supporting prevention, treatment, and control of disease. Besides, it strives to attain universal health coverage by equipping all with access to essential medicines and vaccines, as well as reducing illnesses and deaths caused by hazardous chemicals [23]. However, according to the World Health Organization (WHO), there has been remarkable progress in global health that has seen a reduction in maternal and infant mortality rates and an increase in healthy life expectancy. Progress is needed, however, as some regions record increases in food insecurity despite efforts to speed up the progress towards SDG 3. The benefits associated with reaching Goal 3 have economic implications because the goal is not only linked to health outcomes but also to economic development and the formation of human capital. Research on public health improvements suggests that these improvements aid economic development, as healthy people have the economic ability to earn higher incomes than people who are unemployed [23]. Despite progress in reaching health-related targets, low-income and developed countries are still far from achieving health-related goals. It highlights the need to deal with the challenges and opportunities presented to globally promote good health and well-being, as articulated in Goal 3 of the SDGs.

2.3.4 Goal 4: Quality Education

One of the Sustainable Development Goals is Goal 4, which aims to ensure inclusive and equitable quality education for all. In this context, it is extremely necessary to analyse the progress and the challenges encountered, as well as the future perspectives that would enable quality education globally. Sustainable development cannot be truly achieved without an investment in education, and its priorities include improving access to education and educational outcomes for all. Education for sustainable development is very important, and all countries have committed to achieving the global goals [24]. The aim of SDG 4 is to provide inclusive, good-quality, and low-cost primary and secondary education for all by 2030 to achieve relevant and effective learning outcomes.

2.3.5 Goal 5: Gender Equality

The fifth goal of the Sustainable Development Goals aims for equality and empowerment for women and girls. This goal recognizes the gender-based inequality across the world and that it is critical to eliminate structural hindrances for women's empowerment. However, scholars maintain that economic and educational efforts are not enough to overcome such powerful structural barriers. For this reason, the concept of the goal goes beyond simply representation or minimum standards to include considerations like education, political participation, and work opportunities, strongly affirming that empowerment requires addressing a number of areas. Moreover, the goal emphasizes the relevance of gender mainstreaming in poverty eradication strategies and development approaches to take into consideration the impact of gender inequality as a major cause of poverty and a constraint to development [25].

2.3.6 Goal 6: Clean Water and Sanitation

The United Nations has chosen one goal to encompass, providing clean water and sanitation to all, i.e. Goal 6, from the 17 Sustainable Development Goals. The problem aims at acknowledging the urgent worldwide water crisis, with millions of people not having access to safe drinking water and appropriate sanitation facilities. Goal 6 targets are clean and safe drinking water, declining use of open defecation, integrated management of water resources, an increase in the rates of wastewater treatment, promotion of water recycling, and preventing water wastage [26]. The attainment of these targets faces great challenges, mainly where a large number of populations are not provided with basic drinking water and sanitation facilities [27]. Monitoring of progress towards Goal 6 is therefore vital but difficult because established methods of measuring the quality and use of water and sanitation services involve expense or are elusive. The targets and proposed indicators of Goal 6 put emphasis on safely managed services, which assume that water and sanitation facilities are from an improved source, and they are accessible when needed and do not contaminate. To aid in achieving the targets of Goal 6, the monitoring of the water, sanitation, and hygiene interventions is anticipated to show more cost-effective and efficient ways of improving both global development and sustainability.

2.3.7 Goal 7: Affordable and Clean Energy

SDG 7 is to ensure access to affordable, reliable, sustainable, and modern energy for all. Goal 7 is considerably important in global development; it affects climate change mitigation and provides the path towards sustainable economic growth. For example, Ghana has the problem of moving away from traditional biomass and developing problems in its hydropower sector to achieve universal access to affordable and modern energy services [28]. Efforts to enhance renewable energy potential have seen efforts to diversify the energy supply with alternatives in solar, wind, and small hydro developments, with some challenges such as the 'land grabbing' issue. Moreover, implementing SDG7 also has a role for the European Union (EU), which has acknowledged the necessity of implementing this SDG by developing countries, providing high financial and technical assistance for energy projects aimed at increasing the share of renewable energy and low-carbon technologies [29]. It includes reinforcing policy and regulatory frameworks, building capacity, generating financing, and advocating multi-stakeholder engagement to promote the energy shift towards more abundant and sustainable sources of energy. The achievement of these initiatives is essential for evaluating the advancement in the attainment of Goal 7 and for addressing the energy challenge that the developing countries face in accessing affordable, clean, and reliable energy.

2.3.8 Goal 8: Decent Work and Economic Growth

The path to sustain economic growth, increase productivity and technological innovation for work, and decent employment for all is under Goal 8 of the Sustainable Development Goals. It further provides an indication of how important the issue of decent work is as a fundamental human right and its inclusion as an integral component in the international development policy [30]. However, the SDGs have been criticized for having an extreme encyclopaedic scope and number of 169 targets spanning the 17 goals [31]. In addition, simplification and adaptation of the indicators to the national environment could oversimplify the targets, excluding important qualifiers. The recognition of the right to decent work as a justiciable human right is increasingly being specified, and the Sustainable Development Goals are aimed at promoting full employment and sustainable livelihood by ensuring productive employment and work that is freely chosen. However, these targets can only be attained if there are major changes in policy and national adaptations to allow the indicators to be realistic and relevant within each context.

2.3.9 Goal 9: Industry, Innovation, and Infrastructure

Resilient infrastructure, inclusive and sustainable industrialization, and innovation, which are key to global development, are the object of Goal 9 among the 17 Sustainable Development Goals. It is important to achieve this goal, as it is a key indicator in forecasting future economic performance [32]. It further adds that digitization and better technological infrastructure can help accomplish several Sustainable Development Goals, out of which SDG 3 (Good Health and Well-Being) and SDG 9 are achieved. Also, the research shows that there are trade-offs between Sustainable Development Goals as well as synergies, and the inclusion of Sustainable Development Goals imposes both opportunities and challenges on global businesses. The need for development and innovation that are inclusive in order to realize the Sustainable Development Goals [33]. Instead, they emphasize the role of technology companies in establishing accessibility principles that advance ideas and make life easier for millions across the world with disabilities. The need for inclusive design and development: SDG 9 emphasizes that technology companies should take inclusive design and development as a long-term investment. This emphasizes the value attached to innovation and inclusive infrastructure in promoting sustainable development.

2.3.10 Goal 10: Reduced Inequality

SDG 10 aims to reduce inequality and has sought to address economic, social, and political inequality through its Goal 10. This goal is particularly crucial in low- and middle-income countries to tackle socioeconomic inequalities and non-communicable diseases [34]. The study underlines the activity on social determinants of health and recognizes the concept of health equity from change in internal conditions and the need for universal health protection and ecological development paradigms. This is in line with SDG 10, which seeks to tackle the social determinants of health equity and general inequality in the world. As we envision the future, achieving Goal 10 and creating an inclusive society will require an approach that takes into account health equity and the root of inequality.

2.3.11 Goal 11: Sustainable Cities and Communities

Various targets related to inclusive, safe, resilient, and sustainable urban areas are covered under Goal 11 in the Sustainable Development Goals about sustainable cities and communities. For urban development initiatives to contribute to the generation of outcomes covered by the Sustainable Development Goals and the Paris Agreement, it is crucial to have effective measurement and data disaggregation.

Measuring SDG 11 and its relationship with other urban SDGs constitutes a challenge that only a holistic approach to urban development can overcome [35]. Sustainable Development Goals interlinkage analysis can assist in understanding the relationships among Sustainable Development Goals and guide policy by revealing synergies and trade-offs of different targets and facilitating the development of customized solutions to sustainable urban and rural growth [36].

2.3.12 Goal 12: Responsible Consumption and Production

Sustainable consumption and production patterns are an element of Goal 12 of the 17 Sustainable Development Goals, entitled 'Responsible consumption and production'. This goal is defined as an attempt to eliminate unsustainable patterns of consumption and production. Supporting the long-term well-being of people, communities, and nature conservation through sustainable consumption and production is also to meet the needs of the present as well as future generations. Sustainable consumption and production policies to mitigate trade-offs and manage environmental conservation plans [37]. The study stresses that coherent policy combinations at the global, regional, and national levels are necessary to, in a coordinated way, work towards different Sustainable Development Goals. It also points out the central function of sustainable consumption and production policies in designing integrated policies consistent with the Sustainable Development Goals ambition.

2.3.13 Goal 13: Climate Action

The 13th Sustainable Development Goal aims at addressing the action on climate and its possible effect on the development of the entire globe. Health, livelihoods, and overall sustainability are all at risk from climate change, which has been identified as a major threat to global development. Due to climate change, the intensity and frequency of extreme events are expected to increase, thus threatening public health, poverty reduction, education, as well as other development indicators [38]. Taking measures to lower short-lived climate pollutants (SLCPs) in addition to deep CO_2 cuts raises the chance of achieving the 2 °C or less target and slows the rate of warming and mitigates the growing impacts of climate change. Adaptation to climate change is necessary and may include early warning systems of storms, flood defences, and changes to infrastructure to withstand the worst effects of changes in weather [39]. This provides a new reason to accelerate the fight against climate change and adaptation to its impacts as a necessary means to achieve sustainable development goals.

2.3.14 *Goal 14: Life Below Water*

SDG 14 focuses on the conservation and sustainable use of oceans, seas, and marine resources for sustainable development. The goal is more concerned with the interconnected way that ocean health interplays with social, economic, and environmental matters at large. An important insight is that strong sustainability measures need to be implemented to preserve the livelihood of people residing in the Small Island Developing States (SIDS) and the least developed countries. Furthermore, Marine Protected Areas (MPAs) are recognized as an important strategy towards marine conservation, with specific methods of regulating harvesting, terminating Illegal, Unreported, and Unregulated (IUU) fishing, and destructive fishing [40, 41]. Moreover, acoustic animal tracking technology is considered a promising monitoring tool in the study of pollution effects on aquatic fauna and the long-term effects of climate change on aquatic organisms and fisheries management actions. Animal tracking is in line with the targets of SDG 14 and most substantially help reach its goals of sustainable management and protection of life below water. This relies on the development of transboundary collaborative efforts, as well as global networks, to fully tap animal tracking and achieve SDG 14 targets.

2.3.15 *Goal 15: Life on Land*

Preserving and restoring terrestrial ecosystems and biodiversity are the targets of Goal 15 within the group of 17 Sustainable Development Goals. Achieving this goal is important to ensuring the health of life on land and its ability to sustain. Goal 15 makes ambitious efforts to protect and manage forests sustainably, halt and reverse the desertification process, and reduce biodiversity loss. With the goal of raising global sustainable development, the SDGs have come under scrutiny because of the lack of certainty in targets and measurement. For example, only 29% of the 169 targets were measured as targetable for specific endpoints, timeframes, and the ability to be accurately measured [42]. There are also debates over the level of targets' weighting, links between Goal 15 and other SDGs, or the conservation of multiple species of wild species beyond a few flagship ones [43]. This focusses on the complicated and interdependent nature of sustainable development initiatives, and how it can be measured involves taking into account a wide variety of ecosystems and species.

2.3.16 *Goal 16: Peace, Justice, and Strong Institutions*

Goal 16 aims to promote peaceful and inclusive societies for sustainable development, provide access to justice for all, and build effective, accountable, and transparent institutions at all levels. This is considered one of the critical enablers of sustainable development, able to firmly influence the success of all the SDGs, particularly in conflict or fragile countries. With a broad and transformative scope, the goal represents a crucial element towards achieving the whole 2030 Agenda. Recognizing that Goal 16 can make a huge difference towards the success of the whole set of SDGs is vital, especially when working in a conflict-affected or fragile country. Peace, security, and development are interdependent and necessary ingredients to create sustainable development and the reduction of poverty [44]. Despite this, peace, security, and justice have been the cause of debate and concern by UN member states in the post-2015 development agenda. However, some countries have also voiced their fears that by paying attention to these areas, it could securitize the development agenda and violate national sovereignty.

2.3.17 Goal 17: Partnerships for the Goals

The goal of SDG 17 is focused on the importance of partnerships towards sustainable development. However, a recent study has pointed out the imbalance of effort around the globe and just how this may lengthen later on the North–South divide [45]. Results showed that most partnerships recorded in the UN database were domestic, and that few partnerships registered have partners from low-income countries, whereas almost all have partners in other income categories. Partnership focuses also differed by the income category of the countries partnered, with the idea that the Sustainable Development Goals place inappropriate focus on development rather than sustainability considerations. In addition, the partnerships are made between countries and stakeholders, which could accelerate the achievement of a number of Sustainable Development Goals. In some countries, a systems model analysis showed synergies in interactions within selected Sustainable Development Goals in terms of mutual reinforcement of progress of targets and goals [46]. In addition, trade-offs were observed among other Sustainable Development Goals, showing that achieving one target of these goals would hinder the benefits of the other. It is evidence of the intricate relationships between the Sustainable Development Goals and the need for a broad view of sustainable development.

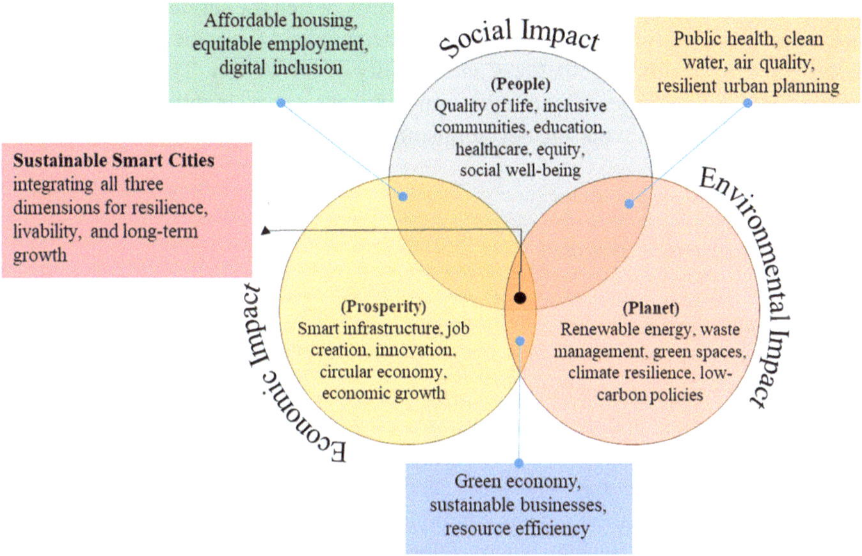

Fig. 2.9 Social, economic, and environmental impact of SDGs in smart cities

2.4 Social, Economic, and Environmental Impact of SDGs in Sustainable Smart Cities

Sustainable Development Goals have a vital role in responding to society, economy, and environment when defining sustainable smart cities, as illustrated in Fig. 2.9. All three aspects are necessary for developing smart cities that will adequately feature computerization, accessibility, and strength, as well as sustainability.

2.4.1 Social Impact

Sustainable Development Goals focus on the importance that the space in urban areas is inclusive and productive in the sense that it should accommodate and meet the essential needs of all the users [47]. The social framework of smart cities depends much on the Sustainable Development Goals. SDG 3 (Good Health and Well-being) and SDG 4 (Quality Education) make goals to ensure that essential services such as healthcare, education, and public safety are available to the urban populations. The digital solutions, such as telemedicine, online learning platforms, and smart policing, are used by smart cities to enhance these aspects. Specifically, Goal 11, "Sustainable Cities and Communities", focuses on this problem because it involves enhancing the safety and accessibility of cities and communities. Improving public transportation, improving the conditions of the housing, and providing basic needs such as health, education, etc., are included in this. When the services of smart cities are moving towards engaging and involving the people of the society, then the smart cities can improve the togetherness and equality of the society. Through the use of technologies such as these digital platforms, citizens can be involved in making the decisions that would favour the development that is required in the urban areas [48].

2.4.2 Economic Impact

Economically, integrative smart cities include steady growth, variety, and stability of the economies [49]. There are several goals to highlight in relation to the mentioned Sustainable Development Goals, where SDG 8 focuses on promoting innovation, decent jobs, and economic growth, as well as SDG 9, which is based on industrialization and infrastructural development. Smart cities apply IoT, AI and big data to enhance economic efficiency and increase productivity [50]. Efficient resource use and operational costs can be lowered for cities in order to establish a sustainable economy. Besides, smart cities have the potential for investments and people's employment in knowledge-based and green economy sectors, which moves the economy to be more vibrant and sustainable. These advancements also enable other areas, such as entrepreneurship, as well as the economic diversification of small businesses through the integration of digital solutions that enable economic resiliency.

2.4.3 Environmental Impact

The environmental component in the framework of Sustainable Development Goals is crucial for the creation of the smart city, which enshrines the principles of rational use of resources and reducing the negative impact on the environment [51]. Target 6.3 of SDG 6 on clean water and sanitation relies on technology for the sustainable provision of water supply and sanitation, while target 7.3 of SDG 7 on affordable and clean energy also emphasizes the use of technology in the provision of clean energy for citizens. Smart city planning has the environmental dimension of SDGs at its centre, seeking, for example, that SDG 13 (Climate Action) and SDG 15 (Life on Land) are focused on sustainable resource management, reducing pollution, and preservation of urban biodiversity. The smart city integrates intelligent technologies and best practices of using available resources for proactive resource utilization for the environment and to minimize the emission of greenhouse gases [52]. Additionally, the smart grid and the utilization of renewable power sources greatly reduce city's emissions. Also, green wedges as part of sustainable urban planning also affect the increase of gradual density of green areas such as green roofs and urban forests. Due to the incorporation of sustainability, smart cities will address climate change's effects and improve future generations' quality of life [53]. Therefore, the main idea is to have an integrative approach to Sustainable Development Goals in smart cities that will implement social equity, economic growth, and environmental sustainability. Incorporating these goals within urban planning and development, cities thus become more inhabitable and more capable of responding to increasing vulnerability, enhancing the chances of sustainable urban development globally. Table 2.5 can be used to categorize and further elaborate on how each of the Sustainable Development Goals can help address all the social, economic, and environmental aspects of sustainable smart cities.

Table 2.5 The description of the social, economic, and environmental aspects of SDGs towards smart cities

SDGs	Social Impact	Economic Impact	Environmental Impact
SDG 1	Provision of basic services, better shelter, and living standards for the poor populations	Promoting employment and quality economic opportunities for the poor	Reducing environmental pressures on poor citizens in urban areas
SDG 2	Improved food security and nutrition for urban populations	Strengthening local food systems and economies	Reduced food waste and sustainable agricultural practices
SDG 3	Access to health care and an improvement in the health of the nation	Properly improved health leads to economic benefits, such as productivity among employees	Enhanced health in society implies less pollution, better air and water quality, and a better urban environment
SDG 4	Enhanced educational opportunities and lifelong learning	Workforce development and economic growth through skilled labour	Promoting environmental education and awareness.
SDG 5	Role of women and gender minorities in city spaces	Economic impact of women's employment	Environmental management through gender mainstreaming
SDG 6	A better health status of the population and well-being	Benefits, which are results of the effective use of water	Protection of water sources and better quality water
SDG 7	Sustainable energy to support the everyday needs of individuals, institutions, and the community	Increased rate of innovation in the energy sector	Reduced rate of greenhouse gas emissions and fossil fuel dependence
SDG 8	Employment opportunities and decent labour conditions	Sustainable economic growth and productivity	Development of climate-neutral occupations and industries
SDG 9	Coordinated and efficient transport systems in the cities	Technological advancement to resilientize the economy	Reduction of the negative impact of infrastructure on the environment
SDG 10	The highest level of social inclusion, eradicating the vivid disparity between new urbanization	Economic perspective involves the improvement of the stability of economic policies to support the growth of the community	Equal share of the resources

(continued)

Table 2.5 (continued)

SDGs	Social Impact	Economic Impact	Environmental Impact
SDG 11	Improved living in urban areas and community resistance	Economic outcomes from the use of efficient urban planning	A clear improvement in the environmental outcome through the use of smart technologies in cities
SDG 12	Promotion of sustainable consumption patterns	Economic efficiency and reduced resource waste	Minimized environmental impact through sustainable practices
SDG 13	Improved public health and disaster resilience	Economic savings from climate adaptation and mitigation	Reduced carbon emissions and enhanced climate resilience
SDG 14	The counselling is to raise awareness about marine ecosystems	The economic benefits of sustainable fisheries, as well as tourism, are used for the conservation of aquatic ecosystems	The efficiency of a country can also be measured using indicators that are part of this goal
SDG 15	Enhanced green spaces and biodiversity in urban areas	Economic opportunities in ecotourism and conservation	Protection of terrestrial ecosystems and biodiversity
SDG 16	Improved governance and reduced urban violence	Economic stability through the rule of law and transparent institutions	Sustainable resource management through effective policies
SDG 17	Engaging the communities and stakeholder collaborations	Partnerships for economic prosperity and business investments	Synergy in conserving the environment and efficient resources

2.5 Role of Technological Advancement in Sustainable Development Goals

The Sustainable Development Goals are the global plan of action for a better and sustainable future. Advanced technologies also apply in the interface between smart cities and the overall achievement of these goals, as illustrated in Fig. 2.10. With more people migrating to urban areas as a result of rural-urban drift, the improved, equitable, and sustainable solutions will be pursued even more [54]. Technologically integrated smart cities are a way of combating these issues and advancing toward a number of the Sustainable Development Goals, such as affordable and clean energy (SDG 7), industry, innovation, and infrastructure (SDG 9), sustainable communities and cities (SDG 11), and climate change (SDG 13). A smart city is a city that applies the old and latest technology, such as IoT, AI, big data logistics, renewable power, and so on. These technologies make it possible to achieve efficient use of resources in a city and, at the same time, lower emission levels and enhance citizens' well-being [55]. For example, via IoT sensors, people living in a particular city are able to get real-time information concerning the quality of the air and water around them, information that the policymakers and city planners can use to enhance

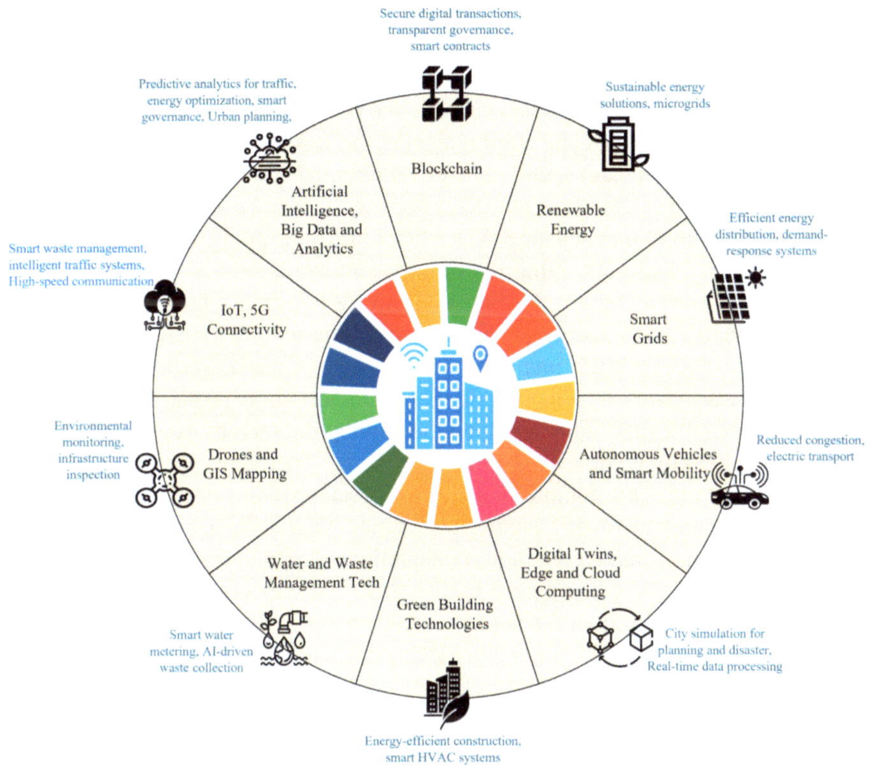

Fig. 2.10 Advanced technologies with their roles in smart cities with SDGs

environmental conservation and, thus, contribute positively to SDG 3 on good health and well-being for all and SDG 6 on clean and available water. In the same regard, smart grids, incorporating AI and IoT into the electrical system, optimize energy use and incorporate renewable energy into the power system; this contributes to SDG 7 of accessing affordable, reliable, sustainable, and clean energy. Big data analytics as an element of smart cities is also applied to solve the issues related to SDG 11, which focuses on making cities sustainable, inclusive, safe, and resilient. Such results help a city avoid blockages, organize traffic better, and ultimately, minimize air pollution output in terms of greenhouse gases. Smart technologies in transport systems, through the use of data, enhance the functioning of public transport, increase the usage and sales of electric cars, and support the establishment of bike sharing [56]. Such steps also help in the reduction of transport-related environmental impacts, including traffic jams and pollution, and assist in improving the transportation systems for all the inhabitants of urban areas, as stipulated in Goal 10 on inequality. Furthermore, such technologies promote innovation and infrastructure for new economic solutions and better services while dealing with SDG 9. Digital tools could then become integrated for the improvement of such services, such as the handling of waste, security services, and disaster control in cities. For example, the use of AI in surveillance that enables the system to identify events and act quickly to prevent such events, or an application that assists in determining the mode of collecting and managing waste in a smart manner to minimize the dumping of waste into the landfill. They also enhance the life standards of living in cities and give birth to new industries and employment in the sectors of technology, leading to economic development and sound industrialization. Another impressive application of smart cities, as it relates to technological advances, may be attributed to its capacity to contribute to climate change (SDG 13). Due to the application of technology in tracking down and controlling the physical infrastructures and usage of cities, CO_2 emissions can be minimized. Smart energy systems make buildings intelligent when it comes to using energy resources, while smart water management systems make water supply facilities intelligent in the usage of water. Vertical farming and hydroponics, as forms of urban agriculture, support food production, cut down emissions from transportation, and feed the population, thereby supporting the UN Sustainable Development Goal 2—Zero Hunger. Therefore, the environmental effects of smart cities are complemented by the social benefits of improving technological aspects and advancing inclusive societies. Digital inclusion strategies deal with a plan towards eradicating digital inequality and making sure that any oppression of conventional use by citizens in that area of low-income range is addressed. This may be relevant to both the gender equality objective of SDG 5 and the decent work and economic growth outlined in SDG 8, since it provides its target population with the necessary tools to flourish in the global digital economy. In addition, smart cities enable the improvement of governance and also public involvement in relation to SDG 16, institutions of peace, justice, and governance. Modern technologies and web-based applications make it easier to engage the public in responsive and effective governance since the interactions are real in nature [57]. Technology-based participatory models of governance enable citizens to be involved in the decision-making processes of governance to ensure that the overall development of urban cities provides for people's desires within the society. The enhancement of technology in smart cities also brings certain effects on partnerships, which are crucial for the achievement of the Sustainable Development Goals (SDG 17).

A fundamental implication of the concept of smart city is that the requisite solutions require action by a broad coalition of governments, non-governmental entities, universities, and civil societies. It means that companies are prepared to invest in smart city programmes, while institutions can provide knowledge and findings. Collaboration between countries is of importance in the sharing of knowledge and technology in order to make cities in the international sphere develop smart solutions. However, there are still some challenges that needs to be considered in smart cities in meeting the Sustainable Development Goals. For technology to be beneficial to all the citizens, the challenges that relate to it should be tackled perfectly. The challenges include data privacy, cybersecurity, and digital inequality. However, smart city technologies are capital and political capital-intensive, thus likely to present difficulties for many resource-scarce cities. It is only critical to rely on long-term strategies and policies so as to make sure that smarter city projects are sustainable and deal with the ever-changing technological advances and risks for sustainable environments in the future. Altogether, technology represents an important characteristic element of smart cities, and it can significantly contribute to the attainment of the Sustainable Development Goals. With the rapid urbanization process, the thing that is going to be crucial for the development of the smart city concept based on a sustainability and inclusion perspective is needed to develop a better future for next generation.

2.5.1 Technological Management in Smart Cities

Smart cities use technological management in order to increase the quality of life in urban areas, as well as optimize the performance of various items of infrastructure for sustainable development. A concept of smart city is centred on bringing together ICT with facilities within cities [55]. This includes the use of sensors, data analytics, and the IoT set up deployed all over the city, which captures data on such aspects as traffic flow, energy consumption, air quality, and public security. In smart cities, technological management aims at achieving the efficient use of available resources in service provision. Big data analytics, therefore, allows the city managers to get information about the dynamics of the city, hence helping to act proactively and better resource the city. Smart grid and renewable energy systems improve the efficiency of energy usage, while a smart water management system helps in the efficient utilization and supply of water. In addition, technological management deals more with the citizens and how they can participate in smart cities. It enables citizens to access government services, report issues, and provide feedback conveniently through digital platforms and mobile applications [58]. The problem of cybersecurity is also important because smart cities are interconnected systems, which means that they may be prone to cyber threats. There must be anti-piracy measures that are put in place to ensure that delicate information is safeguarded well and customer confidence is upheld. Furthermore, the technological management of smart cities is responsive to economic development and creativity. So, in a sense, smart cities can be utilized to make a better environment for business advancement by giving a proper platform for new companies and innovation. Technological management refers to the application of technology towards enhancing urban and city capacities, reliability, and sustainability. It involves the use of technology, governance, policies, and people's engagement in an adequate solution to the various issues emerging from urbanization.

Sustainable Development Goals are particularly significant in how technological management is pursued in smart cities because it mainly focus on endorsing sustainable and innovative urbanization. Several Sustainable Development Goals can be seen as most relevant to this integration to make sure that cities embrace technology towards enhancing the quality of life for people and that they respect the environment. Smart city solutions that relate to Industry, Innovation, and Infrastructure (SDG 9) ensure the use of technologies like IoT, AI, and big data analytics for innovation. These technologies assist in the planning and designing of cities, increase the stability of infrastructures as well, and promote sustainable industrialization. SDG 11 targets sustainable cities and communities, this goal is adopted by technological management in smart cities through efficient use of resources, smart grid and systems in transportation, efficient provision of e-governance and public utility services through an enhanced digital platform, and citizen participation. The SDG 7 goals mainly focus on the provision of clean and environment-friendly energy for all citizens across the globe. Smart cities also use renewable sources of power, the use of smart grid systems to enhance power consumption, and reduce carbon footprint. Furthermore, SDG 13, which relates to climate action, is important as a smart city measures the environmental effects of the proposed technologies and deploys technologies to take real-time decisions and actions that reduce climate change. In general, the Sustainable Development Goals offer a foundation for managing technology in smart cities, guaranteeing that innovations deliver positive impacts on global development goals, leading to the creation of smart cities that are effective, sustainable, and better. Table 2.6 illustrates an understanding of how various technologies map in the realization of Sustainable Development Goals in the context of smart cities. Every technology has its role in the utilization of resources and the delivery of efficient services in urban areas of the region that will ultimately enhance the quality of life for the inhabitants while promoting sustainable development.

Table 2.6 Various technologies used in smart cities in the realization of the SDGs

SDGs	Technologies Used in Smart Cities
SDG 1	Digital payment systems, e-commerce platforms, IoT for resource distribution, AI for job matching and training
SDG 2	IoT for smart agriculture, precision farming, blockchain for supply chain transparency, and food waste management systems
SDG 3	Telemedicine, health monitoring devices, AI for predictive healthcare, data analytics for epidemic control
SDG 4	Online learning platforms, virtual reality for immersive learning, AI tutors, educational data analytics
SDG 5	Mobile apps for women's safety, digital literacy programs, AI for fair recruitment, and platforms for women entrepreneurs
SDG 6	Smart water management systems, IoT sensors for water quality, wastewater treatment technologies, and data analytics
SDG 7	The prominent emerging technologies include smart grids and renewable energy systems such as solar energy, wind energy, energy storage mechanisms, and IoT systems for energy control and monitoring
SDG 8	Automation and robotics, IoT for business efficiency, digital platforms for entrepreneurship, blockchain for jobs
SDG 9	3D printing solutions, intelligent infrastructure systems for maintenance, IoT for maintenance, and artificial intelligence for business process optimization
SDG 10	Digital inclusion initiatives, e-governance platforms, AI for policy analysis, and online education resources

(continued)

Table 2.6 (continued)

SDGs	Technologies Used in Smart Cities
SDG 11	Smart transportation systems, IoT for urban management, renewable energy integration, data analytics for planning
SDG 12	Responsible Consumption and Production can be directly linked to supply chain optimization, utilization of waste-to-energy technologies, and integrated IoT to manage resources and energy on circular economy platforms
SDG 13	Climate monitoring sensors, renewable energy technologies, AI for climate modelling, carbon footprint analysis tools
SDG 14	IoT for marine conservation, autonomous underwater vehicles, and data analytics for ecosystem management
SDG 15	Predicting vegetation, monitoring endangered species, using drones, and developing effective conservation strategies
SDG 16	Blockchain for transparent governance, digital identity systems, and AI for law enforcement support
SDG 17	Irrespective of the type of relation between the two entities, a power partnership can generate new benefits through an innovative approach to cooperation: blockchain, data sharing platforms, and cloud storage for resource sharing

2.6 Importance of SDGs in Sustainable Urban Development

The Sustainable Development Goals establish the foundation that controls how urban regions will develop globally in the future. The new urban agenda makes essential contributions to urban development through its conceptualization in the HABITAT III agreement [35]. The agenda dedicates attention to emphasizing that local government collaboration with existing international targets, particularly the Sustainable Development Goals, forms a partnership. Urban governments maintain an essential role, to support both the Sustainable Development Goals and the Paris Agreement [59]. Initiating major projects always includes numerous challenges that need to be addressed. Determining the precise effects of SDG 11 concerning sustainable cities and communities demands special consideration when performing measurement. Analysing the interactive relationship between SDG 11 and all other Sustainable Development Goals leads to creating successful implementation approaches. The proper implementation requires addressing data fragmentation issues in specific datasets. The data disaggregation process stands as a vital component because it assists policymakers, along with stakeholders, to obtain clear analyses of Sustainable Development Goals progress assessment. When data is segmented into specific groups, assessments become more effective because the analysis reveals vital information about current implementation limitations for creating focused responses. Local governments play a central and vital role in supporting the entire range of 17 SDGs, specifically focusing on achieving SDG 11. It prioritizes sustainable urban development as its essential framework; therefore, different approaches must be implemented to achieve its targets. The system implements several programs for public transportation prioritization while working on carbon emissions reduction through waste management strategies. Progressive, transformative, and sustainable actions enable cities to create a sustainable path toward the future. Effectively monitoring the impact, breaking down data, and implementing localized SDG programs represent necessary steps for generating sustainable, successful cities that will benefit future generations [60]. Several cities are actively working on implementing the Sustainable Development Goals at a local level.

2.6.1 Objectives of SDG 11 in Sustainable Smart Cities

The Sustainable Development Goals within the framework of smart cities come with multiple complicated, specific objectives. The Sustainable Development Goals tackle distinct objectives that cover sustainability in the environment, along with economic growth, social fairness, and political participation for all. To achieve main goal, the Sustainable Development Goals make the promotion of sustainable cities a vital objective. The goal stands as essential when considering that urban areas are expected to accommodate more than 70% of global residents by 2030 [61]. The worldwide production of greenhouse gas emissions stems mainly from urban areas, which maintain a 70% contribution. SDG 11 expressly shows that one must develop sustainable urban locations that both welcome swift metropolis expansion and reduce environmental deterioration [62]. The Sustainable Development Goals recognize ICTs to play an essential part in constructing smart cities. Cities that successfully merge ICT frameworks develop into intelligent and resource-effective locations that provide increased life standards for citizens. The IoT functions as an essential basis to fulfil sustainability needs in smart cities. The implementation of IoT devices allows cities to distribute resources effectively and track environmental conditions for maximizing operational efficiency in sustainable urban growth. The introduction of the new urban agenda creates an important platform to handle the extensive effects of SDG 11. This approach enables a thorough investigation of the relationship between sustainable cities and urban citizenship at different levels. Smart city policies must first understand what citizens need and hope for, because it enables better policy development that suits their requirements. The core of citizen-centric urban transformation depends on enabling broad-based participation by individuals in making urban decisions, which leads to sustainable and liveable urban environments. By implementing the Sustainable Development Goals, smart city initiatives obtain direction toward sustainable development practices while focusing on effective urban change that benefits citizens [63]. The Sustainable Development Goals establish a framework that enables future smart city development whose features combine sustainability with social justice and political equality, together with economic soundness and environmental protection. Sustainable urban development toward resident well-being and low ecological impact can be achieved by following these identified goals.

2.6.2 Alignment of Smart City Initiatives with SDGs

The integration of ICTs in smart cities creates opportunities to reach the United Nations Sustainable Development Goals by dealing with social, economic, and environmental problems [64]. Smart city solutions demonstrate effectiveness through better citizen outcomes and economic development while simultaneously protecting environmental resources. The systematic measurement of alternative data should involve all social backgrounds of citizens, while public-private partnerships create the necessary framework for fulfilling SDG-11. The collaborations between private and public stakeholders ensure that every interest group gets represented while developing a sustainable environment throughout the city. ICT plays a fundamental role in smart city projects because it enables the management of social and environmental impacts, which support the Sustainable Development Goals agenda. Technology implementation allows smart cities to achieve goals of social inclusion, with environmental sustainability. Smart city applications enable key features that decrease climate pollution through more effective energy controls, together with reduced waste and improved public transportation networks.

Big data, ICT technologies, and IoT, as well as TripleRM tools for managing risks, resources, and resilience, are identified as the essential components to construct a sustainable smart city. These innovations allow the cities to make data-driven decisions for efficient monitoring and management of the city resources and achieve long-term sustainability. Therefore, these insights illustrate the great opportunity that smart city initiatives exploit to achieve the Sustainable Development Goals' overall agenda of sustainable development. With the proper use of smart city technologies and close cooperation between the public and private sectors, cities boost their performance towards the realization of the Sustainable Development Goals. SDG 11 emphasizes the importance of sustainable cities, and so the smart city initiatives are promoting urban sustainability and overcome global issues [65]. Urban smartness can enhance economic diversification at the urban level through the use of smart infrastructure and citizen engagement to contribute to the creation of an innovative, economically prosperous, and socially well-being-friendly urban environment. Initial initiatives for smart cities shall inspire the future cities that are intelligent, sustainable, and inclusive communities.

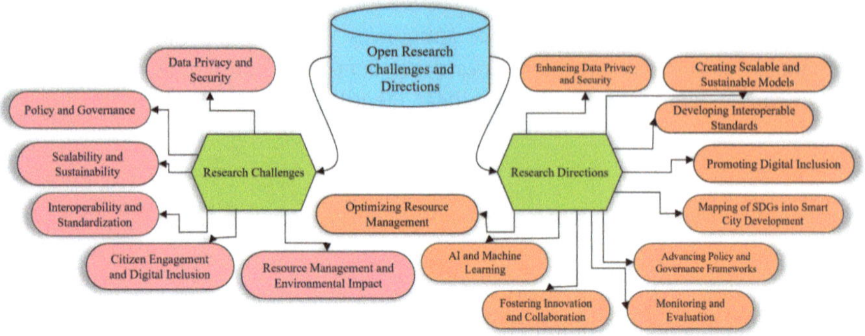

Fig. 2.11 Open research challenges and directions in sustainable smart cities with SDGs

2.7 Open Research Challenges and Research Directions

A smart urban environment is a well-developed model that is realized by managing and facilitating the services of cities with the help of information technology. This approach is closely related to the United Nations Sustainable Development Goals, which define the main global objectives and tasks for tackling the issues of human development by 2030. While the leaders of smart cities attempt to attain the Sustainable Development Goals, multiple research issues and open questions arise. The following sections discuss these challenges and offer suggestions for future research with the aim of promoting the sustainable development of smart cities as illustrated in Fig. 2.11.

2.7.1 Open Research Challenges

2.7.1.1 Data Privacy and Security

The strength of smart cities lies in the data acquisition from the users, IoT, and sensors. Protecting personal information is still a great concern that many individuals face [66]. At the same time, many cases of data loss and hacking have led to the need for an adequate system of cybersecurity. However, the question of how much information should be provided to the public regarding their database and to what extent citizens' rights will be protected when it comes to information released for the purposes of innovation remains debated.

2.7.1.2 Interoperability and Standardization

It has been observed that a smart city includes a number of technologies and systems that are required to interconnect with one another. The propagation of systems and reluctance to make those systems compatible means that there may be less time and resources regarding the introduction of various technologies [67]. Establishing the universal standards and protocols is rather important for the consistent operation of smart city systems.

2.7.1.3 Scalability and Sustainability

The fundamental issue of piloting technologies lies in expanding those solutions, which work successfully on a small scale, to large cities [68]. In research, the models have to be adaptable and can be scaled up or scaled down according to the respective urban system.

2.7.1.4 Citizen Engagement and Digital Inclusion

The engagement of the citizens to enhance the realization of the respective goals in smart cities is critical [69]. However, there is always a digital divide that prevents some special groups of people from having equal access to the aspects of a smart city. More specifically, research should focus on how the process of increasing the population's digital literacy and creating equal opportunities for participation should be implemented.

2.7.1.5 Resource Management and Environmental Impact

Smart cities have the goal of effective utilization of resources, but the consequences of such usage of technologies still face challenges. It is also necessary to carry out further studies in order to evaluate the life cycle effect of smart technologies and find ways to reduce energy consumption and carbon emissions while increasing the usefulness of the technology in its application [70].

2.7.1.6 Policy and Governance

For any integration of a smart city project, it is very crucial to have an effective governance framework in place [71]. The policy-making process is rather challenging, especially because of the dynamic technologies in the present world. More research needs to be carried out so as to design and propose policies and regulations for implementation that can be changed to suit future technological and social environments that may exist in the future.

2.7.2 *Open Research Directions*

2.7.2.1 Enhancing Data Privacy and Security

Future work has to be strengthened with more encryption algorithms, a database with distributed features, and secure data mining. It is therefore important to regulate and set ethical standards and policies for the use of data in smart cities [72].

2.7.2.2 Developing Interoperable Standards

There is a need to promote standards among governments, collaborating industries, and academicians to work together to make them interoperable [73]. This implies the need for more research in finding open-source platforms and protocols to enable better integration of such diverse technologies in the smart city system.

2.7.2.3 Creating Scalable and Sustainable Models

Research on the economic and environmental feasibility of these models will contribute to the determination of the effectiveness of such systems in relation to costs as well as their impact on the environment. More research needs to be done to define smart city models that can be relatively more singularized based on the environment, but at the same time offer the ability to add or incorporate more modules as required [68].

2.7.2.4 Promoting Digital Inclusion

More research should be conducted examining the various means by which digital literacy should be promoted and made easily available, such as training that can be conducted in the community, and inexpensive technologies should be implemented [74]. Approximately, measures that relate to addressing the gap that exists in the distribution of ICT should be encouraged. Thus, it is important to incorporate the concept of the ID into the smart city projects to make them available to every person.

2.7.2.5 Optimizing Resource Management

The most interesting research-related field should be related to the automation and, in particular, the intelligent monitoring of the resources [70]. Research on the effects of the external environment on smart technologies, as well as measures to enable the attainment of environmentally friendly smart cities, will be valuable.

2.7.2.6 Advancing Policy and Governance Frameworks

New methodologies capable of adapting to the technologies and the society are presented by the adaptive models. There is a need to extend the knowledge base concerning the policies in smart cities, as well as the extent to which citizens engage in policy-making and the role they play in the management of smart cities [75]. It is therefore important to establish procedures for ethical decision making and reporting in smart city projects as well.

2.7.2.7 Artificial Intelligence and Machine Learning

AI/ML has immense possibilities in the smart city operations and procedures [76]. The general conversation that emerges from the existing study should be about the use of research to achieve smart maintenance to avoid machine failure, intelligent traffic control, and smart energy consumption. Other issues that should also be discussed include the concept of ethical decision-making and the handling of bias in AI/ML.

2.7.2.8 Mapping of SDGs into Smart City Development

Such scholarship should focus on identifying strategies that have to be incorporated into smart city frameworks and procedures. This is because research carried out on different aspects of working relationships, coherence, and tensions between the multiple Sustainable Development Goals will assist in developing well-rounded, smart city strategies for achieving the intended sustainable development outcomes [77].

2.7.2.9 Innovation and Collaboration

It is crucial to establish the right environment needed for the stakeholders to work efficiently and interdependently [78]. There is a need to consider the impact of innovation hubs, incubators, and other collaborative platforms as a subject for further research for smart city advancement. This will be highly valuable since cooperation with universities allows for launching cross-sectoral and interdisciplinary projects focused on the development of smart city solutions.

2.7.2.10 Monitoring and Evaluation

The need to create tools for monitoring and evaluation to determine the effectiveness of smart strategies becomes imperative [79]. This means that more effort should be directed towards developing indices that help in evaluating the impact of smart cities on the achievement of the Sustainable Development Goals and the quality of life.

The relationship between smart cities and the Sustainable Development Goals is both a research concern and a great opportunity. All these need to be supplemented by new technologies as well as social, economic, and environmental contexts and impacts. From the outlined research directions, scholars and practitioners get an opportunity to work towards the creation of smart cities that are not only technologically sustainable but also scientifically sustainable, citizen-inclusive, and development goals-oriented. Fortunately, smart cities are becoming more advanced day by day, and hence, more research is needed towards the development of smart cities so that these can effectively play a constructive role in enhancing the quality of life in this world.

2.8 Summary

Smart cities and Sustainable Development Goals constitute a complex landscape with the potential to create barriers to realizing this potential. Moreover, the study included the demand for inclusive data measurement from urban citizens, matters of urban governance that enable the achievement of a wide array of Sustainable Development Goals, and the capacity of Smart Cities to address UN goals for sustainable development. However, it indicated the absence of environmental impacts in the social justice elements of ICT and a deadline for the combined consideration of social and environmental issues for Smart City projects. These perspective identify the ongoing evolution and challenges in the smart cities and Sustainable Development Goals context in terms of identifying the scope of further research and the importance of focusing on research from a general perspective of social, environmental, and governance factors.

References

1. "Transforming our world: The 2030 agenda for sustainable development | Department of Economic and Social Affairs." Accessed: Apr. 22, 2025. [Online]. Available: https://sdgs.un.org/2030agenda
2. M.A. Ahad, S. Paiva, G. Tripathi, N. Feroz, Enabling technologies and sustainable smart cities. Sustain. Cities Soc. **61**, 102301 (2020). https://doi.org/10.1016/J.SCS.2020.102301
3. D.K. Das, Digital technology and AI for smart sustainable cities in the global south: A critical review of literature and case studies. Urban Sci. **9**(3), 72 (2025). https://doi.org/10.3390/urbansci9030072
4. Z. Mohammadzadeh, H.R. Saeidnia, A. Lotfata, M. Hassanzadeh, N. Ghiasi, Smart city healthcare delivery innovations: A systematic review of essential technologies and indicators for developing nations. BMC Health Serv. Res. **23**(1), 1180 (2023). https://doi.org/10.1186/s12913-023-10200-8

5. T. Makkonen, T. Inkinen, Inclusive smart cities? Technology-driven urban development and disabilities. Cities **154**, 105334 (2024). https://doi.org/10.1016/J.CITIES.2024.105334
6. G. Dzhunushalieva, R. Teuber, Roles of innovation in achieving the sustainable development goals: A bibliometric analysis. J. Innov. Knowl. **9**(2), 100472 (2024). https://doi.org/10.1016/J.JIK.2024.100472
7. N. Donthu, S. Kumar, D. Mukherjee, N. Pandey, W.M. Lim, How to conduct a bibliometric analysis: An overview and guidelines. J. Bus. Res. **133**, 285–296 (2021). https://doi.org/10.1016/j.jbusres.2021.04.070
8. S.E. Bibri, J. Krogstie, Smart sustainable cities of the future: An extensive interdisciplinary literature review. Sustain. Cities Soc. **31**, 183–212 (2017). https://doi.org/10.1016/j.scs.2017.02.016
9. H. Ahvenniemi, A. Huovila, I. Pinto-Seppä, M. Airaksinen, What are the differences between sustainable and smart cities? Cities **60**, 234–245 (2017). https://doi.org/10.1016/j.cities.2016.09.009
10. C. Moreno, Z. Allam, D. Chabaud, C. Gall, F. Pratlong, Introducing the '15-Minute City': Sustainability, resilience and place identity in future post-pandemic cities. Smart Cities **4**(1), 93–111 (2021). https://doi.org/10.3390/smartcities4010006
11. Z. Allam, Z.A. Dhunny, On big data, artificial intelligence and smart cities. Cities **89**, 80–91 (2019). https://doi.org/10.1016/j.cities.2019.01.032
12. S. Nizetic, P. Solic, D. Lopez-de-Ipina Gonzalez-de-Artaza, L. Patrono, Internet of things (IoT): Opportunities, issues and challenges towards a smart and sustainable future. J. Clean. Prod. **274**, 122877 (2020). https://doi.org/10.1016/j.jclepro.2020.122877
13. E. Ismagilova, L. Hughes, Y.K. Dwivedi, K.R. Raman, Smart cities: Advances in research—An information systems perspective. Int. J. Inf. Manag. **47**, 88–100 (2019). https://doi.org/10.1016/j.ijinfomgt.2019.01.004
14. S.E. Bibri, The IoT for smart sustainable cities of the future: An analytical framework for sensor-based big data applications for environmental sustainability. Sustain. Cities Soc. **38**, 230–253 (2018). https://doi.org/10.1016/j.scs.2017.12.034
15. T. Yigitcanlar, M. Kamruzzaman, M. Foth, J. Sabatini-Marques, E. da Costa, G. Ioppolo, Can cities become smart without being sustainable? A systematic review of the literature. Sustain. Cities Soc. **45**, 348–365 (2019). https://doi.org/10.1016/j.scs.2018.11.033
16. Y.A. Fatimah, K. Govindan, R. Murniningsih, A. Setiawan, Industry 4.0 based sustainable circular economy approach for smart waste management system to achieve sustainable development goals: A case study of Indonesia. J. Clean. Prod. **269**, 122263 (2020). https://doi.org/10.1016/j.jclepro.2020.122263
17. M. Kaika, *'Don't call me resilient again!'*: The new urban agenda as immunology … or … what happens when communities refuse to be vaccinated with 'smart cities' and indicators. Environ. Urban. **29**(1), 89–102 (2017). https://doi.org/10.1177/0956247816684763
18. S. Borchardt, G.B. Vignola, D. Buscaglia, M. Maroni, *Mapping EU policies with the 2030 agenda and SDGs-fostering policy coherence through text-based SDG mapping*. https://doi.org/10.2760/110687
19. F. Twum Barimah, S.K. Diko, The impact of United Nations millennium development goals on Sub-Saharan Africa. **3**(7) (2013) [Online]. Available: www.iiste.org
20. W. Leal Filho, S.K. Tripathi, J.B.S.O.D. Andrade Guerra, R. Giné-Garriga, V. Orlovic Lovren, J. Willats, Using the sustainable development goals towards a better understanding of sustainability challenges. Int. J. Sustain. Dev. World Ecol. **26**(2), 179–190 (2019). https://doi.org/10.1080/13504509.2018.1505674
21. J.D.B. Gil, P. Reidsma, K. Giller, L. Todman, A. Whitmore, M. van Ittersum, Sustainable development goal 2: Improved targets and indicators for agriculture and food security. Ambio **48**(7), 685–698 (2019). https://doi.org/10.1007/s13280-018-1101-4
22. I. Madzorera et al., Food systems as drivers of optimal nutrition and health: Complexities and opportunities for research and implementation. Curr. Dev. Nutr. **5**, nzab062 (2021). https://doi.org/10.1093/CDN/NZAB062

23. A. Takian, A. Raoofi, H. Haghighi, COVID-19 pandemic: The fears and hopes for SDG 3, with focus on prevention and control of noncommunicable diseases (SDG 3.4) and universal health coverage (SDG 3.8), in *COVID-19 and the sustainable development goals*, (2022), pp. 211–234. https://doi.org/10.1016/B978-0-323-91307-2.00014-6
24. M. Saini, E. Sengupta, M. Singh, H. Singh, J. Singh, Sustainable development goal for quality education (SDG 4): A study on SDG 4 to extract the pattern of association among the indicators of SDG 4 employing a genetic algorithm. Educ. Inf. Technol. (Dordr) **28**(2), 2031–2069 (2023). https://doi.org/10.1007/s10639-022-11265-4
25. J. M. Denney, "Brief 11: Gender and the Sustainable Development Goals: Moving Beyond Women as a 'Quick Fix' for Development," 2015. [Online]. Available: https://scholarworks.umb.edu/cgs_issue_brief_series
26. S. Singh, R. Jayaram, Attainment of water and sanitation goals: A review and agenda for research. Sustain. Water Resour. Manag. **8**(5), 146 (2022). https://doi.org/10.1007/s40899-022-00719-9
27. L. Andres, K. Boateng, C. Borja-Vega, E. Thomas, A review of in-situ and remote sensing technologies to monitor water and sanitation interventions. Water (Basel) **10**(6), 756 (2018). https://doi.org/10.3390/w10060756
28. M. Acheampong et al., Is Ghana ready to attain sustainable development goal (SDG) number 7?—A comprehensive assessment of its renewable energy potential and pitfalls. Energies (Basel) **12**(3), 408 (2019). https://doi.org/10.3390/en12030408
29. "ACCESS TO CLEAN AND EFFICIENT ENERGY IN DEVELOPING COUNTRIES: THE NEED FOR EU ACTION TO IMPLEMENT SDG7." [Online]. Available: www.epc.eu
30. G. Macnaughton, D. F. Frey, "Decent work, human rights, and the millennium development goals," HeinOnline, 2010. [Online]. Available: https://heinonline.org/HOL/License
31. G. Allegre, "THE IMPERATIVE OF SUSTAINABILITY: Economic, social, environmental," 2019. [Online]. Available: https://sciencespo.hal.science/hal-03403311v1
32. O. Gulseven, A. Elmi, O. Bataineh, The divergence between industrial infrastructure and research output among the GCC member states. Int. J. Busi. Appl. Sci. **9**(2), 2471–8858 (2020)
33. L. Advisor et al., "Redefining leadership in the age of the SDGs accelerating and scaling up delivery through innovation and inclusion Phumzile Mlambo-Ngcuka"
34. L.W. Niessen et al., Tackling socioeconomic inequalities and non-communicable diseases in low-income and middle-income countries under the sustainable development agenda. Lancet **391**(10134), 2036–2046 (2018). https://doi.org/10.1016/S0140-6736(18)30482-3
35. F. Caprotti et al., The new urban agenda: Key opportunities and challenges for policy and practice. Urban Res. Pract. **10**(3), 367–378 (2017). https://doi.org/10.1080/17535069.2016.1275618
36. G. Baffoe et al., Urban–rural linkages: Effective solutions for achieving sustainable development in Ghana from an SDG interlinkage perspective. Sustain. Sci. **16**(4), 1341–1362 (2021). https://doi.org/10.1007/s11625-021-00929-8
37. M. Obersteiner et al., Assessing the land resource–food price nexus of the sustainable development goals. Sci. Adv. **2**(9), e1501499 (2016). https://doi.org/10.1126/sciadv.1501499
38. A. Haines, M. Amann, N. Borgford-Parnell, S. Leonard, J. Kuylenstierna, D. Shindell, Short-lived climate pollutant mitigation and the sustainable development goals. Nat. Clim. Chang. **7**(12), 863–869 (2017). https://doi.org/10.1038/s41558-017-0012-x
39. D.E. Mthembu, G. Nhamo, Aligning SDG 13 with South Africa's development agenda: Adaptation policies and institutional frameworks. Jamba-J. Disas. Risk Stud. **14**(1), 1155 (2022)
40. B. Neumann, K. Ott, R. Kenchington, Strong sustainability in coastal areas: A conceptual interpretation of SDG 14. Sustain. Sci. **12**(6), 1019–1035 (2017). https://doi.org/10.1007/s11625-017-0472-y
41. J. Alós et al., Toward a decade of ocean science for sustainable development through acoustic animal tracking. Glob. Chang. Biol. **28**(19), 5630–5653 (2022). https://doi.org/10.1111/gcb.16343

42. B. Ferdouz Vuilliomenet Cochran, N. A. Brunsell, et al., "Biophysical indicators of sustainability for climate change mitigation and adaptation," 2015
43. C. A. Tisdell, "India's biodiversity conservation responses to the UN's sustainable development goals: are they adequate?" 2019, [Online]. Available: http://www.uq.edu.au/economics/PDF/staff/Clem_Tisdell_WorkingPapers.pdf
44. H. Koff, C. Maganda, The EU and the human right to water and sanitation: Normative coherence as the key to transformative development. Eur. J. Dev. Res. **28**(1), 91–110 (2016). https://doi.org/10.1057/ejdr.2015.77
45. M. Blicharska, C. Teutschbein, R.J. Smithers, SDG partnerships may perpetuate the global North–South divide. Sci. Rep. **11**(1), 22092 (2021). https://doi.org/10.1038/s41598-021-01534-6
46. C.C. Anderson, M. Denich, A. Warchold, J.P. Kropp, P. Pradhan, A systems model of SDG target influence on the 2030 agenda for sustainable development. Sustain. Sci. **17**(4), 1459–1472 (2022). https://doi.org/10.1007/s11625-021-01040-8
47. P. Mishra, G. Singh, *Sustainable smart cities* (Springer, Cham, 2023). https://doi.org/10.1007/978-3-031-33354-5
48. P. Mishra, P. Thakur, G. Singh, Enabling technologies for IoT based smart city, in *2021 Sixth international conference on image information processing (ICIIP)*, (IEEE, Nov. 2021), pp. 99–104. https://doi.org/10.1109/ICIIP53038.2021.9702673
49. P. Mishra, G. Singh, Introduction: Importance of sustainable smart city, in *Sustainable smart cities*, (Springer, Cham, 2023), pp. 1–37. https://doi.org/10.1007/978-3-031-33354-5_1
50. P. Mishra, G. Singh, Internet of things for sustainable smart city, in *Sustainable smart cities*, (Springer, Cham, 2023), pp. 75–96. https://doi.org/10.1007/978-3-031-33354-5_4
51. J. Mensah, Sustainable development: Meaning, history, principles, pillars, and implications for human action: Literature review. Cogent Soc. Sci. **5**(1) (2019). https://doi.org/10.1080/23311886.2019.1653531
52. S. Karatzimas, Smart cities' actions, performance and reporting practices on climate change challenges: An exploratory analysis in a sample of awarded smart cities. Cities **153**, 105270 (2024). https://doi.org/10.1016/J.CITIES.2024.105270
53. C. García Fernández, D. Peek, Smart and sustainable? Positioning adaptation to climate change in the European smart city. Smart Cities **3**(2), 511–526 (2020). https://doi.org/10.3390/smartcities3020027
54. P. Mishra, G. Singh, Enabling technologies for sustainable smart city, in *Sustainable smart cities*, (Springer, Cham, 2023), pp. 59–73. https://doi.org/10.1007/978-3-031-33354-5_3
55. J. Wu, S. Guo, H. Huang, W. Liu, Y. Xiang, Information and communications technologies for sustainable development goals: State-of-the-art, needs and perspectives. IEEE Commun. Surv. Tutor. **20**(3), 2389–2406 (2018). https://doi.org/10.1109/COMST.2018.2812301
56. "The Role of Smart Bike-sharing Systems in Urban Mobility," 2009
57. U. Sivarajah, Z. Irani, V. Weerakkody, Evaluating the use and impact of Web 2.0 technologies in local government. Gov. Inf. Q. **32**(4), 473–487 (2015). https://doi.org/10.1016/J.GIQ.2015.06.004
58. M. Adnan, M. Ghazali, N.Z.S. Othman, E-participation within the context of e-government initiatives: A comprehensive systematic review. Telemat. Inform. Rep. **8**, 100015 (2022). https://doi.org/10.1016/J.TELER.2022.100015
59. L. Gomez-Echeverri, Climate and development: Enhancing impact through stronger linkages in the implementation of the Paris agreement and the sustainable development goals (SDGs). Philos. Trans. R. Soc. A Math. Phys. Eng. Sci. **376**(2119), 20160444 (2018). https://doi.org/10.1098/rsta.2016.0444
60. P. Jones, D. Comfort, A commentary on the localisation of the sustainable development goals. J Public Aff **20**(1) (2020). https://doi.org/10.1002/pa.1943
61. "World urbanization prospects." Accessed: Apr. 23, 2025. [Online]. Available: https://population.un.org/wup/

62. H. Vaidya, T. Chatterji, SDG 11 sustainable cities and communities, in *Actioning the global goals for local impact. science for sustainable societies*, (Springer, Singapore, 2020), pp. 173–185. https://doi.org/10.1007/978-981-32-9927-6_12
63. A. Sharifi, Z. Allam, S.E. Bibri, A.R. Khavarian-Garmsir, Smart cities and sustainable development goals (SDGs): A systematic literature review of co-benefits and trade-offs. Cities **146**, 104659 (2024). https://doi.org/10.1016/J.CITIES.2023.104659
64. F.-F. Rothe, L. Van Audenhove, J. Loisen, ICT for development and the novel principles of the sustainable development goals. Third World Q. **43**(6), 1495–1514 (2022). https://doi.org/10.1080/01436597.2022.2060202
65. U. Sengupta, U. Sengupta, SDG-11 and smart cities: Contradictions and overlaps between social and environmental justice research agendas. Front. Sociol. **7** (2022). https://doi.org/10.3389/fsoc.2022.995603
66. E. Ismagilova, L. Hughes, N.P. Rana, Y.K. Dwivedi, Security, privacy and risks within smart cities: Literature review and development of a smart city interaction framework. Inf. Syst. Front. **24**(2), 393–414 (2022). https://doi.org/10.1007/s10796-020-10044-1
67. J. Koo, Y.-G. Kim, Interoperability requirements for a smart city, in *Proceedings of the 36th annual ACM symposium on applied computing*, (ACM, New York, 2021), pp. 690–698. https://doi.org/10.1145/3412841.3441948
68. A.d.M. Del Esposte et al., Design and evaluation of a scalable smart city software platform with large-scale simulations. Futur. Gener. Comput. Syst. **93**, 427–441 (2019). https://doi.org/10.1016/J.FUTURE.2018.10.026
69. B. Stelzle, A. Jannack, T. Holmer, F. Naumann, A. Wilde, J.R. Noennig, Smart citizens for smart cities, in *Internet of things, infrastructures and mobile applications. IMCL 2019. Advances in intelligent systems and computing*, (Springer, Cham, 2021), pp. 571–581. https://doi.org/10.1007/978-3-030-49932-7_54
70. A.H. Sodhro, S. Pirbhulal, Z. Luo, V.H.C. de Albuquerque, Towards an optimal resource management for IoT based green and sustainable smart cities. J. Clean. Prod. **220**, 1167–1179 (2019). https://doi.org/10.1016/J.JCLEPRO.2019.01.188
71. W. Castelnovo, G. Misuraca, A. Savoldelli, Smart cities governance. Soc. Sci. Comput. Rev. **34**(6), 724–739 (2016). https://doi.org/10.1177/0894439315611103
72. M.H. Panahi Rizi, S.A. Hosseini Seno, A systematic review of technologies and solutions to improve security and privacy protection of citizens in the smart city. Internet of Things **20**, 100584 (2022). https://doi.org/10.1016/J.IOT.2022.100584
73. A. Brutti et al., Smart city platform specification: a modular approach to achieve interoperability in smart cities, in *The internet of things for smart urban ecosystems. Internet of things*, (Springer, Cham, 2019), pp. 25–50. https://doi.org/10.1007/978-3-319-96550-5_2
74. O. Kolotouchkina, L. Ripoll González, W. Belabas, Smart cities, digital inequalities, and the challenge of inclusion. Smart cities **7**(6), 3355–3370 (2024). https://doi.org/10.3390/smartcities7060130
75. Z.R.M.A. Kaiser, Smart governance for smart cities and nations. J. Econ. Technol. **2**, 216–234 (2024). https://doi.org/10.1016/J.JECT.2024.07.003
76. Z. Ullah, F. Al-Turjman, L. Mostarda, R. Gagliardi, Applications of artificial intelligence and machine learning in smart cities. Comput. Commun. **154**, 313–323 (2020). https://doi.org/10.1016/J.COMCOM.2020.02.069
77. S. Ebrahiem, A.O. El-Kholei, G. Yassein, Socially sustainable smart cities: Mapping the research trends by co-word analysis. Open House Int. **49**(2), 315–339 (2024). https://doi.org/10.1108/OHI-10-2022-0247
78. E. Näslund, F. Strömberg, Open data within a smart city initiative a case study exploring how collaboration can foster innovation within a smart city initiative
79. M. Waghmare, S. Singhal, Monitoring and evaluation framework for inclusive smart cities in India. Dev. Pract. **32**(2), 144–162 (2022). https://doi.org/10.1080/09614524.2021.1907535

Chapter 3
Innovative Solutions for Smart Cities with Internet of Things

Contents

P. Mishra, G. Singh, *Sustainable Smart Cities 2.0*,
https://doi.org/10.1007/978-3-032-01102-2_3

3.1 Introduction

The advancement in technology through the Internet of Things (IoT) has now made it possible for the city to enhance the quality and management of people's lifestyles is referred to as 'smart cities' [1, 2]. With more people shifting to urban areas, new vision-oriented cities turn into a potential solution for handling the complex problems of a modern world, such as traffic jams, power and light usage, pollution, and security. By integrating IoT, it is possible to connect all sectors in a city to enable continuous exchange of data through the internet and its subsequent use to support relevant decision-making processes [3]. In its generic meaning, a smart city is an innovative idea of utilizing information and communication technologies (ICT) to improve the performance of cities [4]. The IoT has a highly important role in this integration, as it interconnects almost all the devices and sensors spread across the smart city infrastructures, including traffic signals, waste management mechanisms, energy networks, and public transportation. These IoT devices create immense quantities of data that reveal valuable information about city operations to the city's planners and administrators. It means that the accessibility of this data and its processing in real-time promote efficiency in resource usage, quality of public services, and, generally, resilience of the cities. Transportation and mobility are among the primary and popular domain where IoT has been integrated with the smart city concept. Intelligent Transport System (ITS) implements the transportation system environment by the use of IoT to control traffic flow, minimize congestion, and improve the public transport system. Cities could place IoT sensors and cameras at intersections in order to control traffic signals in accordance with the real-time traffic situation, which reduces traffic congestion and emissions [5]. Also, IoT contributes to smart parking systems, which involve the use of sensors to identify available parking spaces and direct the driver to the available spaces, hence saving a lot of time searching for a parking space and the pollution resulting from it. Public transport systems also benefit from the IoT with tracking, monitoring, and analysing information to enhance the provision of services to passengers. Energy management is essentially another important sector where IoT innovations play such a significant role [6]. Smart grids based on IoT mean better management of electricity consumption since distributors are given control of usage through IoT [7]. Smart appliances can easily pick up the variation within demand and thus help in controlling the wastage of energy as well as in the stability of the grids. Similarly, smart meters enable consumers to see how they use energy in a more detailed manner, thus ensuring that they adopt conservation measures. Besides, smart grids of solar and wind energy integrated by IoT technologies lead to a more effective balance between supply and demand and contribute to creating a more sustainable and stronger system of energy.

Management of waste has always been a concern, especially in areas that are urbanized, and with the help of IoT, some solutions have been developed. Smart waste bins that detect the amount of waste in them and provide information on status to the waste collection services for efficient routing and scheduling of the waste collection vehicles. It not only helps control collection routes to be more efficient in terms of fuel usage but also eliminates unnecessary spillages of waste that pollute and make cities dirty. Furthermore, IoT technologies can play a role in closely monitoring and controlling waste processing plants to enhance the effective conversion of waste into energy. Security and safety are two factors that are essential in smart cities, and the use of IoT technologies offers significant solutions for these factors [8]. Through the use of cameras and sensors that are connected to the internet, surveillance systems provide real-time monitoring of the public spaces to help prevent crime as well as respond to cases of disasters and other calamities on time. These systems can be integrated with artificial intelligence (AI) algorithms to give out an alert when certain activities thought to be unlawful are identified, thus allowing the police to take action. In addition, IoT enhances disaster management as IoT devices are capable of collecting essential real-time data about conditions of the environment and participating in the formulation of timely reactions to evacuate from areas that may experience floods, air pollution, water deficiency, or earthquakes.

Another important application of IoT innovations is in water management [9]. Smart water networks use IoT devices to keep track of the quality of the water supply, as well as the pressure and the rate of flow of water in the network. This real-time data is useful in detecting leakages and anomalies, and the integrity of the water supply would be maintained. Likewise, the IoT enhances control of irrigation systems applied in the green areas of cities to consume less water while maintaining the green appearance and the health of the plants. Healthcare services of a smart city are also improved with the help of IoT inventions [2]. Telemedicine and remote monitoring enabled by IoT allow patients to contact doctors from home, reducing pressure on healthcare centres and ensuring timely access to medical assistance. [10]. Several wearables and health devices are interconnected with IoT to track vital signs and other health factors; they inform clinicians if the situation becomes critical. Both of these goals contribute to improved patient satisfaction, but also helps in the prevention of new cases of arising complications, thereby reducing costs incurred through admissions and emergency response. Thus, the IoT will remain crucial for the formation of modern city strategies, as it offers a more sustainable and effective solution to the problems that cities face nowadays. Figure 3.1 illustrates the overview of the IoT-based smart cities.

Fig. 3.1 The overview of an IoT-enabled smart city

3.2 Current Research Trends of IoT in Smart Cities

This study investigates the trends, contributions, and impact of research in the Internet of Things with smart cities from 2015 to 2025 through a bibliometric analysis of Scopus data. The Bibliometrix R package, along with Origin software, was utilized to analyse and depict the data. This approach ensures that researchers are presented with the most current research developments in the field, enabling them to stay informed on emerging trends and advancements. In March 2025, a bibliometric analysis was conducted utilizing the Scopus core collection database to thoroughly investigate the concept and understanding of smart cities. Ultimately, 5327 articles were exported and saved in BibTeX format before being included and examined in the bibliometric study. The primary information, syntax and specific criteria used in screening for this study are shown in Table 3.1.

3.2.1 Annual Scientific Production and Top 10 Most Published Journals/Cited Papers

Figure 3.2 shows the annual publication trends from 2015 to 2025, highlighting the growth in research output. The early years show a steady increase in publications, rising from 2015 (55 doc) to 2022 (799 doc), indicating a growing interest in the

Table 3.1 Summary of data characteristics and main information of the bibliometric study

Description	Specific Standard Requirements/Results
Research database	Scopus
Search query	("smart") AND ("cities") AND ("internet") AND ("of") AND ("things")
Time span	2015–2025
Document type	Articles
Language	English
Data export format	BibTeX
Documents	5327
Annual growth rate %	12.84
Document average age	3.83
Average citations per doc	34.65
Keywords plus (ID)	18,962
Author's keywords (DE)	11,726
Authors	14,592
Authors of single-authored docs	295
Single-authored docs	319
Co-authors per doc	4.18
International co-authorships %	39.95

field. There is a slight decrease in the number of documents in 2023 (782), followed by a small increase in 2024 (791). In 2025 (184), the number continues to rise and may continue increasing until the end of the year.

Figure 3.3 presents the highest number of articles in the IoT-based smart city research domain published in "IEEE Internet of Things Journal" (348 docs), "IEEE Access" (336 docs), "Sensors" (170 docs), and "Sensors (Switzerland)" (152 docs), which indicates a higher output of articles compared to the other journals. This highlights that these four journals focus primarily on the IoT in the field of smart city research. In addition, some other journals, such as the "Future Generation Computer Systems" (121 docs), "Sustainability (Switzerland)" (98 docs), "Sustainable Cities and Society" (93 docs), "International Journal of Advanced Computer Science and Applications" (70 docs), "Applied Sciences" (69 docs), and "Electronics (Switzerland)" (68 docs), are in this domain.

Table 3.2 displays the top ten most cited papers from the study, which presents both the ranking and authors with their published works. This section presents information about authors, article titles, and their corresponding publications in journals with documentation for publication years and citation frequency. The ranking focuses on engineering field smart cities articles to demonstrate their total citation (TC) counts and annual citation rates and normalized total citation (NTC) data, which collectively signify their impact in the field.

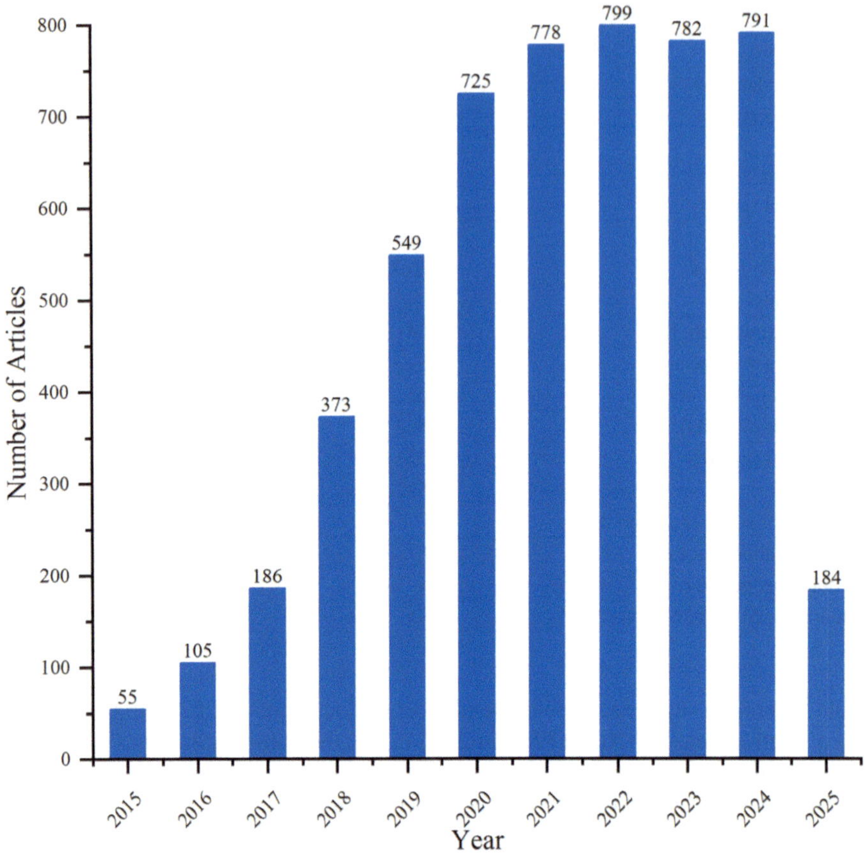

Fig. 3.2 Yearly trend of publication numbers on IoT-based smart cities

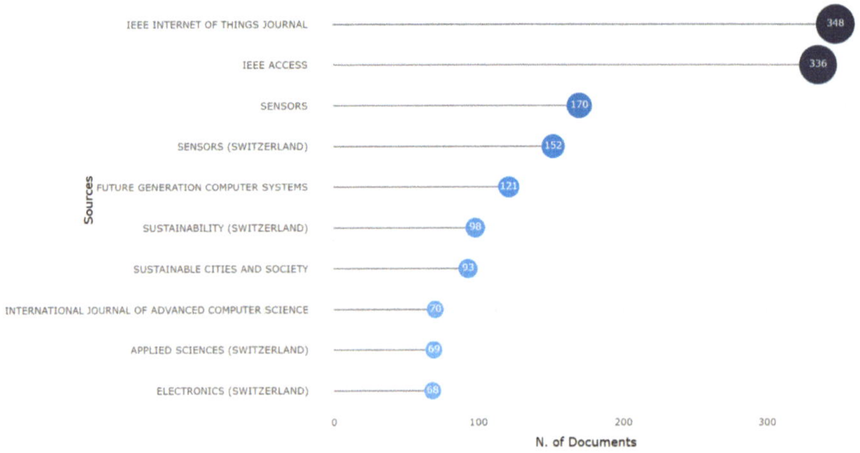

Fig. 3.3 Most relevant sources with number of articles on IoT-based smart cities

Table 3.2 Most globally cited documents in IoT-based smart city research domain

Rank	Author's name	Paper Title	Journal's Name	TC	TC/ Year	NTC
1	Shi et al. (2016) [11]	Edge Computing: Vision and Challenges	IEEE Internet of Things Journal	6033	603.30	88.03
2	Lin et al. (2017) [12]	A Survey on Internet of Things: Architecture, Enabling Technologies, Security and Privacy, and Applications	IEEE Internet of Things Journal	2214	246.00	40.47
3	Khan and Salah (2018) [13]	IoT Security: Review, Blockchain Solutions, and Open Challenges	Future Generation Computer Systems	1942	242.75	38.98
4	Botta et al. (2016) [14]	Integration of Cloud Computing and Internet of Things: A Survey	Future Generation Computer Systems	1907	190.70	27.83
5	Raza et al. (2017) [15]	Low Power Wide Area Networks: An Overview	IEEE Communications Surveys & Tutorials	1560	173.33	28.52
6	Fuller et al. (2020) [16]	Digital Twin: Enabling Technologies, Challenges, and Open Research	IEEE Access	1432	238.67	40.10
7	Centenaro et al. (2016) [17]	Long-Range Communications in Unlicensed Bands: The Rising Stars in the IoT and Smart City Scenarios	IEEE Wireless Communications	931	93.10	13.58
8	Hashem et al. (2016) [18]	The Role of Big Data in Smart City	International Journal of Information Management	864	172.80	32.03
9	Kumar et al. (2019) [3]	Internet of Things Is a Revolutionary Approach for Future Technology Enhancement: A Review	Journal of Big Data	854	85.40	12.46
10	Farahani (2018) et al. [19]	Towards Fog-Driven IoT eHealth: Promises and Challenges of IoT in Medicine and Healthcare	Future Generation Computer Systems	829	207.25	44.47

3.2.2 *Sankey Diagram (Three-Field Plot and Thematic Map)*

In Fig. 3.4, the Sankey diagram illustrates the three-field plot thematic progression to visualize the flow and the distribution in scientific research. In this case, keywords, an abstract, and a title are added, acting as indicators to highlight the most influential research in the subject. These keywords indicate vital concepts and themes that stand out in academic discussions of the IoT-based smart city research domain. Figure 3.5 categorizes various IoT-based smart city themes based on their development and relevant degrees. In the top right quadrant (motor

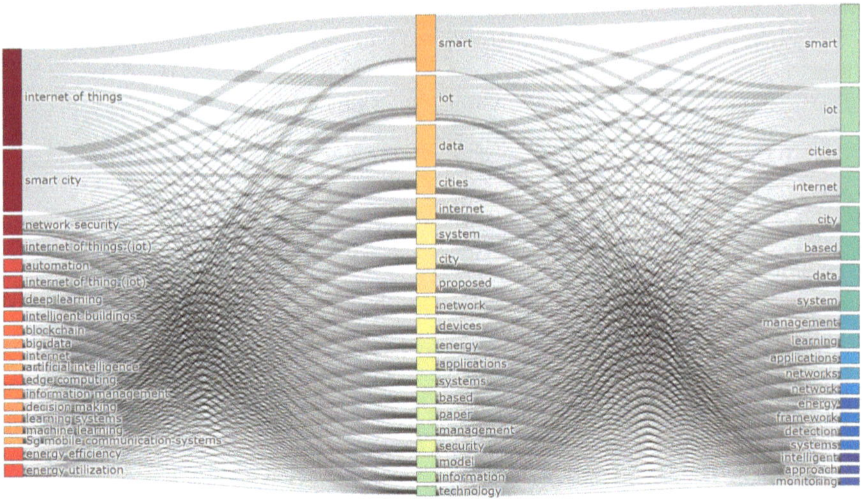

Fig. 3.4 Three-field plot (thematic development) using the Sankey diagram in IoT-based smart cities

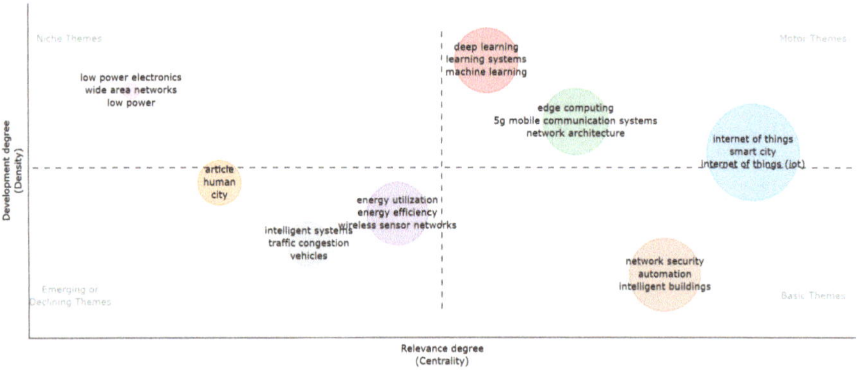

Fig. 3.5 Thematic map in IoT-based smart city research domain

themes), which represents themes that are highly developed and highly relevant, terms like “deep learning”, “learning systems”, “machine learning”, “edge computing”, “5G mobile communication systems”, “network architecture”, “smart city”, and “internet of things” appear. In the left quadrants (niche themes), there are representations such as “low power electronics”, “wide area networks”, and “low power”. This indicates a strong focus on these areas, suggesting they are central to IoT-based smart city research and practice. This quadrant reflects the areas where significant advances and investments are made, highlighting their

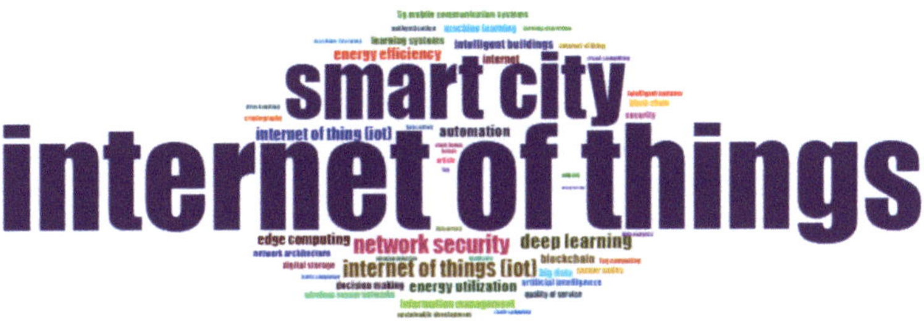

Fig. 3.6 Co-occurrence keyword trends in IoT-based smart city research domain

critical role in a smart city. In contrast, the bottom left quadrant, which represents themes that are either emerging or declining and have lower relevance, includes terms like "human", "city", "energy urbanization", "energy efficiency", "wireless sensor networks", "intelligent systems", "traffic congestion", and "vehicles". Additionally, terms like "network security", "automation", and "intelligent buildings" are the foundational information for further analysis and serve as a base for more advanced analyses. These themes might be considered foundational, but must be emphasized more in IoT-based smart cities. This visual categorization provides valuable insights for educators, curriculum developers, and researchers, helping them understand which areas are gaining traction and which might need more attention or innovation.

3.2.3 Co-occurrence Keyword Trends and Network Visualization

Figure 3.6 depicts the most prominent themes and keywords related to research in IoT-based smart cities. The most significant and central terms, such as "internet of things", "smart city", "network security", "internet of things (IoT)", and "deep learning", indicate the core focus of the research on how researchers grasp and alter their conceptual knowledge, particularly within smart city contexts.

An illustration in Fig. 3.7 displays word occurrence frequencies from the smart city research network. Research focuses mainly on the "Internet of Things", as this term appears 3337 times throughout the literature. Research on smart cities indicates the importance of "smart city" based on its 2099 word occurrences. Additionally, "Network Security" (576), "Internet of Things (IoT)" (555), and "Deep Learning" (500) combinations occupy crucial importance in emerging technology fields. The data shows that smart city research revolves primarily around several major subjects.

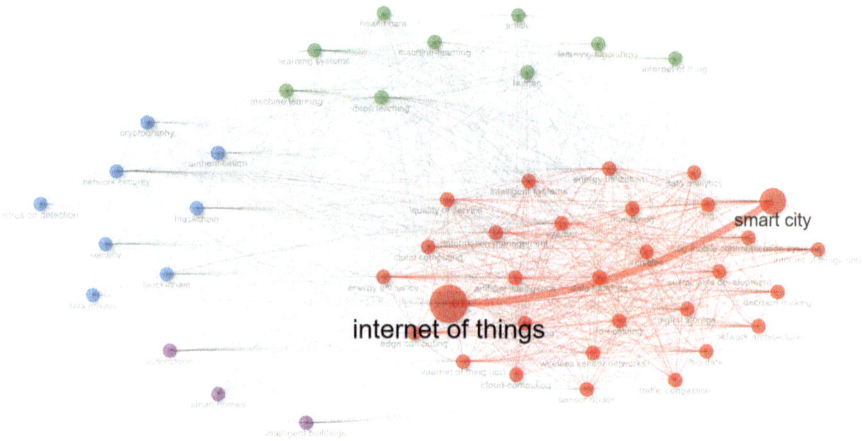

Fig. 3.7 Co-occurrence network of words in IoT-based smart city research domain

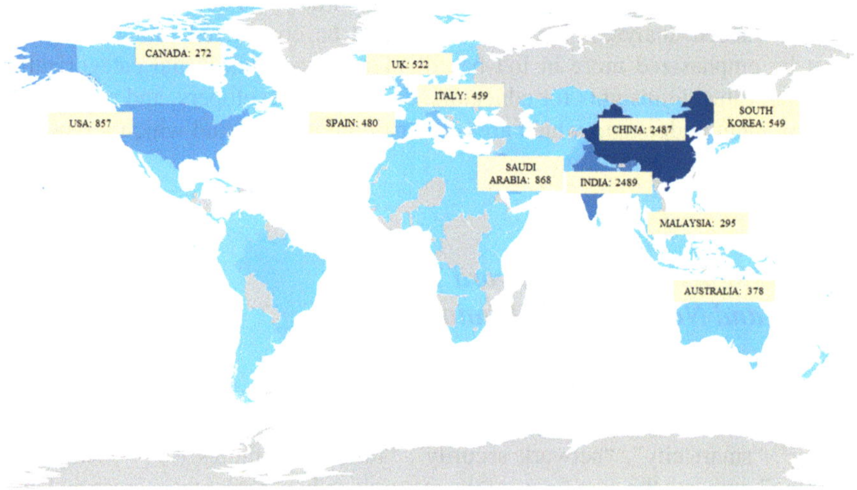

Fig. 3.8 Top 10 countries' scientific production of various countries based on frequency

3.2.4 Countries' Scientific Production

The scientific publication frequencies of various countries in the world are represented in Fig. 3.8. Scientific contributions from India stand at 2489, which demonstrates their powerful role in research development. The growing scientific power of China displays itself through its 2487 scientific publications, which rank second after India. The research hub of Saudi Arabia produces 868 scientific papers among its total contributions. Numerous publications emerge from the USA and South

Korea, together with the UK, given their research output comprises 857, 549, and 522 papers, respectively. Academic and industrial research sectors of Spain and Italy produce approximately 480 and 459 scholarly works, respectively. Australia, Malaysia, and Canada have smaller publication numbers, but their contributions amount to totals of 378, 295, and 272 research items, respectively.

3.2.5 Mapping IoT-Based Smart City Research to Sustainable Development Goals (SDGs)

Sustainable Development Goals, their targets (bubble series), and the percentage (bar graph) of corresponding keywords detected in the text (calculated as the ratio of keywords in one goal to the total number of keywords detected) are depicted in Fig. 3.9. For SDG mapping, the study utilized the SDG Mapper tool from the "Knowledge Base for the Sustainable Development Goals", provided by the European Commission [20]. This tool uses bubble charts to depict SDG interconnections and relevance among bibliometric records. The size of each bubble correlates with the proportion of associated keywords found, which is calculated as the number of goal-specific keywords divided by the total keyword count. As such, SDG 11 is marked by 33 keyword occurrences. This is followed by SDG 12, with 17 occurrences, and SDG 6, with 10

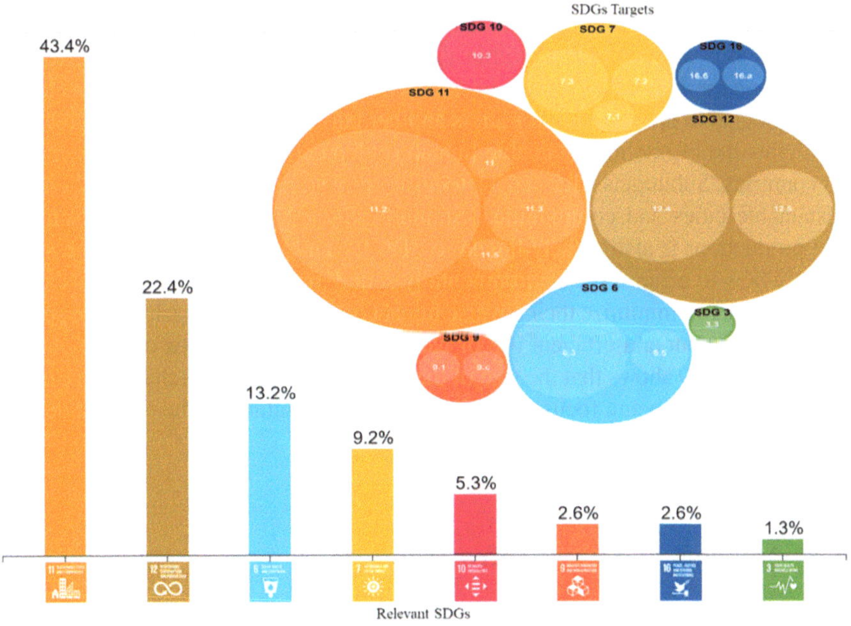

Fig. 3.9 Mapping IoT-based smart city research to relevant SDGs and targets

Table 3.3 SDGs ranking, targets, and relevance in IoT-based smart city research domain with SDGs

SDGs (Rank)	Description	Targets	No. of Occurrence		Relevance
SDG 11	Sustainable cities and communities	11.2	25	33	43.4%
		11.3	6		
		11.5	1		
		11	1		
SDG 12	Responsible consumption and production	12.4	11	17	22.4%
		12.5	6		
SDG 6	Clean water and sanitation	6.3	8	10	13.2%
		6.5	2		
SDG 7	Affordable and clean energy	7.3	4	7	9.2%
		7.2	2		
		7.1	1		
SDG 10	Reduced inequalities	10.3	4	4	5.3%
SDG 9	Industry, innovation, and infrastructure	9.1	1	2	2.6%
		9.c	1		
SDG 16	Peace, justice, and strong institutions	16.a	1	2	2.6%
		16.6	1		
SDG 3	Good health and well-being	3.3	1	1	1.3%

occurrences, underlining their relative prominence in the research database. Within SDG 11, there are several key subtargets that the research addresses: Subtarget 11.2 is shown as a distinct bubble within the larger SDG 11 circle, indicating focused research on the Internet of Things in smart cities. Subtarget 11.3, which aims to provide sustainable urbanization for all, is another bubble within the SDG 11 area. This highlights the efforts in sustainable human settlement planning and management in all countries. Subtargets 11.5 and 11 focus on protecting human beings and creating sustainable cities and communities. Similarly, SDG 12 (Responsible consumption and production) is also closely linked to SDG 11 and suggests that research should explore how to achieve the environmentally sound management of chemicals and all waste in order to minimize their adverse impact on humans and to achieve the sustainable management and efficient use of natural resources within the context of smart cities. Figure 3.9 shows that the size of the SDG 6 bubble implies that advancements in smart cities are being focused on clean water and sanitation. The detailed description of other relevant SDGs is depicted in Table 3.3.

3.3 Internet of Things Infrastructure

The Internet of Things is revolutionizing industries by making smarter, connected devices collect, process, and share huge amounts of data [21]. The infrastructure of IoT systems, as constituted by the sensors, connectivity, and network integration,

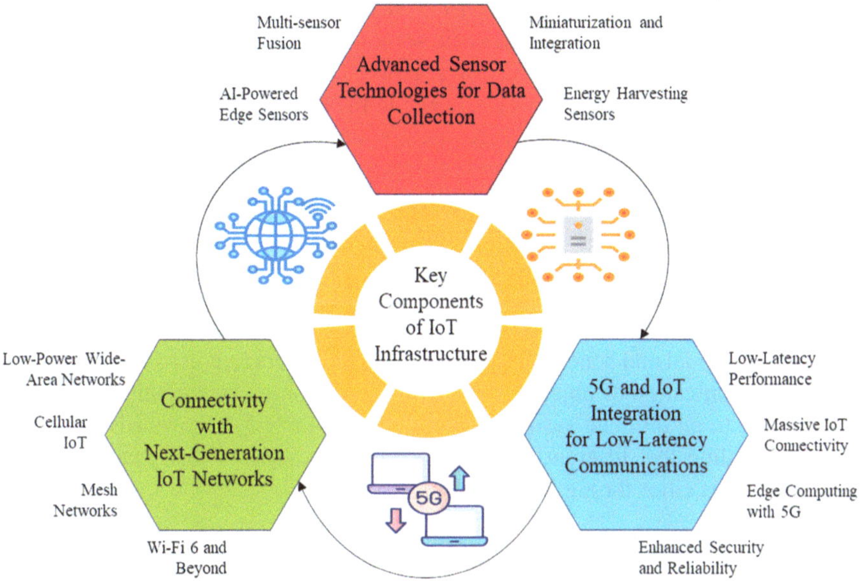

Fig. 3.10 The key components of IoT infrastructure

plays a central role in the success of IoT systems. The three key aspects of the IoT infrastructure, namely, advanced sensor technologies for data collection, connectivity and other next-generation IoT networks, and integration of 5G and other wireless cellular network technologies for low-latency transmission, are depicted in Fig. 3.10. These elements together comprise the backbone of IoT ecosystems, allowing real-time decision making and automation of a wide range of applications.

3.3.1 *Advanced Sensor Technologies for Data Collection*

The IoT is based on sensors, which are used to collect important data that are then transmitted and analysed to drive intelligent actions. There are sensors that convert physical parameters like temperature, pressure, motion, light, and sound into digital signals, and IoT systems read these digital signals. Advanced sensor technologies are able for more granular data collection across product lines, all the way from consumer devices to industrial systems, and facilitating smarter decision making [22]. It is essential for each class of sensor to have certain uses that would make efficient, sustainable management of a city possible. Miniaturization of sensors is one of the most typical trends in sensor technology [23]. Micro Electro Mechanical Systems (MEMS) technologies have enabled the production of very tiny and high-precision sensors. Devices such as smartphones, wearables, industrial machinery, and medical implants, which require the utmost in terms of space efficiency and power efficiency, can place these sensors in them. In autonomous vehicles, data from cameras, radar, lidar, and ultrasonic sensors are fused for better decision making and situational awareness. However, powering billions of sensors needed for large-scale deployment is a major challenge in IoT data collection. Energy harvesting sensors provide a solution where sensors are able to harvest ambient energy, such as solar, thermal, or vibrational energy, to power the sensor [24]. The main advantage of this technology is that it does not require replacing the batteries too often, particularly in remote or difficult-to-access areas. Yet, another advancement in sensor technology comes from embedding AI in sensors, or 'edge AI' [25]. It can preprocess data locally so that significantly less raw data have to be sent to the cloud for processing. In applications such as security and surveillance, real-time decision making is essential and beneficial. For example, AI cameras are used for local detection and classification of objects or behaviours and communicate only the relevant information to an ultimate system for a follow-up. Table 3.4 shows advanced sensors that outline the different technologies in smart sensors that enable the efficiency of smart cities through IoT applications.

3.3.2 *Connectivity with Next-Generation IoT Networks*

It is equally important to be able to transmit the data across the networks. For IoT systems to work using device-to-device communication, the supporting networks have to be robust, scalable, and reliable. To meet such needs, the next generation of IoT networks is evolving toward providing efficient coverage, higher bandwidth, and better energy efficiency. There are various connectivity options available, and each has its strengths and appropriate use cases as follows:

3.3.2.1 Low-Power Wide-Area Networks (LPWANs)

One of the largest boosts in IoT connectivity is Low-Power Wide-Area Networks (LPWANs). Long Range (LoRa) and Sigfox are technologies that belong to the LPWAN category, which offers communication over long distances with low power consumption, suitable for battery-powered IoT devices [26]. These networks are very useful for large-scale deployments in smart cities, agriculture, and industrial monitoring, where thousands of devices have to communicate over large distances without the need for frequent battery recharge.

3.3.2.2 Cellular IoT

More reliable, secure, and higher-bandwidth connections have made cellular networks, especially using LTE-M (Long Term Evolution for Machines) and NB-IoT (Narrowband IoT), extremely popular for IoT applications. The advantage of

Table 3.4 Advanced sensor technology via IoT in smart city applications

Topic	Description	Purpose in Smart Cities
Miniaturization and integration of sensor technology	Small-sized sensors that can be integrated onto other electrical equipment and the human environment. It aims at minimizing the size, power demands, and cost	Small sensors may be concealed in extensive urban structures and interfaces to monitor data influx comprehensively without causing inconvenience. This leads to better control over the operations of city spaces, traffic patterns, pollutants, and the usage of resources to improve the ability to manage a city proactively
Multi-sensor fusion	To ensure the data accrued is of higher accuracy and credibility, fusion of sensors is done to collect information accrued from different sensors. This entails the combination of data from various sensors, including visual cameras, LIDAR, and environmental sensors	It improves traffic flow, public safety, and the ability to monitor the environment in order to make better decisions to improve a city and its management
Energy harvesting sensors	These sensors are self-powered by means of energy that is utilized from surrounding environment, like solar, thermal, kinetic energy, etc. This leads to a minimized need for battery replacements and low maintenance expenses	Energy-harvesting sensors are beneficial to IoT networks in smart cities, pertaining to energy consumption and the need for an external power source. It is to work especially for the streetlights and water management, and many other facilities, in which the frequency of maintenance is high
AI-enabled sensors	AI-enabled peripheral sensors that analyse data at the sensor level by employing predictive models to curb the need to send much data to a server. Such sensors are capable of making real-time decisions on the data that has been processed	The use of AI sensors in smart cities provides a more efficient way of responding faster to tasks like adaptive traffic signals, maintenance, and safety

cellular IoT is that it is already available everywhere, and the devices can connect to the existing cellular infrastructure [27]. LTE-M supports mobile applications such as asset tracking and fleet management, whereas NB-IoT is better suited for stationary, low bandwidth use cases like environmental monitoring.

3.3.2.3 Mesh Networks

Mesh networks can be used as a good alternative when devices are deployed near each other [28]. A mesh network is a decentralized network where devices communicate with each other and relay data amongst one another by means of multiple hops. Usually, smart home and building automation applications use mesh networking protocols such as Zigbee and Z-Wave. The reason is that they provide resilience and redundancy, as well as the ability to scale well in high-density IoT device scenarios.

3.3.2.4 Wi-Fi 6 and Beyond

To implement IoT, traditional Wi-Fi networks have made advances with the introduction of Wi-Fi 6 (802.11ax) and Wi-Fi 7 (802.11be) [29]. The next generation of Wi-Fi standards provides more bandwidth, better power usage, and more efficient performance in crowded areas. For example, Wi-Fi 6 includes a feature called Target Wake Time (TWT), which lets IoT devices set the times when communicate with the network and, thereby, decrease power consumption. With IoT presence in the homes, enterprises, and public spaces getting larger, Wi-Fi will take up an even bigger role, especially in IoT applications that need high-speed data transmission, like video surveillance and augmented reality.

3.3.3 5G and IoT Integration for Low-Latency Communications

Since IoT devices and applications are getting more and more advanced, there is a growing demand for ultra-reliable and low-latency communication. 5G technology integration then comes into the picture as 5G networks are able to offer real-time communication and support huge device connectivity due to high speed, bandwidth, and low latency [30]. It reduces latency down to 1 millisecond, which is very important in IoT use cases such as autonomous driving, remote surgery, and industrial automation. Apart from low latency, 5G also accommodate a massive number of IoT devices: 1 million devices per square kilometre. Due to this, 5G is suitable for densely populated areas and wide-scale IoT deployment. It is used in smart cities, where thousands of sensors, cameras, and other devices must communicate simultaneously. Also, 5G's network slicing feature gives service providers network capabilities to create virtual networks within a network for individual IoT applications to ensure that each use case gets the required bandwidth and latency levels. 5G also brings enhanced security features, which are important for IoT deployments. Since IoT devices frequently work in sensitive environments, like healthcare, transport, and critical infrastructure, securing the transmission of data is crucial, and 5G networks include more powerful encryption protocols and virtual network slices for more secure operation of IoT devices as well as protection of IoT data. Furthermore, 5G and edge computing are integrated to provide even more capability for IoT because some of the data does not need to be sent to the cloud for processing [31]. Edge computing refers to processing data closer to the source, on the edge of the network, for reduced latency and bandwidth consumption. For example, real-time on-site analytics can be performed in a smart factory so that the machines can correct the processes in a mere few milliseconds before having to send information back and forth to a remote cloud server. It streams 5G innovation and edge computing in order to make faster decisions and use resources and networks in the most efficient way. Therefore, the IoT infrastructure is progressing rapidly due to advancements in sensor technology, next-generation connectivity solutions, and 5G integration for low-latency communication. These advancements leads to more efficient systems, be scalable and be real time, and they will certainly continue to reshape the future of industries as well as the lives of people all over the world.

3.4 Urban Mobility Management with IoT Applications

The area of IoT applications in urban mobility management reduces the need for travel on city transport systems and render it more secure and ecological [32]. IoT enables smarter management of traffic, fewer congestions, and a better experience in public transit rides by using the real-time data from the connected sensors, vehicles, and infrastructure. Figure 3.11 illustrates the urban mobility management with IoT applications.

3.4.1 IoT-Enabled Intelligent Traffic Management Systems

The use of IoT devices for urban mobility management, such as IoT-based intelligent traffic management systems, addresses some of the major challenges modern cities face, including traffic congestion [33]. ITS involves the use of ICTs aimed at improving urban transport systems. ITS brings even greater potential when it comes to use in combination with the IoT [34]. Traffic management systems that are dispersed generate data from IoT networks of embedded roadway sensors and linked automobiles to accumulate authentic time traffic statistics. Speeds, time, and/or congestion levels, as obtained by the sensors, are

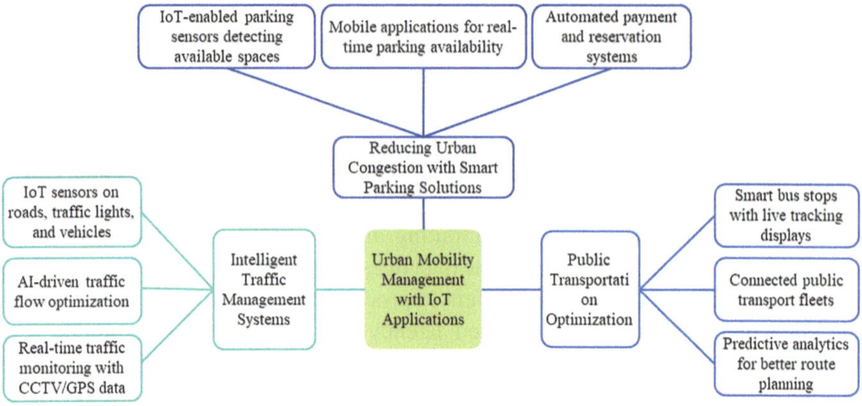

Fig. 3.11 Urban mobility management with IoT application

mostly of vehicle speeds on certain routes. V2I is the use of wireless means for a vehicle to send location details, speed, and direction [35]. The collected information about the traffic density is then referred to at the traffic control centre for the purpose of evaluating how the traffic signals' timings and speed limits need to be changed to improve the traffic volume. Systems that reduce speeds near areas of traffic congestion and increase speeds in zones of low traffic density can be taken up by traffic managers to achieve smooth traffic flow. Messages on signs, signalized intersections, and other information can also ideally redirect automobiles from traffic-prone areas. Smart traffic management systems in the use of IoT offer consumers' current traffic status through websites, mobile applications, and in-car devices. The drivers can get real-time notifications for traffic, accidents, and other recommendations for a better route. The information can be useful in traffic control and improvement of public transport systems, including appropriate schedules, bus frequencies, and routes. A clustering model with IoT and big data analytics makes it possible for near-term predictive queueing, past and current traffic data are analysed for probable impending traffic conditions and probable corrective measures to be taken before the congestion occurs. The IoT-based smart traffic control is useful in enhancing the movement of traffic within the urban streets to cause less congestion, take less time, consume less fuel and energy, and help avoid losses resulting from reduced time on the road.

3.4.2 IoT-Enabled Public Transportation Optimization

The development of public transport plays a crucial role in controlling traffic in the city and avoiding congestion [36]. However, many issues are relevant to transit systems, such as reliability, productivity, security issues, and managing devices and plans. It has been realized that the incorporation of IoT technologies is among the most efficient ways of enhancing public transportation. This makes it possible for the transit agencies to monitor the buses, enhance strict following of the schedule, change frequencies depending on the demand, and inform the users of the buses' predicted time of arrival through apps. Vehicle health monitoring IoT applications use the auto-installed sensors to provide data on engine status, such as oil condition and tire pressure. This implies that failures are avoided and there is always an ability to foresee maintenance needs, which helps to reduce unnecessary time and cost. It is also possible to report a problem or ask for help via an onboard IoT connection for riders. For the installation of smart ticketing, the implementation of IoT, in place of electronic tickets, to facilitate fare collection and boarding control [37]. By connecting the IoT transit fleet systems to the connected transport infrastructure, buses can be caused to move safely and efficiently. There is transit signal priority, which gives buses at intersections a green light or changes the signals to favour buses. The data analytics also assist planning decisions within infrastructure investment decisions, route changes, and service expansion. Use of the IoT for mobility optimization is central in developing efficiency in the public transport systems that are vital in the development of smart cities of the future [38].

3.4.3 Reducing Urban Congestion with Smart Parking Solutions

Elimination of congestion is a great concern in developed and developing cities due to rapid growth in population and economy [39]. Smart parking has emerged as an effective use of technology in parking management of a city and has become an advanced application of the IoT in urban centres [40]. These advanced technologies, not only improve parking but also support the complex concept of smart cities. Smart parking is the application of IoT that helps in tracking parking and space availability instantly. It consists of these sensors that are connected to a central system, which informs the users about the available spaces through mobile applications. This real-time data helps drivers save time for searching a free parking bay, thus reducing the traffic hold-up and vehicle exhaust fumes. This way, the citizens of cities would cut down on driving as often, which consequently results in less abrasion of the roads, less fuel consumption, and less air pollution. In addition, smart parking systems can also have a dynamic parking fee structure that may vary with the utilization rates. This may consist of higher fees during periods of high demand to discourage overstaying at the parking lot and frequent turnover, while cheaper rates during off-peak periods would clearly attract more clients. This eliminates the necessity of having different prices for parking spaces and is also convenient in handling car traffic. Accessibility with the public transport system is another feature that is under the smart parking system. Smart parking solutions, just applying the IoT technologies, bring the benefits and create opportunities for cities to open a path towards smart cities [41]. Most cities employ ever-increasing parking resources and affect urban transport network effectiveness, which presents an opportunity to better enhance the quality of life in urban centres.

3.5 Monitoring and Managing Resources Using IoT

IoT introduces the real-time tracking of the use of resources by connecting devices to the internet for data-sharing purposes [42]. Smart sensors creates a permanent system to monitor usage performance alongside operating conditions. This data is sent over the internet to cloud solutions, whereby analysis yields discriminative information. It provides the intelligent tracking of resources and optimized decisions, and prevents the management from being reactive. There are various resources where IoT can be used for monitoring and management, like air and water quality monitoring, recycling and waste management, and the smart grid for efficient resource distribution, as illustrated in Fig. 3.12 in various aspects.

3.5.1 Air and Water Quality in Real Time with IoT Sensors

The IoT development enables the substantial connection of sensors for delivering real time, continuous monitoring of air and water quality [43]. For air quality, integrated sensors can be installed on streetlights, buildings, and even other fixed

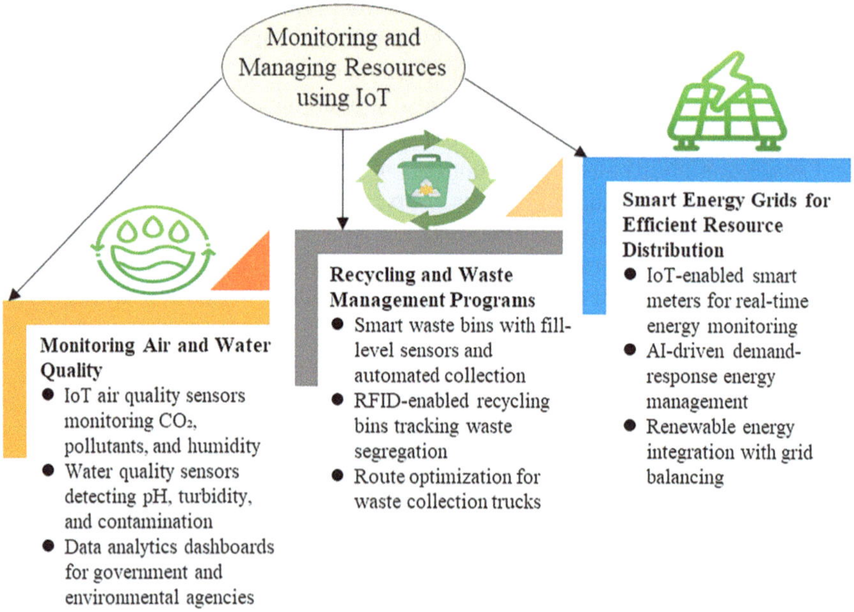

Fig. 3.12 Monitoring and managing resources using IoT

infrastructures in order to evaluate parameters such as particulate matter, carbon monoxide, sulfur dioxide, and nitrogen dioxide. Likewise, pH, turbidity, and dissolved oxygen levels that are measured along the length of the river can be used to monitor water quality. The IoT platforms can take in large amounts of data from the sensors and look at the readings to identify a violation or a threshold. In the long term, the sensor data helps in making improved policies on emission standards, implementation, and health recommendations. Hyperlocal data also helps to address various issues using specific remediation measures. As more and more organizations implement IoT solutions at a larger scale, it is possible to deploy millions of sensors for monitoring the city's environment [44].

3.5.2 Recycling and Waste Management on the Basis of IoT

IoT provides excellent opportunities for recycling and waste management innovation [45]. Integrated and connected waste bins, commonly referred to as 'smart' bins, may have features that allows to identify or even quantify the items that are disposed [46]. This allows for measurements and different types, from which the papers, plastics, and glasses, as well as metallic materials, are separated. Real-time bin utilization level means that route optimization algorithms can facilitate the pick-up and transport processes. RFID tags follow recyclables through the supply chain from used to disposal and deliver a comprehensive view instead of aggregate weight [47]. As investments, in-built compressors can optimize the storage volumes available between pick-up cycles. At the collection points, additional techniques such as the use of robots whereby there is robotic sorting still enhances the sorting of the waste into the right category. Integrated IoT platforms also have machine learning (ML) intelligence to identify other information, like patterns of consumption for areas, peak waste rate, violation of compliance, and so on for proper systemic planning for change.

3.5.3 Smart Energy Grids for Efficient Resource Distribution

The use of IoT in the deployment of electricity grid modernization is vital in the management of energy distribution [48]. Smart meters have two-way communication and transmit to the provider the consumption rate at the individual unit level [49]. Aggregate demands are used for tracking fluctuations in high and low points of areas under a certain grid provision. Working with historical data, it is possible to achieve accurate load forecasting. A metering and distributing apparatus, like grid sensors that are used to track equipment health and transformers, predicts maintenance requirements before the occurrence of a failure. Real-time meter data also capture irregularities in power usage that may point to theft of electrical power. IoT also allows for load distribution centred at intelligent devices for smooth control of electricity usage [50]. It means that instead of limiting green energy, devices can directly connect with renewable sources when production is higher than storage. Consumers also gain more control over costs through decisions about when to use the facility. The growth of renewable energy-based supplies in the grid improves both power stability and economic system stability.

3.6 IoT Security and Privacy Considerations

Through IoT technology, everyday objects and physical items become connected to the internet in order to collect data and receive remote commands. With more and more connected IoT devices, security become a big challenge. Just like with any other software solutions, there are some security threats inherent to IoT devices. It was observed that a large number of IoT consumer devices operating in the market hardly have basic security features and can easily be hacked. The attackers can take over control of the targeted devices because a lack of strong authentication protocols and insecure transmission of data. Loosely connected IoT devices are vulnerable to being hacked and turned into botnets to execute a variety of cyberattacks.

Hardware makers provide no software updates for the device's use throughout its service life and security support. This increases the attack surface as new threats comes. Security and privacy of users is another important factor where data-collecting IoT devices are installed in homes, vehicles, wearables, and cities. The security and privacy by design concepts need to be implemented by the manufacturers of products at all stages of the life cycle. Last, increasing users' awareness of security measures and individuals' privacy, insisting on higher security, and promoting clarity. Solutions to overcome these issues are discussed below and illustrated in Fig. 3.13.

3.6.1 Securing IoT Data Transactions Through Blockchain

This is significantly more important in the case of smart cities because of the IoT data transactions and the combinations of multiple and interconnected systems in cities. Smart cities are characterized by a large number of IoT devices; these devices produce a large amount of data that has to be transmitted and processed safely to provide services necessary for the functioning of a city. Blockchain, on the other hand, offers a reliable solution to these challenges by bringing about a decentralized system for IoT data exchange. In the smart city, adopting blockchain technology is particularly important to handle the information generated from IoT devices, where all the data are recorded, and only provided to the authorized users [51]. This is significant in fields such as traffic control, power, water supply, security, and other areas where the accuracy of data cannot be compromised. Through the distributed arrangement of data in a peer-to-peer network,

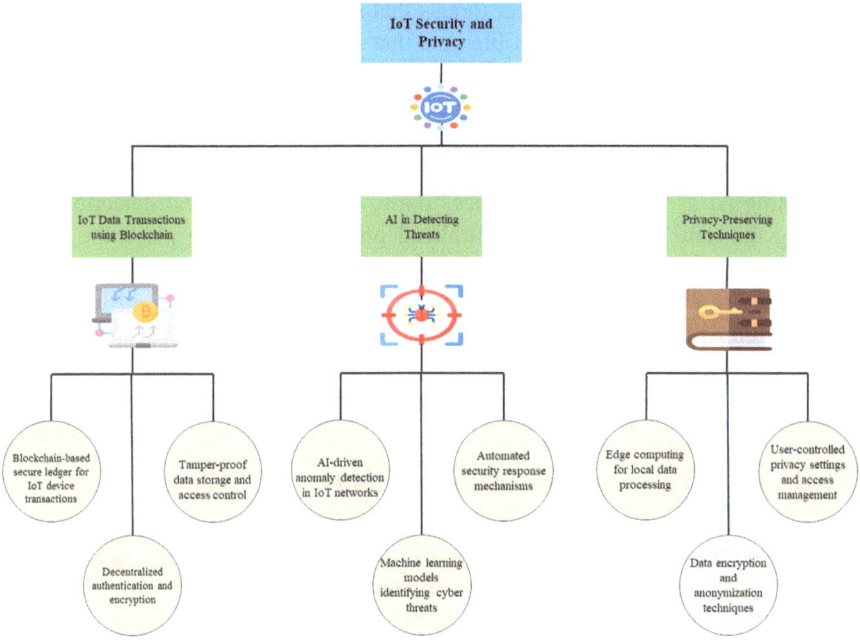

Fig. 3.13 IoT security and privacy solutions

blockchain eliminates the possibility of failure. Every transaction that occurs when IoT devices are communicating on a blockchain platform has to be signed and sealed through the use of cryptographic principles. Transparency is also created in blockchain, and distrust between different devices is excluded, since all actions on the network are visible to participants. There are leading blockchain platforms, such as Hyperledger, that are well-equipped with solutions for the protection of applied IoT device communications. The combination of blockchain and IoT in smart cities can provide increased openness and accountability to the operations. Blockchain technology for IoT data transactions is a significant component in smart cities, especially if the future of smart cities is to adopt efficient and sustainable ecosystems [52].

3.6.2 *The Role of AI in Detecting Threats*

AI applications become even more important in the case of threat detection because of number of various interconnected devices and systems in smart cities [53]. A smart city includes several IoT devices not only for traffic, safety control, and public utilities but also for the energy grid and the environment. This leads to the formation of several data points that interconnect with one another and make hacking even possible. This enables AI to detect threats that may disrupt the normal functioning of a city. IoT creates a stream of big data that AI systems analyses to see patterns that may indicate threats to IoT cybersecurity [54]. In particular, with the help of ML algorithms that are based on normal traffic flows, AI is capable of identifying strange data flows, unauthorized attempts to enter suspicious actions. One more important advantage of applying AI in threat detection is its much higher speed. Integrating AI as a part of IoT devices improves real-time threat detection while limiting the data incoming for centralized processing. It also plays the role of strengthening and increasing the robustness and flexibility of city infrastructures, protecting the digital environment behind the modern urban quality of life. In this way, the implementation of smart cities sustain a protected context that supports experimenting and enhancing people's quality of life.

3.6.3 IoT Privacy-Preserving Techniques

Privacy-preserving techniques are important in smart cities since the use of IoT in such cities is widespread, and the devices used are connected [55]. Differential privacy enables the grouping of individual data in such a way that pattern analysis is possible, but individuals' identities are kept anonymous through the introduction of random noise on the data. This ensures that data does not need to be transmitted to a central server for training, which drastically reduces the possibility of data breaches in the server. User privacy has to be preserved as more IoT applications collect personal information. Modern IoT platforms demand security elements that include control systems for access as well as encryption methods for data protection and anonymous functionality to defend confidentiality. Federated learning is another promising AI training technique, using which IoT models are trained without uploading user data to the central server and keeping it on devices [56]. Smart cities also seek to protect privacy by taking privacy into account during the early stages of technology and making user control and transparency significant factors of IoT applications. Implementing a number of these privacy-preserving approaches will be pertinent as smart cities grow in the future. However, the use of IoT technologies must be ethical in order to maintain customers' trust [57].

3.7 Scalability and Interoperability Challenges in IoT Systems

Large-scale implementation of IoT networks through thousands of sensors and devices made by various manufacturers produces difficulties for IoT approaches to scale effectively. Interoperability is a challenge because there are no standards, application programming interfaces, and message infrastructures that enable two or more platforms to exchange data. The solution requires standardized approaches for needs, modular and scalable designs, and better bandwidth with computing power. Proper undertaking of the activities requires planning and beginning with governance frameworks for future expansion and integration. Various scalability and interoperability challenges in IoT are illustrated in Fig. 3.14.

3.7.1 *Interoperable IoT Standards and Protocols*

The successful development of smart cities depends on IoT integration, specifically for the integration of IoT devices with associated systems. The various systems for IoT have to have a similar platform so they can be integrated and share information easily. There are a number of standard bodies and alliances that are working systematically to make such interoperable standards and protocols. One of these is the one Machine-to-Machine (M2M) standard dedicated to IoT vertical service layer and protocols to establish conformity and interconnectivity in smart city solutions to directly address the vast quantity and various impacted devices and sensors [58]. Another significant standard is the cross-platform compatibility that represents IoT device connectivity level interoperability and includes networking protocols. The service range contains wireless services through WiFi and Bluetooth, alongside cellular IoT

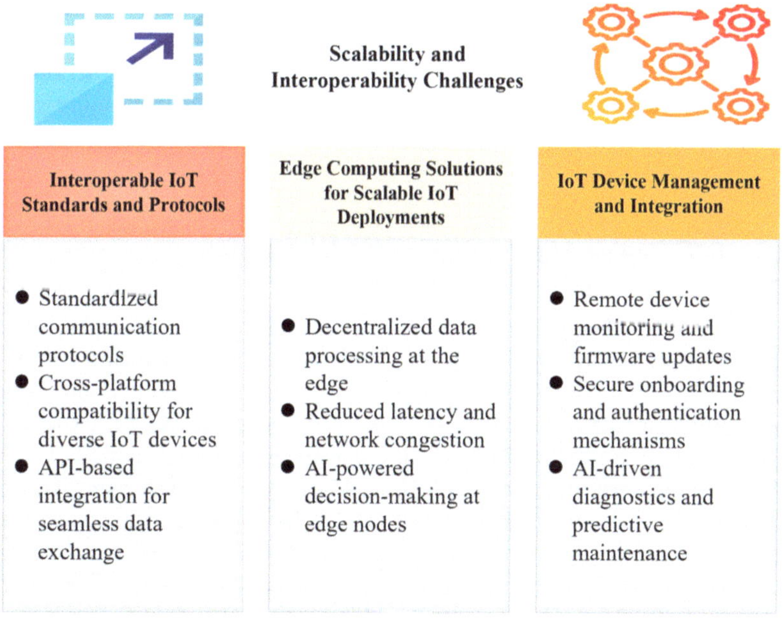

Fig. 3.14 The scalability and interoperability challenges

standards that incorporate interoperability with NB-IoT networks offered directly by cellular IoT providers [59]. At the application layer, the standard message protocols like MQTT, AMQP, CoAP, and WebSocket offer standard messaging interfaces for IoT resource-limited devices with features such as publish-subscription. Therefore, further efforts are necessary to establish and implement open, aligned IoT protocols to build real smart city environments.

3.7.2 Edge Computing Solutions for Scalable IoT Deployments

When a large-scale IoT network with sensors and devices is being installed for managing the infrastructures of a smart city, edge computing becomes vital for real-time data processing [60]. The advantages of edge computing over existing approaches of communicating with cloud data centres over the internet are enormous. Some of the important areas for low latency are traffic management, smart city safety, and surveillance. Edge computing solutions involve the use of small-scale data centres or micro data centres that can be located, where possible, at base stations of cellular carrier networks or IoT aggregation points. For example, real-time processing of data produced by smart cameras that monitors city regions, such as video streams, on local edge servers are able to detect risks without having to send large volumes of video data to a cloud platform. When IoT expands to large geographical areas, edge computing becomes the game-changer that allows IoT at scale for smart city officials. Data localization is achieved while data volumes are mitigated, helping to prevent operational issues within these databases. As organizations advance the application of edge solutions with IoT in focus, dedicated edge options for IoT implementations promotes even more efficient creation of new smart city solutions.

3.7.3 IoT Device Management and Integration

The smart city applications are dependent on IoT devices to collect data and perform functions. Secure onboarding and authentication mechanisms are the key strength in the IoT framework in order to get maximum benefits with security, compatibility, and dependability [61]. Thus, a language like MQTT has to be used to ensure that different devices from different manufacturers communicate effectively at a high level of complexity [62]. Also, such a middleware platform is needed to be able to effectively integrate large volumes of data from disparate and isolated city systems. As thousands of sensors and other devices should be placed all over the city, it is crucial to develop a remote monitoring and management system that will allow control of the device's health state, identify the faults, and firmware/software updates if needed. In a less latency and bandwidth-intensive setup, AI-driven devices need to be backed by a resilient mobile switching centre and scalable cloud infrastructure solutions, with the help of edge computing. Some of the security policies that should be put in place during the IoT device life cycle include password protection, encoding of data, limited physical access to the device, and penetration testing. From the standpoint of security and compatibility standards of IoT devices, smart cities strategically offer intelligent services to residents.

3.8 Emerging Trends and Innovations in IoT for Smart Cities

Most cities are being enhanced by IoT devices, sensors, and networks. Environmental monitors, along with traffic lamps and intelligent street lamps, are getting real-time data to improve services. Shifts in new generation AI, the fifth generation of internet-connected devices, and the advancement in computational edge will enable for improved data analytics from the IoT devices. IoT has applicability in intelligent transportation systems and smart cities, controlling traffic, and efficient energy usage. Figure 3.15 illustrates the emerging trends and innovations in IoT for smart cities.

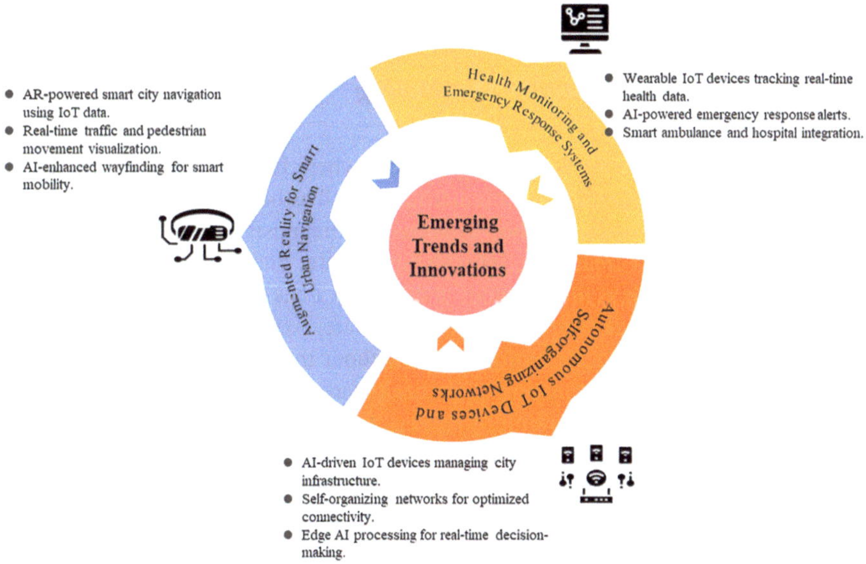

Fig. 3.15 Emerging trends and innovations in IoT for smart cities

3.8.1 Integration of IoT with Augmented Reality

IoT means a system that comprises devices with sensors, software, and connectivity to collect and share information. Augmented reality (AR) is digital information and the digital objects that are overlaid into the real environment in real time [63]. It indicated that these two technologies, when integrated, can produce immersive solutions and open up new opportunities [64]. For example, in a smart factory, plant managers could wear AR glasses that give real-time data on temperature, pressure, and vibration levels right on a see-through screen on the equipment. Also, the AR system could inform the workers on repair procedures by providing actual instructions on the surface of the machines. As applied in smart homes and buildings, AR can offer homeowners a better way to control connected IoT products. An AR dashboard is capable of recognizing all smart devices in the environment and showing the current state, such as whether lights are on, temperature, electricity used, etc. In the smart cities, IoT and AR allow city planners to see data about real-time traffic and the movement of pedestrians and navigation of the route map. On-ground customers can get necessary augmented information about the monuments, word map, and special offers, which are directly imposed on the actual site that the tourist is perceiving. This combination allows an immediate decision or action based upon available data [65]. As both these technologies mature, enterprise IoT-driven AR experiences will be possible across many different industries, from manufacturing to healthcare. It will bring increased effectiveness, measurability, and usability to the technology.

3.8.2 IoT-Enabled Health Monitoring

The IoT technology is one of the biggest areas where it is implemented in healthcare through real-time tracking of biometrics as well as health status with the use of wearables, remote sensors, etc. [66]. This enables diagnosis of health complications in an early stage and treatment, individualized care, and optimization of health. Wearable devices such as fitness bands and smart watches are embedded with built-in sensors that can track heart rate, blood pressure, oxygen levels, sleep patterns, and so on. These send health information without the use of cables to the apps, which in turn send the information to the healthcare service provider and analyse it for any irregularities. It also reminds citizens to take medication on time with the help of sensors. Currently, wearables that are used to monitor the chronic disease biomarkers are in the process of being developed into IoT skin patches. Other connected devices, such as weight scales, also allow people to monitor themselves against certain health targets. These wearables generate useful information for the healthcare providers, which enables early intervention. With telehealth using IoT, doctors are able to have online consultations with patients from rural areas and those who cannot be mobilized. It enhances compliance post-discharge and home-based healthcare services. IoT integration with electronic health records (EHR) leads to big data analysis to support health management programs from insurance companies and governmental entities [67]. The remote, real-time health monitoring that is controlled by IoT provides clients with even more synchronous and prognostic care.

3.8.3 Autonomous IoT Devices and Self-Organizing Networks

With the IoT technology in smart cities, there has been an increase in the number of linked devices. These devices carry sensors and actuators that allow them to gather information, reason, and perform functions without any human input. When more devices are used, complications arise regarding how best to handle or categorize them. This is where self-organizing networks come into the picture. Self-organizing networks are of significance to standardization and its subsequent implementation allows users in telecommunications networks to meet autonomously. Self-organizing networks can be defined as those associated with IoT devices that can reorganize themselves based on the existing situation of the network [68]. It relies on ICT possibilities such as AI, ML, and software-defined networking to demonstrate autonomic features. The devices are capable of searching for each other, self-organizing in terms of resource sharing, and performing time and data synchronization to make sure that they can work and communicate with each other [69]. Similarly, self-guided aerial vehicles can also plan and work conjointly to navigate through ambiguous geography for search and rescue operations, choosing optimum teams or groups. The self-organization that the algorithms power relies on the local interactions in which each device only shares information with its neighbours. This localized behaviour gives an optimum behaviour of the network at the network level. This also enhances scalability, as other nodes can be incorporated without readjustment. This will help wide applications such as dynamic topology reconstructions, distributed control, and reliable connectivity for multiple smart home applications, industries, and smart cities. The provided plug-and-play abilities will substantially improve the approaches to the utilization of vast, decentralized IoT systems.

3.9 Summary

Smart cities and new applications on urban infrastructure, mobility, and resource management are also transformed with the Internet of Things. The 5G communication is utilized in developing cities with advanced sensor technologies and the next generation of IoT networks to deploy real-time data, optimize decisions, reduce traffic congestion, and distribute resources for more seamless urban mobility through intelligent traffic management and public transportation optimization. IoT is part of its tasks, such as providing environmental and resource monitoring for continuous assessment of air and water qualities, waste management efficiency and smart energy grids required for sustainable urban expansion. It includes the role of blockchain, AI-driven threat detection, as well as privacy-preserving techniques in addressing the security and privacy concerns to make IoT solutions secure and resilient against cyber threats. While IoT systems face challenges such as standards and interoperability, edge computing and good device management solutions shows that system deployments are becoming more scalable and robust and are only growing in such direction. Emerging trends of integration of IoT in augmented reality, health monitoring, self-organizing, and autonomous networks are initiating new ways of living in urban spaces. Eventually, smart cities that embrace these innovations enable them to deliver on this promise of helping to meet the changing needs of their inhabitants towards building on a sustainable, connected future.

References

1. S.P. Mohanty, U. Choppali, E. Kougianos, Everything you wanted to know about smart cities: The internet of things is the backbone. IEEE Cons. Electron. Magaz. **5**(3), 60–70 (2016). https://doi.org/10.1109/MCE.2016.2556879
2. P. Mishra, G. Singh, Internet of things for sustainable smart city, in *Sustainable smart cities*, (Springer, Cham, 2023), pp. 75–96. https://doi.org/10.1007/978-3-031-33354-5_4
3. S. Kumar, P. Tiwari, M. Zymbler, Internet of things is a revolutionary approach for future technology enhancement: A review. J. Big Data **6**(1), 111 (2019). https://doi.org/10.1186/s40537-019-0268-2
4. A.P. Rodriguez Müller, Making smart cities 'smarter' through ICT-enabled citizen coproduction, in *Handbook of smart cities*, (Springer, Cham, 2021), pp. 1–21. https://doi.org/10.1007/978-3-030-15145-4_63-1
5. M.Q. Kheder, A.A. Mohammed, Real-time traffic monitoring system using IoT-aided robotics and deep learning techniques. Kuwait J. Sci. **51**(1), 100153 (2024). https://doi.org/10.1016/J.KJS.2023.10.017
6. C. Stephen Ball, D. Degischer, IoT implementation for energy system sustainability: The role of actors and related challenges. Util. Policy **90**, 101769 (2024). https://doi.org/10.1016/J.JUP.2024.101769
7. P. Pawar, K. Vittal, Design and development of advanced smart energy management system integrated with IoT framework in smart grid environment. J. Ener. Stor. **25**, 100846 (2019). https://doi.org/10.1016/J.EST.2019.100846
8. P.M. Rao, B.D. Deebak, Security and privacy issues in smart cities/industries: Technologies, applications, and challenges. J. Ambient. Intell. Humaniz. Comput. **14**(8), 10517–10553 (2023). https://doi.org/10.1007/s12652-022-03707-1
9. A. Alshami, E. Ali, M. Elsayed, A.E.E. Eltoukhy, T. Zayed, IoT innovations in sustainable water and wastewater management and water quality monitoring: A comprehensive review of advancements, implications, and future directions. IEEE Access **12**, 58427–58453 (2024). https://doi.org/10.1109/ACCESS.2024.3392573
10. P. Mishra, G. Singh, Smart healthcare in sustainable smart cities, in *Sustainable smart cities*, (Springer, Cham, 2023), pp. 195–219. https://doi.org/10.1007/978-3-031-33354-5_9
11. W. Shi, J. Cao, Q. Zhang, Y. Li, L. Xu, Edge computing: Vision and challenges. IEEE Internet Things J. **3**(5), 637–646 (2016). https://doi.org/10.1109/JIOT.2016.2579198
12. J. Lin, W. Yu, N. Zhang, X. Yang, H. Zhang, W. Zhao, A survey on internet of things: Architecture, enabling technologies, security and privacy, and applications. IEEE Internet Things J. **4**(5), 1125–1142 (2017). https://doi.org/10.1109/JIOT.2017.2683200

13. M.A. Khan, K. Salah, IoT security: Review, blockchain solutions, and open challenges. Futur. Gener. Comput. Syst. **82**, 395–411 (2018). https://doi.org/10.1016/j.future.2017.11.022
14. A. Botta, W. de Donato, V. Persico, A. Pescapé, Integration of cloud computing and internet of things: A survey. Futur. Gener. Comput. Syst. **56**, 684–700 (2016). https://doi.org/10.1016/j.future.2015.09.021
15. U. Raza, P. Kulkarni, M. Sooriyabandara, Low power wide area networks: An overview. IEEE Commun. Surv. Tutor. **19**(2), 855–873 (2017). https://doi.org/10.1109/COMST.2017.2652320
16. A. Fuller, Z. Fan, C. Day, C. Barlow, Digital twin: Enabling technologies, challenges and open research. IEEE Access **8**, 108952–108971 (2020). https://doi.org/10.1109/ACCESS.2020.2998358
17. M. Centenaro, L. Vangelista, A. Zanella, M. Zorzi, Long-range communications in unlicensed bands: The rising stars in the IoT and smart city scenarios. IEEE Wirel. Commun. **23**(5), 60–67 (2016). https://doi.org/10.1109/MWC.2016.7721743
18. I.A.T. Hashem et al., The role of big data in smart city. Int. J. Inf. Manag. **36**(5), 748–758 (2016). https://doi.org/10.1016/j.ijinfomgt.2016.05.002
19. B. Farahani, F. Firouzi, V. Chang, M. Badaroglu, N. Constant, K. Mankodiya, Towards fog-driven IoT eHealth: Promises and challenges of IoT in medicine and healthcare. Futur. Gener. Comput. Syst. **78**, 659–676 (2018). https://doi.org/10.1016/j.future.2017.04.036
20. S. Borchardt, G. B. Vignola, D. Buscaglia, M. Maroni, Mapping EU policies with the 2030 agenda and SDGs-fostering policy coherence through text-based SDG mapping, https://doi.org/10.2760/110687
21. S. Munirathinam, Industry 4.0: Industrial Internet of Things (IIOT). Adv. Comput. **117**, 129–164 (2020). https://doi.org/10.1016/bs.adcom.2019.10.010
22. T. Kalsoom, N. Ramzan, S. Ahmed, M. Ur-Rehman, Advances in sensor technologies in the era of smart factory and industry 4.0. Sensors **20**(23), 6783 (2020). https://doi.org/10.3390/s20236783
23. A.B. Frazier, R.O. Warrington, C. Friedrich, The miniaturization technologies: Past, present, and future. IEEE Trans. Ind. Electron. **42**(5), 423–430 (1995). https://doi.org/10.1109/41.464603
24. T. Sanislav, G.D. Mois, S. Zeadally, S.C. Folea, Energy harvesting techniques for internet of things (IoT). IEEE Access **9**, 39530–39549 (2021). https://doi.org/10.1109/ACCESS.2021.3064066
25. R. Singh, S.S. Gill, Edge AI: A survey. Internet of Things Cyber Phys. Syst. **3**, 71–92 (2023). https://doi.org/10.1016/J.IOTCPS.2023.02.004
26. M. Bembe, A. Abu-Mahfouz, M. Masonta, T. Ngqondi, A survey on low-power wide area networks for IoT applications. Telecommun. Syst. **71**(2), 249–274 (2019). https://doi.org/10.1007/s11235-019-00557-9
27. T.H. Moges, D.S. Lakew, N.P. Nguyen, N.N. Dao, S. Cho, Cellular internet of things: Use cases, technologies, and future work. Internet of Things **24**, 100910 (2023). https://doi.org/10.1016/J.IOT.2023.100910
28. E. Di Pascale, I. Macaluso, A. Nag, M. Kelly, L. Doyle, The network as a computer: A framework for distributed computing over IoT mesh networks. IEEE Internet Things J. **5**(3), 2107–2119 (2018). https://doi.org/10.1109/JIOT.2018.2823978
29. D.H. Morais, Wi-Fi 6/6E and Wi-Fi 7 overview, in *5G/5G-Advanced, Wi-Fi 6/7, and Bluetooth 5/6*, (Springer Nature Switzerland, Cham, 2025), pp. 149–179. https://doi.org/10.1007/978-3-031-82830-0_9
30. S.S. Sefati, S. Halunga, Ultra-reliability and low-latency communications on the internet of things based on 5G network: Literature review, classification, and future research view. Trans. Emerg. Telecommun. Technol. **34**(6) (2023). https://doi.org/10.1002/ett.4770
31. Y. Liu, M. Peng, G. Shou, Y. Chen, S. Chen, Toward edge intelligence: Multiaccess edge computing for 5G and Internet of Things. IEEE Internet Things J. **7**(8), 6722–6747 (2020). https://doi.org/10.1109/JIOT.2020.3004500
32. V. Chauhan, M. Patel, S. Tanwar, S. Tyagi, N. Kumar, IoT enabled real-time urban transport management system. Comput. Electr. Eng. **86**, 106746 (2020). https://doi.org/10.1016/J.COMPELECENG.2020.106746

33. S. Bansal, A. Gupta, IoT-enabled intelligent traffic management system, in *IoT based smart applications. EAI/Springer innovations in communication and computing*, (Springer, Cham, 2023), pp. 89–111. https://doi.org/10.1007/978-3-031-04524-0_6
34. X.-G. Luo, H.-B. Zhang, Z.-L. Zhang, Y. Yu, K. Li, A new framework of intelligent public transportation system based on the internet of things. IEEE Access **7**, 55290–55304 (2019). https://doi.org/10.1109/ACCESS.2019.2913288
35. G.A. Ubiergo, W.L. Jin, Mobility and environment improvement of signalized networks through vehicle-to-infrastructure (V2I) communications. Transp. Res. Part C Emerg. Technol. **68**, 70–82 (2016). https://doi.org/10.1016/J.TRC.2016.03.010
36. Y. Wang, Y. Cui, Z. Kong, X. Liao, W. Wang, Design of public transportation system scheduling and optimization in the Internet of Things (IoT) environment. Adv. Eng. Technol. Res. **9**(1), 89 (2023). https://doi.org/10.56028/aetr.9.1.89.2024
37. V.K. Balla, H. Nandibhatla, G. Santhoshi, A.V. Krishna, M. Mohan, P. Purushotham, Smart bus ticketing system with RFID using IoT, in *2024 IEEE 6th International Conference on Cybernetics, Cognition and Machine Learning Applications (ICCCMLA)*, (IEEE, 2024), pp. 297–301. https://doi.org/10.1109/ICCCMLA63077.2024.10871743
38. A.S. Syed, D. Sierra-Sosa, A. Kumar, A. Elmaghraby, Making cities smarter—Optimization problems for the IoT enabled Smart City development: A mapping of applications, objectives, constraints. Sensors **22**(12), 4380 (2022). https://doi.org/10.3390/s22124380
39. M.R. Islam et al., Smart parking management system to reduce congestion in urban area, in *2020 2nd International Conference on Electrical, Control and Instrumentation Engineering (ICECIE)*, (IEEE, 2020), pp. 1–6. https://doi.org/10.1109/ICECIE50279.2020.9309546
40. T. Lin, H. Rivano, F. Le Mouel, A survey of smart parking solutions. IEEE Trans. Intell. Transp. Syst. **18**(12), 3229–3253 (2017). https://doi.org/10.1109/TITS.2017.2685143
41. M. Jemmali, L.K.B. Melhim, M.T. Alharbi, A. Bajahzar, M.N. Omri, Smart-parking management algorithms in smart city. Sci. Rep. **12**(1), 6533 (2022). https://doi.org/10.1038/s41598-022-10076-4
42. M.B. Mollah, M.A.K. Azad, A. Vasilakos, Secure data sharing and searching at the edge of cloud-assisted internet of things. IEEE Cloud Comput **4**(1), 34–42 (2017). https://doi.org/10.1109/MCC.2017.9
43. V. Garrido-Momparler, M. Peris, Smart sensors in environmental/water quality monitoring using IoT and cloud services. Trends Environ. Analyt. Chem. **35**, e00173 (2022). https://doi.org/10.1016/J.TEAC.2022.E00173
44. F. Montori, L. Bedogni, L. Bononi, A collaborative internet of things architecture for smart cities and environmental monitoring. IEEE Internet Things J. **5**(2), 592–605 (2018). https://doi.org/10.1109/JIOT.2017.2720855
45. A. Lakhouit, Revolutionizing urban solid waste management with AI and IoT: A review of smart solutions for waste collection, sorting, and recycling. Results Eng. **25**, 104018 (2025). https://doi.org/10.1016/J.RINENG.2025.104018
46. J. Kim, A. Manna, A. Roy, I. Moon, Clustered vehicle routing problem for waste collection with smart operational management approaches. Int. Trans. Oper. Res. **32**(2), 863–887 (2025). https://doi.org/10.1111/itor.13282
47. V. SrirengaNachiyar, R. Chandralekha, R. Kasi Raman, V. Krishnameera, A. Guna, S. Harini Shriram, Smart bin system based on IoT for effective food waste management, in *2025 4th International Conference on Sentiment Analysis and Deep Learning (ICSADL)*, (IEEE, 2025), pp. 525–530. https://doi.org/10.1109/ICSADL65848.2025.10933108
48. S.M.A.A. Abir, A. Anwar, J. Choi, A.S.M. Kayes, IoT-enabled smart energy grid: Applications and challenges. IEEE Access **9**, 50961–50981 (2021). https://doi.org/10.1109/ACCESS.2021.3067331
49. D.B. Avancini, J.J.P.C. Rodrigues, R.A.L. Rabêlo, A.K. Das, S. Kozlov, P. Solic, A new IoT-based smart energy meter for smart grids. Int. J. Energy Res. **45**(1), 189–202 (2021). https://doi.org/10.1002/er.5177
50. K. Sharma, A. Malik, W. Abdelfattah, I. Batra, M. Shabaz, A smart electricity consumption management framework using internet of things (IoT) to optimize electricity demand. Measure. Sens. **33**, 101218 (2024). https://doi.org/10.1016/J.MEASEN.2024.101218

51. U. Majeed, L.U. Khan, I. Yaqoob, S.M.A. Kazmi, K. Salah, C.S. Hong, Blockchain for IoT-based smart cities: Recent advances, requirements, and future challenges. J. Netw. Comput. Appl. **181**, 103007 (2021). https://doi.org/10.1016/J.JNCA.2021.103007
52. M.S. Rahman, M.A.P. Chamikara, I. Khalil, A. Bouras, Blockchain-of-blockchains: An interoperable blockchain platform for ensuring IoT data integrity in smart city. J. Ind. Inf. Integr. **30**, 100408 (2022). https://doi.org/10.1016/J.JII.2022.100408
53. M.E.E. Alahi et al., Integration of IoT-enabled technologies and artificial intelligence (AI) for Smart City scenario: Recent advancements and future trends. Sensors **23**(11), 5206 (2023). https://doi.org/10.3390/s23115206
54. I.H. Sarker, AI-enabled cybersecurity for IoT and smart city applications, in *AI-driven cybersecurity and threat intelligence*, (Springer Nature Switzerland, Cham, 2024), pp. 121–136. https://doi.org/10.1007/978-3-031-54497-2_7
55. D. Torre, A. Chennamaneni, A. Rodriguez, Privacy-preservation techniques for IoT devices: A systematic mapping study. IEEE Access **11**, 16323–16345 (2023). https://doi.org/10.1109/ACCESS.2023.3245524
56. L.U. Khan, W. Saad, Z. Han, E. Hossain, C.S. Hong, Federated learning for internet of things: Recent advances, taxonomy, and open challenges. IEEE Commun. Surv. Tutor. **23**(3), 1759–1799 (2021). https://doi.org/10.1109/COMST.2021.3090430
57. M.H. Panahi Rizi, S.A. Hosseini Seno, A systematic review of technologies and solutions to improve security and privacy protection of citizens in the smart city. Internet of Things **20**, 100584 (2022). https://doi.org/10.1016/J.IOT.2022.100584
58. V.S. Chakravarthi, *Internet of Things and M2M communication technologies* (Springer, Cham, 2021). https://doi.org/10.1007/978-3-030-79272-5
59. A. Hazra, M. Adhikari, T. Amgoth, S.N. Srirama, A comprehensive survey on interoperability for IIoT: Taxonomy, standards, and future directions. ACM Comput. Surv. **55**(1), 1–35 (2023). https://doi.org/10.1145/3485130
60. S. Algarni, F.E.A. El-Samie, Energy-efficient distributed edge computing to assist dense internet of things. Fut. Internet **17**(1), 37 (2025). https://doi.org/10.3390/fi17010037
61. S. Maksuti et al., Automated and secure onboarding for system of systems. IEEE Access **9**, 111095–111113 (2021). https://doi.org/10.1109/ACCESS.2021.3102280
62. S. Lakshminarayana, A. Praseed, P.S. Thilagam, Securing the IoT application layer from an MQTT protocol perspective: Challenges and research prospects. IEEE Commun. Surv. Tutor. **26**, 2510 (2024). https://doi.org/10.1109/COMST.2024.3372630
63. S. Dargan, S. Bansal, M. Kumar, A. Mittal, K. Kumar, Augmented reality: A comprehensive review. Arch. Comput. Methods Eng. **30**(2), 1057–1080 (2023). https://doi.org/10.1007/s11831-022-09831-7
64. K.N. Qureshi, A. Alhudhaif, R.W. Anwar, S.N. Bhati, G. Jeon, Fully integrated data communication framework by using visualization augmented reality for internet of things networks. Big Data **9**(4), 253–264 (2021). https://doi.org/10.1089/big.2020.0282
65. A.A. Syahidi, K. Arai, H. Tolle, A.A. Supianto, K. Kiyokawa, Augmented reality in the Internet of Things (AR + IoT): A review. IJICS (Int. J. Inform. Comput. Sci.) **5**(3), 258 (2021). https://doi.org/10.30865/ijics.v5i3.3341
66. M.S.H. Talpur et al., Illuminating healthcare management: A comprehensive review of IoT-enabled chronic disease monitoring. IEEE Access **12**, 48189–48209 (2024). https://doi.org/10.1109/ACCESS.2024.3382011
67. A. Ahmed, R. Xi, M. Hou, S.A. Shah, S. Hameed, Harnessing big data analytics for healthcare: A comprehensive review of frameworks, implications, applications, and impacts. IEEE Access **11**, 112891–112928 (2023). https://doi.org/10.1109/ACCESS.2023.3323574
68. N. Li, X. Liu, Research on self-organization and adaptive strategy of the internet of things sensor networks. IEEE Access **12**, 66569–66579 (2024). https://doi.org/10.1109/ACCESS.2024.3399537
69. S. Maltseva, V. Kornilov, V. Barakhnin, A. Gorbunov, Self-Organization in Network Sociotechnical Systems. Complexity **2022**(1) (2022). https://doi.org/10.1155/2022/5714395

Chapter 4
Advanced Artificial Intelligence for Sustainable Smart Cities

Contents

P. Mishra, G. Singh, *Sustainable Smart Cities 2.0*,
https://doi.org/10.1007/978-3-032-01102-2_4

4.1 Introduction

Smart cities have recently gained attention due to the potential to enhance sustainability, efficiency, and the quality of life for inhabitants. One of the most important aspects is the application of advanced artificial intelligence (AI) technologies in smart cities as part of sustained utilization of resources, with the intent to improve people's quality of life [1]. AI is on the rise in smart city projects because of its appeal in handling big data, making predictions, and supporting decision-making. The population living in urban areas has risen at a considerably high rate, with many challenges following these developments, such as the high emission of greenhouse gases, pollution, traffic jams, and wastage of resources. To overcome these challenges, smart city concepts have come into existence to involve the use of technology and data to enhance the residents' quality of life in urban centres. Fundamental to smart city initiatives is sustainability; to satisfy needs and ensure renewable resources in the future [2]. First and foremost, advanced AI technologies are essential in reaching the goal of cities by offering the right information for the proper functioning of numerous systems in urban areas. Energy efficiency is one of the most remarkable advantages of AI, which contributes to sustainable smart cities [3]. AI can also be widely used to help analyse the data from sensors, meters, to manage energy use across buildings and transportation, and much more. For example, predictive maintenance are used to diagnose equipment faults before they cause failures that result in energy loss. Furthermore, integrated applications such as smart grids that rely on AI are also capable of managing supply demands in relation to energy at variable rates that lead to efficient management of energy, as well as an reduction in greenhouse gas emissions [4]. One of the most important concerns of using AI and big data for sustainable smart cities is the transportation system [5]. The intelligent transport systems (ITS) and self-driven cars helps in traffic management and emissions by providing a better way of traffic management and routing. They also assist in the optimization of the available resources depending on the expected demand for public transport and the reduction of traffic incidents due to installed intelligent transport systems. In addition, through the use of AI technologies, recycling and methods of improving waste management are enhanced and improved [6]. By analysing waste data, more effective waste management policies are made, and the principles of circular economies can be learned and applied. Waste sorting is still effective in recycling processes since it can sort different materials correctly. It can also support sustainable smart cities by helping with decision-making for inhabitants of cities and enhancing their standard of living [7]. The information from different sources, an AI system can gain an understanding of data and pass the relevant information to the urban planners/politicians and various stakeholders to enable them to make better decisions. In addition, AI enhances public services, especially healthcare and education, by analysing the distribution of resources and meeting the requirements of users.

The state of the art in modern sustainable smart cities is primarily based on machine learning (ML), deep learning (DL), and natural language processing

(NLP). With the help of ML, DL methods are trained on a large number of datasets to recognize patterns and then make certain predictions, which helps in various applications such as energy optimization, intelligent transportation, and waste management. In particular, AI and more specifically, ML has enabled DL using neural networks to be able to enhance many different AI applications for smart cities. DL has abilities of analysing satellite images and identifying the places of deforestation, water pollution, etc., in the sphere of environmental monitoring. NLP can therefore also be employed to support the development and running of sustainable smart cities for the purpose of making sense of big data in the form of text, such as tweets, reviews, and commentaries [8]. In particular, NLP algorithms are used to investigate needs and attitudes contained in the massive amounts of textual data to help urban planners and policymakers get a more detailed idea of the issues affecting residents of large cities. Other AI methods being considered for application in the sustainable smart city incorporates computer vision, reinforcement learning, and robotics.

Figure 4.1 illustrates the AI-driven conceptual framework for sustainable smart cities. It shows how sustainable smart city applications are built using AI and highlights the flow from data to smart city services. First, data comes from Internet of Things (IoT) devices, smart sensors, environmental sources, and citizen feedback. This data feeds into an AI solution (like AI-driven energy grids and waste categorization) and explainable AI (XAI) for transparency. Technologies like DL and reinforcement learning power the AI models. The final application includes smart energy management, AI-powered waste management, AI-driven weather forecasting, green mobility, and AI-driven governance. As a result, the future application of advanced AI technologies for sustainable smart cities is optimistic because new techniques as well as application development are fast. Also, the application of AI to the IoT is making people's lives even more connected and sensitive to people's needs as another category of development. Therefore, advanced AI technologies are becoming more and more essential in the creation and functioning of efficient smart cities. The role that AI can play in the context of urbanization and necessary improvements to the quality of life of city inhabitants may include enabling smart decision making and enhancing or optimizing the urban systems. While the AI already has promising applications in smart cities, there is great potential for future development and its positive impact on sustainability and future-proofing of smart cities.

4.2 Current Research Trends in Artificial Intelligence for Smart Cities

This study investigates the trends, contributions, and impact of research on artificial intelligence in smart cities from 2015 to 2025 through a bibliometric analysis of Scopus data. The Bibliometrix R package, along with Origin software, was utilized to analyse and depict the data. This approach ensures that researchers are

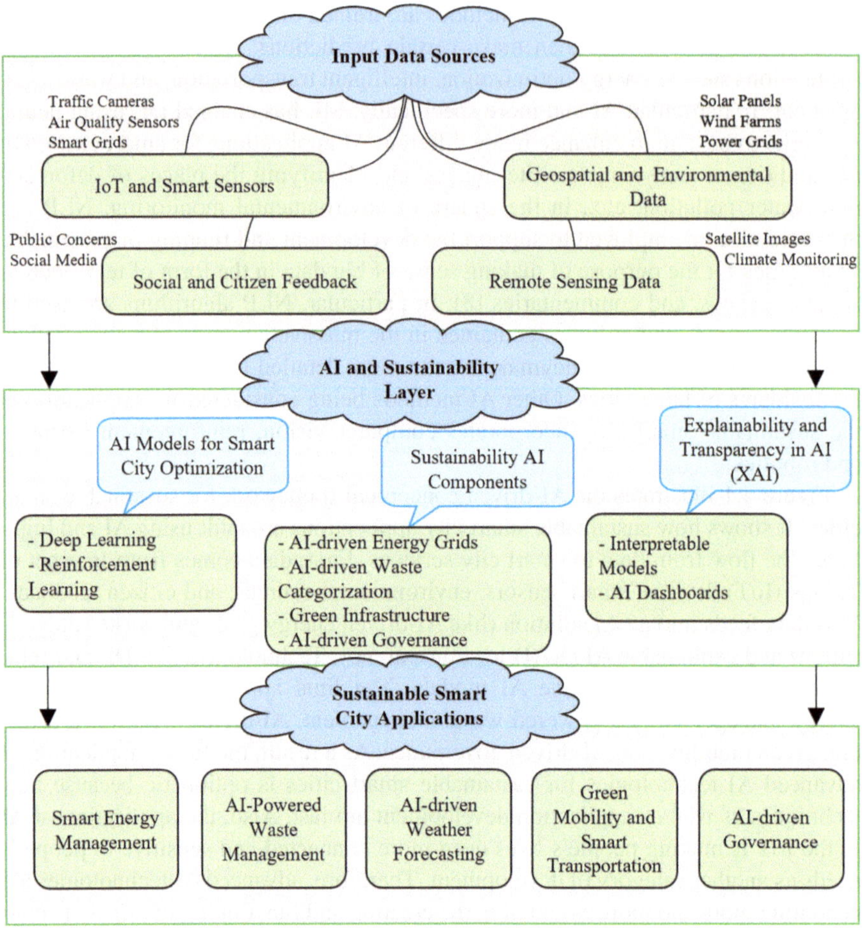

Fig. 4.1 AI-driven conceptual framework for sustainable smart cities

presented with the most current research developments in the field, enabling to focus on emerging trends and advancements. In March 2025, a bibliometric analysis was conducted utilizing the Scopus core collection database to thoroughly investigate the concept, impact, and implementation of AI in smart cities. Ultimately, 1514 articles were exported and saved in BibTeX format before being included and examined in the bibliometric study. The primary information, syntax, and specific criteria used in screening for this study are shown in Table 4.1.

Table 4.1 Summary of data characteristics and main information of the bibliometric study

Description	Specific Standard Requirements/Results
Research database	Scopus
Search query	(“smart”) AND (“city”) AND (“artificial”) AND (“intelligence”) AND (“machine”) AND (“learning”) AND (“deep”) AND (“learning”)
Document type	Articles
Language	English
Data export format	BibTeX
Timespan	2015:2025
Documents	1514
Annual growth rate %	17.46
Document average age	3.13
Average citations per doc	21.9
Keywords plus (ID)	6780
Author’s keywords (DE)	3634
Authors	4775
Authors of single-authored docs	130
Single-authored docs	221
Co-authors per doc	3.78
International co-authorships %	30.91

4.2.1 Annual Scientific Production and Top 10 Most Published Journals/Cited Papers

Figure 4.2 shows the annual publication trends from 2015 to 2025, highlighting the growth in research output. The early years show a steady increase in publications, rising from 2015 (18 doc) to 2018 (91 doc), indicating a growing interest in the field. There is a slight decrease in the number of documents in 2019 (76), followed by an increase from 2020 (107) to 2024 (361). In 2025 (90), the number continues to rise and may continue increasing until the end of the year. Figure 4.3 presents the highest number of articles in the AI-driven smart city research domain published in “Lecture Notes in Networks and Systems” (50 docs), “IEEE Access” (46 docs), “Communications in Computer and Information Science” (30 docs), and “ACM International Conference Proceeding Series” (24 docs), which indicates a higher output of articles compared to the other journals. In addition, some other journals, such as the “Advances in Intelligent Systems and Computing” (23 docs), “Sustainability (Switzerland)” (23 docs), “Lecture Notes in Computer Science (Including Subseries Lecture notes in Artificial Intelligence and Lecture Notes in Bioinformatics)” (20 docs), “Sensors” (20 docs), “Lecture Notes in Electrical Engineering” (19 docs), and “Arts-Based Coaching: Using Creative Tools to Promote Better Self-Expression” (17 docs), are in this domain. Table 4.2 displays the top ten most cited papers from the study, which presents both the ranking and authors with their published works. This section presents information about authors,

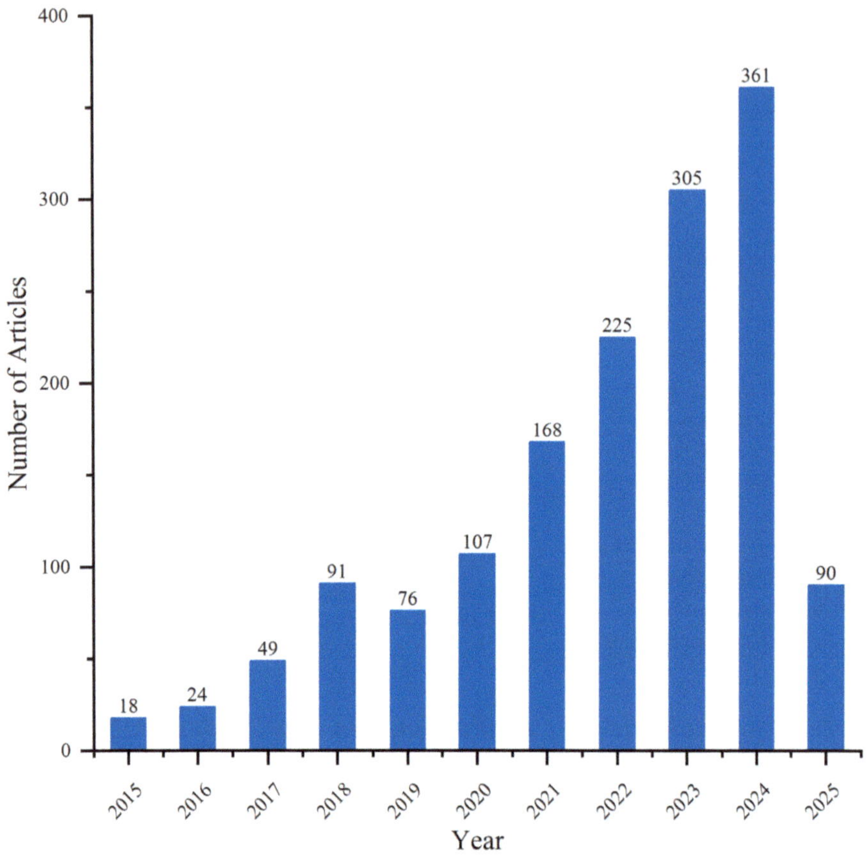

Fig. 4.2 Yearly trend of publication numbers in artificial intelligence in smart cities

article titles, and their corresponding publications in journals with documentation for publication years and citation frequency. The ranking focuses on engineering field smart city articles to demonstrate their total citation (TC) counts, annual citation rates, and normalized total citation (NTC) data, which collectively signify their impact in the field.

4.2.2 *Sankey Diagram (Three-Field Plot and Thematic Map)*

In Fig. 4.4, the Sankey diagram illustrates the three-field plot thematic progression to visualize the flow and the distribution in scientific research. In this case, keywords, an abstract, and a title are added, acting as indicators to highlight the most influential research in the subject. These keywords indicate vital concepts and themes that stand out in academic discussions of artificial intelligence in the smart city domain. The diagram's keywords are sourced from earlier published studies and

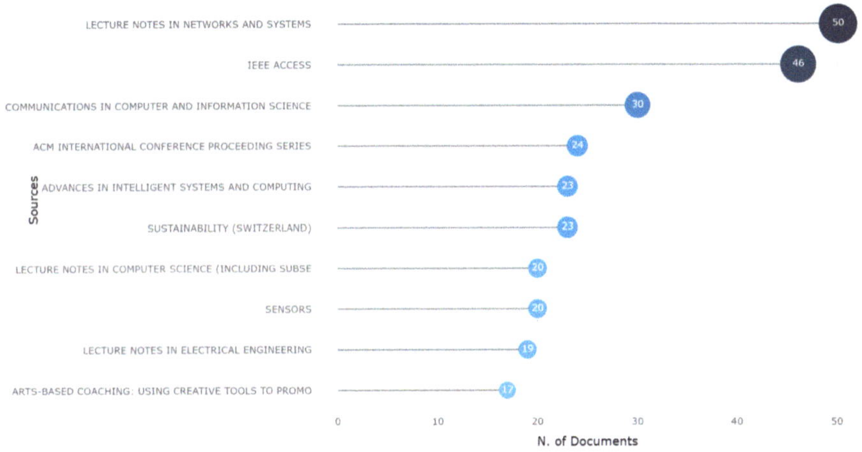

Fig. 4.3 Most relevant sources with the number of articles

were then categorized thematically. Figure 4.5 categorizes various smart city themes based on development and relevant degrees. The centre of the thematic map contains a green bubble, which represents medium-level development alongside moderate relevance for "intelligent systems" and "traffic congestion", along with "adversarial machine learning". The subjects present themselves as promising mixed areas that unite basic AI principles with specific functional challenges in a real-world environment. The central location seems to indicate these areas might develop into essential research fronts that deliver extensive outcomes in the future. In the left quadrants (niche themes), there are mainly representations of "network security", "5G mobile communication systems", and "embedded systems". This indicates a strong focus on these areas, suggesting they are central to artificial intelligence in smart city research and practice. This quadrant reflects the areas where significant advances and investments are made, highlighting their critical role in the smart city. Additionally, terms like "learning systems", "big data", "machine learning", "smart city", and "artificial intelligence" are foundational information for further analysis. These themes might be considered foundational but must be emphasized more in smart city research. This visual categorization provides valuable insights for educators, curriculum developers, and researchers, helping them understand which areas are gaining traction and which might need more attention or innovation.

4.2.3 Co-occurrence Keyword Trends and Network Visualization

Figure 4.6 depicts the most prominent themes and keywords related to research in conceptual change. The most significant and most central terms such as "smart city", "artificial intelligence", "learning systems", "internet of things", and "machine

Table 4.2 Most globally cited documents in the smart city research domain with artificial intelligence

Rank	Author's name	Paper Title	Journal's Name	TC	TC/ Year	NTC
1	Sarker (2021) [9]	Machine learning: Algorithms, real-world applications, and research directions	SN Computer Science	2801	560.20	56.14
2	Fuller et al. (2020) [10]	Digital twin: Enabling technologies, challenges, and open research	IEEE Access	1458	243.00	26.08
3	Khan et al. (2021) [11]	Federated learning for Internet of Things: Recent advances, taxonomy, and open challenges	IEEE Communications Surveys & Tutorials	827	165.40	16.58
4	Diro and Chilamkurti (2018) [12]	Distributed attack detection scheme using a deep learning approach for the Internet of Things	Future Generation Computer Systems	750	93.75	19.36
5	Rahman et al. (2018) [13]	Predicting electricity consumption for commercial and residential buildings using deep recurrent neural networks	Applied Energy	601	75.13	15.51
6	Sivanathan et al. (2019) [14]	Classifying IoT devices in smart environments using network traffic characteristics	IEEE Transactions on Mobile Computing	537	76.71	11.83
7	Ullah et al. (2020) [15]	Applications of artificial intelligence and machine learning in smart cities	Computer Communications	516	86.00	9.23
8	Zhang and Tao (2021) [16]	Empowering things with intelligence: a survey of the progress, challenges, and opportunities in artificial intelligence of things	IEEE Internet of Things Journal	502	100.40	10.06
9	Minerva et al. (2020) [17]	Digital twin in the IoT context: A survey on technical features, scenarios, and architectural models	Proceedings of the IEEE	494	82.33	8.84
10	Tian and Pan (2015) [18]	Predicting short-term traffic flow by a long short-term memory recurrent neural network	IEEE International Conference on Smart City/SocialCom/ SustainCom (SmartCity)	464	42.18	11.09

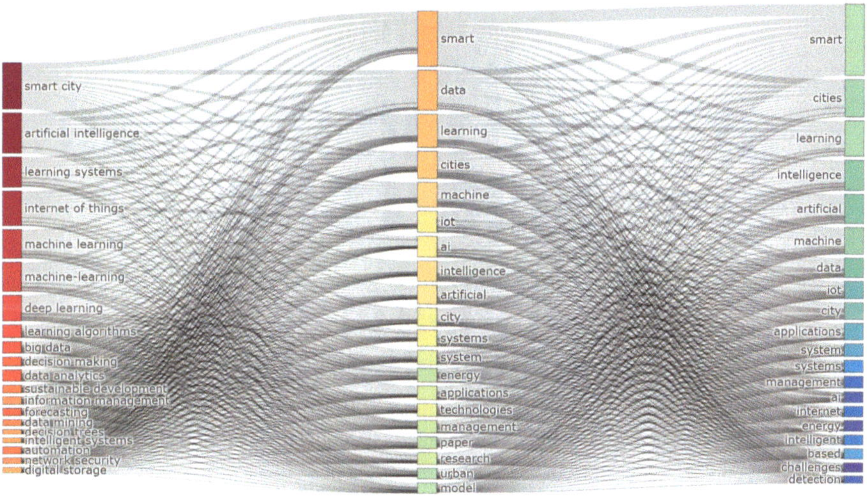

Fig. 4.4 Three field plots (thematic development) using a Sankey diagram

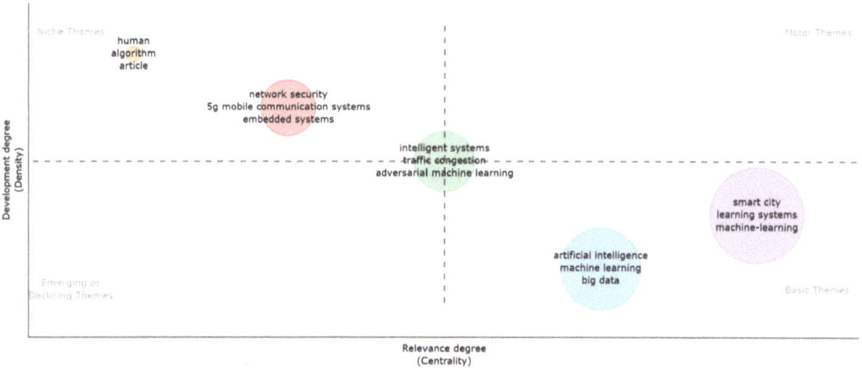

Fig. 4.5 Thematic map in the smart city research domain with artificial intelligence

learning", indicate the core focus of the research: how researchers grasp and alter their conceptual knowledge, particularly within smart city contexts.

An illustration in Fig. 4.7 displays word occurrence frequencies for the artificial intelligence in the smart city research network. Research focuses mainly on "smart city", as this term appears 515 times throughout the literature. Research on smart cities indicates the importance of "artificial intelligence" based on its 481-word occurrences. Additionally, "learning systems" (370), "internet of things" (338), "machine learning" (330), "deep learning" (265), "learning algorithms" (147), big data (126), and data analytics (101) combinations occupy crucial importance in emerging technology fields. The data shows that smart city research revolves primarily around several major subjects.

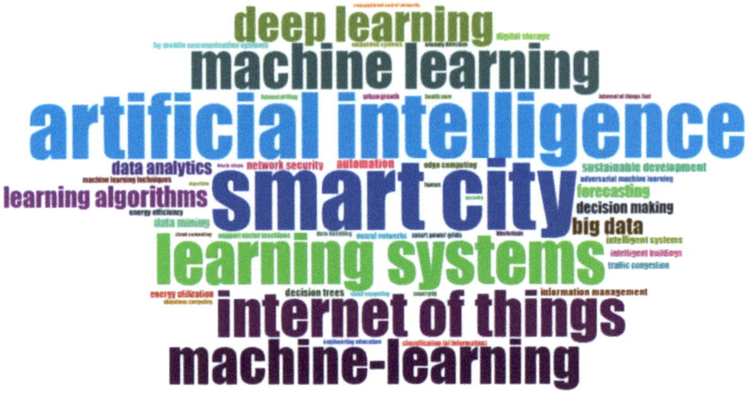

Fig. 4.6 Co-occurrence keyword trends in the smart city domain with artificial intelligence

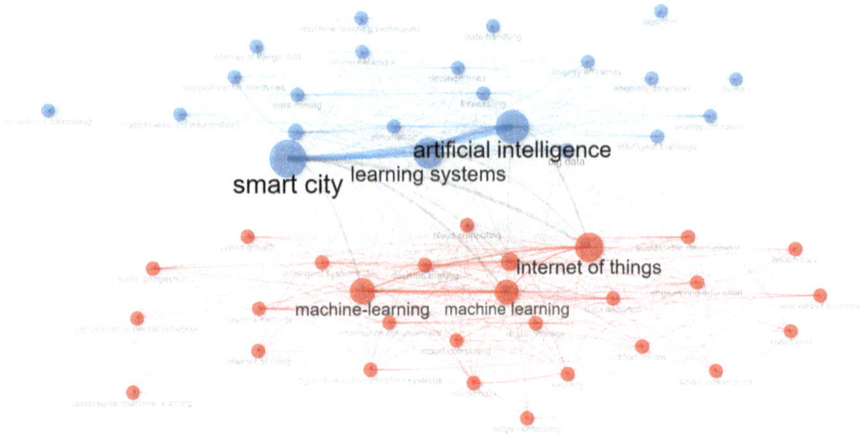

Fig. 4.7 Co-occurrence network of words in the smart city research domain with artificial intelligence

4.2.4 Countries' Scientific Production

The scientific publication frequencies of various countries in the world are represented in Fig. 4.8. Scientific contributions from India stand at 963, which demonstrates the powerful role in research development. The growing scientific power of China displays itself through its 322 scientific publications, which rank second after India. The research hub of the USA produces 253 scientific papers among its total contributions. Numerous publications emerge from Saudi Arabia, the UK, together with Italy, given that their research output comprises 213, 156, and 96 papers, respectively. Academic and industrial research sectors of South Korea and Australia produce approximately 85 and 76 scholarly works, respectively. Morocco and

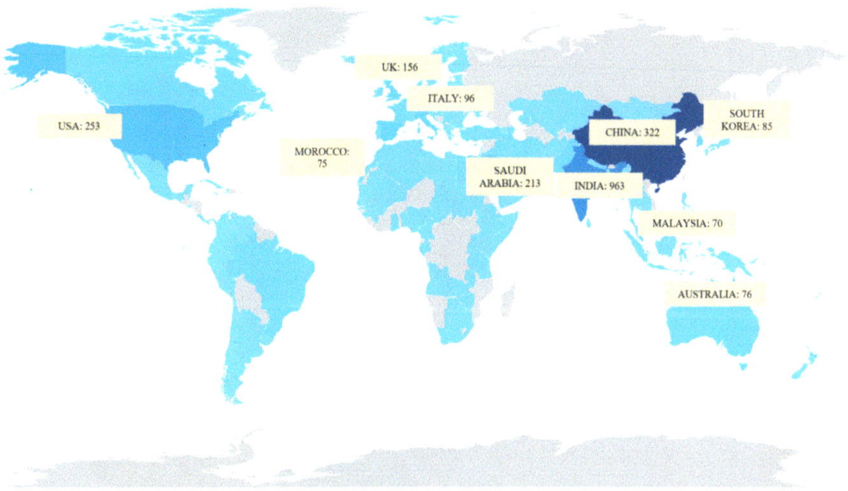

Fig. 4.8 Top 10 countries' scientific production of various countries based on frequency

Malaysia have smaller publication numbers, but their contributions amount to totals of 75 and 70 research items, respectively.

4.2.5 Mapping Smart City Research to Sustainable Development Goals (SDGs)

Sustainable Development Goals, their targets (bubble series), and the percentage (bar graph) of corresponding keywords detected in the text (calculated as the ratio of keywords in one goal to the total number of keywords detected) are depicted in Fig. 4.9. For SDG mapping, the study utilized the SDG Mapper tool from the "Knowledge Base for the Sustainable Development Goals", provided by the European Commission [19]. This tool uses bubble charts to depict SDG interconnections and relevance among bibliometric records. The size of each bubble correlates with the proportion of associated keywords found, which is calculated as the number of goal-specific keywords divided by the total keyword count. As such, SDG 11 is marked by 44 keyword occurrences. This is followed by SDG 6, with 15 occurrences, and SDG 7, with 13 occurrences, underlining their relative prominence in the research database. Within SDG 11, there are several key subtargets that the research addresses: Subtarget 11.3 is shown as a distinct bubble within the larger SDG 11 circle, indicating focused research on the role of artificial intelligence in smart cities. Subtarget 11.2, which aims to provide access to safe, affordable, accessible, and sustainable transport systems for all, is another bubble within the SDG 11 area. This highlights the efforts in research to improve public transport with artificial intelligence in a smart city. Subtarget 11.6 focuses on reducing the adverse per

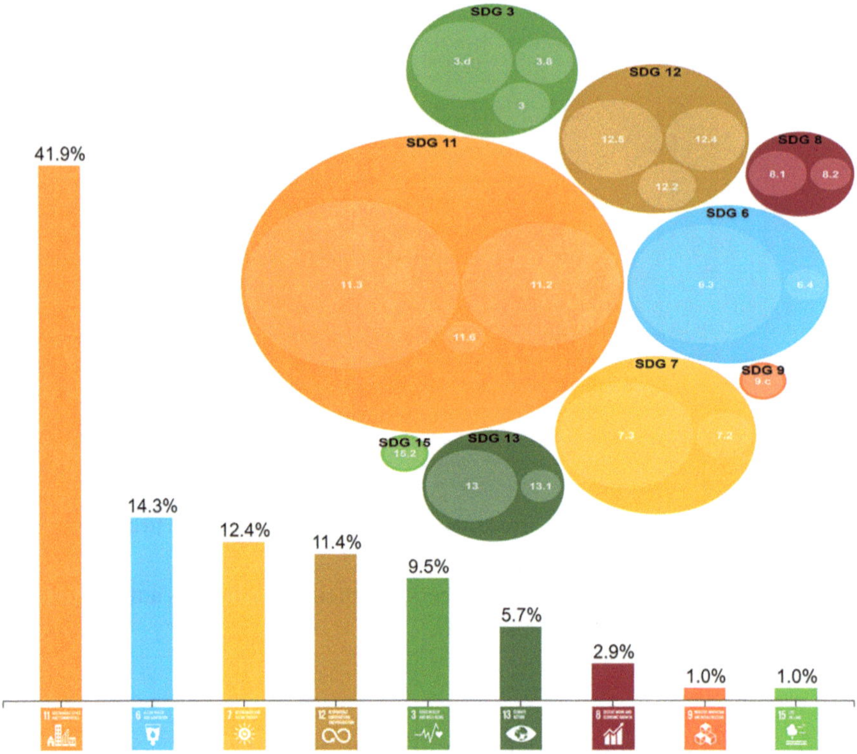

Fig. 4.9 Mapping AI-driven smart city research to relevant SDGs and targets

capita environmental impact of cities in smart cities solutions. Figure 4.9 shows that the size of the SDG 6 bubble implies that advancements in smart cities are being focused on clean water and sanitation that prioritize sustainability. Furthermore, the SDG 7 bubble implies that cities are being focused on affordable and clean energy that prioritizes sustainability. Similarly, SDG 12 (Responsible consumption and production) is also closely linked to SDG 11 and suggests that research should explore how to achieve the environmentally sound management of chemicals and all waste in order to minimize their adverse impact on humans and to achieve the sustainable management and efficient use of natural resources within the context of smart cities. The detailed description of other relevant SDGs is depicted in Table 4.3.

4.3 Urban Analytics and Decision Support Systems with AI

Urban analytics and decision support systems with AI are relatively new and emerging disciplines concerned with applying data analysis, AI, and decision sciences. With the growth of smart cities and new urban development at the national

Table 4.3 SDGs ranking, targets, and relevance in the AI-driven smart city research domain with SDGs

SDGs (Rank)	Description	Targets	No. of Occurrence		Relevance
SDG11	Sustainable cities and communities	11.3	28	44	41.9%
		11.2	15		
		11.6	1		
SDG 6	Clean water and sanitation	6.3	14	15	14.3%
		6.4	1		
SDG 7	Affordable and clean energy	7.3	11	13	12.4%
		7.2	2		
SDG 12	Responsible consumption and production	12.5	6	12	11.4%
		12.4	4		
		12.2	2		
SDG 3	Good health and well-being	3.d	6	10	9.5%
		3.8	2		
		3	2		
SDG 13	Climate action	13	5	6	5.7%
		13.1	1		
SDG 8	Decent work and economic growth	8.1	2	3	2.9%
		8.2	1		
SDG 9	Industry, innovation, and infrastructure	9.c	1	1	1.0%
SDG 15	Life on land	15.2	1	1	1.0%

and global level, the emerging point of interest is how to facilitate the management and planning. Urban analytics and decision support systems with AI, giving a new way to address these challenges. The main focus of urban analytics is about using data for discovering, modelling, and optimizing the components of cities, such as infrastructure, health, transportation, and economy. Hence, by collecting and processing an enormous amount of data from various sources, including satellite images, social media, sensors, and government databases, urban planners, policymakers, and researchers analyze the trends, patterns, and correlations, and then make proper decisions [20]. Furthermore, urban analytics are applied to traffic forecasting, energy usage, or optimal delivery of public services, such as city sanitation and education. Decision support systems are complex software systems that help in the decision-making process since they provide access to data, present data in an understandable format, and even make suggestions. The following description explains that the urban planning and management processes rely on urban analytics and decision support systems using AI. These systems combine the power of data and the application of advanced technologies to help decision makers make more solution-based decisions that solves the problems of cities. As more and more people shift to urban centres and these areas become denser and more complex, the deployment of urban analytics and decision support systems with AI will continue to be increasingly important, which means that in the future, cities will only become even better places to live for all.

4.3.1 Advanced AI Algorithms for Urban Data Analytics

Smart cities are founded on the usage of existing modern technologies in enhancing the lives of citizens and building an efficient urban society. Such changes can be seen among the strategic goals, in which the application of AI and data analytics for improving urban processes and resource management [21]. The use of AI in urban data analysis could be dated back to data availability, computational development, and the emergence of interest in utilizing the resources in decision making for an enhanced intelligent urban environment. Specifically, urban AI-driven data refers to the variety of data, such as sensor data, social media data, traffic data, energy consumption data, and much more. These data sources are about the urban systems, and there are advantages in trying to understand patterns, trends, and abnormalities in the data that benefit decision making. Also, there are ML, DL, and NLP, which are AI algorithms capable of processing large amounts of data to identify previously hidden patterns and correlations. There is a clear need to further enhance the use of AI algorithms in order to enhance the collection of data in urban areas [22]. Firstly, it can be employed to build prognostic models that predicts prospective urban issues. Secondly, it also helps to reveal unused or unnecessarily arranged resources, save energy, and enhance the maintenance of the infrastructure used in production processes and services. Additionally, by increasing the utilization of an AI recommendation system, the benefits of environmentally friendly transport, such as cycling or walking, can be recommended to residents. Furthermore, AI algorithms are useful in identifying the parts of the system that require more energy and those parts that do not use energy efficiently. Thus, smart cities will ensure a correct distribution of energy demand and supply, minimize waste, and minimize CO_2 emissions. NLP algorithms are utilized to analyse data posted on social media or received from citizens, so that city authorities get an idea of how the public feels about certain issues or how they can address the citizens' issues [23]. Besides, these algorithms used in improving the emergency response systems due to the fact of forecasting the effects of natural disasters and other calamities. Hence, AI algorithms offer a great perspective for processing big urban data and creating smart cities. These algorithms gives meaning to large volumes of data, answer important questions, and equip the decision makers with effective tools to build more humane and efficient habitats in cities.

4.3.2 AI-Driven Decision Support Systems for Urban Planning

A smart city uses ICT to optimize the delivery of urban services for the purposes of boosting the quality of life of its people. A major use of smart city technologies is the AI-based decision support systems for urban planning. AI-based decision support systems are systems that utilize AI methods and tools to support the process of evaluating and treating big data and provide advice regarding decision making. These systems have become an essential part of smart cities because of extremely fast processing and handle enormous quantities of data while making intelligent predictions. Use of AI in the development of decision support systems can greatly enhance realistic urban planning by offering city planners and decision makers actual data on the effectiveness and productivity of various planning scenarios and their possible impact on the planning environment. AI-based decision support systems are used to analyse traffic information that are used in traffic control and organizing public transport routes [24]. Environmental applications of AI-driven decision support systems include surveillance of the quality of air, water, soil, land utilization, and control of wastes. Based on data, such as from smart meters and other devices, AI-driven decision support systems are implemented to reduce energy utilization [25]. It projects the energy demand and incorporate renewable energy resources, alongside making recommendations needed to minimize energy usage in smart cities to improve sustainability. Public safety can benefit from AI-driven decision support systems because of the ability of the system to analyse security threats, detect irregularities, and forecast criminal activities. Therefore, AI-synergized decision support systems for urban planning will become even more important in the future as urbanization progresses and technology advances.

4.4 Machine Learning-Based Predictive Maintenance for Infrastructure

Turning to the constantly evolving concept of smart cities, one of the more crucial elements is the integration of the ML-based predictive maintenance solution for critical city infrastructure [26]. The usage of the ML-based predictive maintenance saves time and money, but also removes possible disasters. The use of ML by organizations supports identifying degradation levels of infrastructures such as bridges, roads, traffic signals, or energy networks and makes precise forecasts on condition, thus improving the strain to be able to efficiently schedule maintenance. It employs the previous data to determine traffic flow patterns in order to decide on the repair of roads during times when traffic is light. The early detection of the faults in the bridges also helps ensure that mishaps, as well as expensive, exigent repairs, are avoided. In energy grids, predictive maintenance is really useful for identifying possible spots of vulnerability so as to avoid blackouts and hence improve energy efficiency and reliability. This is not just about cost savings; however, ML-based predictive maintenance made clear that it is about making smart city systems stronger [27]. From this perspective, maintenance has value in the sense that it reduces operational costs, controls environmental effects, secures public safety, and improves the quality of life of the populace. Maintaining and preserving the infrastructure of smart cities to ensure sustainability and development requires innovations, such as predictive maintenance.

4.4.1 Machine Learning Models for Predictive Maintenance

As the technology advances towards smart cities, which would optimize the use of technology to enhance efficiency and the quality of the environment and citizens' well-being, the use of ML models for prediction or predictive maintenance has become relevant. Predictive maintenance, which provides an instance of the application of ML models in real manufacturing, helps to reduce downtime, maintenance costs, and the occurrence of machine failures. These models develop the probability of equipment failure based on the past and present use, and thus, the maintenance workers can be prepared to prevent or reduce the incidence of failure. Some of the ML models that have been used with positive results concerning the field of predictive maintenance include decision trees, support vector machines, artificial neural networks, and deep learning.

4.4.1.1 Decision Trees

A decision tree is one of the most efficient algorithms for approaching the problem of equipment failure prediction. This model works with the construction of a tree, where the decisions taken rely completely on the input data that has to be achieved. The resulting data is then analysed in the tree to make a decision on whether any particular piece of equipment is likely to fail. Decision trees are easily understandable due to graphical representation and are easily interpretable, that is why they are suitable for predictive maintenance [28]. Random forest is a fundamental ensemble learning methods that uses decision trees as its base models and combines multiple trees to reduce overfitting. An individual decision tree might have high variance, which is controlled by Random Forest since it combines the results of many trees that were built on different samples of data, as well as using different subsets of features. The illustration of a random forest is represented in Fig. 4.10. This not only increases the prediction accuracy but also ranks features, which can be useful in the case of classifiers and regressors.

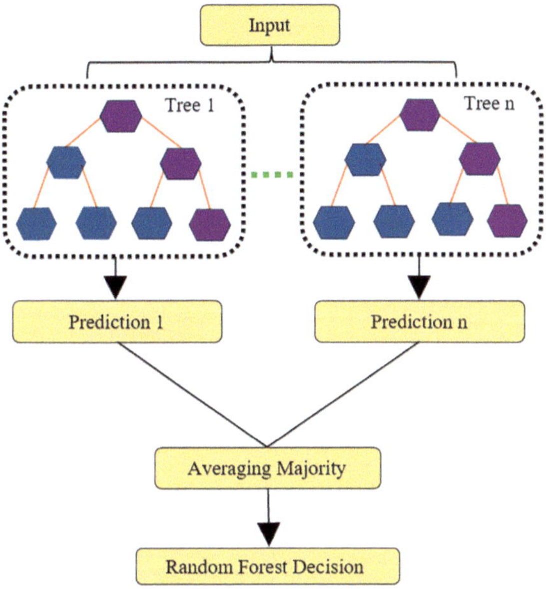

Fig. 4.10 Random forest algorithm

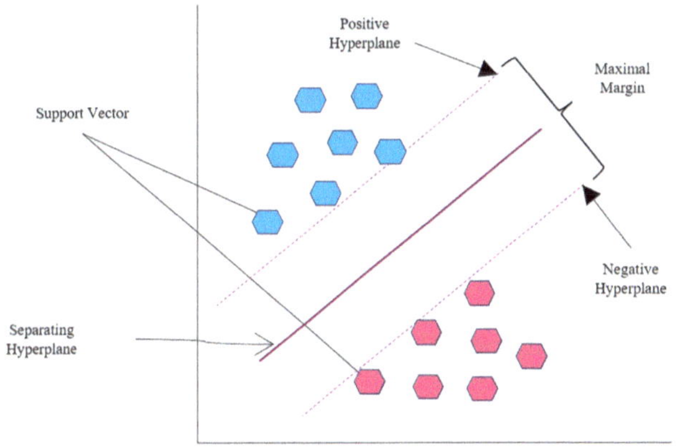

Fig. 4.11 Support vector machine algorithm

4.4.1.2 Support Vector Machines (SVMs)

This algorithm is another widely used ML model that has been successfully applied to determine the efficient scheme of predictive maintenance [29]. It operates under the principle of identifying the most appropriate separation plane between different classes of data, in this case, the normal working condition and the ones likely to cause equipment failure, as depicted in Fig. 4.11. The other key applications of SVMs include high-dimensional data and linear and non-linear problems as well.

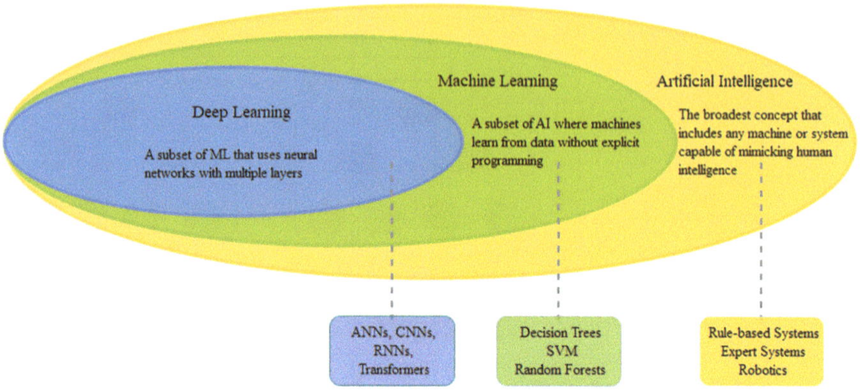

Fig. 4.12 Deep learning, machine learning, and artificial intelligence with key techniques

4.4.1.3 Deep Learning

Deep Learning as a form of a broader category, where 'machine learning' itself is a part of a larger category called artificial intelligence, as represented in Fig. 4.12. There are further ways in which it breaks down ML; one of those techniques is DL, which works with deep neural networks to learn from large and complex data sets for predictive maintenance [30]. One of the characteristics of the models is that there are layers of interconnected nodes, which allows the models to learn some abstract features of the data at an increasingly abstract or higher level. DL has a lot of potential in the field of predictive maintenance, as it will be capable of managing multiple input-output mappings without imposing restrictions on the complexity type and input-output that have to be derived from the raw data.

4.4.1.4 Artificial Neural Networks

It is worth noting that artificial neural networks are a class of ML models that are based on the structure of the human brain. Figure 4.13 represents the model of artificial neural networks. These models come with an inbuilt ability to learn from past data and make new predictions later. Artificial neural networks have been successfully applied to predictive maintenance since they deal well with non-linear interactions between the different variables [31].

Particularly, in smart cities, there are many advantages to applying ML models for the prediction of maintenance. Through condition monitoring, the maintenance personnel are able to recognize equipment failure before it happens and repair it when the demand in the city is low, hence reducing the risk of eventual interruptions. Furthermore, predictive maintenance directly contributes to covering the city's needs in terms of optimizing the budget for maintenance works and choosing

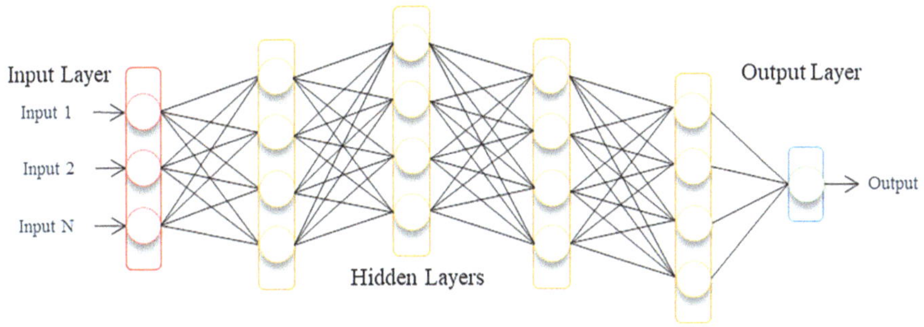

Fig. 4.13 Artificial neural network model

the right priorities. The use of ML models for carrying out predictive maintenance has been identified as a key component in the integration of smart cities.

4.4.2 Monitoring Health System Using Deep Learning Techniques

Health monitoring systems play an important role in ensuring the health of city inhabitants, as well as in the formulation of health policies in order to enhance the health of citizens [32]. These systems monitor different parameters, including patient status, disease incidence, waiting lists, and patients' compliance with medications. The introduction of DL approaches in health monitoring systems has become one of the interesting developments for smart cities to enhance public health. DL algorithms can provide big data to cities to effectively monitor the dispensation of healthcare and prescribe interventions for residents, by which the quality of residents' lives can be greatly improved. DL belongs to that category of supervised learning techniques that uses several layered neural networks to process data. By employing DL algorithms, it is possible to monitor large amounts of data within short periods of time with high accuracy. Some of the most promising applications of DL in health monitoring are disease prediction, telemedicine, personalized medicine, sentiment analysis, and anomaly detection. Disease prediction using DL models is an advanced way of improving the existing healthcare systems [33]. The care of people's health is an important part of the development of smart city technologies, where technology has been adopted to ease human lives. DL models, due to proficiency in handling a large amount of data in unstructured format, can be very useful in predicting epidemics and in managing healthcare systems efficiently. These models process data from patients' records, their social media activity, environment, as well as wearable devices. Thus, analysing such patterns and correlations with the help of DL algorithms predict incidents of disease outbreaks in advance. Figure 4.14 explains how DL models are used in healthcare systems. Input sources like electronic health records, wearable sensors, medical imaging, social

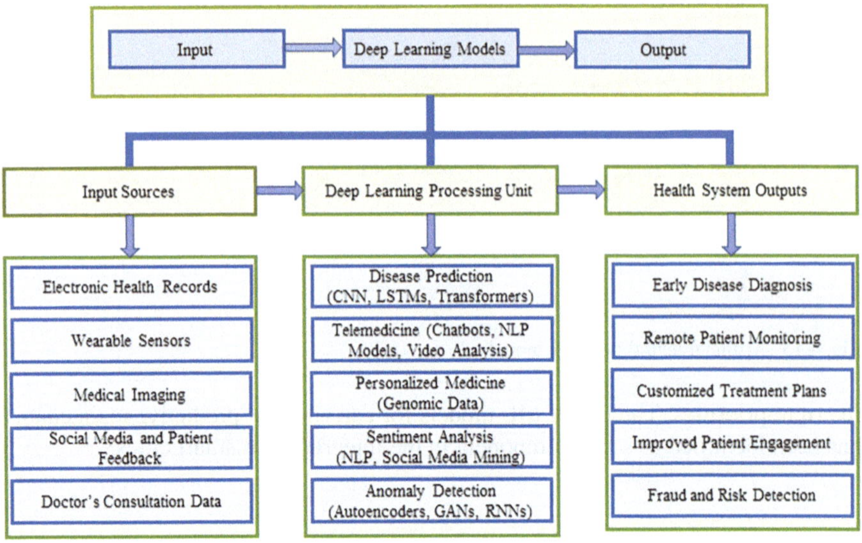

Fig. 4.14 Deep learning framework for healthcare systems

media feedback, and doctor consultation data are fed into DL processing units. These units perform tasks like disease prediction, telemedicine support, personalized medicine, sentiment analysis, and anomaly detection. As a result, healthcare systems achieve outputs like early disease diagnosis, remote patient monitoring, customized treatment plans, better patient engagement, and fraud detection. The flow diagram shows how DL transforms raw medical data into actionable healthcare solutions. Telemedicine for smarter cities, incorporating advanced DL, is slowly and gradually establishing itself as one of the most important branches of health care. The use of telemedicine in smart cities helps in the equal distribution of health care, since all places are provided with adequate medical services regardless of geographical location [34]. In the context of modern urbanization, combining telemedicine and a DL model can be viewed as the development of the healthcare system, bringing the focus to the patient. Personalized medicine is a revolutionary concept in healthcare and deals with the handling of each person on the basis of patient genes, biological makeup, and surroundings. In smart cities, which are equipped with technology and manifold data systems, DL models are vital in helping to implement the concept of personalized medicine [35]. The features of DL allow for the analysis of various kinds of information flows to monitor the health status of a person in real time. The treatment procedure formulated by these models includes genetic imperatives, the physical and social environments to which patients are exposed, and patients' behaviours. It has become increasingly necessary to utilize DL models in self-supervised sentiment analysis, and possibly useful for surveillance of communication platforms such as social media to gauge the public opinion regarding specific health policies [36]. The application of DL feeds reveal anomalies within various health datasets and include potential problems within hospital functioning, patient treatment, or equipment performance. Anomaly detection is the trending technique, with nothing but a radical shift towards interpretive health data analysis to trigger warning impulses of irregularity [37].

The algorithms are particularly well suited to detect patterns with small distances to normal patterns in large, complex health datasets, analyse large amounts of data and detect multilevel infrastructures. This ability is important in ensuring timely ward checks, patient treatment progress, and general efficient performance of equipment used in the hospitals. This helps in increasing the efficiency of the hospital and thus the standard of care by revealing the areas where appropriate change is required.

4.5 AI Applications in Traffic Management and Optimization

The urban centres use information technology to enhance almost all aspects of city operations, including traffic management and optimization [38]. The primary focus of the traffic systems revolution is AI and its diverse uses in making systems more efficient and sustainable [39]. Advanced signal systems, ML traffic models, basic traffic control, and intelligent traffic signals are the essential AI-driven applications in traffic management, as shown in Fig. 4.15. Enhancing traffic signal control is an important way that traffic management can benefit from AI applications. In contrast

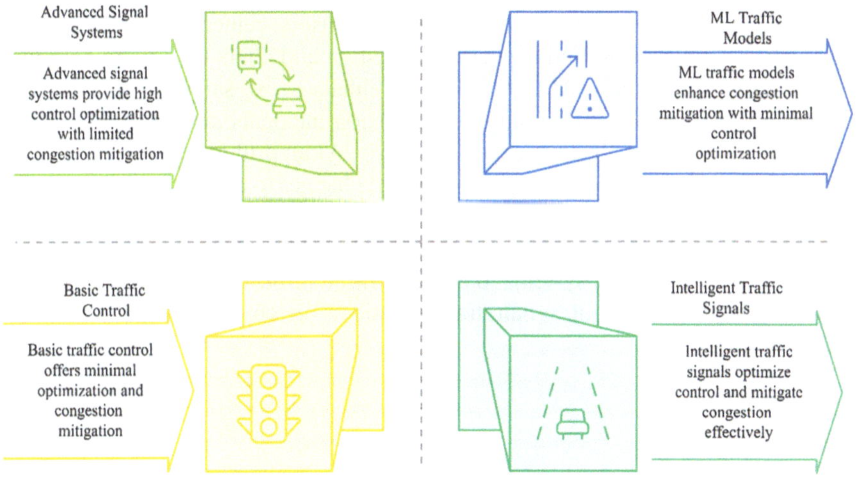

Fig. 4.15 Artificial intelligence applications in traffic management

with active systems that adapt to current real-world traffic conditions, traditional traffic signals work on predetermined timing plans. AI uses real-time data to adjust signal timings, allowing the maximum number of vehicles to pass through an intersection while minimizing waiting times. AI algorithms also predicts the occurrence of congestion and timely alter the signal cycle so as to avert congestion. Another essential feature where AI is used in traffic systems is in analysing and handling incidents. Predictive analytics with an AI application helps with the timing of buses and trains so that whenever they arrive, they will arrive on time [40]. It not only helps in managing the operation of the public transport system but also lowers the number of drivers on the road, helping to reduce traffic jams. In addition, AI can be employed in the strategic decision-making process of the construction of transport networks. The experience gained in providing support for the infrastructure projects puts the expenses where it is needed in the future of the city's growth and development. The applications of AI in traffic control and signal optimization tasks are indispensable for smart cities' solutions. Additionally, AI is being implemented in the mitigation of congestion using ML-based traffic modelling. Eventually, AI involvement in traffic contributes to creating better-quality, efficient, and sustainable cities.

4.5.1 AI-Driven Traffic Control and Signal Optimization

The idea of smart cities is getting popularized, where modern technologies like traffic control and signalization based on AI are used to regulate the traffic flows in the city. The ultimate goal of using AI in a traffic control system is to use data collected from the operational traffic and the signal system to provide real-time traffic control [41]. Advanced signal control, utilizing various sensors, cameras, and data analysis tools, studies traffic flow and expected congestion and alters the timing of signals accordingly. Additionally, AI traffic controls enhance road safety since the system analyses incidents of collision and near-collision incidents and adjusts the traffic light signals to reduce the incidences. Similarly, AI algorithms identify new traffic patterns like construction or bumpers and make changes to the existing signalization to lower the impact on drivers. Therefore, AI-assisted traffic control and signal management are important components of smart cities [42]. Taking advantage of the potential of the AI-based systems to analyse traffic information, such technologies enhance the efficiency of urban transportation, thereby enhancing safety and sustainability.

4.5.2 *Mitigation of Congestion Using ML-Based Traffic Modelling*

There is a challenge today in every big city around the world, and that is congestion [43]. This has led to significant system constraints, particularly in areas with higher population densities, where more people live in towns and the number of vehicles is higher. While traditional cities face such traffic problems as traffic jams, smart cities are capable of solving such problems using more advanced ML algorithms to predict and avoid traffic jams. These models contribute to the improvement of the traffic operation, which consequently improves the quality of life of citizens in urban areas. The traffic modelling based on the applications of the ML algorithm involves traffic patterns, traffic flow history, weather conditions, public events, and road constructions [44]. Combined with predictive modelling, smart cities determine congestion-prone areas within the city and then take measures such as routing, speed humps, and traffic signal adjustments to relieve congested areas. ML algorithms are used for real-time traffic monitoring to solve traffic issues and to control in an efficient way, reducing travel time and thus fuel consumption. Also, ITS can be coupled with these ML-driven architectures to improve its performance criteria. Furthermore, smart cities practicing the application of ML-based traffic modelling could help drivers select the fastest routes for traffic movement. Also, it decreases congested traffic events, as well as time spent and distance covered by car, which leads to a reduction in the emission level to improve air quality. With the help of such highly developed ML solutions, smart cities can redesign their traffic management systems, which, in turn, will contribute to citizens' overall quality of life.

4.6 Monitoring and Management of the Environment Using Artificial Intelligence

The integration of AI in the realization of sustainable monitoring and management of the environment specific to smart cities has proven to be more and more imperative, seeing that population density and city expansion continue to rise [45]. AI has assisted cities in the efficient use of resources, hence enhancing the health of the public and the natural environment [46]. A major implementation area of AI applications in smart cities is the building of an effective and efficient environmental assessment framework. Various sensors, cameras, and other IoT devices measure air quality, noise level, or traffic intensity. Moreover, the field of AI helps with waste and recycling, which, contributes to decreasing the pollution level of cities. By using big data analysis, smart cities are able to minimize waste and sort the waste to avoid causing harm to the environment. Figure 4.16 illustrates the enhanced environmental quality with AI-driven management and monitoring.

4.6.1 Data Analytics and AI-Powered Environmental Sensors

The environmental sensors, also referred to as IoT sensors, constantly collect and send information such as air and water quality, noise, temperature, humidity, and other environmental data [47]. It has become an essential supplement to smart cities as they offer information on the conditions of the external environment, allowing local authorities to make decisions and consequently introduce corrective measures to enhance the quality of life of those living in smart cities. Business analysis is the act of scrutinizing large quantities of unstructured information with a view to identifying important patterns and trends. The support of data analytics is important to smart cities because they involve multiple IoT devices, such as environmental sensors, and assist decision-makers in decision-making based on an understanding of the effects between multiple urban ecosystems [48]. Environmental sensors integrated with AI are expected to bring innovation by providing an optimal approach to some of the most challenging urban issues. Moreover, AI algorithms forecast possible environmental concerns, including a decline in air quality [49] or water pollution, and then the government takes necessary measures before the problem intensifies. In addition, these sensors are applied in combination with other smart infrastructures of an urban environment, including traffic lights and public transport, to enhance the mobility within cities, thus cutting down on the effects of the negative impacts of transportation from one place to another.

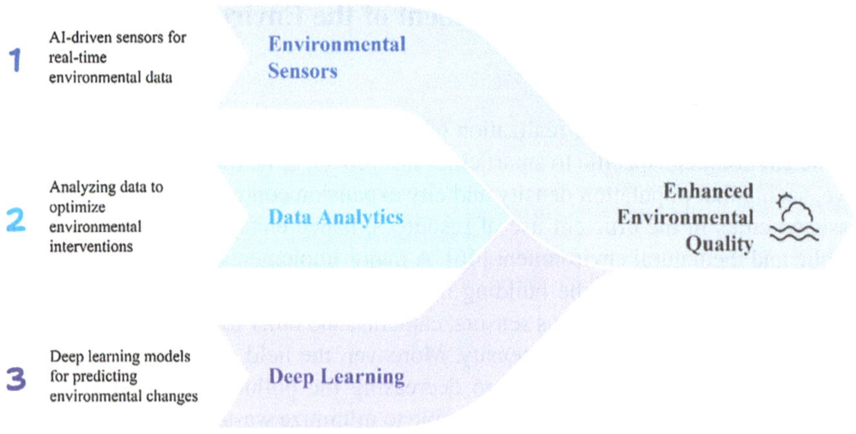

Fig. 4.16 AI-driven environment monitoring and management

4.6.2 *Predicting and Managing Air and Water Quality Using Deep Learning*

With the continuing development of smart cities, air and water quality have emerged as critical parameters that can be responded to and controlled because of direct impacts on the health of citizens and the stability of the environment. With the help of recent developments in DL, there are great possibilities in predicting and controlling the quality of air and water in smart cities [50]. Therefore, the DL algorithms effectively deal with big data that cannot be interpreted by humans in real time and build early warning systems and make efficient decisions rapidly. Multiple interconnected layers or nodes (neurons) that are used in algorithms to forecast air and water quality indicators are based on neural networks being used for data processing and pattern recognition. These networks can, in fact, be trained on millions of past readings and thus give accurate predictions about future air and water quality. Convolutional neural networks are used to analyse the image data from satellite imagery and weather patterns to evaluate and predict air quality, and similarly, recurrent neural networks could be used to analyse time series data from air and water quality sensors to predict trends and high points in the events. Cities deploy reinforcement learning in a city environment to optimize the traffic conditions and reduce vehicle exhaust emissions, thereby enhancing the air quality. In addition, DL algorithms are used for effective water resources to estimate the effects of climate change on water supply and design efficient water supply chain networks that would result in less wastage and pollution [51, 52]. Another application of DL in smart cities is the early warning system for air and water pollution events. Real-world event detection is achieved by DL models, which, when applied in real-time data streams, can signal threats to city administrators such as industrial mishaps, algal poisoning, or adverse weather changes. With such early forecasts, the models are useful in reducing the effects of such occurrences on public health as well as the environment.

4.7 Intelligent Cities and Explainable Artificial Intelligence (XAI)

Globalization has made a huge revolution in the world, and this technology will definitely play a role in the future world. The enhanced rate of urbanization, together with critical issues of the environment, shows that smart cities have developed. Smart city, one of the most significant concepts the world is focusing on, is an intelligent city that is defined to mean actual cities that would provide efficient, safe, and sustainable living environments through the utilization of hi-tech solutions [53]. Intelligent cities come under the broader concept of smart cities, which came into existence in the late 1990s and early twenty-first century. Intelligent cities use revolutionary tools such as AI, IoT, and big data to increase the efficiency of the various sectors, improve the quality of services delivered to the public, and increase the standards of living of the inhabitants. Some of the intelligent cities that the world has today include Copenhagen, Singapore, and Barcelona, with the intelligent solutions that have been implemented in these cities [54], like intelligent transport systems, intelligent energy systems, intelligent water systems, intelligent waste management systems, etc. As the intelligent city applications advance in the implementation of AI, there is a rising demand for improving the interpretability and explainability of those systems. However, that is where XAI comes into the picture. In the context of a smart city, XAI is one important building block in the provision of transparent, interpretable, and understandable AI-based decisions [55]. Figure 4.17 presents a conceptual framework for integrating AI in intelligent cities.

The four components of the framework for applying AI in urban environments. It includes input data sources such as IoT devices, social media, and environmental data. An AI model is used for decision making, including DL and reinforcement learning. XAI techniques to improve AI transparency and interpretability, and intelligent city applications, ranging from traffic management to smart governance. In the transportation system, AI is used in the exploitation of traffic patterns in order to increase traffic flow [56]. By including XAI, the citizens shall be able to know how the AI reaches the result, which shall enable to alter decision making and enhance the system. Likewise, XAI are implemented in energy management systems where the use of AI is for managing renewable resources and minimizing energy consumption [57]. By tuning XAI, it becomes possible for the city officials to know the factors that the AI has used in making a decision, hence guiding them to make the right decisions, hence equal utilization of the energy.

Thus, an intelligent city is the next stage in building smart cities that will enrich people's lives with the help of technologies [58]. The future advancement of technology is definitely going to rely on the aspect of explainable AI, especially in the state making of interpretability and understandability of such systems in intelligent cities. The advantages of XAI in intelligent cities are huge, including public acceptance of smart city technologies and promoting innovations. XAI systems present unique challenges as intelligent cities develop further; thus, cooperation from policymakers, city officials, and technologists will be critical in the implementation of

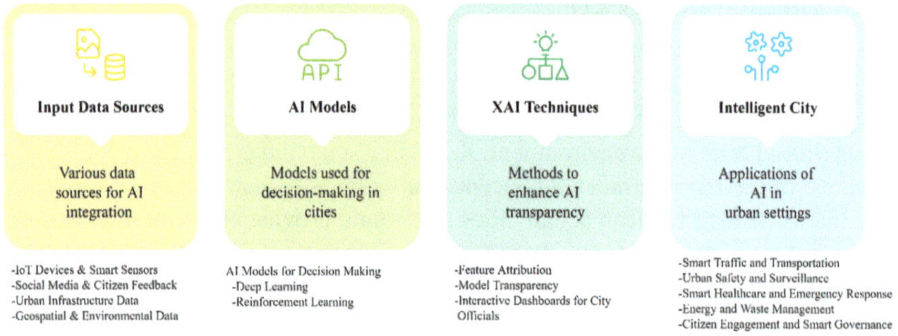

Fig. 4.17 Intelligent cities and explainable AI

XAI systems in these intelligent environments [59]. The use of information technology in various aspects of urban life helps develop better-built environments, making life easier for all city inhabitants.

4.8 Open Research Challenges and Research Directions

The integration of advanced AI into the concept of sustainable smart cities holds immense potential. AI enhances urban living by optimizing resources, improving governance, and ensuring sustainability. However, its application in smart cities presents several challenges that researchers and practitioners must address, as discussed below and depicted in Fig. 4.18.

4.8.1 Data Privacy and Security

Sustainable smart cities rely on AI-driven systems processing vast amounts of data from IoT devices, sensors, and citizen interactions. Ensuring data privacy is challenging, as sensitive information must be protected against unauthorized access and

Fig. 4.18 Open research challenges and research directions on AI-driven sustainable smart cities

breaches. Compliance with regulations like the General Data Protection Regulation requires robust data governance. Additionally, securing data from cyberattacks is critical, as vulnerabilities lead to compromised services. Balancing the need for real-time interpretations with encryption and anonymization methods is complex. The reliance on interconnected systems increases the risk of cascading failures. Addressing these challenges requires privacy-preserving AI techniques and resilient cybersecurity frameworks for urban environments [60].

4.8.2 Data Quality and Integration

Data quality and integration are critical challenges in utilizing advanced AI for sustainable smart cities. Urban datasets are often incomplete, inconsistent, or biased, reducing the accuracy of AI models. Integrating data from diverse sources, such as IoT devices, sensors, and public records, is complex due to differences in formats, standards, and reliability [61]. These challenges hinder real-time decision-making and limit the scalability of AI solutions. Addressing these issues requires developing robust data cleaning methods, standardization frameworks, and efficient integration techniques. Enhancing data quality and interoperability is essential to ensure AI-driven insights align with the goals of smart, sustainable urban development.

4.8.3 Real-Time Data Processing

Real-time data processing is critical for advanced AI in sustainable smart cities, but poses several challenges. The massive influx of data from IoT devices, sensors, and citizen interactions creates a need for scalable and efficient systems [62]. High computational demands, latency issues, and resource constraints hinder real-time decision-making. Ensuring data privacy and security while processing sensitive information adds complexity. Additionally, integrating heterogeneous data streams from diverse sources remains a significant obstacle. Addressing these challenges requires innovations in edge computing, distributed AI frameworks, and energy-efficient algorithms to enable timely, secure, and sustainable data processing for smart city applications.

4.8.4 Scalability and Interoperability

Scalability and interoperability are critical hurdles in implementing AI for sustainable smart cities [63]. The growing population and dynamic urban environments demand AI systems that can scale seamlessly while maintaining performance. Current systems often struggle to process increasing data volumes or adapt to new technologies without significant upgrades. Interoperability challenges arise due to diverse devices, platforms, and standards used across city infrastructures, hindering seamless communication and integration. Standardized protocols, modular AI architectures, and scalable algorithms are essential to address these challenges. Overcoming these issues will ensure robust, adaptable AI systems that support the long-term evolution of smart city ecosystems.

4.8.5 Environmental and Sustainability Concerns

Advanced AI in sustainable smart cities presents environmental challenges, including high energy consumption during model training and deployment. Large-scale AI systems often rely on energy-intensive data centres, contributing to carbon emissions. Additionally, the lack of eco-friendly lifecycle management for AI technologies, from production to disposal, exacerbates environmental harm [64]. Balancing AI's computational demands with sustainability goals is critical, as is addressing the waste generated by obsolete devices and systems. Researchers must focus on developing energy-efficient AI models, utilizing renewable energy sources, and promoting circular economy principles to minimize the ecological footprint of AI-driven smart city solutions.

Future research must prioritize privacy-preserving AI techniques, scalable and explainable models, and lightweight architectures that minimize energy consumption. Advancing green AI initiatives and circular economy practices can ensure AI aligns with sustainability goals. Collaborative, interdisciplinary approaches are essential, promoting partnerships between AI researchers, urban planners, and policymakers. Citizen-centric co-design and the creation of equitable AI solutions bridge the digital divide and enhance societal acceptance. By addressing these challenges, AI empowers sustainable smart cities to thrive in the face of urbanization and climate change.

4.9 Summary

The integration of advanced artificial intelligence into sustainable smart cities marks a significant step toward enhancing urban living through optimized resource management, improved infrastructure, and robust decision-making. Key advancements in urban analytics, such as AI-driven decision support systems and advanced data analytics algorithms, enable more informed urban planning and policy formulation. Predictive maintenance powered by machine learning and deep learning ensures infrastructure durability and efficiency, while AI applications in traffic management address congestion and improve mobility. Additionally, environmental monitoring systems employ AI and DL to manage air and water quality, facilitating healthier urban environments. However, challenges, like data privacy, algorithmic bias, energy consumption, and scalability, hinder widespread implementation. Addressing these requires innovative solutions, including explainable AI to promote transparency and trust, lightweight models for resource-constrained cities, and environmentally friendly AI technologies aligned with sustainability goals. Future research must emphasize cross-disciplinary collaboration, citizen-centric designs, and equitable access to AI technologies to ensure inclusivity. By overcoming these challenges and employing AI's full potential, sustainable smart cities can thrive, addressing urbanization challenges while promoting economic growth, environmental sustainability, and social equity. This vision necessitates continued exploration of open research challenges and emerging directions to build intelligent and resilient urban ecosystems.

References

1. J.P. Ferreira dos Santos, C.A. de Matos, A. Groznik, The role of artificial intelligence in smart city systems usage: Drivers, barriers, and behavioural outcomes. Technol. Soc. **81**, 102867 (2025). https://doi.org/10.1016/J.TECHSOC.2025.102867
2. P. Mishra, G. Singh, Introduction: Importance of sustainable smart city, in *Sustainable smart cities*, (Springer, Cham, 2023), pp. 1–37. https://doi.org/10.1007/978-3-031-33354-5_1
3. D. Rodríguez-Gracia, M. de las Mercedes Capobianco-Uriarte, E. Terán-Yépez, J.A. Piedra-Fernández, L. Iribarne, R. Ayala, Review of artificial intelligence techniques in green/smart buildings. Sustain. Comput. Inform. Syst. **38**, 100861 (2023). https://doi.org/10.1016/J.SUSCOM.2023.100861
4. N.D. Noviati, S.D. Maulina, S. Smith, Smart grids: Integrating AI for efficient renewable energy utilization. Int. Trans. Artif. Intell. (ITALIC) **3**(1), 1–10 (2024). https://doi.org/10.33050/italic.v3i1.644
5. A. Nikitas, K. Michalakopoulou, E.T. Njoya, D. Karampatzakis, Artificial intelligence, transport and the smart city: Definitions and dimensions of a new mobility era. Sustainability **12**(7), 2789 (2020). https://doi.org/10.3390/su12072789
6. B. Fang et al., Artificial intelligence for waste management in smart cities: A review. Environ. Chem. Lett. **21**(4), 1959–1989 (2023). https://doi.org/10.1007/s10311-023-01604-3
7. P. Mishra, G. Singh, Artificial Intelligence for sustainable smart cities, in *Sustainable Smart Cities*, (Springer, Cham, 2023), pp. 119–142. https://doi.org/10.1007/978-3-031-33354-5_6
8. L. Dong, Y. Liu, Frontiers of policy and governance research in a smart city and artificial intelligence: An advanced review based on natural language processing. Front. Sustain. Cities **5** (2023). https://doi.org/10.3389/frsc.2023.1199041
9. I.H. Sarker, Machine learning: Algorithms, real-world applications and research directions. SN Comput. Sci. **2**(3), 160 (2021). https://doi.org/10.1007/s42979-021-00592-x
10. A. Fuller, Z. Fan, C. Day, C. Barlow, Digital twin: Enabling technologies, challenges and open research. IEEE Access **8**, 108952–108971 (2020). https://doi.org/10.1109/ACCESS.2020.2998358
11. L.U. Khan, W. Saad, Z. Han, E. Hossain, C.S. Hong, Federated learning for internet of things: Recent advances, taxonomy, and open challenges. IEEE Commun. Surv. Tutor. **23**(3), 1759–1799 (2021). https://doi.org/10.1109/COMST.2021.3090430
12. A.A. Diro, N. Chilamkurti, Distributed attack detection scheme using deep learning approach for Internet of Things. Futur. Gener. Comput. Syst. **82**, 761–768 (2018). https://doi.org/10.1016/j.future.2017.08.043
13. A. Rahman, V. Srikumar, A.D. Smith, Predicting electricity consumption for commercial and residential buildings using deep recurrent neural networks. Appl. Energy **212**, 372–385 (2018). https://doi.org/10.1016/j.apenergy.2017.12.051

14. A. Sivanathan et al., Classifying IoT devices in smart environments using network traffic characteristics. IEEE Trans. Mob. Comput. **18**(8), 1745–1759 (2019). https://doi.org/10.1109/TMC.2018.2866249
15. Z. Ullah, F. Al-Turjman, L. Mostarda, R. Gagliardi, Applications of artificial intelligence and machine learning in smart cities. Comput. Commun. **154**, 313–323 (2020). https://doi.org/10.1016/j.comcom.2020.02.069
16. J. Zhang, D. Tao, Empowering things with intelligence: A survey of the Progress, challenges, and opportunities in artificial intelligence of things. IEEE Internet Things J. **8**(10), 7789–7817 (2021). https://doi.org/10.1109/JIOT.2020.3039359
17. R. Minerva, G.M. Lee, N. Crespi, Digital twin in the IoT context: A survey on technical features, scenarios, and architectural models. Proc. IEEE **108**(10), 1785–1824 (2020). https://doi.org/10.1109/JPROC.2020.2998530
18. Y. Tian, L. Pan, Predicting short-term traffic flow by long short-term memory recurrent neural network, in *2015 IEEE International Conference on Smart City/SocialCom/SustainCom (SmartCity)*, (IEEE, 2015), pp. 153–158. https://doi.org/10.1109/SmartCity.2015.63
19. S. Borchardt, G. B. Vignola, D. Buscaglia, & M. Maroni, "Mapping EU policies with the 2030 agenda and SDGs-fostering policy coherence through text-based SDG mapping", (2023) https://doi.org/10.2760/110687
20. V. Funda, E. Francke, Artificial intelligence-powered decision support system for operational decision-making in the ICT department of a selected African university. Afr. J. Sci. Technol. Innov. Dev. **16**(5), 689–701 (2024). https://doi.org/10.1080/20421338.2024.2376916
21. T.H. Son, Z. Weedon, T. Yigitcanlar, T. Sanchez, J.M. Corchado, R. Mehmood, Algorithmic urban planning for smart and sustainable development: Systematic review of the literature. Sustain. Cities Soc. **94**, 104562 (2023). https://doi.org/10.1016/J.SCS.2023.104562
22. H.M.K.K.M.B. Herath, M. Mittal, Adoption of artificial intelligence in smart cities: A comprehensive review. Int. J. Inform. Manag. Data Insigh. **2**(1), 100076 (2022). https://doi.org/10.1016/J.JJIMEI.2022.100076
23. A. Hodorog, I. Petri, Y. Rezgui, Machine learning and natural language processing of social media data for event detection in smart cities. Sustain. Cities Soc. **85**, 104026 (2022). https://doi.org/10.1016/J.SCS.2022.104026
24. M. Shulajkovska, M. Smerkol, G. Noveski, M. Bohanec, M. Gams, Artificial intelligence-based decision support system for sustainable urban mobility. Electronics (Basel) **13**(18), 3655 (2024). https://doi.org/10.3390/electronics13183655
25. N. Ma, Artificial intelligence-driven decision support systems for sustainable energy management in smart cities. Int. J. Adv. Comput. Sci. Appl. **15**(9) (2024). https://doi.org/10.14569/IJACSA.2024.0150953
26. G. Sekar, J. Raja, Smart city infrastructure: Predictive maintenance using machine learning. IET Conf. Proc. **2024**(37), 387–392 (2025). https://doi.org/10.1049/icp.2025.0943
27. E. Florian, F. Sgarbossa, I. Zennaro, Machine learning-based predictive maintenance: A cost-oriented model for implementation. Int. J. Prod. Econ. **236**, 108114 (2021). https://doi.org/10.1016/J.IJPE.2021.108114
28. S. Kaparthi, D. Bumblauskas, Designing predictive maintenance systems using decision tree-based machine learning techniques. Int. J. Qual. Reliab. Manag. **37**(4), 659–686 (2020). https://doi.org/10.1108/IJQRM-04-2019-0131
29. R. Langone, C. Alzate, B. De Ketelaere, J. Vlasselaer, W. Meert, J.A.K. Suykens, LS-SVM based spectral clustering and regression for predicting maintenance of industrial machines. Eng. Appl. Artif. Intell. **37**, 268–278 (2015). https://doi.org/10.1016/J.ENGAPPAI.2014.09.008
30. O. Serradilla, E. Zugasti, J. Rodriguez, U. Zurutuza, Deep learning models for predictive maintenance: A survey, comparison, challenges and prospects. Appl. Intell. **52**(10), 10934–10964 (2022). https://doi.org/10.1007/s10489-021-03004-y
31. J. Krenek, K. Kuca, P. Blazek, O. Krejcar, D. Jun, Application of artificial neural networks in condition based predictive maintenance, in *Recent Developments in Intelligent Information and Database Systems*, Studies in Computational Intelligence, vol. 642, (Springer, Cham, 2016), pp. 75–86. https://doi.org/10.1007/978-3-319-31277-4_7

32. P. Rajan Jeyaraj, E.R.S. Nadar, Smart-monitor: Patient monitoring system for IoT-based healthcare system using deep learning. IETE J. Res. **68**(2), 1435–1442 (2022). https://doi.org/10.1080/03772063.2019.1649215
33. S. Xie, Z. Yu, Z. Lv, Multi-disease prediction based on deep learning: A survey. Comput. Model. Eng. Sci. **128**(2), 489–522 (2021). https://doi.org/10.32604/cmes.2021.016728
34. S.Y. Kadum et al., Machine learning-based telemedicine framework to prioritize remote patients with multi-chronic diseases for emergency healthcare services. Netw. Model. Analy. Heal. Inform. Bioinform. **12**(1), 11 (2023). https://doi.org/10.1007/s13721-022-00407-w
35. S. Zhang, S.M.H. Bamakan, Q. Qu, S. Li, Learning for personalized medicine: A comprehensive review from a deep learning perspective. IEEE Rev. Biomed. Eng. **12**, 194–208 (2019). https://doi.org/10.1109/RBME.2018.2864254
36. M.M. Agüero-Torales, J.I. Abreu Salas, A.G. López-Herrera, Deep learning and multilingual sentiment analysis on social media data: An overview. Appl. Soft. Comput. **107**, 107373 (2021). https://doi.org/10.1016/J.ASOC.2021.107373
37. Y.M. Galvão, L. Castro, J. Ferreira, F.B. de Lima Neto, R.A. de Araújo Fagundes, B.J.T. Fernandes, Anomaly detection in smart houses for healthcare. SN Comput. Sci. **5**(1), 136 (2024). https://doi.org/10.1007/s42979-023-02480-y
38. R. Abduljabbar, H. Dia, S. Liyanage, S.A. Bagloee, Applications of artificial intelligence in transport: An overview. Sustainability **11**(1), 189 (2019). https://doi.org/10.3390/su11010189
39. E. Behdadfar, AI-driven urban traffic optimization: A case study of tehran's metropolitan traffic management system, in *8th International Conference On Civil Engineering, Architecture And Urban Planning With An Approach To Urban Infrastructure Development March 15, 2025 Barcelona - Spain*, (2025) [Online]. Available: https://www.researchgate.net/publication/389316781
40. N. Singh, K. Kumar, A review of bus arrival time prediction using artificial intelligence. Wiley Interdiscip. Rev. Data Min. Knowl. Discov. **12**(4) (2022). https://doi.org/10.1002/widm.1457
41. A.A. Ouallane, A. Bahnasse, A. Bakali, M. Talea, Overview of road traffic management solutions based on IoT and AI. Proc. Comput. Sci. **198**, 518–523 (2022). https://doi.org/10.1016/J.PROCS.2021.12.279
42. S.K. Jagatheesaperumal, S.E. Bibri, J. Huang, J. Rajapandian, B. Parthiban, Artificial intelligence of things for smart cities: advanced solutions for enhancing transportation safety. Comput. Urban Sci. **4**(1) (2024). https://doi.org/10.1007/s43762-024-00120-6
43. J. Lu, B. Li, H. Li, A. Al-Barakani, Expansion of city scale, traffic modes, traffic congestion, and air pollution. Cities **108**, 102974 (2021). https://doi.org/10.1016/J.CITIES.2020.102974
44. M. Nama et al., Machine learning-based traffic scheduling techniques for intelligent transportation system: Opportunities and challenges. Int. J. Commun. Syst. **34**(9) (2021). https://doi.org/10.1002/dac.4814
45. M. Al-Raeei, The smart future for sustainable development: Artificial intelligence solutions for sustainable urbanization. Sustain. Dev. **33**(1), 508–517 (2024). https://doi.org/10.1002/sd.3131
46. W. Liu, Y. Xu, D. Fan, Y. Li, X.F. Shao, J. Zheng, Alleviating corporate environmental pollution threats toward public health and safety: The role of smart city and artificial intelligence. Saf. Sci. **143**, 105433 (2021). https://doi.org/10.1016/J.SSCI.2021.105433
47. S.E. Bibri, A. Alexandre, A. Sharifi, J. Krogstie, Environmentally sustainable smart cities and their converging AI, IoT, and big data technologies and solutions: an integrated approach to an extensive literature review. Energ. Inform. **6**, 9 (2023). https://doi.org/10.1186/s42162-023-00259-2
48. A.M. Shahat Osman, A. Elragal, Smart cities and big data analytics: A data-driven decision-making use case. Smart Cities **4**(1), 286–313 (2021). https://doi.org/10.3390/smartcities4010018
49. S. Subramaniam et al., Artificial intelligence technologies for forecasting air pollution and human health: A narrative review. Sustainability **14**(16), 9951 (2022). https://doi.org/10.3390/su14169951
50. L. Zhu, Z.J.B.M. Husny, N.A. Samsudin, H.P. Xu, C. Han, Deep learning method for minimizing water pollution and air pollution in urban environment. Urban Clim. **49**, 101486 (2023). https://doi.org/10.1016/J.UCLIM.2023.101486

51. N. Sarkar, R. Gupta, P.K. Keserwani, M.C. Govil, Air quality index prediction using an effective hybrid deep learning model. Environ. Pollut. **315**, 120404 (2022). https://doi.org/10.1016/J.ENVPOL.2022.120404
52. K.P. Wai, M.Y. Chia, C.H. Koo, Y.F. Huang, W.C. Chong, Applications of deep learning in water quality management: A state-of-the-art review. J. Hydrol. (Amst) **613**, 128332 (2022). https://doi.org/10.1016/J.JHYDROL.2022.128332
53. K. Kourtit, City intelligence for enhancing urban performance value: A conceptual study on data decomposition in smart cities. Asia Pac. J. Reg. Sci. **5**(1), 191–222 (2021). https://doi.org/10.1007/s41685-021-00193-9
54. C.D. Carpentiere, A. Messeni Petruzzelli, L. Ardito, Success factors in smart mobility: a new framework and implications for the EuroMed context from case study of New York, Copenhagen, Singapore, Bari and Barcelona. EuroMed. J. Busi. (2024). https://doi.org/10.1108/EMJB-01-2024-0015
55. A.R. Javed, W. Ahmed, S. Pandya, P.K.R. Maddikunta, M. Alazab, T.R. Gadekallu, A survey of explainable artificial intelligence for smart cities. Electronics **12**(4), 1020 (2023). https://doi.org/10.3390/electronics12041020
56. A.V.S. Madhav, A.K. Tyagi, Explainable Artificial Intelligence (XAI): Connecting artificial decision-making and human trust in autonomous vehicles, in *Proceedings of Third International Conference on Computing, Communications, and Cyber-Security*, Lecture Notes in Networks and Systems, vol. 421, (Springer, Singapore, 2023), pp. 123–136. https://doi.org/10.1007/978-981-19-1142-2_10
57. D.P. Panagoulias, E. Sarmas, V. Marinakis, M. Virvou, G.A. Tsihrintzis, H. Doukas, Intelligent decision support for energy management: A methodology for tailored explainability of artificial intelligence analytics. Electronics (Basel) **12**(21), 4430 (2023). https://doi.org/10.3390/electronics12214430
58. A.R. Javed et al., Future smart cities: Requirements, emerging technologies, applications, challenges, and future aspects. Cities **129**, 103794 (2022). https://doi.org/10.1016/J.CITIES.2022.103794
59. S. Tong, S. Li, Explainable artificial intelligence for urban planning: Challenges, solutions, and future trends from a new perspective. Int. J. Adv. Comput. Sci. Appl. **15**(7) (2024). https://doi.org/10.14569/IJACSA.2024.0150777
60. M. Padhiary, P. Roy, D. Roy, The future of urban connectivity, in *Sustainable Smart Cities and the Future of Urban Development*, (IGI Global, 2024), pp. 33–66. https://doi.org/10.4018/979-8-3693-6740-7.ch002
61. K. Ahmad, M. Maabreh, M. Ghaly, K. Khan, J. Qadir, A. Al-Fuqaha, Developing future human-centered smart cities: Critical analysis of smart city security, data management, and ethical challenges. Comput. Sci. Rev. **43**, 100452 (2022). https://doi.org/10.1016/J.COSREV.2021.100452
62. Z. Lifelo, J. Ding, H. Ning, Qurat-Ul-Ain, S. Dhelim, Artificial intelligence-enabled Metaverse for sustainable smart cities: Technologies, applications, challenges, and future directions. Electronics (Basel) **13**(24), 4874 (2024). https://doi.org/10.3390/electronics13244874
63. A. Hammad and R. Abu-Zaid, "Applications of AI in decentralized computing systems: harnessing artificial intelligence for enhanced scalability, efficiency, and autonomous decision-making in distributed archit applications of AI in decentralized computing systems: harnessing artificial intelligence for enhanced scalability, efficiency, and autonomous decision-making in distributed architectures," (2024). [Online]. Available: https://www.researchgate.net/publication/386219179
64. A.A. Faris et al., Green IoT for eco-friendly and sustainable smart cities: Future directions and opportunities. Mob. Netw. Appl. **28**(1), 178–202 (2023). https://doi.org/10.1007/s11036-021-01790-w

Chapter 5
Blockchain Technology in Sustainable Smart Cities

Contents

P. Mishra, G. Singh, *Sustainable Smart Cities 2.0*,
https://doi.org/10.1007/978-3-032-01102-2_5

5.1 Introduction

Blockchain technology has become a new, powerful tool to create a scientific basis for the creation of eco-smart cities [1]. Internet of Things (IoT) and different smart devices are involved in the operation of smart cities, wherein data processing, storage, and protection are important parameters to address [2]. Blockchain technology provides a reliable and secure solution by sharing and storing data in the blockchain [3]. Using blockchain technology enhances transparency, security, and efficiency within the concept of smart cities. By its consensus, it guarantees the validity of the data adopted from various sensors and devices so as to optimize decision-making and resources. The application of blockchain technology minimizes interconnectivity; intermediate models are not necessary to improve the entire productivity of the city [4]. For example, smart contracts solve the issue of energy usage by controlling exactly how much energy is needed for any specific task, and if there is energy that the system does not need, it can be restored back to the power company. Similarly, blockchain is also used in planning cities to make them more transparent and to record all data and transactions [5]. Thus, the use of blockchain in the sphere of public services also works more efficiently, decreases the level of fraud, and increases the effectiveness of resource distribution. The use of blockchain in the provision of public transport services may help in tracking the transactions in line with the needs of the clients by accomplishing geographical monitoring and payment processing in real-time [6]. Therefore, the application of blockchain in smart sustainable cities can improve several aspects, including energy, planning, public services, data management, financial transactions, cybersecurity, privacy, etc., as depicted in Fig. 5.1. Making use of smart contracts in the distributed infrastructure enables the establishment of a secure, efficient, and transparent system to manage the huge volumes of data emanating from smart cities, hence resource optimality and sustainable development [7].

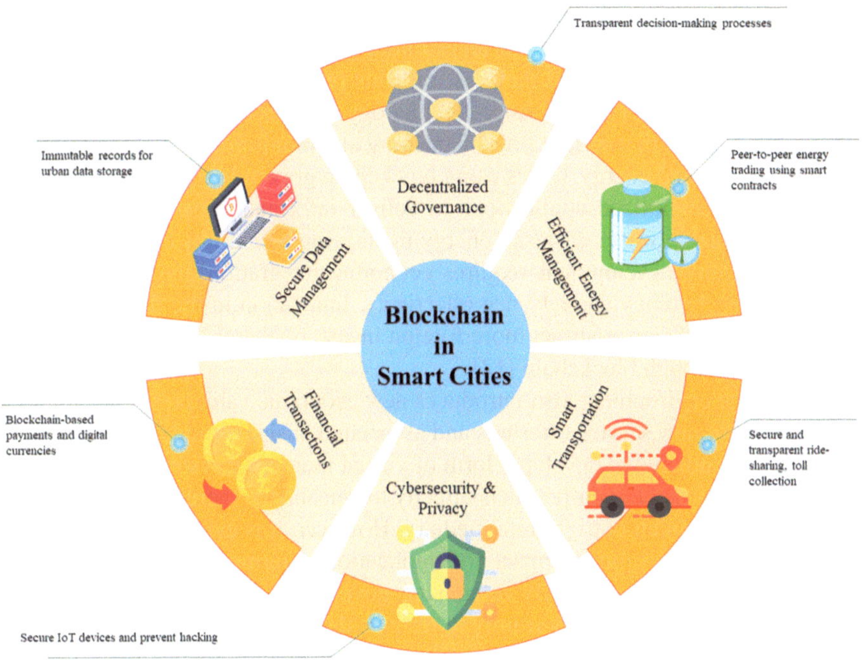

Fig. 5.1 Various aspects of blockchain in smart cities

5.1.1 *Role of Blockchain in Urban Development*

Urban development that aims to infuse smart city practices as sustainable, innovative, and effective has emerged as an interesting field, which involves different stakeholders, including the government, planners, developers, and the community. As a highly technological society is emerging around the world, the need to implement high-tech solutions in the urban environment becomes even more in demand. Undoubtedly, the technology that has the greatest potential in forming future cities is blockchain [8]. With transparency, this technology offers to serve as a solution across various fields, industries, and smart cities. For the purpose of analysing the benefits of blockchain in constructing sustainable smart cities, more consideration should be given to its effectiveness in aspects of governance, infrastructure, resources, and economy. The possibility of recording various transactions or data without the participation of an authoritative centre is the key idea of using blockchain [9]. When applied to the development of urban societies, practical implementation improves the efficiency of government functions by ensuring the records remain accurate. A blockchain technology approach to land registry and property data deliver improvements and reduce instances of fraud. The implementation of distributed registry technology is vital in improving the management of infrastructures in smart cities due to its opportunities in increasing the transparency of supply chains. Since blockchain is an open database, it develop networks that help track, maintain, and update infrastructure such as utilities, transport, and public amenities in real-time. Power management is an essential issue in sustainable smart cities due to increasing populations and the pressures this presents them with in the world's growing environment [10]. Advanced technology, such as blockchain, improves the usage of resources in a certain firm by tracking and monitoring the consumption of utilities, waste, and energy. Cities using blockchain govern IoT devices, smart contracts, and decentralized applications to deliver information-based solutions that ensure sustainable waste disposal, energy management, and low carbon emissions [11]. Smart city development requires economic integration and growth, which blockchain technology can advance effectively. Innovation in the financial system and increased efficiency attract more foreign investors, thereby making the system more secure through blockchain [12].

That same environment also introduces new economic value, resulting from the deployment of new service offerings and associated business models. For example, blockchain is utilized to develop a form of virtual marketing system that will allow local producers to sell their products directly to consumers, thereby providing more local benefits than an outright intermediary. However, it is evident that blockchain has a diverse and profound impact and application in terms of urban development for sustainable smart cities. By refocusing governance, infrastructure, resources, and economy, there is an opportunity to reshape the future of cities with the help of blockchain. So, blockchain technology finds the freedom to revolutionize conventional urbanization for smarter, sustainable cities of tomorrow.

5.1.2 Importance of Blockchain in Smart Cities

Smart cities are huge integrated systems where data needs to be exchanged between sectors such as transportation, utilities, and public services. Blockchain is a technology that achieves a reliable, open, and distributed network for recording and exchanging information on transactions. Thus, it creates a perfect opportunity to build and sustain a secure network for executing smart city functions. Blockchain technology, an innovative product from the basement of cryptography, has gone through a phase of evolution and has become one of the elements of several industries. It is therefore of growing relevance in smart city applications and has possible transformative impacts on these environments. Smart city, therefore, refers to a city that applies Information and Communication Technology (ICT) to enhance the efficiency of services delivered by the government, the welfare of citizens, and the advancement of the economy of a given city. As it has been revealed that all these systems are closely interrelated, it becomes especially important to use secure, open, and effective approaches to data handling and sharing. In smart city applications, blockchain technology also contributes to the development of innovative solutions aimed at improving the living standards of urban populations [13]. Blockchain technology assists in developing a decentralized identity system where record-keeping of personal data can be safely completed without misusing personal information. It also opens the prospect of applying this technology across industries, including energy management, supply chain, and logistics, designing smarter, sustainable, and resilient urban spaces. The significance of blockchain in smart cities is based on the fact that it has the ability to transform the manner in which cities are governed, constructed, and run. Based on the prospects that smart cities will be developed and expanded in the future, the application of blockchain technology in the construction of smart cities will become more and more important [14].

5.2 Current Research Trends of Blockchain in Smart Cities

This study investigates the trends, contributions, and impact of research in smart cities with blockchain technology from 2016 to 2025 through a bibliometric analysis of Scopus data. The Bibliometrix R package, along with Origin software, was utilized to analyse and depict the data. This approach ensures that researchers are presented with the most current research developments in the field, enabling them to stay informed on emerging trends and advancements. In March 2025, a bibliometric analysis was conducted utilizing the Scopus core collection database to thoroughly investigate the concept and understanding of smart cities with blockchain technology. Ultimately, 2478 articles were exported and saved in BibTeX format before being included and examined in the bibliometric study. The primary information, syntax, and specific criteria used in screening for this study are shown in Table 5.1.

5.2.1 Annual Scientific Production and Top 10 Most Published Journals/Cited Papers

Figure 5.2 shows the annual publication trends from 2016 to 2025, highlighting the growth in research output. The early years show a steady increase in publications, rising from 2016 (3 doc) to 2023 (474 doc), indicating a growing interest in the field. There is a slight decrease in the number of documents in 2024 (464). In 2025 (116), the number continues to rise and may continue increasing until the end of the year. Figure 5.3 presents the highest number of articles in the blockchain-based smart city research domain published in "Lecture Notes in Networks and Systems" (72 docs), "IEEE Access" (52 docs), "IEEE Internet of Things Journal" (49 docs), and "Lecture Notes in Computer Science (including subseries Lecture Notes in Artificial Intelligence and Lecture Notes in Bioinformatics)" (49 docs), which indicates a higher output of articles compared to other journals. In addition, some other journals, such as the "ACM International Conference Proceeding Series" (48 docs),

Table 5.1 Summary of data characteristics and main information of the bibliometric study

Description	Specific Standard Requirements/Results
Research database	Scopus
Search query	("smart") AND ("cities") AND ("blockchain")
Document type	Articles, conference papers, book chapters, etc.
Language	English
Data export format	BibTeX
Timespan	2016:2025
Documents	2478
Annual growth rate %	50.1
Document average age	3.16
Average citations per doc	21.61
Keywords plus (ID)	7776
Author's keywords (DE)	4243
Authors	6190
Authors of single-authored docs	157
Single-authored docs	371
Co-authors per doc	3.6
International co-authorships %	31.15

"Communications in Computer and Information Science" (42 docs), "Advances in Intelligent Systems and Computing" (35 docs), "Sensors" (33 docs), "Sustainability (Switzerland)" (32 docs), and "Lecture Notes in Electrical Engineering" (26 docs), are in this domain.

Table 5.2 displays the top ten most cited papers from the study, which presents both the ranking and authors with their published works. This section presents information about authors, article titles, and their corresponding publications in journals with documentation for publication years and citation frequency. The ranking focuses on engineering-field smart cities articles to demonstrate their total citation (TC) counts, annual citation rates, and normalized total citation (NTC) data, which collectively signify their impact in the field.

5.2.2 Sankey Diagram (Three-Field Plot and Thematic Map) in Blockchain with Smart City

In Fig. 5.4, the Sankey diagram illustrates the three-field plot thematic progression to visualize the flow and the distribution in scientific research. In this case, keywords, an abstract, and a title are added, acting as indicators to highlight the most influential research papers on the subject. These keywords indicate vital concepts and themes that stand out in academic discussions of the smart city domain. The diagram's keywords are sourced from earlier published studies and were then thematically organized by a specific researcher. Figure 5.5 categorizes various smart

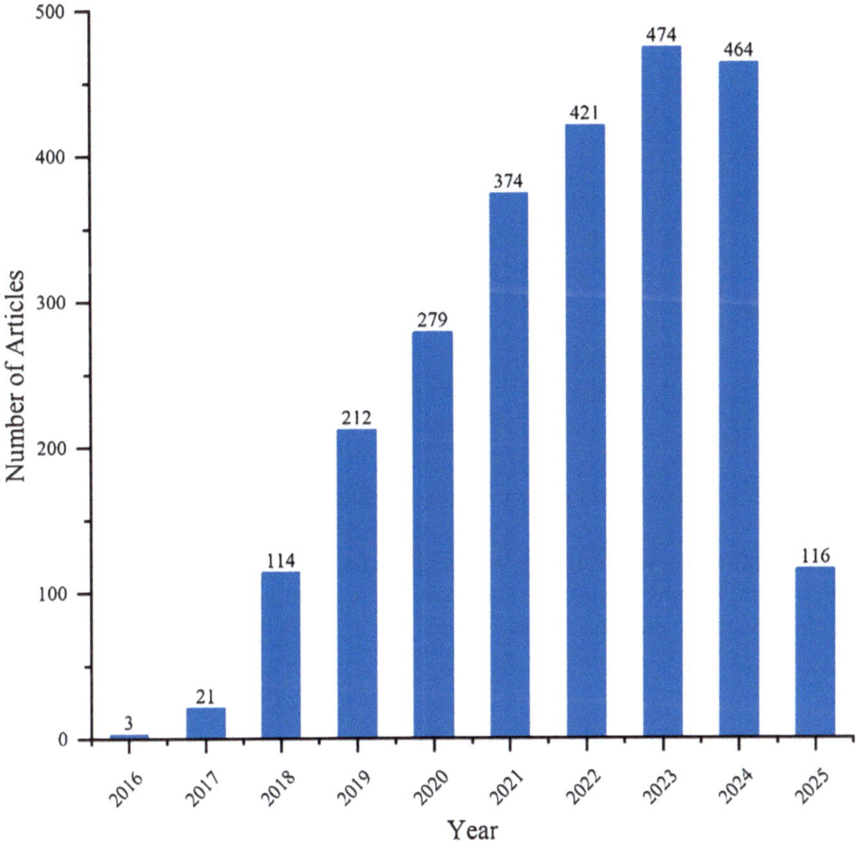

Fig. 5.2 Yearly trend of publication number in smart city research domain with blockchain technology

city themes based on their implementation of blockchain technology and relevant degrees. In the top right quadrant (motor themes), which represents themes that are highly developed and highly relevant, terms like "blockchain" and "smart city" appear. In the left quadrant (niche themes), there are "power market", "energy trading", and "electric vehicles". This indicates a strong focus on these areas, suggesting they are central to smart city research and blockchain technology. This quadrant reflects the areas where significant advances and investments are made, highlighting their critical role in blockchain in smart cities. In contrast, the bottom left quadrant, which represents themes that are either emerging or declining and have lower relevance, includes terms like "intelligent systems", "vehicles", and "intelligent transportation systems". Additionally, terms like "deep learning", "energy utilization", and "machine learning" are the foundational information for further analysis and serve as a base for more advanced analyses. These themes might be considered foundational but must be emphasized more in smart cities. This visual

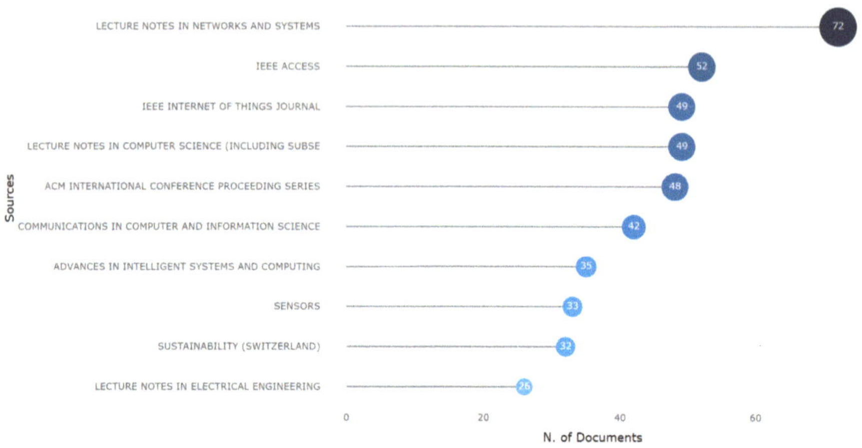

Fig. 5.3 Most relevant sources with the number of articles on blockchain-enabled smart cities

categorization provides valuable insights for educators, curriculum developers, and researchers, helping them understand which areas are gaining traction and which might need more attention or innovation.

5.2.3 Co-occurrence Keyword Trends and Network Visualization

Figure 5.6 depicts the most prominent themes and keywords related to research in blockchain technology in smart cities. The most significant and central terms, such as "blockchain", "smart city", "internet of things", "network security", and "security", indicate the core focus of the research on how researchers grasp and alter their conceptual knowledge of the blockchain technology within smart city contexts.

An illustration in Fig. 5.7 displays word occurrence frequencies from the blockchain technologies in the smart city research network. Research focuses mainly on "blockchain", as this term appears 2245 times, and "smart city" appears 853 times throughout the literature. Research on blockchain in smart cities indicates the importance of "internet of things" based on its 690-word occurrences. Additionally, "network security" (289), "security" (192), "digital storage" (172), "authentication" (136), "information management" (134), and smart contract (133) combinations occupy crucial importance in emerging technology fields. The data shows that blockchain and smart city research revolves primarily around several major subjects.

Table 5.2 Most globally cited documents in the smart city research domain with blockchain technology

Rank	Author's name	Paper Title	Journal's Name	TC	TC/ Year	NTC
1	Khan and Salah (2018) [15]	IoT Security: Review, Blockchain Solutions, and Open Challenges	Future Generation Computer Systems	1954	244.25	26.36
2	Dwivedi et al. (2019) [16]	A Decentralized Privacy-Preserving Healthcare Blockchain for IoT	Sensors	716	102.29	17.32
3	Dagher et al. (2018) [17]	Ancile: Privacy-Preserving Framework for Access Control and Interoperability of Electronic Health Records Using Blockchain Technology	Sustainable Cities and Society	700	87.50	9.44
4	Stoyanova et al. (2020) [18]	A Survey on the Internet of Things (IoT) Forensics: Challenges, Approaches, and Open Issues	IEEE Communications Surveys & Tutorials	630	105.00	15.97
5	Xie et al. (2019) [19]	A Survey of Blockchain Technology Applied to Smart Cities: Research Issues and Challenges	IEEE Communications Surveys & Tutorials	589	84.14	14.25
6	Hammi et al. (2018) [20]	Bubbles of Trust: A Decentralized Blockchain-Based Authentication System for IoT	Computers & Security	586	73.25	7.90
7	Mistry et al. (2020) [21]	Blockchain for 5G-Enabled IoT for Industrial Automation: A Systematic Review, Solutions, and Challenges	Mechanical Systems and Signal Processing	501	83.50	12.70
8	Li et al. (2019) [22]	Blockchain in the Built Environment and Construction Industry: A Systematic Review, Conceptual Models, and Practical Use Cases	Automation in Construction	480	68.57	11.61
9	Banerjee et al. (2018) [23]	A Blockchain Future for Internet of Things Security: A Position Paper	Digital Communications and Networks	460	57.50	6.21
10	Ahad et al. (2020) [24]	Enabling Technologies and Sustainable Smart Cities	Sustainable Cities and Society	434	72.33	11.00

5.2.4 Countries' Scientific Production

The scientific publication frequencies of various countries in the world are represented in Fig. 5.8. Scientific contributions from India stand at 1479, which demonstrates their powerful role in research development. The growing scientific power of China displays itself through its 790 scientific publications, which rank second after

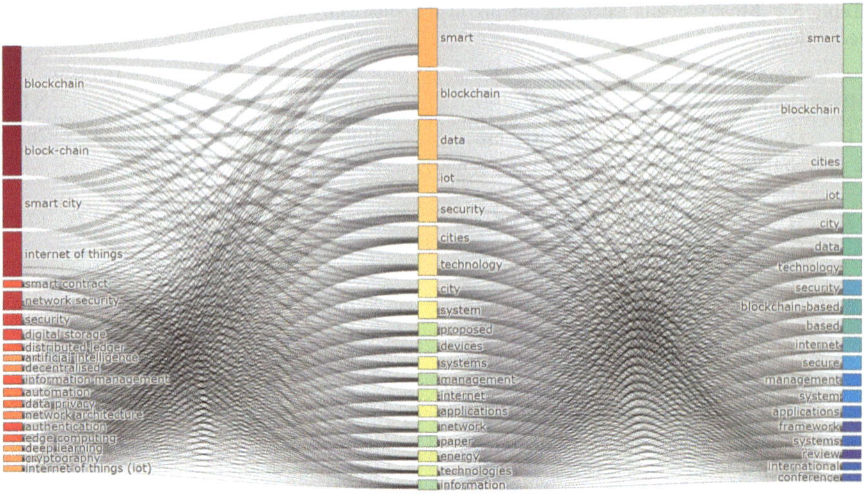

Fig. 5.4 Three field plots (thematic development) using a Sankey diagram in blockchain with smart cities

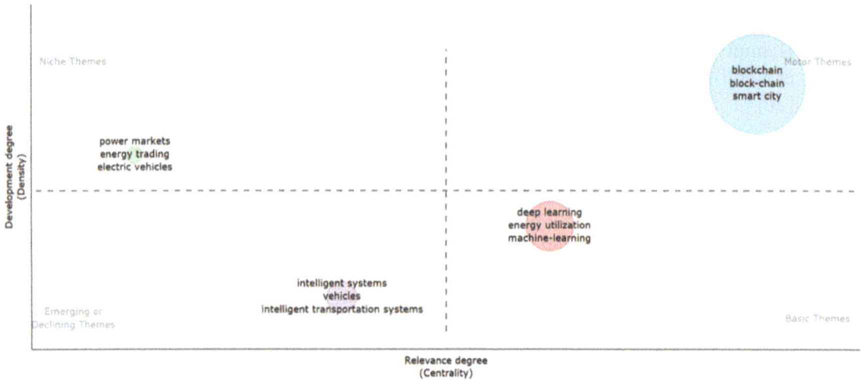

Fig. 5.5 Thematic map of blockchain technology in the smart city research domain

India. The research hub of the USA produces 430 scientific papers among its total contributions. Numerous publications emerge from Saudi Arabia and the UK, together with Australia, as their research output comprises 247, 224, and 169 papers, respectively. Academic and industrial research sectors in Italy and South Korea, produce approximately 161 and 160 scholarly works, respectively. Canada and Malaysia have smaller publication numbers, but their contributions amount to totals of 127 and 104 research items, respectively.

Fig. 5.6 Co-occurrence keyword trends in the smart city domain and sustainable development goals

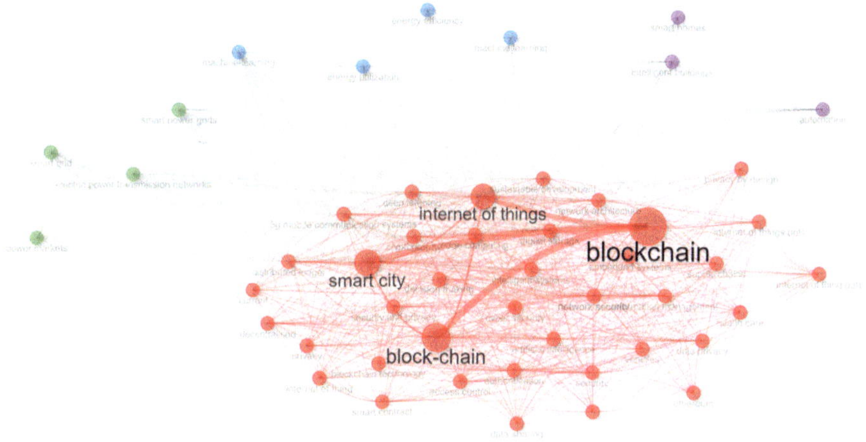

Fig. 5.7 Co-occurrence network of words in the smart city research domain with blockchain

5.2.5 Mapping Blockchain-Based Smart City Research to Sustainable Development Goals (SDGs)

Sustainable Development Goals, their targets (bubble series), and the percentage (bar graph) of corresponding keywords detected in the text (calculated as the ratio of keywords in one goal to the total number of keywords detected) are depicted in Fig. 5.9. For SDG mapping, the study utilized the SDG Mapper tool from the “Knowledge Base for the Sustainable Development Goals”, provided by the European Commission [25]. This tool uses bubble charts to depict SDG

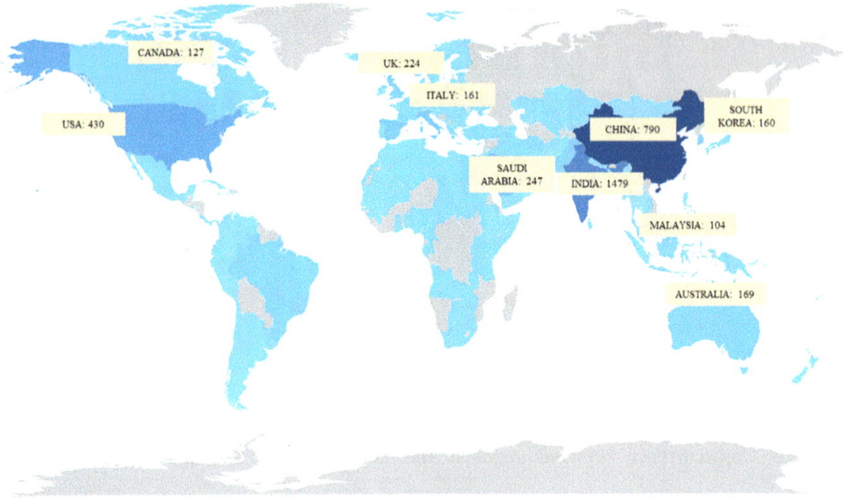

Fig. 5.8 Top 10 countries' scientific production of various countries based on frequency

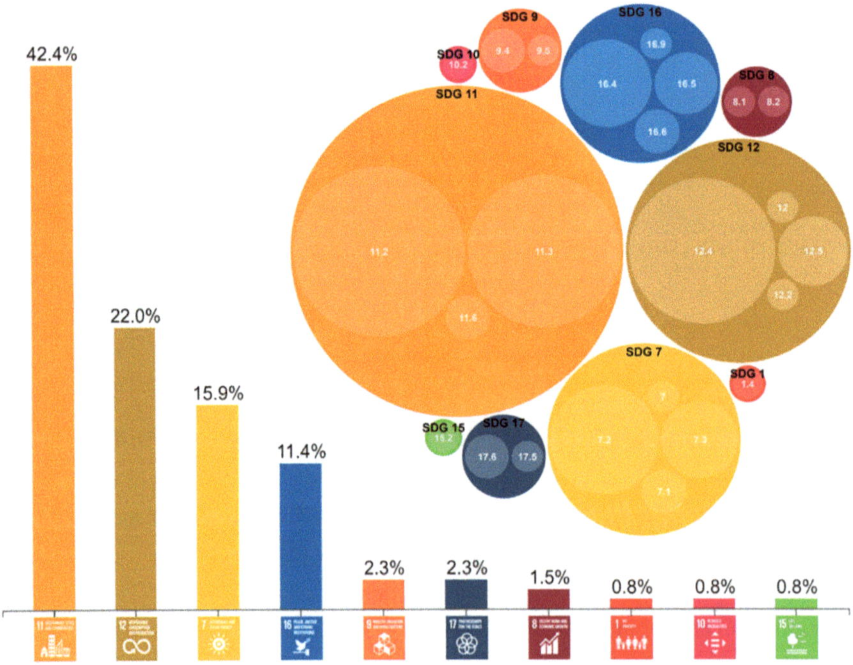

Fig. 5.9 Mapping blockchain-based smart city research to relevant SDGs and targets

interconnections and relevance among bibliometric records. The size of each bubble correlates with the proportion of associated keywords found, which is calculated as the number of goal-specific keywords divided by the total keyword count. As such, SDG 11 is marked by 56 keyword occurrences. This is followed by SDG 12, with 29 occurrences, and SDG 7, with 21 occurrences, underlining their relative prominence in the research database. Within SDG 11, there are several key subtargets that the research addresses: Subtarget 11.2 is shown as a distinct bubble within the larger SDG 11 circle, indicating focused research on smart cities. It is mainly focused on providing safe, accessible, and sustainable smart transportation. Subtarget 11.3, which aims to enhance sustainability, is another bubble within the SDG 11 area. This highlights the efforts in research to improve the cities. Subtarget 11.6 focuses on reducing the adverse per capita environmental impact of cities in smart cities solutions. Similarly, SDG 12 (Responsible consumption and production) is also closely linked to SDG 11 and suggests that research should explore how to achieve the environmentally sound management of chemicals and all waste in order to minimize their adverse impact on humans and to achieve the sustainable management and efficient use of natural resources within the context of smart cities. Furthermore, the size of the SDG 7 bubble implies that advancements in smart cities are being focused on affordable and clean energy that prioritizes sustainability. The detailed description of other relevant SDGs is depicted in Table 5.3.

5.3 Blockchain Technology Fundamentals

Blockchain technology has disrupted the methods and patterns that are used to conduct business and communicate through the internet. This new generation technology makes a good basis for a secure, decentralized, and approved form of transaction, eliminating the use of third parties like banking institutions, which reduces the speed of carrying out transactions. Blockchain is the system that supports all cryptocurrencies that exist today, including Bitcoin [26], Ethereum, and others. This section focuses on the three major aspects of blockchain technology: distributed ledger technology, cryptography, consensus mechanisms, and public, private, and consortium blockchains, as illustrated in Fig. 5.10.

5.3.1 Distributed Ledger Technology

Distributed Ledger Technology (DLT) is a digital system for recording transactions of assets in multiple computers or nodes in a network. This is a constituent of blockchain and guarantees that everyone has an identical, updated, and synchronized view of the data [27]. It has been identified to offer an improved, secure, and simplified method of performing transactions without the use of agents. The biggest benefit of DLT is that it decentralizes the system, making it hard for one entity to alter

Table 5.3 SDGs ranking, targets, and relevance in blockchain-based smart city research domain with SDGs

SDGs (Rank)	Description	Targets	No. of Occurrence		Relevance
SDG 11	Sustainable cities and communities	11.2	30	56	42.4%
		11.3	24		
		11.6	2		
SDG 12	Responsible consumption and production	12.4	22	29	22.0%
		12.5	5		
		12.2	1		
		12	1		
SDG 7	Affordable and clean energy	7.2	12	21	15.9%
		7.3	6		
		7.1	2		
		7	1		
SDG 16	Peace, justice, and strong institutions	16.4	8	15	11.4%
		16.5	4		
		16.6	2		
		16.9	1		
SDG 9	Industry, innovation, and infrastructure	9.4	2	3	2.3%
		9.5	1		
SDG 17	Partnerships for the goals	17.6	2	3	2.3%
		17.5	1		
SDG 8	Decent work and economic growth	8.1	1	2	1.5%
		8.2	1		
SDG 1	No poverty	1.4	1	1	0.8%
SDG 10	Reduced inequalities	10.2	1	1	0.8%
SDG 15	Life on land	15.2	1	1	0.8%

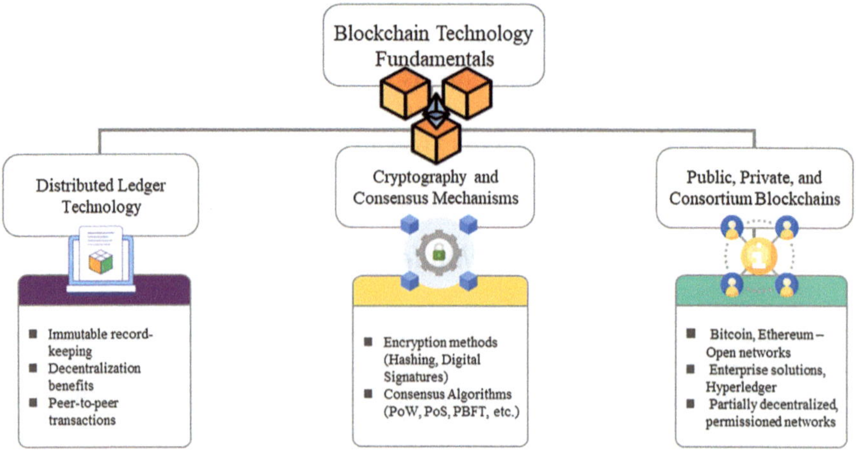

Fig. 5.10 The fundamentals of blockchain technologies

or contaminate the information. However, the data is spread among many computers, and any changes to the data are made, recorded, and validated by those in the network. This implies that any transaction that is posted on a blockchain cannot be changed or modified, making it very secure. DLT, as evidenced by blockchain, is changing the outlook towards sustainable smart cities through a secure, efficient, and transparent management of city systems [28]. In the context of smart cities, DLT's decentralization, immutability, and interoperability offer transformative solutions to the complex challenges of urban sustainability. It is the modern concept of cities that relates to the use of information and communication technology (ICT) to create efficient infrastructure for services and human life in delivering services to the inhabitants sustainably.

These objectives may be supported by DLT in the crucial aspects: distribution of energy grids, smart transport systems, and government transparency. Building on the principle of handling transactions safely in a secure manner, DLT helps to make records safe and free from any form of alteration, hence increasing the level of trust among the stakeholders when it comes to collaborative management of the city. Energy management is one of the most significant areas in which DLT provides service delivery in sustainable smart cities [29]. The smart grid that is based on cryptocurrency enables the inhabitants to trade electricity between themselves, for example, using the excess energy produced from solar panels. Consequently, through the establishment of smart contracts in transactions, energy transactions are made efficient, and, as consumers embrace efficiency, there is sustainable energy consumption [30]. Blockchain helps bring the different modes of transport onto a single platform of transport, hence enhancing availability and minimizing traffic jams. Regarding that, decentralized systems could enable operations of ride-sharing services, bike rentals, and public transportation in the most efficient way, with fewer emissions involved [31]. The level of visibility over the use and maintenance of the vehicles also helps in enhancing the asset life while reducing the wastage of resources. Another potential application is in waste management—optimization of recycling, disposal, and other related processes with the help of DLT systems. Similarly, with the smart city concept, the principles of governance also use DLT. Thus, by introducing the blockchain solution, citizens get an opportunity to engage in decision-making processes within local governments [32]. Furthermore, DLT has possibilities for integration across multiple sectors, and that is conducive to the achievement of sustainable goals. Combining several cities with blockchain systems will create a mutually integrated urban environment, energy, transportation, and even waste disposal. This makes sure that optimum use is made of the available resources, as well as ensures sustainability in the development of the infrastructure.

5.3.2 *Cryptography and Consensus Mechanisms*

There are two very important aspects when it comes to safeguarding a blockchain network and also making sure it is efficient in cryptography and consensus [33, 34]. Cryptography is one of the most significant components of blockchain, as it is responsible for handling the security of the network. It helps to protect the identities of the transacting parties and the transactions they perform. In cryptography, different processes or techniques are used to encode or encrypt the information in such a manner that the data cannot be intercepted or altered by unauthorized people. As for the security of transactions in the context of blockchain, technologies such as hashing, digital signatures, and encryption are used [35]. Hashing is an operation of taking large data and formatting it into a fixed-length portion of characters and strings, making it almost impossible to determine the original data. Digital signatures are used as proof that the transaction is original, while encryption is used to safeguard the contents as they are transmitted through the network. A consensus mechanism is a specific procedural way commonly implemented in a blockchain system to approve an action with no centralized control. Its mechanism ensures that all nodes on the network agree to a consensus on any transaction and ledgers. Consensus algorithms are used in blockchain and are implemented through respective consensus models. Proof of work (PoW) and proof of stake (PoS) are two popular consensus algorithms [36]. The technology that facilitates building sustainable smart cities is built on cryptography and consensus mechanisms [37]. This is because cryptography provides privacy and confidentiality to data transmitted through the networks in the city through encryption. This security is important to guard the rights of publicity of each citizen while enabling the adoption of digital transactions, which are relevant in smart city functions like energy, transport, and service delivery. However, consensus mechanisms, which operate at the basis of a decentralized system, make it possible to reach an acceptable consensus as to the content of a shared ledger without using centres of coordination. Such decentralization enhances the data credibility because no individual or company can manipulate the data. This efficiency is geared towards the achievement of the smart city environmental goals. In other words, the smart cities' digital environment relies on cryptography and consensus mechanisms to attain secure performance.

5.3.3 *Public, Private, and Consortium Blockchains*

Public blockchains, private blockchains, and consortium blockchains, are distinguished by specific features [38]. Bitcoin and Ethereum are public blockchains; they are just peer-to-peer networks that anyone can join [26]. The architecture of these blockchains is open source, which means that anyone can join the network and transact and that no permission is required. There are many properties that make public blockchains highly transparent, secure, and practically tamper-proof. Public blockchains play a crucial role in the creation of a sustainable smart city, since they ensure decentralized, transparent, and secure solutions. The aspect of sharing resources in an efficient way, and this makes the data of the stakeholders effective and reliable. It also involves smart contracts in automating, efficiently and transparently, the delivery of services in the affairs of a municipality. Furthermore, the use of blockchain-based platforms also has the added advantage of increasing traceability for sustainable products in a supply chain. Decentralization of decision making in a public chain stimulates community engagement for sustainable solutions and, therefore, stimulates the creativity and adaptability of smart cities. A private blockchain is always a permissioned network; only the approved parties are allowed to join the network, as well as enter into transactions with others. These blockchains are utilized by large organizations to manage their intra-organizational affairs more effectively and to increase security and efficiency [39]. Decentralized business networks are more centralized, operating at faster and cheaper rates compared to public blockchains, making them more appropriate. Private blockchains help in rolling out efficient and sustainable ideas of smart cities through secure management of data. Through the use of private blockchains, smart cities are able to have controlled decentralization that is necessary for the data integrity and privacy that are always required when collecting data from citizens. Blockchains enhance resource utilization, reduce energy costs, and improve operational efficiency.

Consortium blockchains are considered to be an intermediate between public and private blockchains, with the best elements of both incorporated [40]. These blockchains are private networks that are operated and managed by two or more organizations or entities. They have been designed to offer the same features of a private blockchain, like control, security, and the efficiency of a public blockchain, along with the same transparency and trust as a public blockchain typically offers. However, they imply massive coordination of participants, and hence, they are not fully flexible and not entirely decentralized like other blockchain networks. They play a central role in sustainable smart cities' development, since they bring together multiple parties, which include government, utility firms, and inhabitants of a city [38]. These blockchains are less open, as they provide controlled access and privacy, and are suitable for managing sensitive information from urban areas. They harness the IoT devices to drive smart grids, waste management, and transportation systems for sustainability. Blockchain technology is expected to have a greater impact throughout its different sectors and industries as it progresses through its growth cycle.

5.4 Blockchain-Enabled Smart Cities

Smart cities based on blockchain democratize the living environment through a decentralized, transparent, and highly secure platform. Blockchain smart cities employ the IoT, distributed ledgers for record keeping, and smart contracts to make more efficient decisions. Therefore, the use of blockchain technology in smart cities leads to sustainability, economic development, and social inclusion without compromising the privacy of citizens and continual innovation. With the advancement of technology, the idea of the IoT, where cities are developed and transformed into smart and more resilient systems, has risen immensely. Figure 5.11 illustrates blockchain technology across various applications of smart cities.

Fig. 5.11 Blockchain technology in various applications of smart cities

5.4.1 Blockchain for Government Security and Transparency

Smart cities have been developed to enhance the standards of living in urban environments through the application of ICT across different domains, including education, health, transport, energy, etc. These cities capture, analyse, and disseminate enormous quantities of data. However, data security and data transparency are still issues of concern. Therefore, blockchain provides a crucial solution for managing smart cities through the establishment of safe, clear, and efficient blockchain infrastructures. Blockchain technology uses cryptography to originate details and disperse information throughout the system, making sure that it is available and unalterable. The utilization of blockchain technology in smart cities helps in governmental security [41]. Blockchain technology creates a strong foundation for keeping government data and processes secure. One of the beneficial application areas of blockchain is its usage in government departments like law enforcement or intelligence communities, where large volumes of information can be securely stored and exchanged between departments and agencies in an unmodified manner. For example, it can be applied to help and safeguard election processes. By building a clear and easily traceable record, citizens can check the authenticity of voting results [38]. In addition, blockchain may be adopted by governments to enhance the handling of papers, birth certificates, driver's licenses, and even passports.

Cryptographic security is helpful in smart cities by improving government security and increasing the clarity of governmental processes. Due to the underlying mechanism, blockchain has all the potential to develop transparency, decentralization of information, and high-level data security to reinforce trust among citizens, businesspeople, and government administrative units. Therefore, the application of blockchain in smart cities bring about a level of security and trust for city governance.

5.4.2 *Waste Management via Blockchain Integration*

Garbage disposal poses a major problem as the population has risen exponentially, and there is increased urbanization in society. Today, smart cities, with the help of new innovative technologies like blockchain, seek to implement efficient and sustainable waste management. Recently, it was stated that the implementation of blockchain with waste management systems has numerous advantages that potentially transforms waste disposal in smart cities. Blockchain technology solves many of the issues associated with the traditional waste management system [38]. Moreover, using blockchain technology facilitates the traceability of various waste materials and their handling at the end of their lifecycle. Smart cities can use blockchain in a specific way to create a database for tracking waste materials and their further management. This provides an aid in increasing efficiency in the collection and sorting of recyclable materials, management of waste by reducing landfill disposals, and uptake of the circular economy. Blockchain can also be applied to promote public participation in waste management at a greater level [42]. It also increases the effectiveness of services that deal with waste in that it increases the capacity of the community. Further, it can help solve waste-to-energy management, where waste can be turned into electricity or fuel to support the green agenda of smart city systems. However, the use of blockchain in smart cities' waste management is more of a challenge given the scalability and interoperability of the blockchains involved, as well as governance [43]. Cities need to create the proper systems and provide the suitable specifications, which means engaging fabrication, local authorities, waste management companies, and technology suppliers. Therefore, smart cities promote improved, long-lasting solutions in waste management with the help of blockchain technology to enhance environmental standards and the general well-being of urban inhabitants.

5.4.3 *Public Transportation Systems Through Blockchain*

Blockchain systems have the potential to significantly enhance transportation in smart cities by making it faster, more secure, and transparent [44]. Blockchain is an open, distributed digital ledger that ensures secure and efficient recording of organizational transactions. Through the usage of blockchain, smart cities completely transform how they address operational issues, tracking, and service delivery of public transportation systems for populace. The advantage of using blockchain in public transportation is that this technology provides strong protection. This is especially the case for cards designed for payment and access control for such services as transport tickets and passes. The existing use of blockchain in public transportation systems is also a solution for ticketing and payment processing [45]. Blockchain can also be applied to the enhancement of the public transport system in a smart city as well. Through accessing traffic and parking information as well as transportation data from various sources, the specific areas of congestion, parking difficulties, and public transport demand are accurately identified by city authorities [46]. The information obtained from the analysis is applied to the optimization of routing, scheduling, and resource management in order to enhance transportation services. Therefore, the integration of blockchain technology in public transportation systems is a great opportunity in smart cities. Increased security, reforms in ticketing and payment systems, improvements in transportation services, and collaboration among stakeholders raise the use of blockchain in transportation, making systems more efficient and user-friendly. With the progress of the smart cities concept and the application of new technologies, the importance of blockchain in public transport will increase and involve many benefits for citizens and city authorities.

5.4.4 Trading and Managing Energy with Blockchain Solutions

Energy is one of the most important resources that smart cities should coordinate since it influences nearly all spheres of human activity. Based on the growing importance of decentralized digital platforms for energy trading and management in smart cities, the concept of blockchain deserves special attention [47]. Another application is the development of Peer-to-Peer (P2P) energy trading markets, where producers and consumers of green energy transact directly and sell excess energy to others, instead of depending on large centralized energy companies [48]. In addition, a blockchain-based energy management system enables demand response programs, in which electricity usage is varied in response to the prevailing availability of electricity. As earlier stated, blockchain may be used as a way of encouraging consumers to decrease energy demand during peak hours in order to cut costs and, at the same time, encourage the use of renewable energy. Smart contracts are employed to bring more transparency and security to energy trading by allowing blockchain technology to record all energy trades and usage in an unalterable register. With such transparency and accountability, fraudulent activities may be dealt with effectively, and there is a high likelihood of developing higher levels of trust among participants in the energy market. Smart cities involved in energy trading and management through blockchain technology have a wide range of benefits, such as relative energy efficiency, reduced costs on energy management and infrastructure, and better resilience of the energy grid [19].

5.5 Applications of Blockchain in Smart Cities

5.5.1 *Smart Healthcare*

The integration of blockchain technology in smart healthcare systems in smart cities provides a new, innovative solution to address the handling of health data [19]. Blockchain's distributed and unalterable database provides strong solutions for several issues that affect current systems, data privacy, data sharing, and protection. Blockchain helps bring efficiency and security to the delivery of healthcare because it is, in fact, a key part of smart city technology that involves the integration of technology and a large flow of data. It ensures that records of patients can be shared securely between healthcare providers, patients, and insurance companies so that the records remain easy to obtain but fully protected against unauthorized access. Smart contracts are used to automate the claims and insurance reimbursement processes so that transactions are done quickly and effectively. Such an action not only saves on the costs required to employ many administrative officers but also provides the patients' quicker access to services. Decentralization is made possible by blockchain and ensures that the huge amount of health data, such as the genomic data, is integrated and analysed as well [49]. Moreover, blockchain is used in clinical trials without compromising the security or integrity of the data through proper tracking and promoting new treatments. As blockchain solves the existing healthcare problems, it guarantees the protection of data, increases the efficiency of work with a patient, and creates the basis for a new step in the development of smart cities [50]. Thus, blockchain has a great possibility to transform smart healthcare for smart cities.

5.5.2 *Smart Transport*

Smart transportation systems in smart cities use blockchain technology for effectiveness and security [19]. In urban transport networks, blockchain technology are used to bring significant changes because of its infrastructure. The main application is in the facilitation of secure payment systems. Blockchain ensures that passengers are able to switch to the use of tokens or cryptocurrencies as a means of paying for the services of public transport, hence eliminating the use of cash. Smart contracts have the ability to manage the fares so that every transaction will be affected immediately and be transparent to the service providers as well as the users. Also, supply chain management is made more transparent through this technology. As vehicle maintenance and logistics are provided with an authentic record of vehicle repairs, thus making vehicles safer and more reliable. From the perspective of traffic management, blockchain enables one to build distributed systems of real-time information exchange. This makes it possible to poll traffic information from various sources, like sensors and GPS devices, and transfer the information in the network without the necessity of a central coordinator, besides ensuring traffic data authenticity and minimizing traffic intensity. Moreover, blockchain technology is also applied to the Vehicle-to-Everything (V2X), which enables any vehicle to share information with other vehicles and physical infrastructures [51]. This is essential in providing control for motion and improving security in the organization of transport systems. Additionally, the blockchain assists car-sharing applications and services by helping to automate the process of identifying dishonest participants and their attempts to provide false information on the vehicle and driver and optimizing the process of fair compensation. Thus, blockchain in smart transportation systems in smart cities can be viewed as a potential way to achieve the commonly stated goals of making transport systems more sustainable, efficient, and safe for smart cities of the future, with higher adoption of sustainable development concepts [52].

5.5.3 Smart Energy

Smart energy systems, as part of smart cities, are an improvement that has emerged through the implementation of blockchain technology [53]. When it comes to overcome major issues of sustainable development with an aim to help manage resources to enhance the quality of life for urban citizens, blockchain becomes one of the most important innovations of smart cities, designed to change energy systems [19]. One of the potential applications is peer-to-peer (P2P) energy trading, which is the trading of energy and electricity between various consumer parties. In blockchain, the trading between buyers and sellers is done directly with no interference of third parties, and hence it is applicable for energy. In addition, blockchain helps to improve transparency in energy consumption as well as production indicators, which then be used in monitoring and billing. Smart contracts refer to an automated system where the terms of a contract are coded and are self-executing, thereby lowering regulatory compliance and overhead costs. It plays a significant role in the control of distributed energy resources (DERs) and microgrids otherwise associated with smart cities, because these smart systems need real-time data handling for optimal performance. The application of smart energy systems that rely on blockchain in smart city management systems enables the improvement of energy efficiency, reduction of carbon emissions, and creation of viable trading systems for the economy [54]. The use of blockchain in energy transactions, therefore, enhances a more secure, transparent, and thereby sustainable and resilient urban energy market, which is the focus of smart cities.

5.5.4 Smart Governance

Smart governance in smart cities can be brought in through the application of blockchain technology to improve transparency, efficiency, and citizen participation [19]. It is used in smart governance for effective management of the administrative practices and the provision of public services so that corruption can be eliminated through the removal of unnecessary bureaucratic barriers. It is adopted in the management of land registrations in ways that the ownership of the property is recorded effectively, hence effectively reducing negative activities such as fraud and disputes. This also helps to involve property business transactions as it enhances the confidence of the public in governmental institutions. Furthermore, it can be assumed that through the use of blockchain, voting systems is made secure and can be considered as verifiable platforms for voting exercises [55]. Smart cities also demonstrate how blockchain is used to create a system that openly displays a clean record of the election in progress and immediately shows if a specific event is fraudulent or not, and enhances the voting process. It also applies to participatory budgeting, where the citizens can directly comment on issues relating to the budgets of the municipality, making it more democratic. Smart governance with the aid of blockchain imposes control over political leadership, increases optimization of the provided services, and increases the amount of citizens' trust in the government. The successful incorporation of blockchain systems in the development of smart cities are a real solution that will enable the development of optimal and perfect governance systems for the public [56].

5.5.5 *Smart Infrastructure*

Smart infrastructure in smart cities are improved by adopting blockchain technology that increases efficiency, transparency, and security across various industries. The application of blockchain makes it possible to manage smart cities' issues, such as data management, resource management, and citizen participation, effectively [57]. In infrastructure, one of the most useful applications of blockchain is in utility and energy. They also manage the application of decentralized energy traders; residents trade excess energy with one another with no help from the mediator through decentralized energy grids. This not only decreases the expenses on energy but also promotes green energy sources. These increase the reliability of such transactions, since information technology can only be executed in accordance with certain conditions set for smart contracts.

5.5.6 *Smart Security*

Blockchain solutions are projected to improve the security of smart cities due to their possibilities of decentralized, transparent, and secure record keeping. Employing the idea of the IoT and integrating systems within cities, these cities become increasingly subject to cyber threats. Blockchain technology minimizes the chances of a centralized accumulation of a large amount of data, which makes blockchain suitable for smart security systems [58]. The distribution of data in blockchain makes it impossible for the system to have a single weak link, thus greatly strengthening smart cities' immunity against attacks. Application of smart security for identity has one of its main foundational elements in the use of blockchain [59]. As more connected devices continue to enter the IoT domain, proper and secure user accounts and access control become necessary. Blockchain helps to create a popular or public identity system that effectively addresses the issue of the restriction of rights through the management of personal data by individuals. This also adds to security and makes using the services within the smart city easier to access because access can be easily managed. Moreover, since blockchain maintains records of all activities within its ledgers, it is possible to easily track all changes made to identity data to prevent unauthorized access or identity fraud. The use of AI can also be adopted alongside blockchain to improve security prediction actions [60]. It also makes the AI algorithms using IoT sensors and other devices able to store data securely and analyse the data for any discrepancies or threats in real time. Smart city plans benefit from blockchain technology in delivering a secure and efficient smart security system. Through applying decentralized and immutable solutions, smart cities become much more resistant to cyber threats, organize identities and use AI as a predictive factor for security, and guarantee the reliability of the key components of smart cities, making them more secure in general.

5.6 Open Research Challenges and Research Directions

With the technological advancement across the world, the challenges and solutions when it comes to development, and particularly urban development, have become important aspects to focus on. The use of blockchain technology is of great benefit in enhancing the urban infrastructural systems through strengthening the areas of quality, transparency, and security, as well as cutting the cost of the operations. Smart cities face many issues due to a centralized system, which can be solved with the help of blockchain technology, with its characteristic features such as decentralization, transparency, and security. But despite this opportunity, there are a number of technological, social, and regulatory challenges that have to be met to achieve it. However, the integration of the blockchain in sustainable smart cities also has the following research challenges and opportunities, as depicted in Fig. 5.12.

5.6.1 Open Research Challenges

5.6.1.1 Scalability and Performance

Cryptography-based public blockchains like Bitcoin and Ethereum face inherent limitations to scalability, such that there is only a limited number of transactions per second, and latency is also high. In a smart city where there are millions of devices

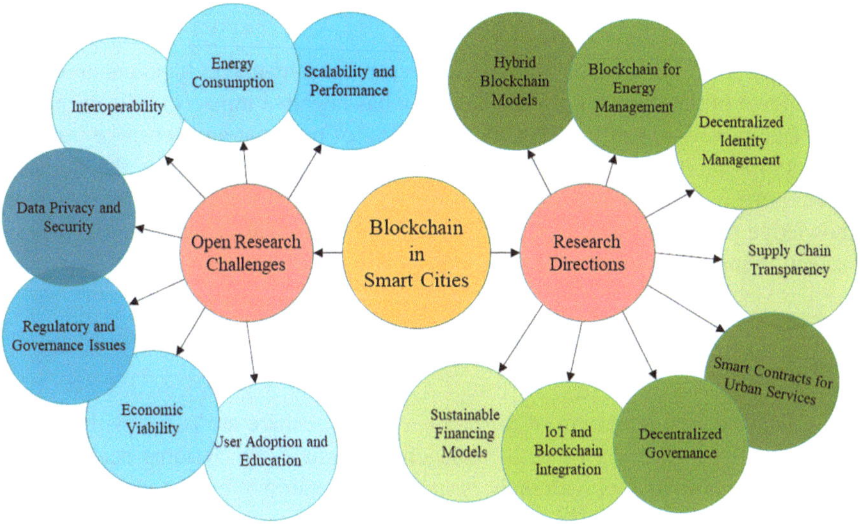

Fig. 5.12 Open research challenges and research directions in blockchain technology-based smart cities

and users, this might be a constraint holding back the blockchain's functioning. It is still a challenge to construct consensus algorithms that will be effective and to optimize the architecture of a blockchain for large numbers of transactions.

5.6.1.2 Energy Consumption

Conventional PoW structures, as employed by Bitcoin, are extremely energy-consuming [61]. This is quite counterproductive to the objective of decreasing carbon footprints. Further work is needed to work out other consensus algorithms, like Proof of Stake or delegated proof of stake. Alternatively, more research on the interconnection of renewables in the blockchain platform might reduce energy issues.

5.6.1.3 Interoperability

Smart cities have various integrated systems as well as various platforms, which include transport, energy, health, and safety. It is equally important to establish compatibility for the successful integration of these systems with blockchain platforms. Standardization of the blockchain and the creation of a set of guidelines for blockchain will ensure compatibility of information and services within various urban infrastructures [62].

5.6.1.4 Data Privacy and Security

Although decentralization accompanied by cryptography makes blockchain provide a high level of data protection, it potentially has negative impacts on its users' privacy. Being smart cities involves the collection of personal and sensitive information; compliance with the privacy regulations is required. To solve the threat of over-reporting bias, further analysis of privacy-preserving technologies like zero-knowledge proof and homomorphic encryption is necessary to incorporate privacy protection into the system without compromising its transparency [19].

5.6.1.5 Regulatory and Governance Issues

Due to the distributed nature of blockchain systems, there are issues with compliance with the legislation or with management. Smart city systems need to have well-defined laws to deal with matters such as ownership of collected data, responsibility, and liability [19]. More research must be done towards the formation of good governance to meet legal and regulatory conditions that would harness blockchain's decentralized benefits.

5.6.1.6 Economic Viability

The application of blockchain technology in a smart city incurs huge initial costs. The long-term gain for stakeholders in areas such as cost savings due to increased efficiency and areas of reduced fraud must also be presented in order to receive the backing of the shareholders. There is a need to employ cost-benefit analysis along with other economic concepts that give the value of blockchain in urban environments [63].

5.6.1.7 User Adoption and Education

For the integration of blockchain technology in smart cities to be efficient, there is a need to have more involvement from the users' adoption and education [64]. Still, there are concerns that may cause the slowdown of the adoption; these are the complications in blockchain systems and the poor knowledge among the general public and policymakers. The awareness campaigns, educational programs, and user-friendly interfaces are the significant areas that require research to close the knowledge gap.

5.6.2 *Open Research Directions*

5.6.2.1 Hybrid Blockchain Models

It may be beneficial to invent consolidated solutions through the integration of the features of both public and private blockchains in order to minimize the scalability, privacy, and interoperability issues [38]. They can provide all the advantages of the public decentralized ledger and the improved speed and privacy of private ledgers at the same time. However, there is still the potential for additional research for hybrid smart city architectures and the corresponding applications.

5.6.2.2 Blockchain for Energy Management

Blockchain also applied in the energy sector since it provides an opportunity for peer-to-peer trading, optimization of energy transmission, and integration of renewable energy. The study of smart grids and microgrid systems, which are based on blockchain technology, ensures that more effective and efficient implementations of the needed energy infrastructure in smart city projects [19, 53].

5.6.2.3 Decentralized Identity Management

More emphasis on the issue of secure and effective management of digital identities for smart city services is essential. Decentralized Identity Management is a feature that can be facilitated by blockchain to offer more security and privacy while avoiding scams [65]. Thus, practical studies of identification methods, self-sovereign identity, and use cases in smart cities are an effective direction.

5.6.2.4 Supply Chain Transparency

Blockchain can be presented as an effective solution for increasing the transparency and giving more opportunities for tracking various flows in urban supply chains regarding nourishment and goods delivery and waste [65]. This, therefore, establishes that studies relating to the application of blockchain in supply chain management enhance conservation of resources, minimize wastage, and promote the usage of sustainable commodities in smart cities [19].

5.6.2.5 Smart Contracts for Urban Services

Through the invention and use of smart contracts, daily urban services in transport, power, safety, and more can be improved [30]. There is a need to investigate the compatibility, validity, and strengths of efficient smart contracts to help in the delivery of services in an urban environment, hence improving the quality of life of citizens.

5.6.2.6 Decentralized Governance

Blockchain technology will help organize decision-making mechanisms for various issues in a smart city, which include making people participate in decision-making processes to render them accountable and transparent [65]. The studies involving blockchain-based voting systems and other forms of casting for the election of leaders and representatives, public engagement platforms, and decentralized autonomous organizations (DAOs) can try to enhance democratic and inclusive urban governance.

5.6.2.7 IoT and Blockchain Integration

In smart cities, IoT is used to relay data from various objects since it deals with connected objects. There are several benefits of using blockchain with IoT, i.e., data can be more reliable due to its ability to provide blockchain and more secure since it allows for the interconnectivity of IoT [66]. Further research into IoT-blockchain systems and architectures, edge computing possibilities for smart cities, and utilization cases is considered to be a very promising field.

5.6.2.8 Sustainable Financing Models

Blockchain is one type of technology that complements smart cities, and one way to apply green bonds or decentralized finance (DeFi) platforms to support sustainable financing [67]. Based on the study of blockchain-based financial instruments and potential effects on sustainable urban development, which can be learned by policymakers as well as investors.

Blockchain promises a significant positive impact on smart cities by providing a means to achieve infinite advancements in medium-density urban areas. However, these potentials are realized only when important challenges in the conduct of research are met and new directions are undertaken. In terms of scalability, energy, interoperability, privacy, governance, cost, and usage, scholars and scientists can open the door towards implementing blockchain in smart cities. Most of these challenges need to be addressed by integrating efforts from academia, industry, and the government in order to realize the full potential of blockchain technology in sustainable urban development.

5.7 Summary

Smart cities with the usage of blockchain technology is characterized as an exciting shift in urban development. This technology is considered one of the fundamental building blocks for the transformation of city governance and service delivery. It provides a useful platform for increasing the level of protection in governmental bodies, as well as increasing public confidence in the implementation of open government. The public records pertaining to the state or municipal affairs are maintained as secure, transparent, and non-alterable through the use of distributed ledgers, which is fundamental to good governance. This leads to cost optimization, reduced fraud in the transport system, and the provision of innovative solutions for mobility, hence increasing the adoption of sustainable urban mobility solutions and systems. These applications, however, are promising but are still limited by scalability issues, compatibility, or transmission with other networks and formal acceptance from the regulatory bodies. Some of the open issues are related to the lack of standards for cross-sector compatibility and the lack of protection of data in distributed systems. As such, more studies should be focused on improving consensus mechanisms, examining the applicability of the hybrid blockchain system and the partnership between government and industry, as well as the university. The concept of a smart city has enormous benefits that are secure, efficient, and more effective through the adoption of blockchain technology. Exploring the opportunities of blockchain and considering the future research prospects, it is identified as a foundation for applying in smart cities when positioned as cities of the future in the context of further digitalization of the world.

References

1. S.A. Bagloee, M. Heshmati, H. Dia, H. Ghaderi, C. Pettit, M. Asadi, Blockchain: The operating system of smart cities. Cities **112**, 103104 (2021). https://doi.org/10.1016/J.CITIES.2021.103104
2. A. Rejeb, K. Rejeb, S. Simske, H. Treiblmaier, S. Zailani, The big picture on the internet of things and the smart city: A review of what we know and what we need to know. Internet of Things **19**, 100565 (2022). https://doi.org/10.1016/J.IOT.2022.100565
3. D. Puthal, N. Malik, S.P. Mohanty, E. Kougianos, G. Das, Everything you wanted to know about the Blockchain: Its promise, components, processes, and problems. IEEE Cons. Electron. Mag. **7**(4), 6–14 (2018). https://doi.org/10.1109/MCE.2018.2816299
4. S. Hakak, W.Z. Khan, G.A. Gilkar, M. Imran, N. Guizani, Securing smart cities through Blockchain technology: Architecture, requirements, and challenges. IEEE Netw. **34**(1), 8–14 (2020). https://doi.org/10.1109/MNET.001.1900178
5. D. Kundu, Blockchain and trust in a smart city. Environ. Urban. ASIA **10**(1), 31–43 (2019). https://doi.org/10.1177/0975425319832392
6. A. Welligton dos Santos Abreu, E.F. Coutinho, C. Ilane Moreira Bezerra, Performance evaluation of data transactions in Blockchain. IEEE Lat. Am. Trans. **20**(3), 409–416 (2022). https://doi.org/10.1109/TLA.2022.9667139
7. H. Taherdoost, Smart contracts in Blockchain technology: A critical review. Information **14**(2), 117 (2023). https://doi.org/10.3390/info14020117
8. J. Rabe, B. Ietto, R. Muth, K. Eisenhut, F. Pascucci, Citizens' engagement in urban development through blockchain: A human-centered design approach, in *2021 IEEE International Conference on Technology Management, Operations and Decisions (ICTMOD)*, (IEEE, 2021), pp. 1–6. https://doi.org/10.1109/ICTMOD52902.2021.9739434

9. M. Baza, N. Lasla, M.M.E.A. Mahmoud, G. Srivastava, M. Abdallah, B-ride: Ride sharing with privacy-preservation, trust and fair payment atop public Blockchain. IEEE Trans Netw Sci Eng **8**(2), 1214–1229 (2021). https://doi.org/10.1109/TNSE.2019.2959230
10. M.G.M. Almihat, M.T.E. Kahn, K. Aboalez, A.M. Almaktoof, Energy and sustainable development in smart cities: An overview. Smart Cities **5**(4), 1389–1408 (2022). https://doi.org/10.3390/smartcities5040071
11. A. Ayub Khan, A.A. Laghari, Z.A. Shaikh, Z. Dacko-Pikiewicz, S. Kot, Internet of Things (IoT) security with Blockchain technology: A state-of-the-art review. IEEE Access **10**, 122679–122695 (2022). https://doi.org/10.1109/ACCESS.2022.3223370
12. S.A. Abdulhakeem, Q. Hu, Powered by Blockchain technology, DeFi (decentralized finance) strives to increase financial inclusion of the unbanked by reshaping the world financial system. Mod. Econ. **12**(01), 1–16 (2021). https://doi.org/10.4236/me.2021.121001
13. P. Mishra, P. Thakur, G. Singh, Sustainable smart city to society 5.0: state-of-the-art and research challenges. SAIEE Afr. Res. J. **113**(4), 152–164 (2022). https://doi.org/10.23919/SAIEE.2022.9945865
14. R.A. Salha, M.A. El-Hallaq, A.I. Alastal, Blockchain in smart cities: Exploring possibilities in terms of opportunities and challenges. J. Data Analy. Inform. Process. **07**(03), 118–139 (2019). https://doi.org/10.4236/jdaip.2019.73008
15. M.A. Khan, K. Salah, IoT security: Review, blockchain solutions, and open challenges. Futur. Gener. Comput. Syst. **82**, 395–411 (2018). https://doi.org/10.1016/J.FUTURE.2017.11.022
16. A.D. Dwivedi, G. Srivastava, S. Dhar, R. Singh, A decentralized privacy-preserving healthcare Blockchain for IoT. Sensors **19**(2), 326 (2019). https://doi.org/10.3390/s19020326
17. G.G. Dagher, J. Mohler, M. Milojkovic, P.B. Marella, Ancile: Privacy-preserving framework for access control and interoperability of electronic health records using blockchain technology. Sustain. Cities Soc. **39**, 283–297 (2018). https://doi.org/10.1016/J.SCS.2018.02.014
18. M. Stoyanova, Y. Nikoloudakis, S. Panagiotakis, E. Pallis, E.K. Markakis, *A Survey on the Internet of Things (IoT) forensics: Challenges, Approaches, and Open Issues* (Institute of Electrical and Electronics Engineers Inc., 2020). https://doi.org/10.1109/COMST.2019.2962586
19. J. Xie et al., A survey of Blockchain technology applied to smart cities: Research issues and challenges. IEEE Commun. Surv. Tutor. **21**(3), 2794–2830 (2019). https://doi.org/10.1109/COMST.2019.2899617
20. M.T. Hammi, B. Hammi, P. Bellot, A. Serhrouchni, Bubbles of trust: A decentralized blockchain-based authentication system for IoT. Comput. Secur. **78**, 126–142 (2018). https://doi.org/10.1016/J.COSE.2018.06.004
21. I. Mistry, S. Tanwar, S. Tyagi, N. Kumar, Blockchain for 5G-enabled IoT for industrial automation: A systematic review, solutions, and challenges. Mech. Syst. Signal Process. **135**, 106382 (2020). https://doi.org/10.1016/J.YMSSP.2019.106382
22. J. Li, D. Greenwood, M. Kassem, Blockchain in the built environment and construction industry: A systematic review, conceptual models and practical use cases. Autom. Constr. **102**, 288–307 (2019). https://doi.org/10.1016/J.AUTCON.2019.02.005
23. M. Banerjee, J. Lee, K.K.R. Choo, A blockchain future for internet of things security: A position paper. Digit Commun. Netw. **4**(3), 149–160 (2018). https://doi.org/10.1016/J.DCAN.2017.10.006
24. M.A. Ahad, S. Paiva, G. Tripathi, N. Feroz, Enabling technologies and sustainable smart cities. Sustain. Cities Soc. **61**, 102301 (2020). https://doi.org/10.1016/J.SCS.2020.102301
25. S. Borchardt, G. B. Vignola, D. Buscaglia, & M. Maroni, Mapping EU policies with the 2030 agenda and SDGs-fostering policy coherence through text-based SDG mapping, https://doi.org/10.2760/110687
26. C. Shen, F. Pena-Mora, Blockchain for cities—A systematic literature review. IEEE Access **6**, 76787–76819 (2018). https://doi.org/10.1109/ACCESS.2018.2880744
27. M. Gorbunova, P. Masek, M. Komarov, A. Ometov, Distributed ledger technology: State-of-the-art and current challenges. Comput. Sci. Inf. Syst. **19**(1), 65–85 (2022). https://doi.org/10.2298/CSIS210215037G

28. M. Zichichi, S. Ferretti, G. D'angelo, A framework based on distributed ledger technologies for data management and services in intelligent transportation systems. IEEE Access **8**, 100384–100402 (2020). https://doi.org/10.1109/ACCESS.2020.2998012
29. A. Papapostolou, I. Andreoulaki, S. Divolis, F. Anagnostopoulos, V. Marinakis, Distributed ledger technology in energy services: The InEExS project objectives and approach, in *2023 14th International Conference on Information, Intelligence, Systems & Applications (IISA)*, (IEEE, 2023), pp. 1–9. https://doi.org/10.1109/IISA59645.2023.10345884
30. T. Bocek, B. Stiller, Smart contracts—Blockchains in the wings, in *Digital Marketplaces Unleashed*, (Springer, Berlin, Heidelberg, 2018), pp. 169–184. https://doi.org/10.1007/978-3-662-49275-8_19
31. S. Khanji, S. Assaf, Boosting ridesharing efficiency through Blockchain: GreenRide application case study, in *2019 10th International Conference on Information and Communication Systems (ICICS)*, (IEEE, 2019), pp. 224–229. https://doi.org/10.1109/IACS.2019.8809108
32. S. Ølnes, J. Ubacht, M. Janssen, Blockchain in government: Benefits and implications of distributed ledger technology for information sharing. Gov. Inf. Q. **34**(3), 355–364 (2017). https://doi.org/10.1016/J.GIQ.2017.09.007
33. Y. Xiao, N. Zhang, W. Lou, Y.T. Hou, A survey of distributed consensus protocols for Blockchain networks. IEEE Commun. Surv. Tutor. **22**(2), 1432–1465 (2020). https://doi.org/10.1109/COMST.2020.2969706
34. S. Zhai, Y. Yang, J. Li, C. Qiu, J. Zhao, Research on the application of cryptography on the Blockchain. J. Phys. Conf. Ser. **1168**, 032077 (2019). https://doi.org/10.1088/1742-6596/1168/3/032077
35. H. Guo, X. Yu, A survey on blockchain technology and its security. Blockchain Res. Appl. **3**(2), 100067 (2022). https://doi.org/10.1016/J.BCRA.2022.100067
36. L.M. Bach, B. Mihaljevic, M. Zagar, Comparative analysis of blockchain consensus algorithms, in *2018 41st International Convention on Information and Communication Technology, Electronics and Microelectronics (MIPRO)*, (IEEE, 2018), pp. 1545–1550. https://doi.org/10.23919/MIPRO.2018.8400278
37. Y. Lv, D. Su, Blockchain security technology based on the asynchronous transmission mode of IoT technology in smart cities. Wirel. Pers. Commun. **126**(3), 1965–1980 (2022). https://doi.org/10.1007/s11277-021-08754-z
38. S. Makani, R. Pittala, E. Alsayed, M. Aloqaily, Y. Jararweh, A survey of blockchain applications in sustainable and smart cities. Clust. Comput. **25**(6), 3915–3936 (2022). https://doi.org/10.1007/s10586-022-03625-z
39. S. Singh, A. Kumar, M. Kathuria, Understanding the public, private and consortium consensus algorithms in blockchain technology. Int. J. Blockchains Cryptocurr. **3**(3), 269 (2022). https://doi.org/10.1504/IJBC.2022.126285
40. X. Chen, S. He, L. Sun, Y. Zheng, C.Q. Wu, A survey of consortium Blockchain and its applications. Cryptography **8**(2), 12 (2024). https://doi.org/10.3390/cryptography8020012
41. A. Alketbi, Q. Nasir, M.A. Talib, Blockchain for government services — Use cases, security benefits and challenges, in *2018 15th Learning and Technology Conference (L&T)*, (IEEE, Feb. 2018), pp. 112–119. https://doi.org/10.1109/LT.2018.8368494
42. A.S.L. França, J. Amato Neto, R.F. Gonçalves, C.M.V.B. Almeida, Proposing the use of blockchain to improve the solid waste management in small municipalities. J. Clean. Prod. **244**, 118529 (2020). https://doi.org/10.1016/J.JCLEPRO.2019.118529
43. S. Biswas et al., Interoperability benefits and challenges in smart city services: Blockchain as a solution. Electronics (Basel) **12**(4), 1036 (2023). https://doi.org/10.3390/electronics12041036
44. K. Abbas, L.A. Tawalbeh, A. Rafiq, A. Muthanna, I.A. Elgendy, A.A. Abd El-Latif, Convergence of blockchain and IoT for secure transportation systems in smart cities. Sec. Commun. Netw. **2021**, 1–13 (2021). https://doi.org/10.1155/2021/5597679
45. S. Kazi, M. Bagasrawala, F. Shaikh, A. Sayyed, Smart E-ticketing system for public transport bus, in *2018 International Conference on Smart City and Emerging Technology (ICSCET)*, (IEEE, 2018), pp. 1–7. https://doi.org/10.1109/ICSCET.2018.8537302

46. H.S. Jennath, S. Adarsh, N.V. Chandran, R. Ananthan, A. Sabir, S. Asharaf, Parkchain: A Blockchain powered parking solution for smart cities. Front. Blockchain **2** (2019). https://doi.org/10.3389/fbloc.2019.00006
47. J. Guo, X. Ding, W. Wu, A Blockchain-enabled ecosystem for distributed electricity trading in Smart City. IEEE Internet Things J. **8**(3), 2040–2050 (2021). https://doi.org/10.1109/JIOT.2020.3015980
48. E.A. Soto, L.B. Bosman, E. Wollega, W.D. Leon-Salas, Peer-to-peer energy trading: A review of the literature. Appl. Energy **283**, 116268 (2021). https://doi.org/10.1016/J.APENERGY.2020.116268
49. E. Gökalp, M.O. Gökalp, S. Çoban, P.E. Eren, Analysing opportunities and challenges of integrated blockchain technologies in healthcare, in *Information Systems: Research, Development, Applications, Education. SIGSAND/PLAIS 2018*, Lecture Notes in Business Information Processing, vol. 333, (Springer, Cham, 2018), pp. 174–183. https://doi.org/10.1007/978-3-030-00060-8_13
50. F. Alam Khan, M. Asif, A. Ahmad, M. Alharbi, H. Aljuaid, Blockchain technology, improvement suggestions, security challenges on smart grid and its application in healthcare for sustainable development. Sustain. Cities Soc. **55**, 102018 (2020). https://doi.org/10.1016/J.SCS.2020.102018
51. R. Jabbar, N. Fetais, M. Kharbeche, M. Krichen, K. Barkaoui, M. Shinoy, Blockchain for the internet of vehicles: How to use Blockchain to secure vehicle-to-everything (V2X) communication and payment? IEEE Sensors J. **21**(14), 15807–15823 (2021). https://doi.org/10.1109/JSEN.2021.3062219
52. R. Jabbar et al., Blockchain technology for intelligent transportation systems: A systematic literature review. IEEE Access **10**, 20995–21031 (2022). https://doi.org/10.1109/ACCESS.2022.3149958
53. P. Zhuang, T. Zamir, H. Liang, Blockchain for cybersecurity in smart grid: A comprehensive survey. IEEE Trans. Industr. Inform. **17**(1), 3–19 (2021). https://doi.org/10.1109/TII.2020.2998479
54. Q. Wang, M. Su, Integrating blockchain technology into the energy sector—From theory of blockchain to research and application of energy blockchain. Comput. Sci. Rev. **37**, 100275 (2020). https://doi.org/10.1016/J.COSREV.2020.100275
55. J. Huang, D. He, M.S. Obaidat, P. Vijayakumar, M. Luo, K.-K.R. Choo, The application of the blockchain technology in voting systems. ACM Comput. Surv. **54**(3), 1–28 (2022). https://doi.org/10.1145/3439725
56. M. Kassen, Blockchain and e-government innovation: Automation of public information processes. Inf. Syst. **103**, 101862 (2022). https://doi.org/10.1016/J.IS.2021.101862
57. S. Li, Application of blockchain technology in smart city infrastructure, in *2018 IEEE International Conference on Smart Internet of Things (SmartIoT)*, (IEEE, 2018), pp. 276–2766. https://doi.org/10.1109/SmartIoT.2018.00056
58. J. Wan, J. Li, M. Imran, D. Li, Fazal-e-Amin, A blockchain-based solution for enhancing security and privacy in smart factory. IEEE Trans. Industr. Inform. **15**(6), 3652–3660 (2019). https://doi.org/10.1109/TII.2019.2894573
59. R. Stephen, A. Alex, A review on BlockChain security. IOP Conf. Ser. Mater. Sci. Eng. **396**, 012030 (2018). https://doi.org/10.1088/1757-899X/396/1/012030
60. O. Fadi, Z. Karim, E.G. Abdellatif, B. Mohammed, A survey on Blockchain and artificial intelligence Technologies for Enhancing Security and Privacy in smart environments. IEEE Access **10**, 93168–93186 (2022). https://doi.org/10.1109/ACCESS.2022.3203568
61. S. Harichandan, S.K. Kar, A. Kumar, A comprehensive review of bitcoin's energy consumption and its environmental implications, in *Machine Learning Approaches in Financial Analytics*, Intelligent Systems Reference Library, vol. 254, (Springer, Cham, 2024), pp. 247–265. https://doi.org/10.1007/978-3-031-61037-0_12
62. N. Alasbali et al., Rules of smart IoT networks within smart cities towards Blockchain standardization. Mob. Inf. Syst. **2022**, 1–11 (2022). https://doi.org/10.1155/2022/9109300

63. T. Jing, L. Zhang, Z. Ma, W. Wang, A cost-benefit secured DAG based analysis of distributed energy generation in urban settings. Sustain. Cities Soc. **107**, 105359 (2024). https://doi.org/10.1016/J.SCS.2024.105359
64. A.R. Sathya, S.K. Panda, S. Hanumanthakari, Enabling smart education system using blockchain technology, in *Blockchain Technology: Applications and Challenges. Intelligent Systems Reference Library*, vol. 203, (Springer, Cham, 2021), pp. 169–177. https://doi.org/10.1007/978-3-030-69395-4_10
65. M. Farnaghi, A. Mansourian, Blockchain, an enabling technology for transparent and accountable decentralized public participatory GIS. Cities **105**, 102850 (2020). https://doi.org/10.1016/J.CITIES.2020.102850
66. M.S. Rahman, M.A.P. Chamikara, I. Khalil, A. Bouras, Blockchain-of-blockchains: An interoperable blockchain platform for ensuring IoT data integrity in smart city. J. Ind. Inf. Integr. **30**, 100408 (2022). https://doi.org/10.1016/J.JII.2022.100408
67. A.N. Turi (ed.), *Financial Technologies and DeFi* (Springer, Cham, 2023). https://doi.org/10.1007/978-3-031-17998-3

Chapter 6
Smart Cities with 6G-Internet of Things

Contents

P. Mishra, G. Singh, *Sustainable Smart Cities 2.0*,
https://doi.org/10.1007/978-3-032-01102-2_6

6.1 Introduction

The concept of smart cities has revolutionized the way urban environments are governed, managed, and developed. Smart cities combine Information Communication Technology (ICT) with conventional industries for the development of a society's welfare of its citizens. This ranges from infrastructure efficiency, environmental standards, health, energy usage, safety, transport, and many other sectors. Most of these advancements have been driven by Internet of Things (IoT); however, 6G technology promises an even greater development of smart cities [1]. The next generation of wireless communication technology is 6G, as it offers internet at very high speed, high reliability, and very low latency. As a result, IoT become even more interconnected and efficient in future due to these enhanced capabilities. For example, intelligent traffic controls will adjust instantly to traffic jams and accident to reduce journey durations and pollution. Monitoring will also be capable of measuring a number of environmental factors such as air and water quality, hence faster decision in pollution and conservation. Additionally, 6G helps to build new autonomous actions through self-driving cars and drones for delivery and emergency services [2]. When the smart grid is implemented, it will help manage electricity consumption to provide the population with constant and efficient electric power

[3]. Smart cities will have more efficient telemedicine and remote monitoring of patients in healthcare, which will enhance the quality of disease treatment and reduce the costs associated with it [4]. Therefore, smart cities in combination with 6G-IoT will improve the future state of urban environment and its inhabitants' quality of life [5]. Computing technologies will offer newer innovative approaches to realizing smart, connected, and sustainable cities of the future.

6.1.1 Impacts of 6G Technology

5G has brought in rich opportunities in the fields of connectivity and smart cities all over the world. As a result of the ultra-high-speed internet, low latency and high system capacity that comes with 5G, cities are now smarter than before [6]. However, the future of smart cities depends on the emergence and application of the next generation communication technology, such as 6G. This application is expected to bring about a radical change in major infrastructures in particular urban systems, leading to enhanced standards of living of people in these areas. Some of the main effects of integrating 6G technology in smart city include the delivery of ultra-reliable and low-latency communication. This enable communication between one smart city subcomponent and another, such as IoT devices, transport, and emergency services. The implication of 6G technology in smart cities consequently accommodate far more connected devices, hence the expansion of IoT use cases. The additional bandwidth would be to process data, analyze, and apply machine learning (ML) real time, in order to help cities adapt to situations and offer a better functionality. Smart transportation in smart cities will be enabled by integration of 6G technology [7]. As this technology will allow for some of the most modern forms of vehicle-to-everything (V2X) communication that include vehicle-to-vehicle (V2V), vehicle-to-infrastructure (V2I), and vehicle-to-pedestrian (V2P) [8]. This should lead to better scalability of traffic control, autonomous vehicles, as well as improved protective abilities for individuals behind the wheel and those affected by automobiles. Further, the technology will enable integration of the public transports which include buses, rail, and trams hence providing efficient and environmentally friendly transport solution for citizens. 6G technology also have the effects on the energy management systems in smart cities [9]. It will also apply to the 6G-supported high volumetric data transfer and real-time analysis to enable smart cities to adapt high efficiency energy consumptions that will contribute to minimum carbon emissions. This is done by the application of big data and artificial intelligence (AI) where its applications include the utilization of energy use and waste disposal in urban regions. Using 6G technology in smart cities will significantly improve safety and emergency services significantly [10]. The technology also facilitates the integration of advanced IoT sensors and drones, which provide essential data and are designed to support disaster response and risk reduction efforts.

6G technologies will achieve a higher connectivity level, better traffic flow management through efficiency, usage of energy, and greater public safety. As a result of the enhancements in its characteristics, such as the speed of its connection, low latency, and huge capability, 6G will make cities smart, connected, and sustainable to a level. Figure 6.1 illustrates the key impact of 6G technology, including advancement in speed, AI and automation, global connectivity, security and privacy, sustainability, and extended reality communication. 6G technology is set to revolutionize communication and connectivity with ultra-high speed, low latency, and AI-driven automation. It enables global connectivity, enhanced security, sustainable networks, and immersive experiences like holographic communication, transforming industries and everyday life. Therefore, its application with 6G technology in smart cities is vital in the advancing direction of the world with respect to wireless communication, from one generation to another, it is evident that 6G wireless communication

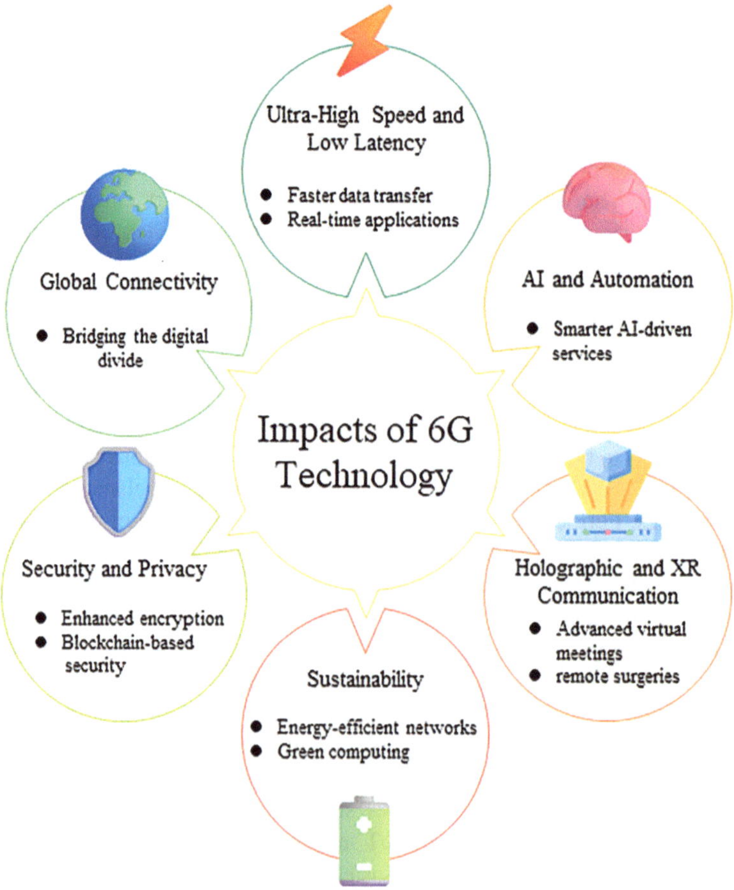

Fig. 6.1 Impact of 6G technology

merged with smart cities [11] contribution in the development of smart cities, is very important for the future.

6.1.2 Role of 6G-IoT for Sustainable Smart Cities

The 6G technology is anticipated to provide high connectivity speeds, extremely low latency, and exponential device density. These capabilities would be required to enable unmediated and synchronous communication, which in turn presents a great opportunity for IoT to implement in urban ecosystems. When integrated with IoT's device communication and huge data handling, this enhancement leads to the development of the 6G-IoT. Figure 6.2 illustrates 6G-IoT-based framework for building sustainable and efficient smart cities by integrating advanced technologies, such as AI, blockchain, edge computing, and high-speed connectivity. It highlights key smart components, like smart infrastructure, mobility, energy, and healthcare aligned with Sustainable Development Goals (SDGs). These components aim to enhance energy efficiency, public safety, and education while reducing pollution and resources consumption. The framework ultimately drives economic growth, reduces environmental impact, and improves citizens' quality of life through optimized city services. It will significantly influence the sustainable development of smart cities through six major pillars.

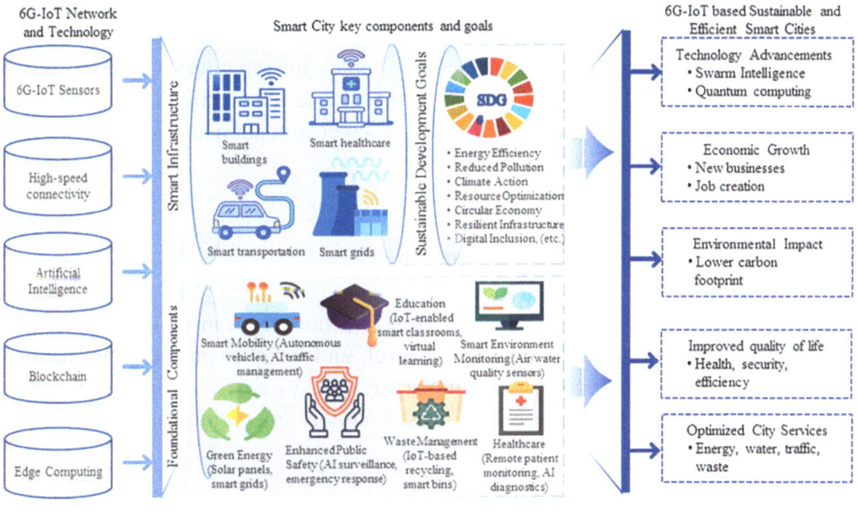

Fig. 6.2 6G-IoT for sustainable smart city framework

6.1.2.1 Green Energy

6G-IoT enables an intelligent energy management within smart city applications by balancing energy need of various sectors [12]. These include smart grids which self-synchronize in response to the demand–supply balance and adjust accordingly, smart buildings that are able to regulate heating, ventilation, and air conditioning (HVAC) from occupants' density, and electric vehicles that are able to transmit charging requirements and availability. All these advancements enable the effective management of energy distribution and formulate ways of cutting on carbon emissions.

6.1.2.2 Smart Transportation

6G-IoT will support machine-to-machine (M2M) communications necessary for successful implementation of the large number of autonomous cars and intelligent transportation systems [13]. It is the ability of communicating between vehicles (V2V) and between vehicles and the infrastructure (V2I) to optimize an appropriate route for a car. In addition, this smart transportation model will greatly reduce emissions and also improve the general safety on roads.

6.1.2.3 Waste Management

The waste levels in smart bins are detected by 6G-IoT-based sensors and optimal routes for collection can be set based on actual information [14]. The special approach of such waste containers also makes waste management more effective, and at the same time, saves fuel and releases fewer greenhouse gases than conventional garbage collection methods.

6.1.2.4 Smart Environment Monitoring

Smart sensors which are featured by 6G-IoT technology can collect information about the quality of air and water, noise level, and weather on the given territory [15]. When this data is incorporated into smart city systems, authorities are able to address pollution sources and equally provide timely interventions, thus creating a healthier environment.

6.1.2.5 Enhanced Public Safety

Based on features such as high reliability and low latency, the case of 6G-IoT enhances the emergency response systems. The identification of occurrence by smart cameras and IoT devices lead to early reporting of events, which makes faster actions and response to the situation as well as immediate mobilization of services [16]. It not only contributes to cutting the rate of accidents and criminal activities, but also assists in making cities ready for disasters.

6.1.2.6 Education and Healthcare

In the future, smart classrooms with 6G-IoT will provide an enriched and customized learning environment through intelligent and innovative technologies [17]. At the same time in health care, operations, surgery, consultations, and constant monitoring of the patients are done remotely in a way that would increase the efficiency of health care and cut down its cost [18].

All the aspects prove the fact that the combination of 6G and IoT is the best model of sustainable smart cities. As for its advanced features, the present system constitutes a rich environment for rational and effective usage of existing resources, the preservation of the natural environment, and the upgrading of people's quality of life.

6.2 Current Research Trends of 6G-IoT in Smart Cities

This study investigates the trends, contributions, and impact of research in smart cities with 6G-IoT from 2019 to 2025 through a bibliometric analysis of Scopus data. The Bibliometrix R package along with origin software were utilized to analyse and depict the data. This approach ensures that researchers are presented with the most current research developments in the field enabling to focus on emerging trends and advancements. In March 2025, a bibliometric analysis was conducted utilizing the Scopus core collection database to thoroughly investigate the concept and understanding of smart cities with 6G-IoT technology. Ultimately, 178 articles were exported and saved in BibTex format before being included and examined in the bibliometric study. The primary information, syntax, and specific criteria used in screening for this study are shown in Table 6.1.

6.2.1 Annual Scientific Production and Top Ten Most Published Journals/Cited Paper

Figure 6.3 shows the annual publication trends from 2019 to 2025, highlighting the growth in research output. The early years show a steady increase in publications, rising from 2019 (1 doc) to 2024 (64 doc), indicating a growing interest in the field. In 2025 (13), the number continues to rise and may continue increasing until the end of the year.

Table 6.1 Summary of data characteristics and main information of the bibliometric study

Description	Specific Standard Requirements/Results
Research database	Scopus
Search query	("smart") and ("cities") AND ("6 g") AND ("iot")
Time span	2019–2025
Document type	Articles
Language	English
Data export format	BibTex
Documents	178
Annual growth rate %	53.34
Document average age	1.99
Average citations per doc	13.25
Keywords plus (ID)	1166
Author's keywords (DE)	589

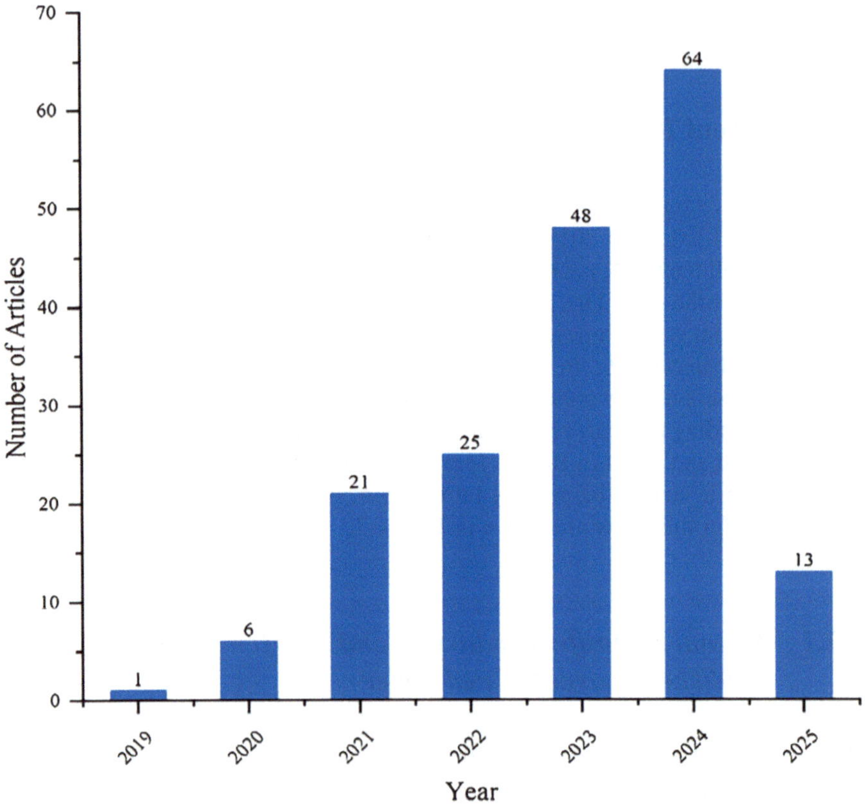

Fig. 6.3 Yearly trend of publication number in smart cities with 6G-IoT

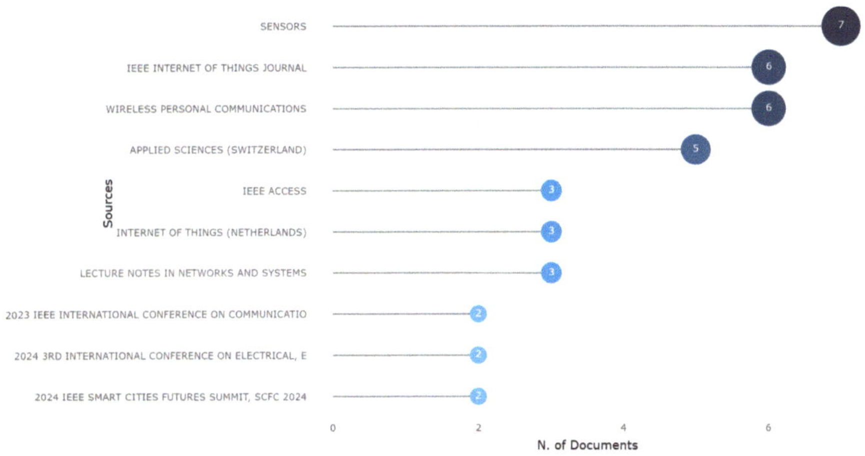

Fig. 6.4 Most relevant sources with number of articles for 6G-IoT smart city

Figure 6.4 presents the highest number of articles in 6G-IoT-based smart city research domain published in "Sensors" (7 docs), "IEEE Internet of Things Journal" (6 docs), "Wireless Personal Communications" (6 docs), and "Applied Sciences (Switzerland)" (5 docs), which indicates a higher output of articles compared to the other journals. In addition, some other journals, such as the "IEEE Access" (3 docs), "Internet of Things (Netherlands)" (3 docs), "Lecture notes in Networks and Systems" (3 docs), "IEEE International Conference on Communications Workshops: Sustainable Communications for Renaissance, ICC Workshops 2023" (2 docs), "3rd International Conference on Electrical, Electronics, Information and Communication Technologies, ICEEICT 2024" (2 docs) and "IEEE Smart Cities Futures Summit, SCFC 2024" (2 docs), are in this domain.

Table 6.2 displays the top ten most cited papers from the study, which presents both the ranking and authors with their published works. This section presents information about authors, article titles, and their corresponding publications in journals with documentation for publication years and citation frequency. The ranking focuses on engineering field smart cities articles to demonstrate their total citation (TC) counts and annual citation rates and normalized total citation (NTC) data which collectively signify their impact in the field.

6.2.2 *Sankey Diagram (Three-Field Plot and Thematic Map)*

Figure 6.5 Sankey diagram illustrates the three-field plot thematic progression to visualize the flow and the distribution in scientific research. In this case, keywords plus, abstract, and title are added, acting as indicators to highlight the most influential research papers in the subject. These keywords indicate vital concepts and

Table 6.2 Most global cited documents in smart city research domain with 6G-IoT technology

Rank	Authors name	Paper Title	Journal's Name	TC	TC/ Year	NTC
1	Javed et al. (2022) [1]	Future smart cities: Requirements, emerging technologies, applications, challenges, and future aspects	Cities	352	88.00	11.75
2	Allam et al. (2021) [19]	Future (post-COVID) digital, smart and sustainable cities in the wake of 6G: Digital twins, immersive realities and new urban economies	Land use policy	194	38.80	5.71
3	Qadir et al. (2023) [13]	Towards 6G Internet of Things: Recent advances, use cases, and open challenges	ICT Express	140	46.67	11.55
4	Kumari et al. (2021) [20]	Amalgamation of blockchain and IoT for smart cities underlying 6G communication: A comprehensive review	Computer Communications	110	22.00	3.24
5	Allam et al. (2022) [21]	Unpacking the '15-minute city' via 6G, IoT, and digital twins: Towards a new narrative for increasing urban efficiency, resilience, and sustainability	Sensors	81	20.25	2.70
6	Chen and Okada (2021) [22]	Toward 6G internet of things and the convergence with RoF system	IEEE Internet of Things Journal	70	14.00	2.06
7	Sharma et al. (2021) [23]	Sustainable smart cities: convergence of artificial intelligence and blockchain	Sustainability	67	13.40	1.97
8	Slalmi et al. (2021) [24]	Toward 6G: Understanding network requirements and key performance indicators	Transactions on Emerging Telecommunications Technologies	63	12.60	1.86
9	Silva and Guerreiro (2020) [25]	On the 5G and Beyond	Applied Sciences	53	8.83	3.98
10	Mishra et al. (2023) [26]	Emerging technologies and design aspects of next generation cyber physical system with a smart city application perspective	International Journal of System Assurance Engineering and Management	48	16.00	3.96

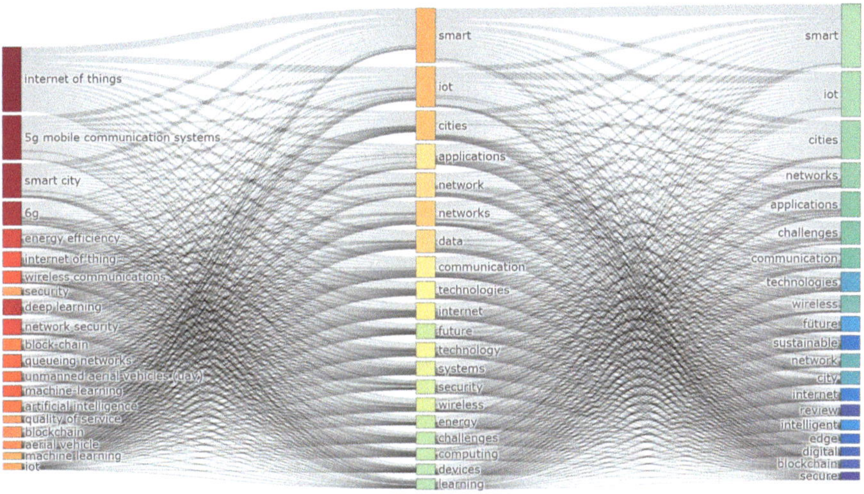

Fig. 6.5 Three field plot (thematic development) using sankey diagram

themes that stand out in academic discussions of the smart city domain with 6G-IoT technology. The diagram's keywords are sourced from earlier published studies and were then thematically organized by a specific researcher. Figure 6.6 categorizes various smart city themes based on 6G-IoT technology and relevant degrees. In the top right quadrant (motor themes), which represents themes that are highly developed and highly relevant, terms like "6g", "smart city", "internet of things" and "5g mobile communication systems". In quadrant left (niche themes) there are represents "internet", "sensors network", "intelligent transportation systems" and "wireless sensor networks". This indicates a strong focus on these areas, suggesting they are central to smart city research and practice with 6G-IoT. This quadrant reflects the areas where significant advances and investments are made, highlighting their critical role in smart city. In contrast, the bottom left quadrant, which represents themes that are either emerging or declining and have lower relevance, includes terms like "benchmarking", "spectrum efficiency", "6g technology", and "cognitive radio". Additionally, the terms like "energy efficiency", "wireless communications", and "unmanned aerial vehicles", are the foundational information for further analysis and serve as a base for more advanced analyses

6.2.3 Co-occurrence Keyword Trends and Network Visualization

Figure 6.7 depicts the most prominent themes and keywords related to research in 6G-IoT with smart city. The most significant and most central terms, such as "6g," "internet of things", "smart cities", "artificial intelligence", and "blockchain"

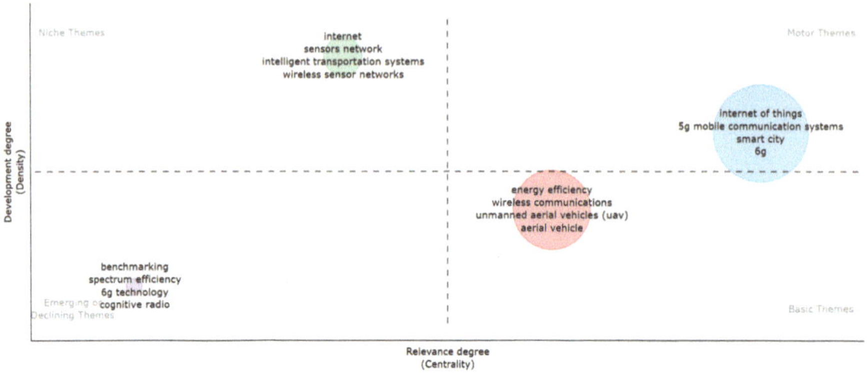

Fig. 6.6 Thematic map of 6G-IoT with smart city

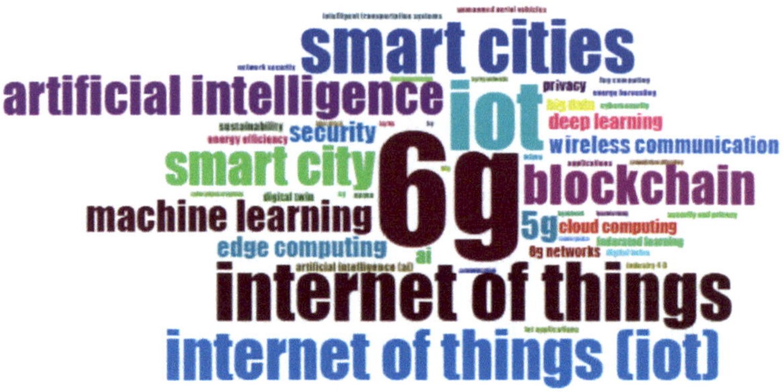

Fig. 6.7 Co-occurrence keyword trends in smart city research domain with 6G-IoT

indicate the core focus of the research on how researchers grasp and alter their conceptual knowledge, particularly within smart city contexts.

An illustration in Fig. 6.8 displays word occurrence frequencies from 6G-IoT technology in smart city research network. Research focuses mainly on "6g" as this term appears 50 times throughout the literature. Research on 6G-IoT with smart cities indicates the importance of "internet of things" based on its 77 word occurrences. The "smart cities" (42) and "artificial intelligence" (18) combinations show the importance and growth in cities. The technologies for security and development are shown through terms like "blockchain" (18) along with "machine learning" (14). The data shows that 6G-IoT technology and smart city research revolves primarily around several major subjects.

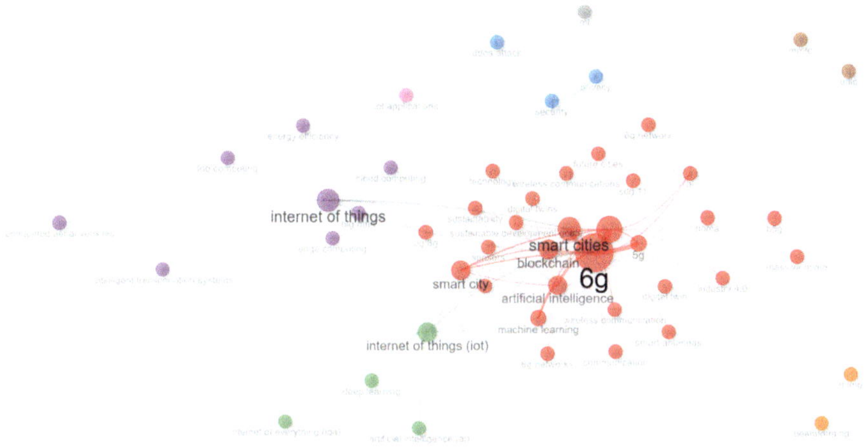

Fig. 6.8 Co-occurrence network of words in smart city research domain

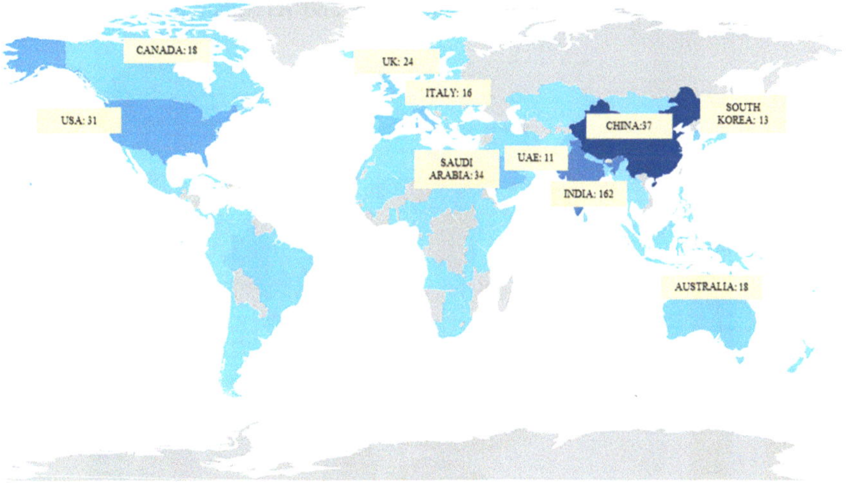

Fig. 6.9 Top ten countries' scientific production of various countries based on frequency

6.2.4 *Countries' Scientific Production*

The scientific publication frequencies of various countries in the world are represented in Fig. 6.9. Scientific contributions from India stand at 162 which demonstrates their powerful role in research development. The growing scientific power of China displays itself through its 37 scientific publications which rank second after India. The research hub of the Saudi Arabia produces 34 scientific papers among its total contributions. Numerous publications emerge from USA and UK together with

Australia given their research output comprises 31, 24, and 18 research papers respectively. Academic and industrial research sectors of Canada and Italy produce approximately 18 and 16 scholarly works between them. South Korea and UAE have smaller publication numbers but their contributions amount to totals of 13 and 11 research items respectively. The measured data indicates that Asian countries significantly influence the global scientific research output.

6.2.5 Mapping 6G- IoT-Based Smart City Research to Sustainable Development Goals (SDGs)

SDGs, their targets (bubble series), and the percentage (bar graph) of corresponding keywords detected in the text (calculated as the ratio of keywords in one goal to the total number of keywords detected) are depicted in Fig. 6.10. For SDG mapping, the study utilized the SDG Mapper tool from the "Knowledge Base for the Sustainable Development Goals," provided by the European Commission [27]. This tool uses bubble charts to depict SDG interconnections and relevance among bibliometric

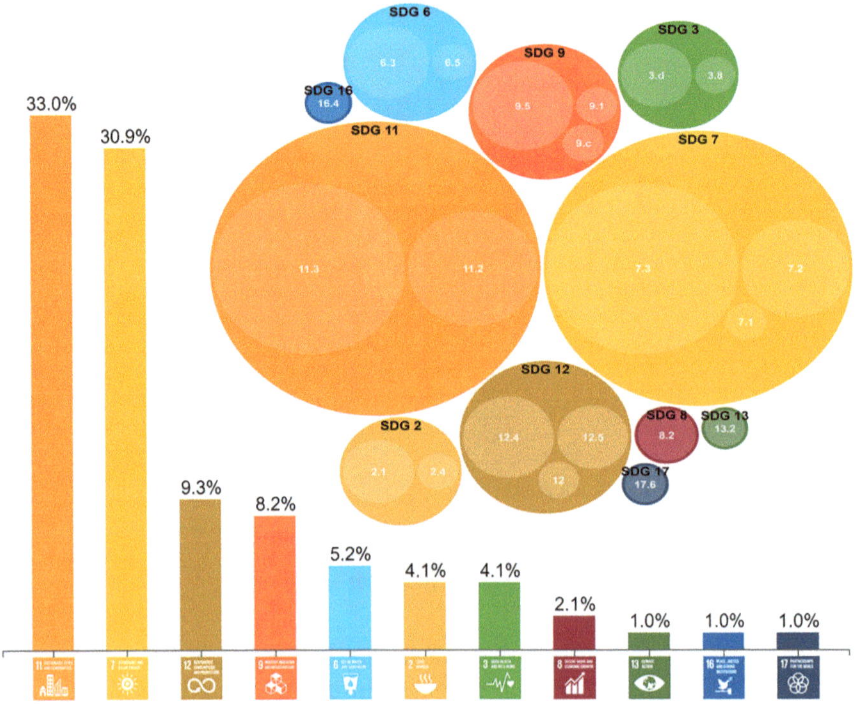

Fig. 6.10 Mapping 6G-IoT-based smart city research to relevant SDGs and targets

records. The size of each bubble correlates with the proportion of associated keywords found, which is calculated as the number of goal-specific keywords divided by the total keyword count. As such, SDG 11 is marked by 32 keyword occurrences. This is followed by SDG 7, with 30 occurrences, and SDG 12, with 9 occurrences, underlining their relative prominence in the research database and so on. Within SDG 11, there are several key subtargets that the research addresses: Subtarget 11.3 is shown as a distinct bubble within the larger SDG 11 circle, indicating focused research on smart city. Subtarget 11.2 focuses on providing the safe, accessible, and sustainable smart transportation, which falls in sustainable cities and communities, is another bubble within the SDG 11 area. This highlights the efforts in research to improving the cities. Figure 6.10 shows that the size of the SDG 7 bubble implies that advancements in smart city are being focussed on affordable and clean energy, which prioritize sustainability. Similarly, SDG 12 (Responsible consumption and production) is also closely linked to SDG 11 and suggests that research should explore how to achieve the environmentally sound management of chemicals and all waste, in order to minimize their adverse impact on humans and to achieve the sustainable management and efficient use of natural resources within the context of smart cities. The detailed description is depicted in Table 6.3.

Table 6.3 SDGs ranking, targets and relevance in 6G-IoT-based smart city research domain with SDGs

SDGs(Rank)	Description	Targets	No. of Occurrence		Relevance
SDG 11	Sustainable cities and communities	11.3	22	32	33.0%
		11.2	10		
SDG 7	Affordable and clean energy	7.3	22	30	30.9%
		7.2	7		
		7.1	1		
SDG 12	Responsible consumption and production	12.4	5	9	9.3%
		12.5	3		
		12	1		
SDG 9	Industry, innovation, and infrastructure	9.5	6	8	8.2%
		9.1	1		
		9.c	1		
SDG 6	Clean water and sanitation	6.3	4	5	5.2%
		6.5	1		
SDG 2	Zero hunger	2.1	3	4	4.1%
		2.4	1		
SDG 3	Good health and well-being	3.d	3	4	4.1%
		3.8	1		
SDG 8	Decent work and economic growth	8.2	2	2	2.1%
SDG 13	Climate action	13.2	1	1	1.0%
SDG 16	Peace, justice, and strong institutions	16.4	1	1	1.0%
SDG 17	Partnerships for the goals	17.6	1	1	1.0%

6.3 Fundamentals of 6G-IoT Technology

The concept of IoT has brought a significant improvement in people's lives, including almost every aspect of smart cities, ranging from smart transportation to smart healthcare, smart automobiles, and beyond. Looking forward into future generations of wireless networks and as the world transitions to the upcoming 6G, IoT application possibilities will increase further. 6G-IoT is outlined to provide high-speed connection and low-end latency to a range of services and applications in the smart city environment. The main purpose of the 6G-IoT technology is to enable high density of terminal connectivity and enable smooth integration. This is done by using technologies like massive MIMO (Multiple Input, Multiple Output), full-duplex communication, and THz (terahertz) frequencies. These technologies will ensure higher data rates for communication and decrease the interconnecting devices response time, which is a crucial factor for ultra-real-time applications such as autonomous cars and intelligent energy distribution systems. Besides, 6G-IoT technology will employ other enhanced networking model including edge computing and network slicing. These technologies will facilitate the establishment of guaranteed network channels to uniquely specified applications leading to increased network functionality. Moreover, the control and the directing of IoT will be more efficient due to the utilization of AI and ML algorithms.

6.3.1 Key Features and Characteristics of 6G

Smart cities are set to benefit hugely with the advancement of the 6G technologies in the further advancement of smart cities. The innovative aspects and parameters of 6G will facilitate a vast number of services and applications that will improve the performance and quality in urban settings. Some of the key features and characteristics of 6G that are particularly relevant to smart cities include the following features:

6.3.1.1 High-Speed Data Transmission

The 6G technology will provide data transmission rate of up to 1Tbps for data exchange [28], which will be important for having a tremendous number of connected devices in smart cities. This will make it possible to have integrated working as well as convenient transfer of information among different city functions. Moreover, 6G technology will be able to reduce the latency down to milliseconds; this will support real-time interaction that is tremendously crucial for specific business applications, including self-driving cars and tele-surgeries. This will also help new applications such as augmented reality (AR) and virtual reality (VR), which will help to provide better user experiences and immersive services in education and training [29]. Moreover, the reliability and efficiency increment with 6G will help IoT and its numerous devices to be integrated and controlled effectively. Thus, smart cities will be capable of improving the use of available resources, improve safety of citizens and offer better services to the population.

6.3.1.2 Ultra-Low Latency

The 6G technology has a latency of less than 100 microseconds [30], which makes it suitable for time-sensitive functions in smart cities including autonomous vehicle operation, efficient traffic control, and response to exigent circumstances. This will help in combining different units of smart cities; the simultaneous exchange of information in self-driving cars makes it easier to bring them individually to the streets without causing any accidents. Also, it will improve the methods of managing the smart grid by providing quick solutions to any shift in the energy demand or supply rate. It will also enhance healthcare expertise through real-time diagnosis and endoscopic operations that will be made possible by the new latency, thus extending the health care outreach.

6.3.1.3 Enhanced Reliability and Security

6G will offer improved security and reliability, safeguard personal and official data and infrastructure networks from cyber threats, and maintain the functionality of the IoT [31]. With that, 6G will also be equipped with enhanced encryption features and quantum security measures due to the increasing incidences of cyber threats. This transfer will improve the reliability of the connection as well as the quality of the connection in a difficult environment. The intelligent networks of 6G will also ensure that security protocols would be more dynamic in nature with a corresponding positive assurance towards the IoT. It ensures that the technology is able to handle large data while at the same time ensuring that privacy of users' data is maintained, thus making it easier for users to develop security in the technology.

6.3.1.4 Ubiquitous Coverage

6G will provide integrated and coherent coverage of the urban application landscape [32] to make a smart city capable of offering high-speed connection and clear communication in all its zones. Besides, full area coverage will be assured in the 6G technology, since it will apply AI and ML algorithms to adjust network resources dynamically. It effectively enhances real-time information processing and analysis, thus driving better the management of cities' physical infrastructure and delivery of public services. Through the transition between 5G and 6G network, continuity and smooth connections will be achieved to meet with the increasing needs in urban areas.

6.3.1.5 Enabling Technologies

6G will include key enabling technologies like AI, big data analytics, and the Internet of Senses [33], major elements for delivering smart cities with interactive environment. In addition, 6G will gain efficiency and security through incorporation of latest technologies such as those in 5G, such as edge computing and quantum communication. Smart city applications require quick responses and involve data processing that must occur close to the data source, a need addressed by edge computing [34]. Quantum communication over certain distance guarantees secure means of protection needed in highly intertwine systems [35]. And, besides, the holographic communication integration will have an option for an immersive user experience, which will transform the human exposure towards the content encountered in the public areas and homes. Each of these technologies will create the basis to develop new, more flexible, and intelligent architectures for urban environments.

6.3.1.6 Sustainable and Green Technology

6G technology will explore methods to reduce harm to environment and enhance the emergence of smart cities that are friendly to the natural environment and use energy effectively [28, 36]. Besides environment preservation, the development of 6G technology will also focus on energy saving communication networks so as to have low effects on carbon footprint [37]. In particular, 6G seeks to enhance renewable energy sources with an intent to advance smart cities that have sustainable infrastructure systems. Besides, through green technology, continuous availability of environmental data for management shall be achieved to promote better decision-making in order to sustain ecosystems. Thus, 6G will promote the development of integrated and resilient societies with the help of technology while preserving the environment.

Therefore, 6G-IoT technology and its principal attributes and modern tendencies will become essential for the further advancement of smart cities. As a global technology supporting high-speed data transmission, ultra-low latency, enhanced security, and coverage, 6G marks the beginning of a new era that has a potential to fundamentally transform urban environments by improving efficiency, quality of life, and sustainability across cities.

6.3.2 *Evolution of IoT from First to Sixth Generations of Wireless Networks*

The evolution of IoT across wireless networks from 1G to 6G highlights transformative technological progress as shown in Fig. 6.11. IoT firstly used the first generation (1G) for communication and then the second generation (2G), third generation (3G), fourth generation (4G), fifth generation (5G), and now it is moving to implement the sixth generation (6G). This has led to the advancement of more complex yet more extensive IoT applications, which has promoted smart technology and connected device infrastructure globally.

6.3.2.1 1G (1980s)—Basic Connectivity

1G, which emerged in the early 1980s, refers to the first generation of cellular technology developed to provide basic connectivity [38]. It is said to be fixed and it mainly consisted of voice communication via radio waves. Some of the data were large and complex while others pertained to basic services such as texting, data transfer or internet connectivity which were previously unavailable. Security was limited, as call interception was not a big challenge at all.

6.3.2.2 2G (1990s)—SMS and Basic Data

The 1990s were considered as the decade when the subsequent evolution of mobile communications called the 2G was mainly characterized by the extension of short message service (SMS) and initial data services. This period marked the era in which people started using SMS as one of the shortest, fastest, and efficient means

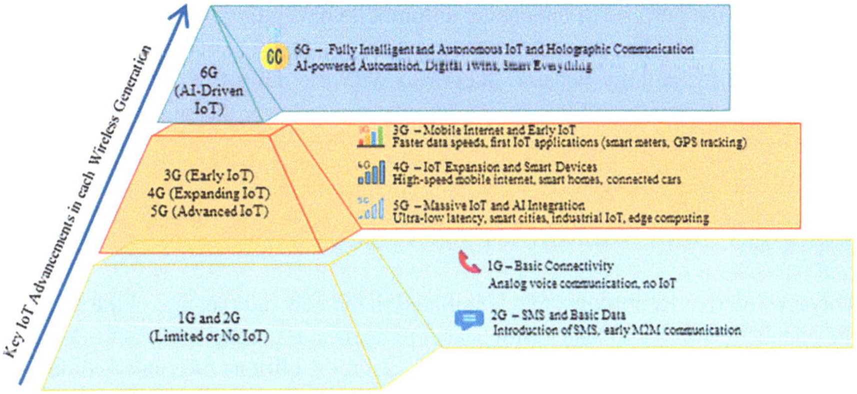

Fig. 6.11 Key IoT advancement in each wireless generation

of sending messages, as well as the sharing of information across the world [38]. Besides, 2G offered the early provision of the M2M communication system through which devices could transmit information to one another freely without the need for a human operator. This technology supported virtually all applications, starting from telemetry, initial stages of remote control, and monitoring of objects, toward creating the world of connected systems that is currently with us. In general, 2G played a crucial role of converting mobile networks from being more of a voice domain to a data domain to an extent.

6.3.2.3 3G (2000s)—Mobile Internet and Early IoT

The transition that occurred around the early 2000s to the 3G effectively boosted the mobile internet offering faster data transfer speeds for browsing and video calling [38]. This technological advancement set the stage for early development of the IoT. 3G made it possible to transmit data from different devices and resulted in the use of smart meters for utilities as well as GPS tracking systems. These early implementations of IoT offered more in terms of process automation and information control in areas like energy and supply chain. It was the time of the appearance of the idea of connected devices that could function and exchange information with each other through the use of mobile networks.

6.3.2.4 4G (2010s)—IoT Expansion and Smart Devices

The use of 4G technology was a significant revolution in the 2010s, which had a huge impact on the change of technology. This advancement enabled mobile internet access for millions of individuals and drove the rapid growth of IoT and smart devices [38]. 4G enabled connectivity of smart homes so that the devices could be controlled and operated efficiently for more convenience and saving energy. There were connected cars that provided effective means for exchanging information in real time for the purpose of enhancing automobile navigation, safety, and entertainment aspects [39]. The period also witnessed the construction of a globalized world, where smart technology became integrated with the daily lives of people which created great progress and laid down the infrastructure for the subsequent generation of mobility technologies.

6.3.2.5 5G (2020s)—Massive IoT and AI Integration

The new internet technology, 5G, is initiated in the new internet age in the 2020s as the connection becomes faster with near zero latency. It greatly benefits the Massive IoT domain and widely facilitates the embedding of billions of connected devices [40]. In smart cities, 5G facilitates efficient resource management and enhanced public services, from traffic control to environmental monitoring. IoT in the industrial sector receives considerable advantages from real-time analysis of gathered data regarding production processes and predictive maintenance. Additionally, AI with the support of edge computing, which performs the information processing closer to the source, improves the decision-making and automation in the 5G network. It is the integration of these technologies that leads to a world of smart systems that will define the destiny in most sectors in smart cities.

6.3.2.6 6G (2030s)—AI-Driven IoT and Holographic Communication

As for the 2030s, 6G is expected to change communication pattern by implementing AI, IoT, and holography. On the one hand, with new generation beyond 5G commonly referred to as the 6G, ubiquitous hyper-speed data transfer and ultra-reliable low-latency communication will be made available to support an intelligent interconnected ecosystem of end-to-end AI-driven automation across the industries [41]. Digital twins will emerge as essential in acting as a live model to support different fields such as smart cities, healthcare, and manufacturing. "Smart everything" will go further with devices continuing with learning and development through the use of AI hence increasing intelligence and functionality. Moreover, 6G also contributes to the realization of immersive holographic communication, as the current virtual interaction is transitioned to the 3D interaction experience [42]. This age will bring a drastic change towards a smarter, connected, and engaged society to transform the ways of living and working.

6.4 6G-IoT Applications in Sustainable Smart Cities

Smart cities operate with optimized performance through the integration of 6G technology. Real-time measurement on energy use, traffic flow, and air quality is enabled by higher integration and data transmission speed. Smart grids enhanced through 6G efficiently manage energies in a way that they minimize wastage and encourage the usage of green energy. Through ITS, appropriate and environmentally sustainable transport solutions are provided in cities. Moreover, new solutions to collect data for 6G-IoT enhance waste management by giving the best routes selection and developing recyclable systems. Improved geometric and solution-based AR and VR improve urban planning, architectural design, and the public's participation. In general, 6G-IoT sets the foundation for making smart sustainable cities. The applications of 6G-IoT in sustainable smart cities are discussed below and illustrated in Fig. 6.12.

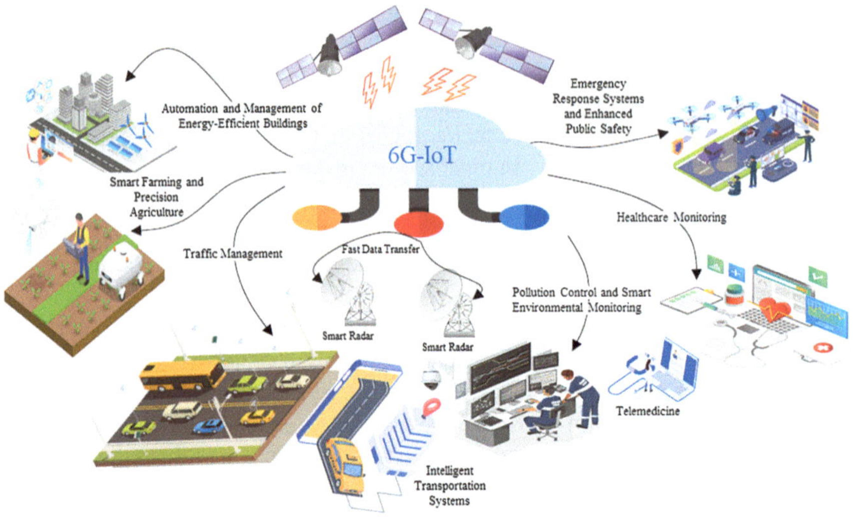

Fig. 6.12 6G-IoT application in sustainable smart cities

6.4.1 Pollution Control and Smart Environmental Monitoring

The sixth generation of wireless technology, which will be ten times faster than the fifth generation, will be beneficial for pollution control [43]. By integrating 6G-IoT, smart cities will have a closed-loop system where sensors, devices, and services are smart and interconnected enough to continuously share relevant data to create the most efficient environment in terms of energy usage, waste management, and air/water quality. Higher data transmission rates, lower delay, and more reliability of 6G networks will enable the continuous monitoring and analysis of environmental parameters. In intelligent waste management, 6G-IoT locate collection schedules, enhance the sorting process, and raise recycling standards [14]. In addressing air pollution problem, 6G-IoT applications offer real-time air pollution information from numerous sensors which are deployed across the city. This would give solution to city authorities to deal with difficulties such as traffic jam, insufficient public transport use, and poor use of electric vehicles. Further, 6G-IoT is utilized for an intelligent distribution of the green areas that are required for coping with the heat island phenomenon. Smart environmental monitoring is one of the visions of the sixth generation of IoT, which will offer an innovative technology for constructing future life and governing cities [44]. Stating the real data on air, water, traffic, energy consumption, etc., smart monitoring systems support city managers and other officials in making the right decision for a better life of residents.

6.4.2 *Traffic Management and Intelligent Transportation Systems*

Traffic management is an important factor in the implementation of smart city, especially in relation to the 6G-IoT. As the world advances toward the next generation of communication, 6G offers an ultra-fast, reliable, and efficient network that enhances traffic control systems. Through, 6G-IoT, there will be full integration of all traffic strategies, for improved real-time communication, data exchange, and smart traffic control [45]. In smart cities, the utilization of 6G-IoT includes traffic management, which will offer traffic insights that will help authorities predict the characteristics of traffic and control it. Moreover, vehicle-to-vehicle (V2V) communication systems and vehicle-to-infrastructure (V2I) communication systems will enable vehicles, traffic signals, and traffic monitoring system to communicate effectively. Intelligent transport systems (ITS) are one of the significant applications of 6G-IoT in sustainable smart cities [46]. This is an innovative technology which provides an improvement in the management and operation of transport through issues of safety, efficiency, and environmental concerns. Through implementing newest wireless communication, IoT, and big data analytics, ITS helps to solve problems of congestion and emissions, also to decrease energy consumption. This is made possible through options such as traffic monitoring in real time, connected car to car and car to infrastructure communications, and self-driving vehicles. ITS is not only implementing better traffic flow and facial recognition to reduce traffic congestion but also enhancing new smart cities to be environmentally friendly for generations to come.

6.4.3 *Automation and Management of Energy-Efficient Buildings*

Automation of energy-efficient buildings in smart cities is one of the innovations that can be embodied in 6G-IoT technology [47]. Environmentally friendly infrastructures form the core of sustainable cities' growth and development. They incorporate many advanced technological facilities to control its usage and wastage of energy efficiently. Smart management is the use of algorithms and AI in the manipulation of information obtained from other smart systems for the optimization of energy usage. This leads to reduction in energy consumption and complements occupant comfort in that the systems adjust and predict the outcome of the user. In addition, 6G-IoT technologies are also capable of facilitating the use of renewable energy power generation sources inside buildings including solar and wind power [48]. Additionally, areas like prediction in maintenance or energy use are applied where it will decrease the environmental impact and operation costs of energy-efficient buildings. The applications of the automation and management of energy-efficient buildings utilizing 6G-IoT technology provides a model for smart and sustainable cities. Hence efficient, sustainable, low-energy, renewable energy-integrated 6G-IoT applications help create green, comfortable, and secure smart urban spaces.

6.4.4 Telemedicine and Healthcare Monitoring

Telemedicine is defined as the practice of medicine at a distance using telecommunication tools. It includes services including the assessment, advice, surveillance, teaching, and therapy. This model promotes patient recovery and enhances access to care, as well as reduce expenditures, time, and environmental effects that accompany in-person sessions [49]. The 6G-IoT ecosystem will be connected with huge number of devices and a high-capacity network to support large amounts of data, application, and services [50]. When it comes to telemedicine and health monitoring, this means accurate, highly dependable, and with minimal delay data exchange between patients, doctors, and IoT gadgets. Telemedicine services will become more interactive and immersive with the help of 6G-IoT technology. Using such advanced systems as audio-video techniques, AR, and VR, the doctor will be able to conduct a complete examination, diagnose the patient, or even make a treatment plan if the doctor and patient aren't at the healthcare facility [51]. Continuous health monitoring will therefore be achieved by users of 6G-IoT devices, within and outside health care facilities. Having such devices means that aspects like the heart rate, blood pressure, as well as oxygen level will be recorded firstly. The collected data will be statistically analyzed with preeminent algorithms in order to define possible health risks and early interferences, and also for proper distribution of health care resources. This will provide fast and efficient access to the patient information required for timely, research-based decisions and actionable interferences to recurrent healthcare costs. Towards a 6G-IoT future, it is key to promote an ecosystem that supports the advancement of telemedicine and healthcare monitoring to build a healthy and sustainable smart city.

6.4.5 Smart Farming and Precision Agriculture

In the modern world where phase of development is characterized by urbanization and population growth at a very higher rate than before, it has become very important that there is provision of adequate and sustainable means of providing food for the growing population [52]. Precision agriculture together with smart farming has come out as possible solutions to the current food security and environmental issues. These techniques will evolve to a new level, when 6G comes along and IoT farming techniques come to sustainable smart cities of the future. Smart farming is the application of innovative technology tools into farming processes for efficiency, effectivity, and sustainable production. Precision agriculture is a technique that incorporates precise application technologies and field information to make decisions which will improve crop productivity with less input. These two concepts are integrated to give a clear framework for the practices of sustainable agriculture in smart cities. The 6G wireless technology and the IoT are anticipated to be the primary support systems of smart cities and proficiency of real-time communication. Through high-speed, low-latency, and reliable 6G networks, all IoT devices, sensors, and machines for smart farming and precision agriculture in sustainable smart cities will be able to effectively connect [53]. Using 6G-IoT smart farming and precision agriculture technologies, farmers are able to make better decisions in respect of crop yields, avoid unnecessary use of water and fertilizers as well as not to have much impact on the external environment. For example, remote sensors can include data on moisture and temperature of the soil as well as nutrient content, and assist farmers to overcome issues of irrigation and fertilization. Some of the complex computations with help of the real-time monitoring and expert systems can help to forecast equipment failures, choose optimal time for the maintenance to minimize the equipment off time or increase its resource life by utilizing ML and AI approach [54]. Moreover, with the process of global urbanization actively ongoing, the provision of the population's food needs with multifunctional, highly productive, environmentally friendly, and relatively resource-saving 6G-IoT-based intelligent and precision farming and agricultural systems will be critical for today's and future generations.

6.4.6 Emergency Response Systems and Enhanced Public Safety

While the world moves closer to smart cities, it becomes crucial to put into practice various technologies that will enhance the lives of the people living there while also helping to safeguard them [55]. Sometimes, such application includes the implementation of 6G-IoT for incorporation in emergency response and improved public security and safety solutions [10]. This innovation addresses the challenges faced by smart cities in achieving high resilience and sustainability, while enabling fast, dependable, and efficient responses to disasters. Emergency response systems are amongst the most important elements of the smart city model since, during critical conditions, the inhabitants' lives should be protected. This enabled smart cities to integrate high-tech emergency response solutions supported by 6G-IoT technology that sense, detect, and respond to emergencies instantaneously. These systems allow combining one or more devices and sensors such as cameras, drones, and environmental sensors to collect data on dangerous conditions, incidents, and disasters

[56]. 6G-IoT helps organizations collect and process large volumes of data within shorter time by providing very high data rates and near-zero latency network. This makes it possible for emergency response teams to gain easy and direct access to the kind of data that needs analyzing in order to make important decisions when disaster strikes. Such systems are also enabled using AI and ML to not only detect but also to specifically predict the probability and the consequences of threat accomplishment in order to prevent threat accomplishment and lower victim costs.

Therefore, advances in 6G-IoT technology intervention in emergency response systems and better mechanisms in smart cities improve the vision of sustainable smart cities. With the help of developing opportunities of extending very high data rates, very low delay, and uninterrupted connections, such smart applications will help the cities to control several disasters and safety of people living in those cities.

6.5 Emerging Technologies and 6G-IoT Integration

The integration of 6G-IoT within smart city is expected to revolutionize living within cities. As connectivity, automation, and intelligence skills grow, so will cities improve in efficiency, sustainability, and equity. AI, ML, VR, edge computing, blockchain, and IoT will support effective and secure data sharing, as illustrated in Fig. 6.13.

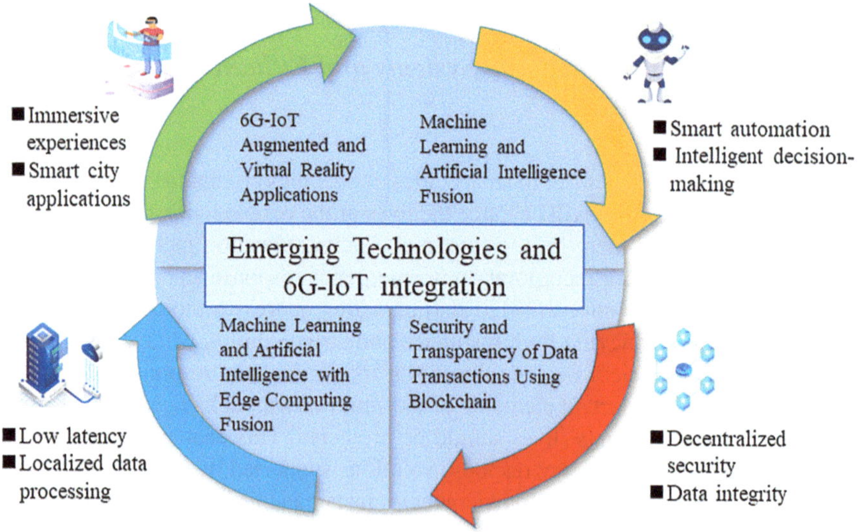

Fig. 6.13 The integration of 6G-IoT and emerging technologies

6.5.1 *Machine Learning and Artificial Intelligence Fusion*

AI and ML are a few of the most cutting-edge technological advancements that have brought quite a lot of change to human life and how humans interact personally with the physical resources [57]. It is the key components in driving the concept of smart cities to meet the needs of the inhabitants through the use of intelligent, efficient, sustainable and integrated infrastructures. ML algorithms allow the smart city to program data and to draw conclusions from existing and potential correlations. These predictions are used by the AI systems to arrive at decisions, automatize processes, and lead infrastructure hence enhancing efficiency. Another advantage for adopting ML and AI in smart cities as a fusion is the improvement in transportation systems. Real-time data from the traffic sensors, cameras and GPS vehicle-based systems are used to develop ML algorithms that learn traffic patterns, minimize traffic and also maximize time efficiency. Another application of the ML and AI in smart cities is in management of energy [58]. Advanced data analysis using ML forecasts energy uses as well as detect areas of wastage and match energy supply to demand. AI plays a role in power management by shutting down or reducing the operation of certain subsystems during periods of high demand. This leads to decreased energy consumption and reduction of greenhouse gasses. Security in the smart city also be increased with ML and AI. Using data from security cameras, social networks, and others, intensive means ML algorithms define threats and prognosis [59]. AI systems can then respond by sending authorities, calling in emergency services, or even take some form of action to stop crime or an accident. The combination of ML and AI with smart city technology has the potential to revolutionize living, working, and urban environment. These technologies are going to help the future generations have cities that are effective, efficient, and most importantly sustainable.

6.5.2 *Security and Transparency of Data Transactions Using Blockchain*

Smart city is not only the focus of technological way but also the center of the digital culture, and among these processes, data transactions are absolutely critical. Since data is produced in a very large quantity in smart cities, the integrity and privacy of data exchanges must be regulated to enhance the effectiveness of the smart city. The blockchain technology is the most appropriate solution to data transaction in smart cities. Bitcoin is a peer to peer, decentralized, distributed digital currency that has a transparent record in a block chain [60]. It is highly secure and sufficiently open for practical realization in the conditions of smart city environment. Data transactions occur in a smart city within a conceptual framework of a smart city where the information flows among two or more different parties, such as government, the other commercial establishments, or a citizen. Sometimes this may contain identity and location data, business and operational data, traffic patterns, or other information about facility energy usage. Smart cities will therefore be able to control for such transactions through implementation of the blockchain technology, which will in turn mean that such transactions will be secure and that there will be no possibility of fraud, misuse or even access to sensitive data [61]. This is a fundamental advantage of using blockchain in smart cities since it is possible to achieve very high levels of transparency in data transaction [62]. Also, through decentralized nature and cryptographic algorithms, the technology protects smart city data transactions on blockchain. That makes it almost impossible for anyone to manipulate the data since the infrastructure of the blockchain is decentralized. For additional augmentation of security and trust in data trade as a part of smart cities, such consensus algorithms like, proof of work and proof of stake can be incorporated. Providing a transparent and well-protected environment for data sharing, blockchain plays a pivotal role in building trust and optimizing organizational process in smart cities, which thus opens the path to more sustainable and future-oriented solutions.

6.5.3 Processing and Analyzing Real-Time Data with Edge Computing

Edge computing has potential to make real-time data processing and analysis and is capable of enhancing performance, efficiency, and scalability of applications in smart cities [63]. In smart cities, real-time data and analytics are most critical in the improvement of the living standards of citizens and efficiency of facilities in smart cities. In traffic management systems, sensors and cameras installed on roads and intersections transmit data to IoT and edge computing platforms, which analyze the information in real time to support decision-making. This means that the traffic lights are intelligent and able to provide the optimum time intervals to prevent traffic jam. Similarly, for monitoring air and water quality, as well as public security and consuming energy, edge computing implementation can be applied in real time. 6G, when combined with edge computing, elevates the concept of smart cities by introducing transformative abilities that drive innovation and disruption [64]. Categorized by its even higher data rates, lower latency, and more connectivity the 6G is expected to support a new generation of IoT devices and applications. This will lead to a tremendous amount of data volumes being created and a need to process and analyze data in real time at the edge. From smart and self-driving cars to delivery drones and from smart glass and contact lenses to AR/VR, 6G and edge computing will serve as a key enabler of smart cities.

6.5.4 6G-IoT Augmented Reality and Virtual Reality Applications

The arrival of 6G technology will enable the creation of an environment that utilizes ultra-fast, reliable, and low-latency connectivity. The main requirement for the next generation of applications such as AR and VR and this revolutionizes the entire concept of smart cities to customers using IoT products and technologies to make the life of urban environments better [65]. The services of 6G-IoT, AR, and VR in smart cities includes enhanced traffic management, improved urban planning and design, immersive educational experiences, enhanced public safety, boosting tourism and cultural experiences, and holistic healthcare. In traffic management, AR helps to deliver navigation and routing information to drivers immediately [66], while VR is used to create realistic simulation of traffic situations [67]. This contribute towards improved urban planning of the smart city hence improving the smart city. Smart cities can provide mission and interactive learning styles to children of different ages due to 6G-IoT-enabled AR and VR applications. Applications in smart cities include monitoring patients, training doctors and nurses, and performing remote health examinations. While VR produce realistic environment with which medical professional practice on, AR can be used to improve telehealth and provide proper diagnostic instruments.

6.6 Open Research Challenges and Research Directions

Among all technological and social transformation, smart cities are progressing due to IoT and wireless communication technologies. With 6G networks, new features are realized, including ultra-low latency, huge connectivity, and edge intelligence towards enabling the growing complexity of urban systems. However, research challenges and opportunities do exist in integrating the two of them (6G and IoT) in smart cities [68]. The open research challenges and potential research on achieving the 6G-IoT and implement to smart cities must be with respect to scalability, security, interoperability, energy efficiency, user-centric design, etc. Figure 6.14 illustrates the open research challenges and research directions in 6G-IoT enabled smart city.

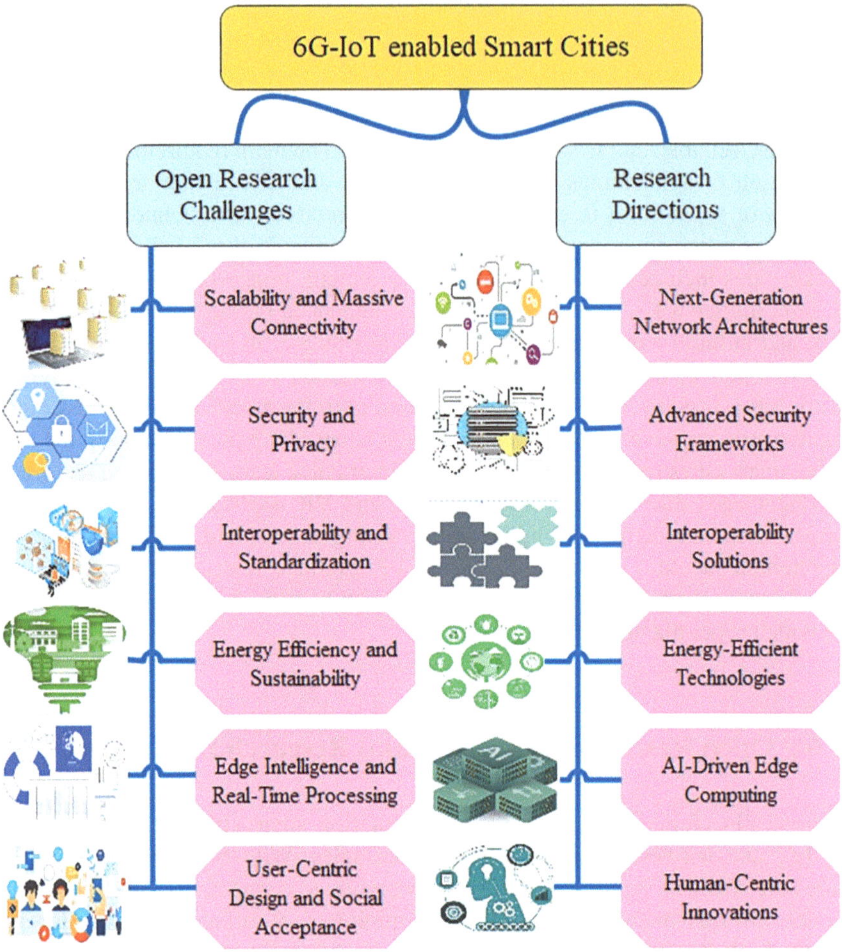

Fig. 6.14 Open research challenges and research directions in 6G-IoT-enabled smart city

6.6.1 Open Research Challenges in 6G-IoT-Enabled Smart City

6.6.1.1 Scalability and Massive Connectivity

The explosive growth of IoT devices brings enormous challenges to the network scalability issue [69]. 6G networks will need to support billions of interconnected devices while ensuring high performance and reliability. It requires efficient resource allocation and management mechanisms to deal with massive data traffic, generated by IoT devices in real time. The challenges on the scalability are handled by the development of advanced multiple access techniques including, non-orthogonal multiple access (NOMA) and the optimization of network slicing towards supporting diverse applications and services [70].

6.6.1.2 Security and Privacy

6G-IoT together opens smart cities to new cyber threats such as data breach, unauthorized access, and denial of service attack. For 6G-IoT networks to ensure secure communication there is need of robust encryption methods, secure authentication protocols, and intrusion detection systems. The collection and processing of an immense amount of personal data gives rise to the privacy concerns. The development of privacy protecting mechanisms (such as differential privacy and federated learning) is essential [71].

6.6.1.3 Interoperability and Standardization

Interoperability in smart cities as a collection of heterogeneous IoT devices and platforms is challenging [72]. For 6G-IoT systems, standardization efforts are required to describe common communication protocols, data formats and application programming interfaces (APIs). Ensuring the standardization, compatibility, and smooth integration between the existing systems and emerging technologies is one of the most important issues for which people must bridge the gap [73].

6.6.1.4 Energy Efficiency and Sustainability

Energy consumption remains a major concern for resource-limited environments and IoT devices, even with the deployment of 6G infrastructure. To reduce the carbon footprint of smart cities, energy-efficient communication protocols, low-power devices, and green network architectures must be developed [12]. Further research on energy harvesting techniques and intelligent power management can make contributions to sustainability.

6.6.1.5 Edge Intelligence and Real-Time Processing

Due to huge volume of data created by IoT devices, processing and analyses on the edge will be highly efficient for reducing latency and bandwidth usage. As an example, integration of edge computing and AI is challenged by computational resource allocation, distributed learning, and model synchronization. Real-time decision-making in dynamic urban environments is a critical research area [74].

6.6.1.6 User-Centric Design and Social Acceptance

User acceptance and trust in the smart cities are the crucial key factors for the 6G-IoT to be successful. Key factors are designing intuitive and user-friendly interfaces, ensuring data transparency, and solving ethical issues. Use of participatory approaches which involve the citizens in smart city solution design and implementation provides for integration into society of smart city solutions [75].

6.6.2 Open Research Directions in 6G-IoT Enabled Smart City

6.6.2.1 Next-Generation Network Architectures

The new generations of networks are crucial for the smart cities, especially considering the near future vision of 6G and IoT devices [76]. iFuture advancements in this field highlight the significance of the following elements. Firstly, 6G networks will provide much higher bandwidths and minimal delay that are vital for processing data in real time in smart city infrastructures. This will eventually lead to the integration of various IoT devices to enhance urban living, improve traffic flow, optimize energy usage, and strengthen public safety, among other areas. Moreover, 6G is predicted to accommodate the vast number of connected devices, to deal with the issues of IoT devices' growth rates. It will make data collection and processing more efficient so that cities make optimal choices for the organization of the urban environment. Furthermore, the application of AI and edge computing will also implement in 6G architectures [77]. The development of the new generation network topologies for 6G as well as IoT offers a great potential for enabling smart cities that would be more connected, intelligent, efficient, and secure.

6.6.2.2 Advanced Security Frameworks

As smart cities are becoming more complex as a result of adopting 6G-IoT, it becomes even more crucial to establish the security framework. These frameworks concern themselves with offering a solution to security to include data integrity, confidentiality, and data availability for a host of IoT appliances. Much focus is on the dynamic methods of protecting systems where AI and ML are used in threat identification and resolution. Further, other decentralized systems like blockchain are also under discussion to improve transparency and reliability of the data transaction. Interoperability of the different devices and systems is also an important consideration to achieve communication compatibility and security. These enhanced security frameworks aim at building an environment that incorporates sustainable and future-proof systems for the complex system of smart cities in the 6G-IoT environment [78].

6.6.2.3 Interoperability Solutions

In the aspects of developing advanced smart cities with 6G-IoT, the issue of Interoperability is highlighted as a major research domain. The interoperability solutions focus on addressing the power of the various systems, devices, or platforms to communicate and interconnect each other in the smart city environment [79]. This research focuses on issues like data fusion, security, and scale of the system. Interoperability thus allows cities to improve on resources usage, offer services to people and provide with new solutions in the urban environment [80].

6.6.2.4 Energy-Efficient Technologies

Another research is the deployment of efficient and sustainable smart city technologies using the 6G-IoT technology. 6G attributes such as low latency, high speed, and massive connection density guarantee the implementation of complex IoT solutions at a large scale. Therefore, there is rising demand of energy-efficient measures for tackling the problem and managing the increased power consumption leading to urban sustainability [81]. It established that application of energy-efficient technologies is capable of enhancing resource utilization in smart city systems to cover any environmental costs resulting from the implementation of smart economy. Thus, with innovative 6G features, smart grids, ITS, and automatic environmental monitoring become feasible in cities using less energy. Experts suggest finding innovative solutions such as ML-based algorithms that will enable dynamic control of energy consumption and development of new and improved batteries to power the IoT devices to maximum lifespan. In 6G, the network slicing concept is further extended to provide application-specific energy-efficient slices for the city verticals. Green communication protocols for interactions and techniques of energy-harvesting are also relevant to the creation of smart cities. These initiatives will not only enhance service delivery in cities but also contribute significantly to energy conservation and the promotion of renewable energy, aligning with environmental sustainability goals.

6.6.2.5 Artificial Intelligence-Driven Edge Computing

AI-based edge computing is one of the most researched domains as part of the smart cities and 6G-IoT technologies development. This makes use of edge devices, which involves the processing of data close to its source to reduce the latency and effective bandwidth which are key when it comes to real-time applications in cities. Through adoption of AI in smart cities, improved and efficient decision-making at edge level is achieved in aspects such as resources, security, and delivery of other public services to its citizens [81]. Edge computing, AI, and 6G-IoT are expected to complement each other in supporting efficient, smart, and connected IoT applications in various innovations on cities services including traffic control, energy supply, and sensing of the environment. Moreover, this integration helps in the formation of "smart", which includes self-driving automobiles, adaptive transportation systems, and so on, and the privacy and security of critical data can be maintained through localized computing. In light of the massive population of cities in the future, near real-time advanced AI-enabled edge computations may be essential to foreseeing smart, adaptive, and inclusive cities for the citizens of the cities.

6.6.2.6 Human-Centric Innovations

Human-centric aspects of innovation in smart cities enabled by 6G-IoT are reshaping the research objectives with a great impact on the society. The main focus is to apply ethical principles in the development of AI technology to promote fairness and equality in smart city goals. Ethical objectives in this research focus in three ways: transparency, accountability, and privacy so they create ethical AI, which rises as a human ethical right to access technologically [82]. Community needs, behavior, and cultural context are analyzed using social sciences in order to find the solutions that are both technological and acceptable by the community. Following ethical AI principles within project planning and deployment will ensure that smart city projects improve public confidence, engagement, and promote the development of robust and sustainable technology cities. This research direction aims at making smart cities fair so that people get benefited brought forward by experts and technological innovation.

The application of 6G and IoT convergence has the ability to impact the smart cities in a positive way enabling various advanced application and services. Realizing this vision, however, presents a number of research challenges: scalability, security, interoperability, energy efficiency, and user-centric design. The research directions discussed are open space for the academic and industrial communities to contribute to the construction of sustainable, secure and inclusive smart city ecosystems. Such challenges must be overcome through continued innovation and collaboration in order to utilize the full potential of 6G-IoT in smart cities.

6.7 Summary

Through convergence of 6G and IoT technologies, smart cities achieve advanced applications and services. 6G-IoT drives sustainable urban development by enabling real-time monitoring, intelligent decision-making, and efficient resource utilization for the future. Applications range from pollution control and traffic management, energy optimization, healthcare, agriculture, to public safety. Other emerging technologies including AI, blockchain, edge computing, and AR/VR enhance the 6G-IoT systems capability. Achieving this vision, however, is impeded by major challenges such as scalability, security, interoperability, energy efficiency, and user-centric design. With all these AI-powered predictive analytics, blockchain-based secure transactions, edge computing for real-time processing, there is a need for interdisciplinary research. The research directions promote the involvement of researchers, policymakers, and industry stakeholders on the creation of sustainable, secure, and inclusive smart cities enabled by 6G-IoT technologies. To get over these challenges and enable the transformative power of this technology, continued innovation is necessary.

References

1. A.R. Javed et al., Future smart cities: Requirements, emerging technologies, applications, challenges, and future aspects. Cities **129**, 103794 (2022). https://doi.org/10.1016/J.CITIES.2022.103794
2. L. Qiao, Y. Li, D. Chen, S. Serikawa, M. Guizani, Z. Lv, A survey on 5G/6G, AI, and robotics. Comput. Electr. Eng. **95**, 107372 (2021). https://doi.org/10.1016/J.COMPELECENG.2021.107372
3. M. Tariq, M. Ali, F. Naeem, H.V. Poor, Vulnerability assessment of 6G-enabled smart grid cyber–physical systems. IEEE Internet Things J. **8**(7), 5468–5475 (2021). https://doi.org/10.1109/JIOT.2020.3042090
4. A. Kumar, M. Masud, M.H. Alsharif, N. Gaur, A. Nanthaamornphong, Integrating 6G technology in smart hospitals: Challenges and opportunities for enhanced healthcare services. Front. Med. (Lausanne) **12** (2025). https://doi.org/10.3389/fmed.2025.1534551
5. P. Mishra, G. Singh, 6G-IoT framework for sustainable Smart City: Vision and challenges, in *Sustainable Smart Cities*, (Springer, Cham, 2023), pp. 97–117. https://doi.org/10.1007/978-3-031-33354-5_5
6. S.K. Rao, R. Prasad, Impact of 5G technologies on Smart City implementation. Wirel. Pers. Commun. **100**(1), 161–176 (2018). https://doi.org/10.1007/s11277-018-5618-4
7. A.V. Jha, B. Appasani, M.S. Khan, S. Zeadally, I. Katib, 6G for intelligent transportation systems: Standards, technologies, and challenges. Telecommun. Syst. **86**(2), 241–268 (2024). https://doi.org/10.1007/s11235-024-01126-5
8. M. Noor-A-Rahim et al., 6G for vehicle-to-everything (V2X) communications: Enabling technologies, challenges, and opportunities. Proc. IEEE **110**(6), 712–734 (2022). https://doi.org/10.1109/JPROC.2022.3173031
9. K.Y. Yap, H.H. Chin, J.J. Klemeš, Future outlook on 6G technology for renewable energy sources (RES). Renew. Sust. Energ. Rev. **167**, 112722 (2022). https://doi.org/10.1016/J.RSER.2022.112722
10. A. Alnoman, A.S. Khwaja, A. Anpalagan, I. Woungang, Emerging AI and 6G-based user localization technologies for emergencies and disasters. IEEE Access **12**, 197877–197906 (2024). https://doi.org/10.1109/ACCESS.2024.3521005
11. M.M. Kamruzzaman, Key technologies, applications and trends of internet of things for energy-efficient 6G wireless communication in smart cities. Energies (Basel) **15**(15), 5608 (2022). https://doi.org/10.3390/en15155608
12. P. Mishra, G. Singh, Energy management systems in sustainable smart cities based on the internet of energy: A technical review. Energies (Basel) **16**(19), 6903 (2023). https://doi.org/10.3390/en16196903
13. Z. Qadir, K.N. Le, N. Saeed, H.S. Munawar, Towards 6G internet of things: Recent advances, use cases, and open challenges. ICT Express **9**(3), 296–312 (2023). https://doi.org/10.1016/J.ICTE.2022.06.006
14. Z. Guo, Y. Shen, C. Chakraborty, F. Alblehai, K. Yu, Industrial 6G-IoT and machine-learning-supported intelligent sensing framework for indicator control strategy in sewage treatment

process. IEEE Internet Things J. **11**(18), 29308–29320 (2024). https://doi.org/10.1109/JIOT.2024.3429341
15. O. Ghosh, B. Kumar, The impact of 6g technology on environmental ecosystems and human life, in *6G Impacts on Natural Habitats and Human Life*, (IGI Global, 2025), pp. 255–290. https://doi.org/10.4018/979-8-3373-2220-9.ch010
16. F. Guo, F.R. Yu, H. Zhang, X. Li, H. Ji, V.C.M. Leung, Enabling massive IoT toward 6G: A comprehensive survey. IEEE Internet Things J. **8**(15), 11891–11915 (2021). https://doi.org/10.1109/JIOT.2021.3063686
17. J.K. Abbas, S.I. Hassan, A.H. Mohsen, A. Hussain, F.H.T. Hussain, Next-gen education with 6G connectivity and intelligent data analytics, in *2024 International Conference on Emerging Research in Computational Science (ICERCS)*, (IEEE, 2024), pp. 1–6. https://doi.org/10.1109/ICERCS63125.2024.10895621
18. S. Nayak, R. Patgiri, 6G communication technology: A vision on intelligent healthcare, in *Health Informatics: A Computational Perspective in Healthcare*, Studies in Computational Intelligence, vol. 932, (Springer, Singapore, 2021), pp. 1–18. https://doi.org/10.1007/978-981-15-9735-0_1
19. Z. Allam, D.S. Jones, Future (post-COVID) digital, smart and sustainable cities in the wake of 6G: Digital twins, immersive realities and new urban economies. Land Use Policy **101**, 105201 (2021). https://doi.org/10.1016/j.landusepol.2020.105201
20. A. Kumari, R. Gupta, S. Tanwar, Amalgamation of blockchain and IoT for smart cities underlying 6G communication: A comprehensive review. Comput. Commun. **172**, 102–118 (2021). https://doi.org/10.1016/j.comcom.2021.03.005
21. Z. Allam, S.E. Bibri, D.S. Jones, D. Chabaud, C. Moreno, Unpacking the '15-Minute City' via 6G, IoT, and digital twins: Towards a new narrative for increasing urban efficiency, resilience, and sustainability. Sensors **22**(4), 1369 (2022). https://doi.org/10.3390/s22041369
22. N. Chen, M. Okada, Toward 6G internet of things and the convergence with RoF system. IEEE Internet Things J. **8**(11), 8719–8733 (2021). https://doi.org/10.1109/JIOT.2020.3047613
23. A. Sharma, E. Podoplelova, G. Shapovalov, A. Tselykh, A. Tselykh, Sustainable smart cities: Convergence of artificial intelligence and Blockchain. Sustainability **13**(23), 13076 (2021). https://doi.org/10.3390/su132313076
24. A. Slalmi, H. Chaibi, A. Chehri, R. Saadane, G. Jeon, Toward 6G: Understanding network requirements and key performance indicators. Trans. Emerg. Telecommun. Technol. **32**(3) (2021). https://doi.org/10.1002/ett.4201
25. M.M. da Silva, J. Guerreiro, On the 5G and beyond. Appl. Sci. **10**(20), 7091 (2020). https://doi.org/10.3390/app10207091
26. A. Mishra, A.V. Jha, B. Appasani, A.K. Ray, D.K. Gupta, A.N. Ghazali, Emerging technologies and design aspects of next generation cyber physical system with a smart city application perspective. Int. J. Syst. Assur. Eng. Manag. **14**(S3), 699–721 (2023). https://doi.org/10.1007/s13198-021-01523-y
27. S. Borchardt, G.B. Vignola, D. Buscaglia, M. Maroni, *Mapping EU Policies with the 2030 Agenda and SDGs-Fostering Policy Coherence through text-based SDG Mapping* (2022). https://doi.org/10.2760/110687
28. T. Huang, W. Yang, J. Wu, J. Ma, X. Zhang, D. Zhang, A survey on green 6G network: Architecture and technologies. IEEE Access **7**, 175758–175768 (2019). https://doi.org/10.1109/ACCESS.2019.2957648
29. Y. Zhang, M.F.A. Bin Azizui, C. Yan, The application of 6G and augmented reality technology in education and training. Wirel. Pers. Commun. (2024). https://doi.org/10.1007/s11277-024-11038-x
30. P. Popovski et al., A perspective on time toward wireless 6G. Proc. IEEE **110**(8), 1116–1146 (2022). https://doi.org/10.1109/JPROC.2022.3190205
31. V. Ziegler, P. Schneider, H. Viswanathan, M. Montag, S. Kanugovi, A. Rezaki, Security and trust in the 6G era. IEEE Access **9**, 142314–142327 (2021). https://doi.org/10.1109/ACCESS.2021.3120143

32. W. Jiang, B. Han, M.A. Habibi, H.D. Schotten, The road towards 6G: A comprehensive survey. IEEE Open J. Commun. Soc. **2**, 334–366 (2021). https://doi.org/10.1109/OJCOMS.2021.3057679
33. T. Bin Ahammed, R. Patgiri, 6G and AI: The emergence of future forefront technology, in *2020 Advanced communication technologies and signal processing (ACTS)*, vol. 2020, (IEEE), pp. 1–6. https://doi.org/10.1109/ACTS49415.2020.9350396
34. L. Zhao, G. Zhou, G. Zheng, C.-L.I.X. You, L. Hanzo, Open-source multi-access edge computing for 6g: Opportunities and challenges. IEEE Access **9**, 158426–158439 (2021). https://doi.org/10.1109/ACCESS.2021.3130418
35. C. Wang, A. Rahman, Quantum-enabled 6G wireless networks: Opportunities and challenges. IEEE Wirel. Commun. **29**(1), 58–69 (2022). https://doi.org/10.1109/MWC.006.00340
36. R. Kumar, S.K. Gupta, H.-C. Wang, C.S. Kumari, S.S.V.P. Korlam, From efficiency to sustainability: Exploring the potential of 6G for a greener future. Sustainability **15**(23), 16387 (2023). https://doi.org/10.3390/su152316387
37. W. Wu, C.-S. Yang, I.-K. Fu, P.-K. Liao, D. Calin, M. Fan, Revisiting the system energy footprint and power efficiency on the way to sustainable 6G systems. IEEE Wirel. Commun. **29**(6), 6–8 (2022). https://doi.org/10.1109/MWC.2022.10003083
38. M. Alsabah et al., 6G wireless communications networks: A comprehensive survey. IEEE Access **9**, 148191–148243 (2021). https://doi.org/10.1109/ACCESS.2021.3124812
39. A. Agarwal, K. Agarwal, The next generation mobile wireless cellular networks—4g and beyond. Am. J. Electr. Electron. Eng. **2**(3), 92–97 (2014). https://doi.org/10.12691/ajeee-2-3-6
40. A. Yıldız, Š. Džakmić, M. Ahmed Saleh, A short survey on next generation 5G wireless networks. Sustain. Eng. Innov. **1**(1), 57–66 (2019). https://doi.org/10.37868/sei.v1i1.93
41. A.A. Ahmed Solyman, K. Yahya, Evolution of wireless communication networks: from 1G to 6G and future perspective. Int. J. Electr. Comput. Eng. (IJECE) **12**(4), 3943 (2022). https://doi.org/10.11591/ijece.v12i4.pp3943-3950
42. X. Shen et al., Toward immersive communications in 6G. Front. Comput. Sci. **4** (2023). https://doi.org/10.3389/fcomp.2022.1068478
43. D. Chauhan et al., Balancing technological innovation and environmental sustainability: A lifecycle analysis of 6G wireless communication technology. Sustainability **16**(15), 6533 (2024). https://doi.org/10.3390/su16156533
44. M.S. Jamil, M.A. Jamil, A. Mazhar, A. Ikram, A. Ahmed, U. Munawar, Smart environment monitoring system by employing wireless sensor networks on vehicles for pollution free smart cities. Proc. Eng. **107**, 480–484 (2015). https://doi.org/10.1016/J.PROENG.2015.06.106
45. J. Logeshwaran, S.K. Patel, O.P. Kumar, F.A. Al-Zahrani, Hybrid optimization for efficient 6G IoT traffic management and multi-routing strategy. Sci. Rep. **14**(1), 30915 (2024). https://doi.org/10.1038/s41598-024-81709-z
46. A. Aghaei Anvigh, Y. Khavan, B. Pourghebleh, Transforming vehicular networks: How 6G can revolutionize intelligent transportation? Sci. Eng. Technol. **4**(1), 80–93 (2024). https://doi.org/10.54327/set2024/v4.i1.127
47. M. Shehab, T. Khattab, M. Kucukvar, D. Trinchero, The role of 5G/6G networks in building sustainable and energy-efficient smart cities, in *2022 IEEE 7th International Energy Conference (ENERGYCON)*, (IEEE, 2022), pp. 1–7. https://doi.org/10.1109/ENERGYCON53164.2022.9830364
48. A. Chel, G. Kaushik, Renewable energy technologies for sustainable development of energy efficient building. Alex. Eng. J. **57**(2), 655–669 (2018). https://doi.org/10.1016/J.AEJ.2017.02.027
49. P. Mishra, G. Singh, Internet of medical things healthcare for sustainable smart cities: Current status and future prospects. Appl. Sci. **13**(15), 8869 (2023). https://doi.org/10.3390/app13158869
50. J.R. Bhat, S.A. Alqahtani, 6G ecosystem: Current status and future perspective. IEEE Access **9**, 43134–43167 (2021). https://doi.org/10.1109/ACCESS.2021.3054833

51. T.K. Vashishth, V. Sharma, K.K. Sharma, B. Kumar, S. Chaudhary, R. Panwar, Virtual reality (VR) and augmented reality (AR) transforming medical applications, in *AI and IoT-Based Technologies for Precision Medicine*, (IGI Global, 2023), pp. 324–348. https://doi.org/10.4018/979-8-3693-0876-9.ch020
52. S. Polymeni, D.N. Skoutas, P. Sarigiannidis, G. Kormentzas, C. Skianis, Smart agriculture and greenhouse gas emission mitigation: A 6G-IoT perspective. Electronics (Basel) **13**(8), 1480 (2024). https://doi.org/10.3390/electronics13081480
53. S. Polymeni, S. Plastras, D.N. Skoutas, G. Kormentzas, C. Skianis, The impact of 6G-IoT technologies on the development of agriculture 5.0: A review. Electronics (Basel) **12**(12), 2651 (2023). https://doi.org/10.3390/electronics12122651
54. T. Ayoub Shaikh, T. Rasool, F. Rasheed Lone, Towards leveraging the role of machine learning and artificial intelligence in precision agriculture and smart farming. Comput. Electron. Agric. **198**, 107119 (2022). https://doi.org/10.1016/J.COMPAG.2022.107119
55. M. Lacinák, J. Ristvej, Smart City, safety and security. Proc. Eng. **192**, 522–527 (2017). https://doi.org/10.1016/J.PROENG.2017.06.090
56. S. Islam, A. Karam, A. Boualem, K.V. Pathan, A.F. Sayed, A. Rahman, 6G-enabled network security in drone technology: Revolutionizing communications and enhancing security applications, in *Nordic e-Infrastructure Tomorrow. NeIC 2024*, Communications in Computer and Information Science, (Springer, Cham, 2025), pp. 28–44. https://doi.org/10.1007/978-3-031-86240-3_3
57. C. Casetti, C.F. Chiasserini, F. Dressler, A. Memedi, D. Gasco, E.M. Schiller, AI/ML-based services and applications for 6G-connected and autonomous vehicles. Comput. Netw. **255**, 110854 (2024). https://doi.org/10.1016/J.COMNET.2024.110854
58. A.T. Mithul Raaj et al., Intelligent energy management across smart grids deploying 6G IoT, AI, and Blockchain in sustainable smart cities. IoT **5**(3), 560–591 (2024). https://doi.org/10.3390/iot5030025
59. N. Maheswaran, S. Bose, T. Shruthi, G. Logeswari, T. Anitha, D. Prabhu, Transformative storage framework for 6G smart city deployments, in *2024 Third International Conference on Electrical, Electronics, Information and Communication Technologies (ICEEICT)*, (IEEE, 2024), pp. 1–6. https://doi.org/10.1109/ICEEICT61591.2024.10718383
60. A. Hughes, A. Park, J. Kietzmann, C. Archer-Brown, Beyond Bitcoin: What blockchain and distributed ledger technologies mean for firms. Bus. Horiz. **62**(3), 273–281 (2019). https://doi.org/10.1016/J.BUSHOR.2019.01.002
61. A. Jahid, M.H. Alsharif, T.J. Hall, The convergence of blockchain, IoT and 6G: Potential, opportunities, challenges and research roadmap. J. Netw. Comput. Appl. **217**, 103677 (2023). https://doi.org/10.1016/J.JNCA.2023.103677
62. X. Shen et al., Blockchain for transparent data management toward 6G. Engineering **8**, 74–85 (2022). https://doi.org/10.1016/J.ENG.2021.10.002
63. L.U. Khan, I. Yaqoob, N.H. Tran, S.M.A. Kazmi, T.N. Dang, C.S. Hong, Edge-computing-enabled smart cities: A comprehensive survey. IEEE Internet Things J. **7**(10), 10200–10232 (2020). https://doi.org/10.1109/JIOT.2020.2987070
64. A. Alawadhi, A. Almogahed, E. Azrag, Towards edge computing for 6G internet of everything: Challenges and opportunities, in *2023 1st International Conference on Advanced Innovations in Smart Cities (ICAISC)*, (IEEE, 2023), pp. 1–6. https://doi.org/10.1109/ICAISC56366.2023.10085007
65. M.Z. Aloudat et al., Metaverse unbound: A survey on synergistic integration between semantic communication, 6G, and edge learning. IEEE Access **13**, 58302–58350 (2025). https://doi.org/10.1109/ACCESS.2025.3555753
66. L. Abdi, F. Ben Abdallah, A. Meddeb, In-vehicle augmented reality traffic information system: A new type of communication between driver and vehicle. Proc. Comput. Sci. **73**, 242–249 (2015). https://doi.org/10.1016/J.PROCS.2015.12.024

67. M. Hasan, D. Perez, Y. Shen, H. Yang, Distributed microscopic traffic simulation with human-in-the-loop enabled by virtual reality technologies. Adv. Eng. Softw. **154**, 102985 (2021). https://doi.org/10.1016/J.ADVENGSOFT.2021.102985
68. P. Mishra, G. Singh, *Sustainable Smart Cities* (Springer International Publishing, Cham, 2023). https://doi.org/10.1007/978-3-031-33354-5
69. Y. Bin Zikria, R. Ali, M.K. Afzal, S.W. Kim, Next-generation internet of things (IoT): Opportunities, challenges, and solutions. Sensors **21**(4), 1174 (2021). https://doi.org/10.3390/s21041174
70. M.B. Shahab, R. Abbas, M. Shirvanimoghaddam, S.J. Johnson, Grant-free non-orthogonal multiple access for IoT: A survey. IEEE Commun. Surv. Tutor. **22**(3), 1805–1838 (2020). https://doi.org/10.1109/COMST.2020.2996032
71. S. Sharma et al., The role of 6G technologies in advancing smart city applications: Opportunities and challenges. Sustainability **16**(16), 7039 (2024). https://doi.org/10.3390/su16167039
72. I. Tsampoulatidis, N. Komninos, E. Syrmos, D. Bechtsis, Universality and interoperability across smart city ecosystems, in *Distributed, Ambient and Pervasive Interactions. Smart Environments, Ecosystems, and Cities. HCII 2022. Lecture Notes in Computer Science*, (Springer, Cham, 2022), pp. 218–230. https://doi.org/10.1007/978-3-031-05463-1_16
73. Y. Mehmood, F. Ahmad, I. Yaqoob, A. Adnane, M. Imran, S. Guizani, Internet-of-things-based smart cities: Recent advances and challenges. IEEE Commun. Mag. **55**(9), 16–24 (2017). https://doi.org/10.1109/MCOM.2017.1600514
74. B. Nathali Silva, M. Khan, K. Han, Big data analytics embedded Smart City architecture for performance enhancement through real-time data processing and decision-making. Wirel. Commun. Mob. Comput. **2017**, 1–12 (2017). https://doi.org/10.1155/2017/9429676
75. P. Mishra, G. Singh, Introduction: Importance of sustainable smart city, in *Sustainable Smart Cities*, (Springer International Publishing, Cham, 2023), pp. 1–37. https://doi.org/10.1007/978-3-031-33354-5_1
76. M. Murroni et al., 6G—Enabling the new smart city: A survey. Sensors **23**(17), 7528 (2023). https://doi.org/10.3390/s23177528
77. K.B. Letaief, Y. Shi, J. Lu, J. Lu, Edge artificial intelligence for 6G: Vision, enabling technologies, and applications. IEEE J Sel Areas Commun **40**(1), 5–36 (2022). https://doi.org/10.1109/JSAC.2021.3126076
78. P. Mishra, G. Singh, 6G-IoT framework for sustainable smart city: vision and challenges. IEEE Cons. Electron. Mag. **13**(5), 93–103 (2024). https://doi.org/10.1109/MCE.2023.3307225
79. D. Lakshmi et al., IoT in power electronics: transforming the future of energy management, in *IoT for Smart Grid*, (Wiley, 2025), pp. 107–125. https://doi.org/10.1002/9781394279401.ch6
80. B. Ahlgren, M. Hidell, E.C.-H. Ngai, Internet of things for smart cities: Interoperability and open data. IEEE Internet Comput. **20**(6), 52–56 (2016). https://doi.org/10.1109/MIC.2016.124
81. M.W.P. Maduranga, V. Tilwari, R.M.M.R. Rathnayake, C. Sandamini, AI-enabled 6G internet of things: Opportunities, key technologies, challenges, and future directions. Telecom **5**(3), 804–822 (2024). https://doi.org/10.3390/telecom5030041
82. B. Gera, Y.S. Raghuvanshi, O. Rawlley, S. Gupta, A. Dua, P. Sharma, Leveraging AI-enabled 6G-driven IoT for sustainable smart cities. Int. J. Commun. Syst. **36**(16) (2023). https://doi.org/10.1002/dac.5588

Chapter 7
Healthcare 5.0: Smart and Connected Healthcare Systems for Sustainable Smart Cities

Contents

P. Mishra, G. Singh, *Sustainable Smart Cities 2.0*,
https://doi.org/10.1007/978-3-032-01102-2_7

7.1 Introduction

Healthcare 5.0 is an advancement of the healthcare systems moving to the smart and connected healthcare for residents in smart cities [1]. The idea behind this concept is the combined use of smart sensors, internet of things (IoT), artificial intelligence (AI), and data analysis to develop individual, anticipatory, and patient-centred medicine, where the medicine is being developed for the individual [2]. When these technologies are incorporated into smart city environments, sustainable and responsive cities that address the growing demand in healthcare. Healthcare 5.0 helps to eliminate health differences and improve the human value in families and communities. This concept of embracing this new model of healthcare points to the possibility of changing the way people maintain health and health systems as the world progressively becomes more urban and globally connected.

7.1.1 The Need for Smart and Connected Healthcare

Medical care is an important segment of any community and is an important component in maintaining and improving the quality of human life in sustainable smart cities [3]. This is particularly the case in the present information age where healthcare has been defined by the emergent advanced and networked systems. These systems apply information and communication technologies (ICT) to expand and advance the established models of providing enhanced, more effective, and innovative solutions [4]. This is because the need for smart and connected healthcare has been driven by the constant need for accessible, affordable, and customized health care services [5]. The future smart and connected health care systems have the potential to address these shortcomings of traditional healthcare. Connected and smart healthcare includes electronic health records (EHRs), telemedicine, wearables, AI, and big data [6]. These technologies are employed in diagnosis, treatment planning, and patient monitoring as well as planning of medication dosages. For example, EHRs provide clinicians with access to electronic information from multiple sources in order to make clinical decisions on time [7]. Telemedicine enables patients to consult with a healthcare practitioner without physical presence at the location, thereby reducing the travel among healthcare [8]. AI as well as big data analytics interventions are crucial components that improve the effectiveness of smart and connected health systems [9]. Diagnostic systems using AI can process a huge amount of big data and assist doctors in making more precise and prompt decisions; while machine learning (ML) models estimate patients' condition and find the people who require intervention as soon as possible [10]. In the world of health care, smart and connected healthcare systems is on the verge of a radical change and the lives of millions of people will be improved. Figure 7.1 illustrates the concept of smart and connected healthcare, integrating technologies like IoT, AI, 5G, and cloud computing to enable real-time diagnostics, remote monitoring, and personalized treatment. It highlights the essential components such as infrastructure and security, key enabling technologies, and the evolution from healthcare 1.0 to 5.0. It also emphasizes the role of data privacy, cybersecurity, and digital tools in modern healthcare systems. However, it is important to look at the difficulties and the hazards that are involved in using such technologies including the risks to patient data, achieving compatibility and developing good understanding of the technologies among the practitioners and patients. As the health care system changes time and again, it is important to stay flexible and be able to meet the society's needs and the continuous change in technology that defines our world. The future of healthcare is defined by smarter and more interconnected systems and helping patients, healthcare providers, and policy makers [11].

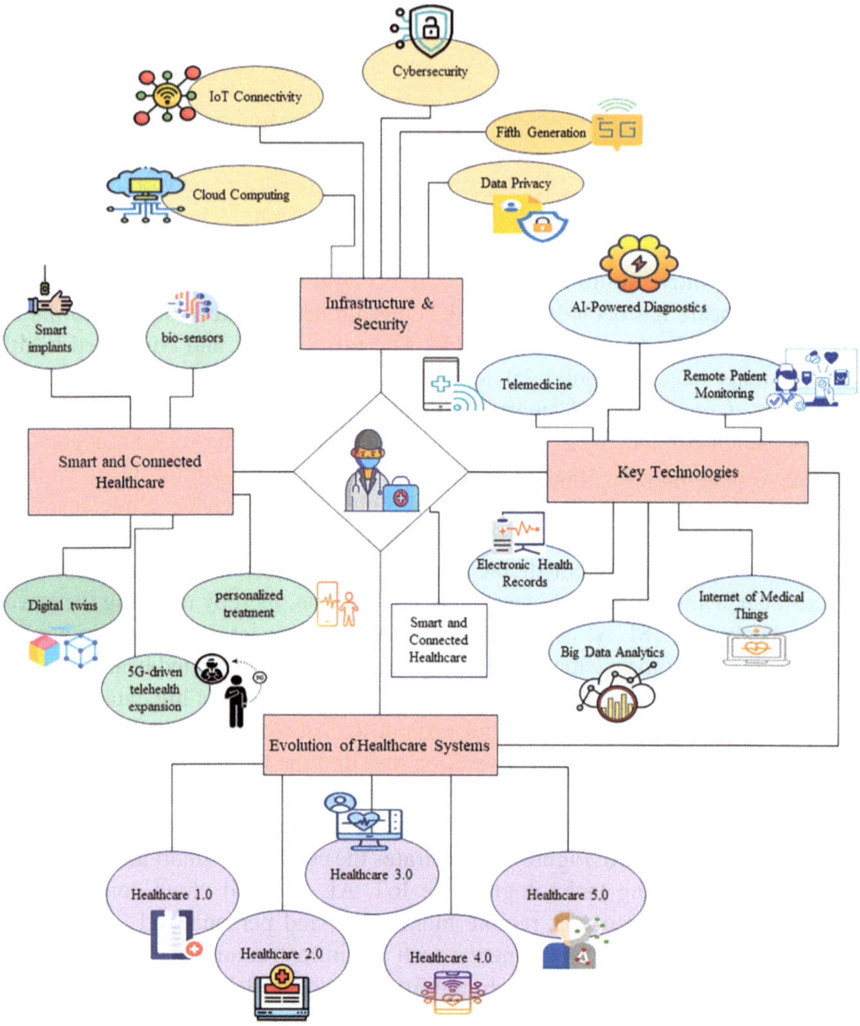

Fig. 7.1 Evolution, technologies, and infrastructure enabling intelligent medical systems

7.1.2 *Role of Technology in Healthcare*

Healthcare has seen major change due to technology as it reaches as far as diagnosis, treatment, patient management and operational efficiency [12]. With advancement in magnetic resonance (MR), computed tomography (CT) scans, wearables devices, and other medical equipment, accurate and early disease detection has become possible. This enables on-time interventions, potentially saving lives by identifying symptoms at an early stage [13]. A growing number of AI and ML algorithms are used to analyse large amounts of medical data in order to help doctors make diagnoses, predict patient outcomes, and provide personalized treatments [14, 15]. Telemedicine has broken the geographical barriers to enable people living in remote areas access high-quality health services. Medical services have been made more accessible and convenient through virtual consultations, remote monitoring, and digital prescriptions. EHRs are another example where efficient information management has enhanced communication among healthcare providers and reduced the risk of errors [16]. Digital systems allow clinicians to make faster decisions based on data that has been synthesized resulting in improved outcome. Robotics and minimally invasive surgical methods have revolutionized procedures conferring increased precision, faster recovery time and decreased risk [17]. Besides, in the field of biotechnology, new frontier is opened in the treatment of genetic disorders and development of personalized therapy based on biological tools, including clustered regularly interspaced short palindromic repeats (CRISPR), for genome editing [18]. Wearable health technology and mobile apps put the power in the hands of people to monitor their own health and create opportunities to practice preventative care and a healthier lifestyle [19]. On the whole, technology plays an ever more important role in modernizing healthcare, increasing effectiveness, precision, and patient outcomes. It addresses global health challenges with significant potential and innovation, leading to a future where high-quality care is accessible to all and the healthcare environment becomes more patient-centered and technology-driven.

7.2 Current Research Trends of Healthcare in Smart Cities

This study investigates the trends, contributions, and impact of research in smart cities with healthcare from 2015 to 2025 through a bibliometric analysis of Scopus data. The Bibliometrix R package along with origin software were utilized to analyse and depict the data. This approach ensures that researchers are presented with the most current research developments in the field enabling them to stay informed on emerging trends and advancements. In March 2025, a bibliometric analysis was conducted utilizing the Scopus core collection database to thoroughly investigate the concept and understanding of healthcare in smart cities. Ultimately, 3143 articles were exported and saved in BibTex format before being included and examined in the bibliometric study. The primary information, syntax, and specific criteria used in the screening for this study are shown in Table 7.1.

Table 7.1 Summary of data characteristics and main information of the bibliometric study

Description	Specific Standard Requirements/Results
Research database	Scopus
Search query	("smart") AND ("cities") AND ("healthcare")
Time span	2015–2025
Document type	Articles, conference paper, book chapter, etc.
Language	English
Data export format	BibTex
Documents	3143
Annual growth rate %	12.4
Document average age	3.45
Average citations per doc	21.71
Keywords plus (ID)	10,876
Author's keywords (DE)	6560
Authors	9108
Authors of single-authored docs	208
Single-authored docs	342
Co-authors per doc	3.78
International co-authorships %	31.88

7.2.1 Annual Scientific Production and Top Ten Most Published Journals/Cited Papers

Figure 7.2 shows the annual publication trends from 2015 to 2025, highlighting the growth in research output. The early years shows a steady increase in publications, rising from 2015 (41 doc) to 2024 (675 doc), indicating a growing interest in the field. In 2025 (132), the number continues to rise and may continue increasing until the end of the year.

Figure 7.3 presents the highest number of articles in healthcare in the smart city research domain published in "IEEE Access" (84 docs), "Lecture notes in Networks and Systems" (77 docs), "Lecture notes in Computer Science (Including subseries lecture notes in Artificial Intelligence and Lecture notes in Bioinformatics)" (47 docs), and "Sensors" (43 docs), which indicates a higher output of articles compared to the other journals. In addition, some other journals, such as the "IEEE Internet of Things Journal" (39 docs), "ACM International Conference Proceeding Series" (37 docs), "communications in Computer and Information Science" (37 docs), "Sustainability (Switzerland)" (32 docs), "Lecture notes in Electrical Engineering" (31 docs), and "Sustainable Cities and Society" (29 docs), are in this domain.

Table 7.2 displays the top ten most cited papers from the study which presents both the ranking and authors with their published works. This section presents information about authors, article titles, and their corresponding publications in journals with documentation for publication years and citation frequency. The ranking focuses on engineering field smart cities articles to demonstrate their total

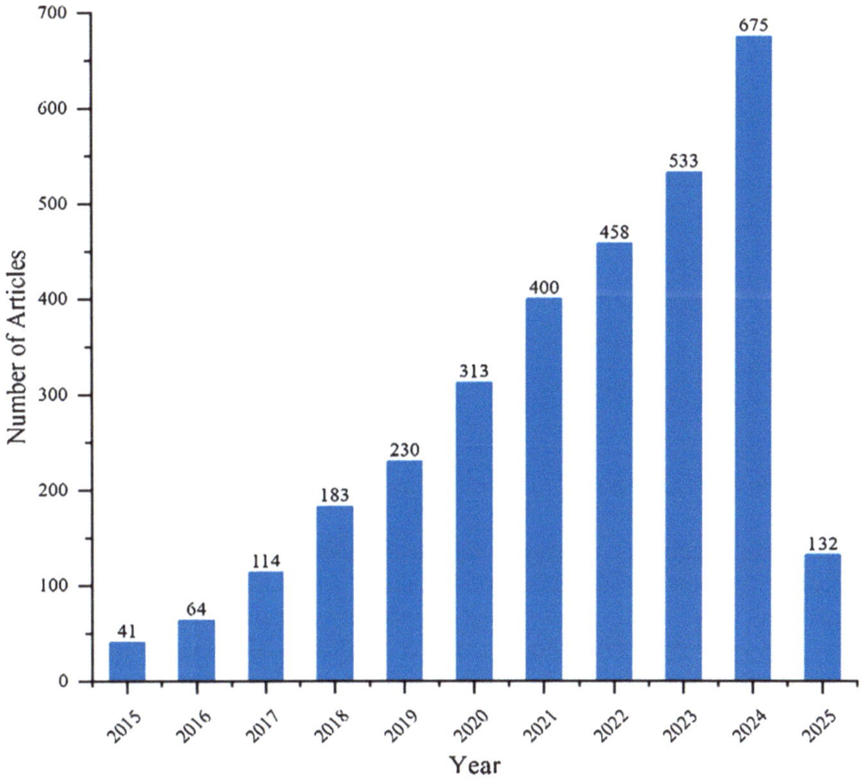

Fig. 7.2 Yearly trend of publication number with healthcare in smart city

citation (TC) counts and annual citation rates and normalized total citation (NTC) data which collectively signify their impact in the field.

7.2.2 Sankey Diagram (Three-Field Plot and Thematic Map)

Figure 7.4 Sankey diagram illustrates the three-field plot thematic progression to visualize the flow and the distribution in scientific research. In this case, keywords plus, abstract, and title are added, acting as indicators to highlight the most influential research papers in the subject. These keywords indicate vital concepts and themes that stand out in academic discussions of the smart city domain. The diagram's keywords are sourced from earlier published studies and were then thematically organized by a specific researcher. Figure 7.5 categorizes various smart city themes based on their healthcare goals and relevant degrees. In the top right quadrant (motor themes), which represents themes that are highly developed and highly relevant, there are terms like "smart city", "internet of things", "cloud computing",

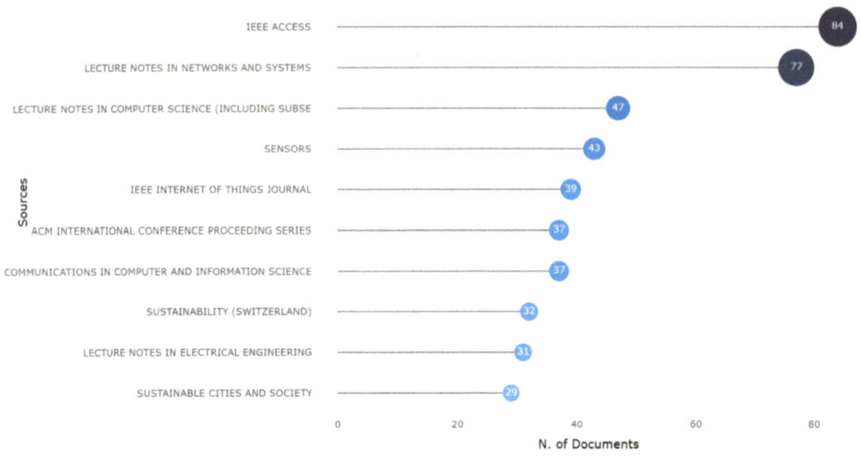

Fig. 7.3 Most relevant sources with number of articles with healthcare in smart city

and "Smart healthcare". Quadrant left (niche themes) represents "blockchain", "security", "privacy", and "federated learning". This indicates a strong focus on these areas, suggesting they are central to smart city research and practice in healthcare. This quadrant reflects the areas where significant advances and investments are made, highlighting their critical role in smart city. In contrast, the bottom left quadrant, which represents themes that are either emerging or declining and have lower relevance, includes terms like "machine learning", "artificial Intelligence", "deep learning", and "big data". Additionally, the terms like "internet of things", "smart cities", and "healthcare" serve as a base for more advanced analyses. These themes might be considered foundational but must be emphasized more in smart cities. This visual categorization provides valuable insights for educators, curriculum developers, and researchers, helping them understand which areas are gaining traction and which might need more attention or innovation

7.2.3 Co-occurrence Keyword Trends and Network Visualization

Figure 7.6 depicts the most prominent themes and keywords related to research in conceptual change. The most significant and most central terms, such as "internet of things," "smart city", "health care", "automation", and "deep learning" indicate the core focus of the research on how researchers grasp and alter their conceptual knowledge, particularly within smart city contexts.

An illustration in Fig. 7.7 displays word occurrence frequencies from the sustainable development goals in smart city research network. Research focuses mainly on "internet of things" as this term appears 1040 times throughout the literature.

Table 7.2 Most global cited documents with healthcare in smart city research domain

Rank	Authors name	Paper Title	Journal's Name	TC	TC/ Year	NTC
1	Sarker (2021) [20]	Machine learning: Algorithms, real-world applications and research directions	SN Computer Science	2801	560.20	73.35
2	Fuller et al. (2020) [21]	Digital twin: Enabling technologies, challenges and open research	IEEE Access	1458	243.00	36.00
3	Nguyen et al. (2023) [13]	Federated learning for internet of things: A comprehensive survey	ACM Computing Surveys	827	165.40	21.66
4	Farahani et al. (2018) [22]	Towards fog-driven IoT eHealth: Promises and challenges of IoT in medicine and healthcare	Future Generation Computer Systems	791	98.88	16.79
5	Dwivedi et al. (2019) [23]	A decentralized privacy-preserving healthcare blockchain for IoT	Sensors	716	102.29	21.70
6	Dagher et al. (2018) [24]	Ancile: Privacy-preserving framework for access control and interoperability of electronic health records using blockchain technology	Sustainable Cities and Society	700	87.50	14.86
7	Xie et al. (2019) [25]	A survey of blockchain technology applied to smart cities: Research issues and challenges	IEEE Communications Surveys & Tutorials	589	84.14	17.85
8	Wu et al. (2021) [26]	Digital twin networks: A survey	IEEE Internet of Things Journal	564	112.80	14.77
9	Shi et al. (2020) [27]	Smart textile-integrated microelectronic systems for wearable applications	Advanced Materials	543	90.50	13.41
10	Du et al. (2016) [28]	Recurrent marked temporal point processes: Embedding event history to vector	Proceedings of the 22nd ACM SIGKDD International Conference on Knowledge Discovery and Data Mining	532	53.20	12.19

Research on healthcare in smart cities indicates the importance of "smart city" and "healthcare" based on its 823 and 485 word occurrences, respectively. Additionally, "automation" (288), "deep learning" (270), "network security" (267) and "block chain" (214) combinations occupies crucial importance in emerging technology in healthcare fields. The data shows that smart city research revolves primarily around

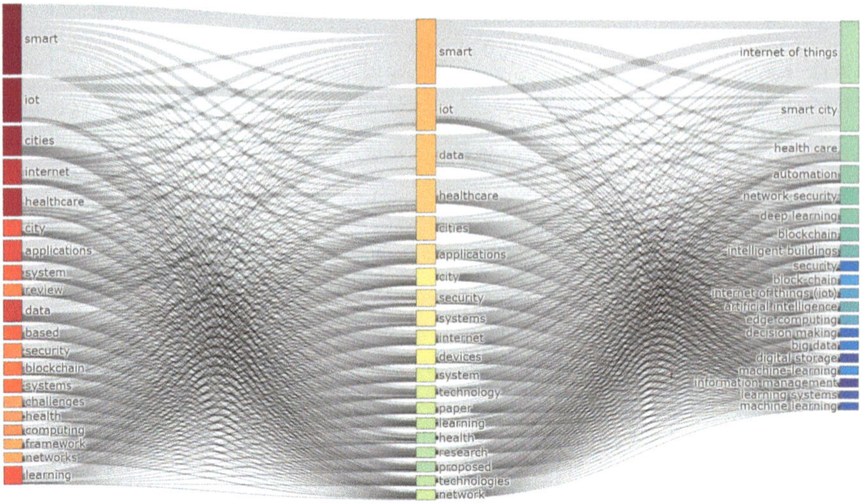

Fig. 7.4 Three field plot (thematic development) using sankey diagram

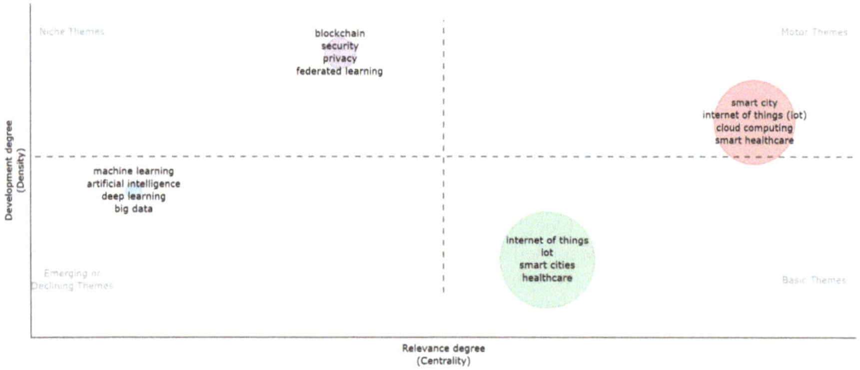

Fig. 7.5 Thematic map of healthcare in smart city research domain

several major subjects. The data shows that healthcare and smart city research revolves primarily around several major subjects.

7.2.4 *Countries' Scientific Production*

The scientific publication frequencies of various countries in the world are represented in Fig. 7.8. Scientific contributions from India stand at 2504, which demonstrates their powerful role in research development. The growing scientific power of China displays itself through its 603 scientific publications which rank second after India. The research hub of the USA produces 480 scientific papers among its total contributions. Numerous publications emerge from Saudi Arabia and UK together

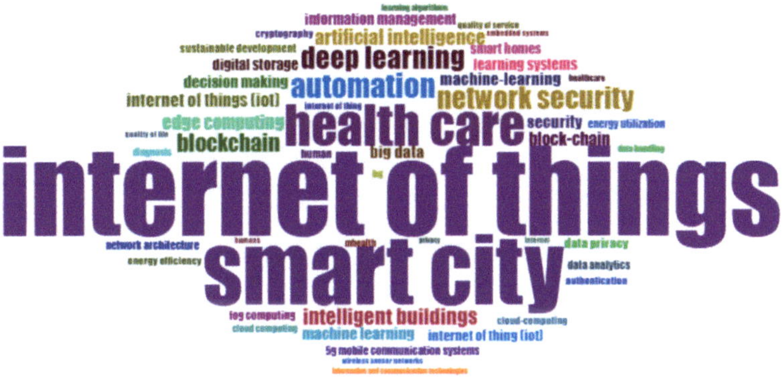

Fig. 7.6 Co-occurrence keyword trends in smart city domain and healthcare sector

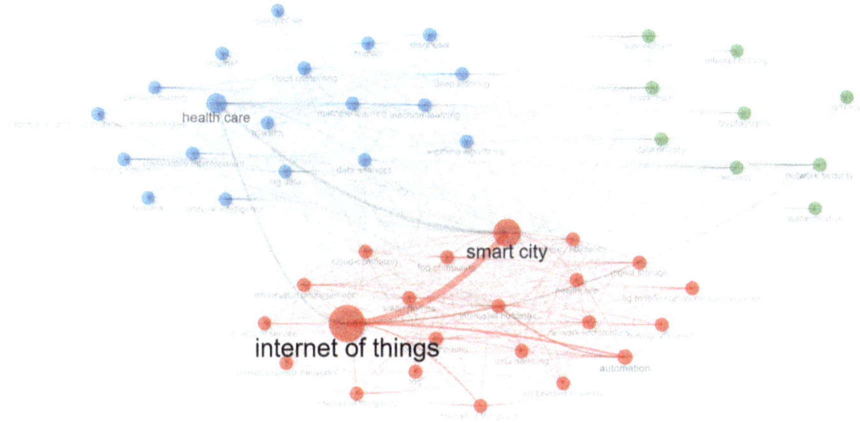

Fig. 7.7 Co-occurrence network of words in smart city research domain with healthcare

with South Korea given their research output comprises 434, 248, and 218 papers respectively. Academic and industrial research sectors of the Italy and Australia produce approximately 191 and 176 scholarly works between them. Malaysia and Spain have smaller publication numbers but their contributions amount to totals of 156 and 122 research items, respectively.

7.2.5 Mapping Healthcare in Smart City Research to Sustainable Development Goals (SDGs)

SDG Goals, its targets (bubble series), and the percentage (bar graph) of corresponding keywords detected in the text (calculated as the ratio of keywords in one Goal to the total number of keywords detected) are depicted in Fig. 7.9. For SDG mapping, the study utilized the SDG Mapper tool from the "Knowledge Base for the

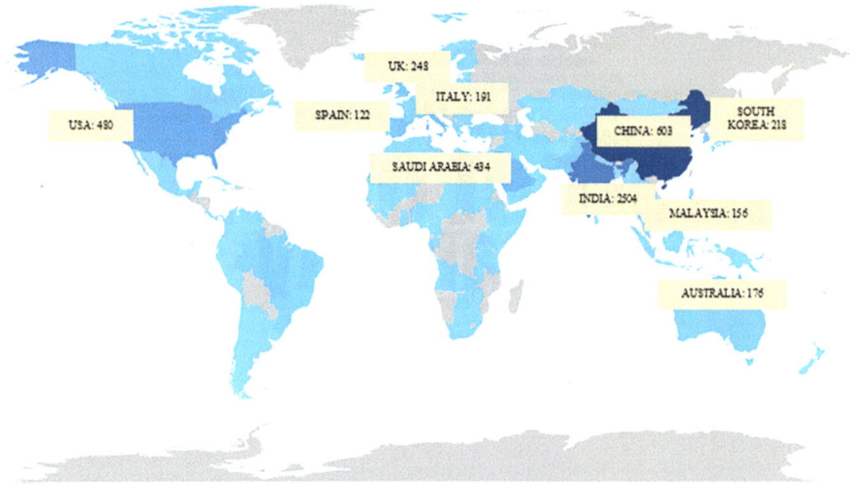

Fig. 7.8 Top ten countries' scientific production of various countries based on frequency

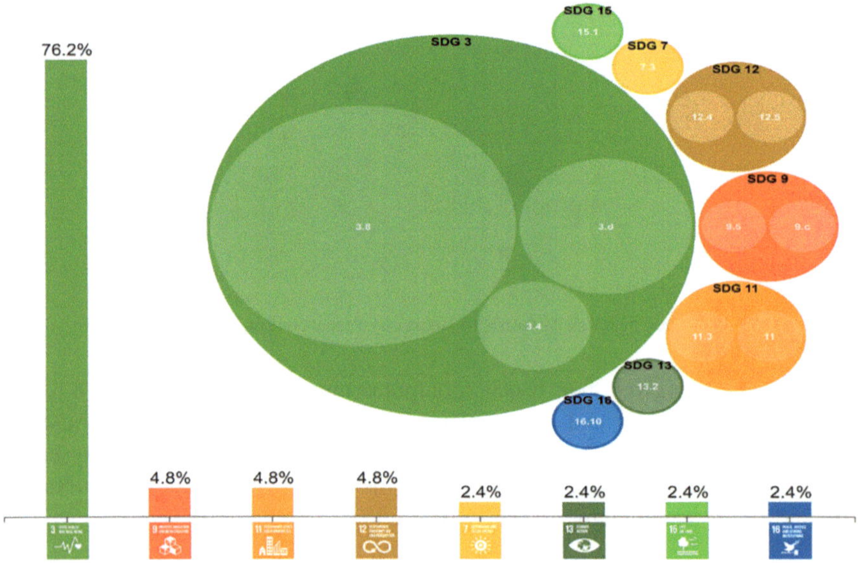

Fig. 7.9 Mapping healthcare in smart city research to relevant SDGs and targets

Sustainable Development Goals," provided by the European Commission [29]. This tool uses bubble charts to depict SDG interconnections and relevance among bibliometric records. The size of each bubble correlates with the proportion of associated keywords found, which is calculated as the number of goal-specific keywords divided by the total keyword count. As such, SDG 3, marked by 32 keyword

Table 7.3 SDGs ranking, targets, and relevance in healthcare-based smart city research domain with SDGs

SDGs (By Rank)	Description	Targets	No. of Occurrence		Relevance%
SDG 3	Good health and well-being	3.8	22	32	76.2
		3.d	7		
		3.4	3		
SDG 11	Sustainable cities and communities	11.3	1	2	4.8
		11	1		
SDG 9	Industry, innovation, and infrastructure	9.5	1	2	4.8
		9.c	1		
SDG 12	Responsible consumption and production	12.4	1	2	4.8
		12.5	1		
SDG 7	Affordable and clean energy	7.3	1	1	2.4
SDG 13	Climate action	13.2	1	1	2.4
SDG 15	Life on land	15.1	1	1	2.4
SDG 16	Peace, justice, and strong institutions	16.10	1	1	2.4

occurrences. This is followed by SDG 11, with 2 occurrences, and SDG 9, with 2 occurrences, underlining their relative prominence in the research database.

Within SDG 3, there are several key subtargets that the research addresses: Subtarget 3.8 is shown as a distinct bubble within the larger SDG 3 circle, indicating focused research healthcare in smart city. Subtarget 3.d, which aims in reduction and management of national and global health risk, is another bubble within the SDG 3 area. This highlights the efforts in research to improving the cities in healthcare sector. Subtarget 3.4 focuses on reducing premature mortality from non-communicable diseases by one-third through prevention and treatment, while also promoting mental health and well-being. Within SDG 11, there are several key subtargets that the research address: Subtarget 11.3 is shown as a distinct bubble within the larger SDG 11 circle, indicating focused research on smart city. Subtarget 11, which aims sustainable cities and communities, is another bubble within the SDG 11 area. This highlights the efforts in research to improving the cities. Similarly, SDG 9 (Industry, innovation, and infrastructure) is also closely linked to SDG 11 and aims to enhance scientific research and affordable access to internet. The detailed description of other relevant SDGs is depicted in Table 7.3.

7.3 Overview of Internet of Medical Things

Internet of Medical Things (IoMT) is a broad term that covers all medical devices and advanced software that uses the internet to share information. IoMT devices transform healthcare services by remote patient management, telemedicine, patient diagnosis, and overall low costs. This way the IoMT market is expected to grow at a fast pace given that there is growing demand in personally integrated medicine and efficient management or chronic diseases.

7.3.1 Types of IoMT Devices

7.3.1.1 Wearable Devices

These are small devices that patients may carry with them and keep a record of their health status anytime. These are smart bands that act as a fitness tracker, smartwatches, as well as biosensors that monitor heart rate, blood pressure, and blood glucose levels [30]. Such devices may aid patients in chronic illnesses in remembering to check their status. IoMT is related to wearable technology since it interconnects these devices to healthcare systems to facilitate the collection and transmission of data in real time between a patient and healthcare providers [31]. It is not limited to wearables but also extends to connected inhalers, insulin pumps, and even smart beds that contain sensors analysing a patient. These devices are capable of generating signals to the healthcare practitioners if there are complications to enable immediate action. Wearable devices combined with IoMT technologies have the potential to enhance patient care by enabling early intervention, continuous monitoring, and improved accessibility.

7.3.1.2 Implantable Devices

Implantable devices have remained one of the best interfaces of technology and medicine with promising potentials in handling both chronic and acute illnesses [31]. For example, modern pacemakers self-monitor the level of physical activity and regulate the rate of heartbeat accordingly to improve the patient's quality of life. Furthermore, advances in technology make some types of insulin pumps that track the glucose level and regulate it and release the insulin accurately as required. Recent developments with implantable cardioverter defibrillators (ICDs) help detect the type of the problem and respond accordingly eliminating the need of giving shocks to the patient [32]. Also, such devices include the facility for wireless connectivity so that the healthcare practitioners directly access the patient information and, if needed, alter the settings on the device and treatment plans without physically visiting the patients. Moreover, it has made it technologically possible to produce comparatively more compact devices that are placed in human bodies with marginal surgical complications and inconveniences to the patient. The advanced implantable technology of the future may contain all what the human body needs from a general medical plan to a specific scheme to solve relatively complex health issues that may require the use of advanced technologies and other health devices [33].

7.3.1.3 In Vitro Diagnostics (IVDs)

The market of IVDs is growing and increasing in demand in recent times due to new technologies for effective and efficient healthcare systems [34]. These diagnostics are not only used in clinical care but are also going more and more often in home care. Through connection of the IVDs to IoT, the flow of data in real time has been made possible making it easier for healthcare facilities to monitor patients and draw conclusion from real-time data collected. The advancement in point-of-care technologies (POCT) to increase access and availability, which has helped in early accurate diagnosis and treatment [35]. These devices are mobile, simple to employ and offer fast outcomes, which ought to be useful in areas of the limited supplies. Thus, new trends in biosensor technology also impact the increased sensitivity and specificity of IVDs to detect various conditions. The information that has been gathered from these diagnostics are fed into AI and ML algorithms to figure out better diagnostics and to predict how the patient may suffer in the future. This application of AI to IVDs is a great advancement towards precision medicine, where patients receive a treatment that is most suitable to them [36]. At the same time, legislative instruments are being adjusted to the innovative developments in order to maintain high safety and effectiveness of new IVD products. Therefore, patients, as well as healthcare providers, are likely to experience improved accuracy in the diagnosis imaging techniques.

7.3.1.4 Telemedicine Devices

Such devices allow distant interactions between the patients and the healthcare providers [37]. The devices are video conference systems, virtual reality headsets, and applications on smart phones. Telemedicine devices help provide healthcare services to people living in districts with limited physician availability or in remote and hard to reach areas. It also includes advanced diagnostic equipment such as digital stethoscopes, portable electrocardiogram (ECG), and patent ductus arteriosus (PDA)-controlled diagnostic ultrasounds. These tools enable one to monitor or examine patients as well as perform some tests without physically examining the patients. Also, telemedicine helps the patient to be engaged and educated with the available applications that include reminders, health tracking, and real-time consultations with health practitioners [38]. This technology is very useful in those illnesses which require check-ups frequently, such as in diabetes or hypertension. Moreover, telemedicine does not only enhance the availability of health care but also enhances the processes that take place within the health care field across the world. Telemedicine devices help provide healthcare services to people living in districts with limited physician availability or in remote and hard-to-reach areas.

7.3.1.5 Medical Imaging Devices

These devices use highly effective imaging techniques to produce pictures of the human body, these are the devices such as X-ray machine, CT scan, and MRI scan [39]. Tremendous progress has been made towards the introduction of image processing systems that are faster and use less radiation exposure. With AI and other advanced applications of ML in medical imaging, scans and related tests continue to achieve high accuracy, often surpassing the human eye, and detect pattern that might go unnoticed by clinicians. Moreover, there is the development of 3D imaging and virtual reality for presenting the possible anatomy of the patient for planning and teaching [40]. Each of these advancements result in a better treatment of patients, better individualized therapy, and more efficient functioning of medical centres. Therefore, these technologies revolutionize diagnostic imaging from being a static tool in the medical practice.

7.3.1.6 Robotic Devices

Technology in the healthcare field is developing robotic devices that enhance patients' well-being [41]. These robots help surgeons in increasing the precision of surgical movements thus allow for procedures that would be impossible to imagine to be conducted manually. Rehabilitation robots are another important category that assist individuals recovering from injuries or surgeries in regaining movement and strength [42]. These robots provide personalized, one-on-one therapy that adapts to patient's progress, ensuring that the exercises performed are accurate, appropriate, and of high equality. A hospital transport robot helps the organization in achieving coordination of resources within the health facility, since sometimes it is required to move medication, samples and equipment from one department to the other in a safe and speedy manner. Furthermore, most of these devices are embedded with sensors as well as AI technologies that allow them to work with much efficiency and precision. This integration of technology enables healthcare providers to provide timely and accurate services to patients. With the advancement of technology, there is even more that robotic-based devices offer to the healthcare system and these are the possibilities for enhancement in future care for patients along with more organizational improvements.

7.3.1.7 Medical Software Applications

In today's practice, medical software applications play critical roles in the improvement of healthcare system. The EHR system consolidates patient records, enabling relevant and accurate information to follow the patient across different healthcare professionals within an organization, thereby reducing the possibility of errors. Clinical Decision Support Systems (CDSS) are tools that utilize algorithms and patient data to assist experts in making decisions based on the information available in a patient's record [43]. Medication management systems support this process by tracking prescriptions and alerting healthcare providers to drug interactions, helping to prevent adverse situation.

7.3.2 IoMT System Architecture

The IoMT system architecture provides assistance in connecting and exchanging the information and data between devices and health care users. The architecture consists of various components, including the following:

7.3.2.1 Sensors and Actuators

IoMT system architecture consists of the sensors and actuators that make it possible to support proper functioning with collection of immediate data and responsive actions [44]. These facilities acquire data through different ways such as measuring the rate of heartbeat, glucose level, and the level of physical activity. It is used in smartwatches, fitness bands, smart clothing, and other wearable technologies that are constantly tracking the physiological characteristics of a patient. Such devices include pacemakers and insulin pumps where the implantable devices are capable of offering direct and improved measure of the human body activities.

7.3.2.2 Communication Networks

Communication system in the IoMT system architecture determines how the various users will be able to share the data in the systems efficiently. These networks facilitate monitoring or the patient data in real time hence enabling doctors intervene at the right time and manner to improve the patience's health [45]. Besides the wireless technologies, which include wi-fi, Bluetooth, and cellular networks and wired technologies including ethernet, there are other communication protocols and standards that are useful to IoMT architecture. One such technology is Zigbee that is characterized by low power consumption, and it is used in connecting medical devices in homes. Near Field Communication (NFC) is another wireless sensor network technology that is used for short ranges where communication may be needed in patients' identification or secure data transfers. Technologies like Low Power Wide Area Network (LPWAN) such as Long-Range Wide Area Network (LoRaWAN) are appealing because they enable communication over a vast distance with limited power consumption and therefore more suitable in supporting remote patient monitoring systems.

7.3.2.3 Data Storage and Management

Reception, storage, processing, and dissemination of data from IoMT patient related devices is a part of this component. It comprises cloud solutions alongside local storage solutions, data integration, analysis, and visualization tools [45]. Furthermore, IEEE 802.15.6 is a WBAN standard targeted towards healthcare application to facilitate the communication between the medical body wearable devices. CoAP (Constrained Application Protocol) is also being used in IoMT more and more as it is lighter protocol being suitable for constrained devices and networks. Moreover, with the advancement of 5G networks, higher speed and slightly delayed internet connection is still essential for health monitoring and telemedicine services. These protocols of communication guarantee that most of the IoMT systems are compatible with many devices and sensors, thereby enriching the information collected and analysed for better healthcare delivery.

7.3.2.4 Data Processing and Analysis

The components of data processing and analysis for establishing predictive models, risk assessment, and creating client-specific treatment plans, these components are helpful in the description of a real-time monitoring system and alert [45]. IoMT devices helps to detect possible health conditions by analysing the data streams collected by them on a constant basis and the healthcare providers intervenes at the right time assuming a preventive measure. For the health care system, the extent of use of big data and AI has the power to enhance the decision-making process in order to improve the accuracy and efficiency of delivery of treatment.

7.3.2.5 User Interfaces

The IoMT devices are interfaced through these interfaces to enable communication from healthcare providers and access to patient information. These include mobile apps, websites and wearable devices that notifies the patients' medical conditions [44]. IoMT can also be connected to smart home devices so that the patients can operate them through voice commands as well as receive notifications through smart home devices. Telemedicine has become a crucial aspect as a number of individuals have access to real-time video conferences with the doctor as a way of seeking treatment. Moreover, EHR systems are connected with IoMT for a smooth inflow of data so that the patient data will be more efficient for clinical diagnosis. This provides flexibility to the health care providers by allowing them to access patient information over the cloud as long as they are authorized. The datasets accumulated by IoMT devices can be analysed by AI-driven analytics platforms to produce information that will be beneficial to healthcare providers to enhance patient care plans.

7.3.2.6 Security and Privacy

This involves the ways of how the patient's information is guarded from being stolen or misused. The necessary security systems for the database include the installation of encryption methods that prevent illegitimate access to information when stored on the servers. Additional security measures such as multi-factor authentication also assure to grant access to the patient data only to the permitted individual [46]. Moreover, carrying out end-user audits and monitoring procedures as routine also assists in identifying threats, and data leakage. Legal requirements include data protection especially regarding new legislations such as Health Insurance Portability and Accountability Act, HIPAA in the United States. The staff should also be trained on how to handle the private information so that there is no leakage of data. The other layer of defence against the threats is the use of firewalls and intrusion detection systems. By incorporating most of these measures, healthcare organizations and facilities will be able to safeguard and secure patient data to reduce cases of breaches, loss of confidentiality, integrity, and availability.

7.4 The Concept of Healthcare 5.0

With the world advancing into the next revolution period, Healthcare 5.0 comes to present a coming evolution of the medical technologies and patient solution-oriented healthcare system. This idea is the continuation of healthcare 4.0 where digital models' development, exchange, and integration were the key concepts. Healthcare 5.0 expands beyond involving the use of human-centric social and technological elements in delivering improved healthcare services. While Healthcare 4.0 was built based on Industrial Internet of Things (IIoT), big data, and AI, Healthcare 5.0 involves these technologies focusses on the human being as a core concept with patient-centric care.

7.4.1 Key Components of Healthcare 5.0

Healthcare 5.0 includes one of the most tremendous transformations in the system and its approach to providing medical care. By combining technology with a patient-centred and patient-sensitive approach, this concept significantly enhances healthcare and positively impacts patients worldwide. The key components of healthcare 5.0 are human-centric technology, customized care delivery, data exchange and interoperability, patient empowerment and engagement, hierarchy of healthcare and ethical considerations and data protection discussed below and illustrated in Fig. 7.10.

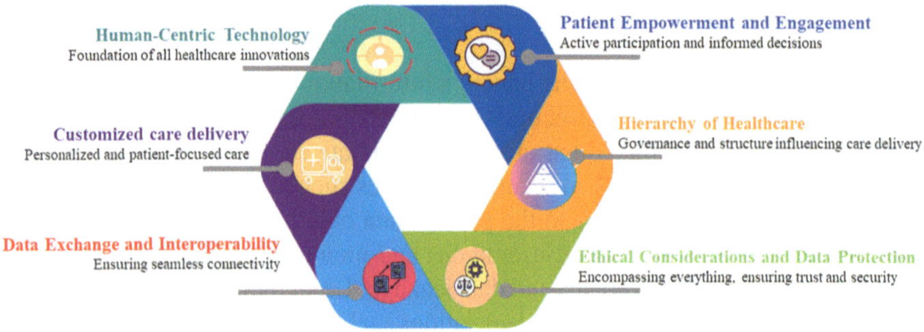

Fig. 7.10 The key component of healthcare 5.0

7.4.1.1 Human-Centric Technology

This is a principle of the Healthcare 5.0 civilization that emphasizes the integration of technology based on human values. This approach guarantees that technology improvement does not suppress but rather supports human aspects of care [47]. AI, ML, and robotics are expected to aid healthcare professionals in providing quality care through enhancing their working tools. AI can take large data set of patient records and analyse them to define health outcomes, and the doctor preserves a personal touch with the patient.

7.4.1.2 Customized Care Delivery

It concerns the personalized treatment of the patients and is referred to as personalized medicine. Genomic, proteomic, and metabolomics studies make it possible for treating physicians to design therapies that are individualized according to the genetic constitution of the patients. It also enhances the effectiveness of the treatment and reduces the side effects that are endured by the patients to, in turn, produce outcomes [48].

7.4.1.3 Data Exchange and Interoperability

Healthcare 5.0 involves an interoperability and sharing of data between systems and platforms used in healthcare. This creates a link between patient's data for the benefit of healthcare providers since all data collected by one health professional is made available to others. Interoperability prevents medical mistakes, guarantees patients' security, and optimizes the healthcare system performance [49]. The integration is made possible by the use of standard data format and communication protocols.

7.4.1.4 Patient Empowerment and Engagement

With a large number of new technologies in digital health, which includes mobile health applications, wearable devices, and telemedicine, the patient has better awareness into their health and collaborate in the development of care management strategies [50]. These tools help inform patients and also provide constant feedback to health status, hence detection and early action to be made. Thus, the data generated by patients are very valuable when it comes to capturing a broader picture of the health situation.

7.4.1.5 Hierarchy of Healthcare

Healthcare 5.0 stands to break this hierarchical order of the world of healthcare away from being based on geographic or financial boundaries. Consequently, through the application of technology, it is possible for healthcare providers to reach out for those who were once considered out of reach due to the increased cost [51]. Telemedicine helps overcome barriers of distant delivery to ensure that patients who cannot be accorded a chance to see qualified practitioners receive quality service.

7.4.1.6 Ethical Considerations and Data Protection

Health care has incorporated various technologies and there is need to address various ethical issues and security of data. Healthcare 5.0 also takes into consideration the ethical impact of such technologies to patients and emphasizes that the use of technology must not intrude into the patient's rights to privacy [52]. With the increasing cases of breach in the health information, the organization needs strong measures of protecting the information. It is imperative that ethical clinical practices and professional standards are clearly defined, and that laws governing the doctor-patient relationship are firmly established.

7.5 The Elements of Communication in Healthcare System

The elements of communication in the healthcare system specifically provides the data transferring process protocol between the various medical devices or systems. In addition to the wireless and wired technologies, there are other communication protocols and technologies that are used to support optimized functioning of healthcare systems as depicted in Fig. 7.11.

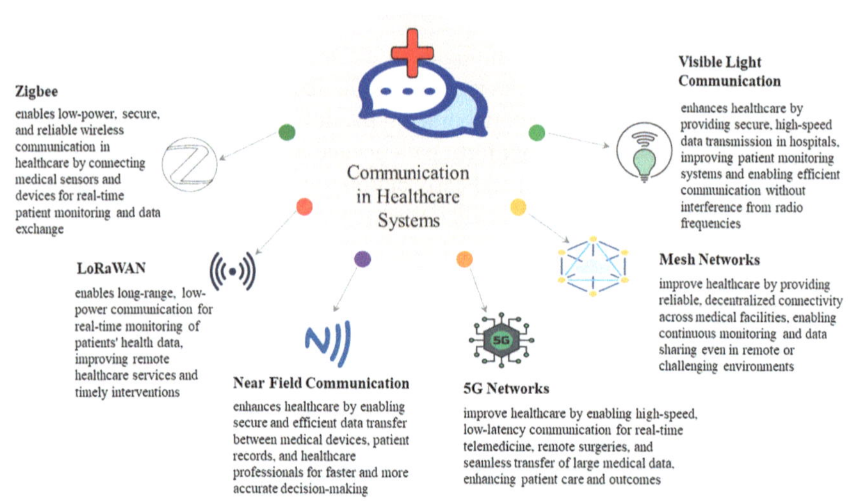

Fig. 7.11 Communication technologies in health systems

7.5.1 Zigbee

This is a wireless personal area network and it operates on low-power and low data transfer rate; it is suitable for use in medical facilities because of its minimal power usage and enhanced security [53]. It is suitable when coupling devices which exchange small amounts of data very often. Zigbee includes increased security measures such as encryption and secure key distribution to ensure that patients' information is not shared to unauthorized individuals. This characteristic is another benefit in health care since Zigbee devices transmit data with increased range and interconnection using other devices in the intermediary [54]. These features hold Zigbee that defines it to be suitable to connect IoT devices in medical premises for patients' safety and medical facilities' optimization.

7.5.2 Long-Range Wide Area Network

This network is specifically designed for high coverage low data rate of wireless communication and hence best suited for remote patient monitoring use cases in which devices have to send data over long distance without the need for frequent recharging or replacement of batteries [55]. It provides an ideal solution for healthcare since the application includes an enhanced patient tracking system particularly in remote monitorization. This is especially applicable in the developing countries that are often located in weak zones and where patients have poor access to health care services. There is flexibility of navigating big distances, which helps in cutting down costs and general management of infrastructures for healthcare solutions in the network. It incorporates the capability of adding more sensors and other devices for monitoring of several other health conditions. Therefore, by applying LoRaWAN, physical healthcare providers are be able to improve efficiency in the health system, since it access real-time data which could help in preventive measures to be taken in order to change the negative health status of a patient.

7.5.3 Near Field Communication (NFC)

NFC technology is adapted for fast communication with devices, which are close physically, typically used for identification of the patient and the usage of the data-logging concepts in the healthcare services [56]. This technology is even progressing healthcare by offering possible means for inter-connection and interaction for devices and systems in fields involving fast and sensitive data transfer. In the healthcare sector, specifically, NFC is essential for fast identification and verification of the patient allowing medical practitioners to access relevant information immediately. This capability enriches patient care to monitor the patient's status in terms of general health status through vital data such as temperature, pulse rate, blood pressure, to be checked frequently. Additionally, NFC enhances the organizational procedure like digital documentation of patients' information and storage of medical equipment in hospitals [57]. Therefore, NFC holds great potential in the healthcare industry primarily because of quality care enhancement, cost efficiency, and patient satisfaction.

7.5.4 5G Networks

The development of the 5G networks is expected to greatly influence the advancement of IoMT since it offers high data rates with low latency [58]. This will promote management of the patient data in real-time manner to support enhanced telemedicine and other IoT health services. Additionally, the increasing number of connected devices in the healthcare sector will be able to share and collect data for comprehensive patient assessment, supported by the enhanced bandwidth and connectivity offered by 5G. The rapid analysis of large volumes of data will enable more accurate diagnostic tools and personalized treatment plans, contributing to development of predictive healthcare analytics. In rural areas, 5G will expand access to telemedicine services, ensuring that patients receive on-time medical attention regardless of their location. Therefore, the widespread implementation of 5G technology is expected to transform patient care and significantly enhance healthcare systems across various regions of the world [58].

7.5.5 Mesh Networks

Hospitals are one of the key areas where these networks are applied, as devices within the network establish their own connections, ensuring reliable coverage without the need for a central router. Mesh networks are especially suitable in hospitals because they preserve the network's connection and signal quality. In such areas that require consistent communication and the use of medical equipment that requires to be communicated to and from a particular pathway, a mesh network helps in sustaining such connections without involving a router [59]. Also, mesh networks pass the data from one end to another due to signal flexibility thus making network more reliable. The self-healing property of the mesh network means that if a certain node is down, the system does not stop as it can recover from the failure quickly. These networks enable several types of medical devices, including wearable health monitors, imaging equipment, it enables the sharing of real-time data. This connectivity is very important for telemedicine practice since it facilitates remote consultations and monitoring, hence improving the quality of services to patients. Thus, the use of mesh networks offers a large, dependable network capable of meeting the needs of most today's healthcare organizations.

7.5.6 Visible Light Communication (VLC)

This is a communication system that employs light to transmit data, it is applied in areas where interference is not allowed in healthcare centres. VLC has enormous potential in the health care sector as it offers reliable and interference-free data communications channel [60]. Also, VLC improves patients' safety through integration with the IoMT, which enables the connection between medical devices and healthcare management. These connections enable the continuous tracking and sharing of data, which has a positive impact on the care of the patients. Moreover, light is used for data transmission in ambient conditions, and this makes the health facilities of high standards to hold sustainable practices. With the emerging trends in technology within the healthcare sector, VLC is being applied to deploy safer, more efficient, and sustainable communication systems.

7.6 Smart and Connected Healthcare as Key Enabler Infrastructure for Sustainable Smart Cities

Smart and Connected Healthcare is the key element in a smart city that is sustainable in terms of public health and infrastructure resources. Technologies such as smart and connected health in healthcare 5.0 will also help in the growth sustainable smart cities due to better health, reduced health care expenditure and improved

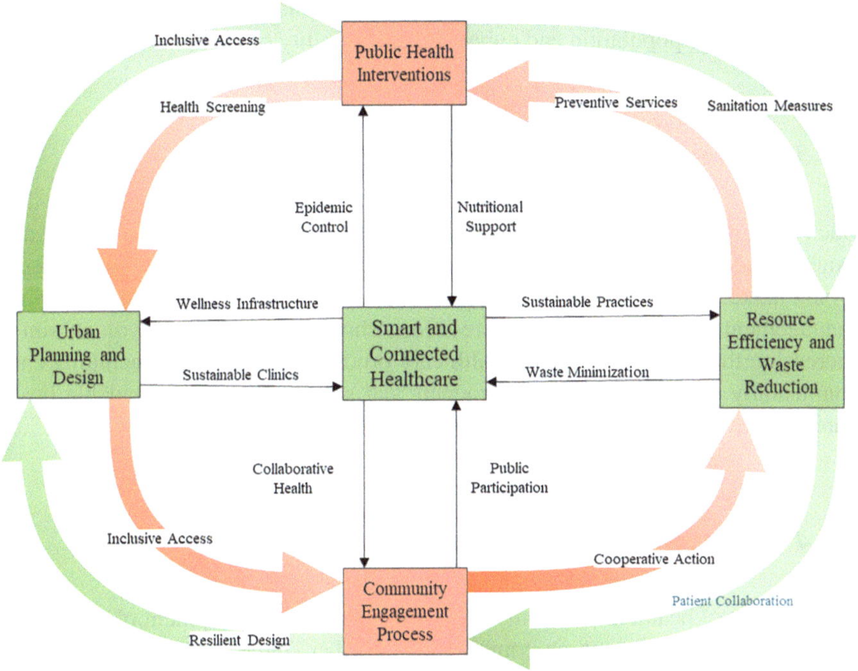

Fig. 7.12 Key enablers of smart and connected healthcare in sustainable smart cities

quality of life. Figure 7.12 illustrates the key enablers of smart and connected healthcare in sustainable smart cities.

7.6.1 Urban Planning and Design

Research in Healthcare system should be directed towards urban and architectural planning so as to develop health by enhancing architectural work [61]. These needs developing clinical structures that are accessible to everyone within the society irrespective of the disability, age, group or financial status. Furthermore, future research should motivate into the concept of contemporary healthcare that involves community health centre that offer health promotion, preventive, diagnostic as well as treatment facilities.

7.6.2 Public Health Interventions

A concept with the integration of smart and connected healthcare systems in smart cities is to implement more progressive public interventions such as early warning systems for epidemics and monitoring of quality of limited natural resources such as air and water [62]. It also contributes to the elimination of diseases, improving the health of the population and enhancing the viability of cities.

7.6.3 Resource Efficiency and Waste Reduction

Such resource efficiencies and waste minimization help healthcare system towards establishment of smart cities. In other words, it comprises of green technologies that would make use of renewable resources and developing a circular health care waste management system [63]. Also, more efforts should be carried out with an aim of increasing the creation of green health care products and services, including environmentally friendly health care products and services like biodegradable health care products and green pharmaceuticals.

7.6.4 Community Engagement Process

The concept of healthcare system and its impact on the development of sustainable smart cities highly depend on communities. Further studies are needed to explore effective ways to engage the community in healthcare decision-making processes, including health promotion and upcoming community health programs. The strategy of healthcare system encourages further enhancement of the healthcare services in delivering job prospects relevant to smart cities' construction [64]. However, the successful realization of benefits of digitization will depend on addressing four key areas; data privacy and security, interoperability and standardization, AI and ML, and personalized, remote, and telemedicine healthcare. Researchers and policymakers must collaborate to establish a foundation for a more sustainable, inclusive, and resilient future city.

7.7 Internet of Medical Things for Healthcare Services and Applications

IoMT has successively become one of the dynamic technologies in the health care sector that gives tools to the doctors and other professional health care providers to tend their patients with personal and most efficient and effective care irrespective of the geographical boundaries of the world. The IoMT links implantable biomedical devices, wearable technology, wireless health applications, and heterogeneous data networks to provide ubiquitous and pervasive monitoring, health assessment, non-invasive diagnostics, treatment delivery, and patient communications. In this section various uses and services of IoMT that have been discussed and illustrated in Fig. 7.13.

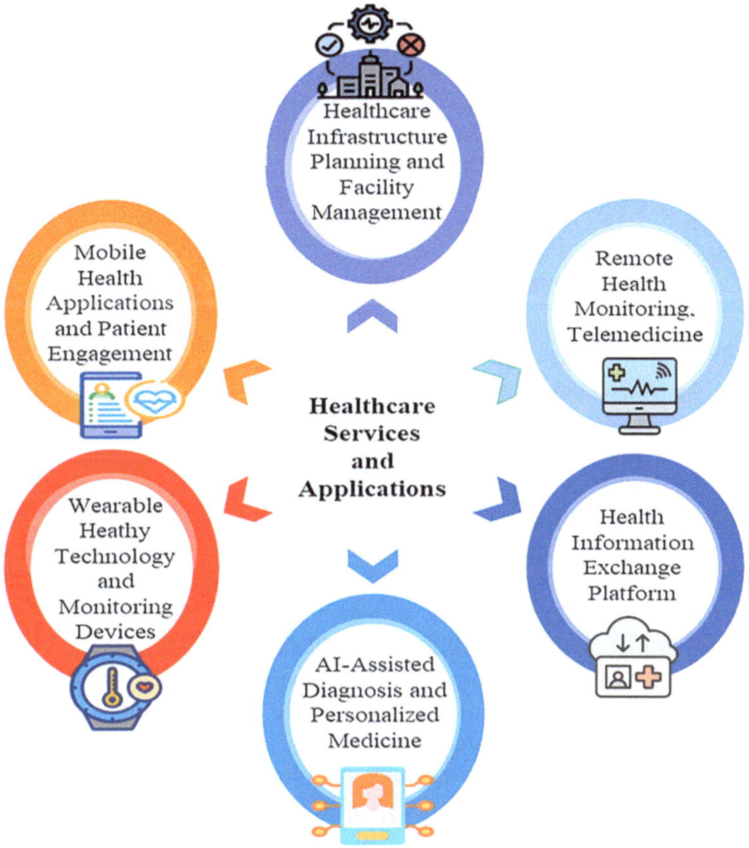

Fig. 7.13 Healthcare services and applications of IoMT

7.7.1 *Remote Health Monitoring and Telemedicine*

The technology involved includes the use of technology to directly reach and treat the patient, as well as other health care professionals. It takes advantage of the IoMT technology composed of connected devices, software applications and telecommunication system to facilitate remote consultation of patients. Telemedicine enables patients in remote or hard-to-reach areas to access a qualified medical practitioner, therefore eliminating visit to the hospital [65]. The strategic application of IoMT in the development of telemedicine systems make it easier for healthcare practitioners to monitor the patient's condition and appropriate on-time follow-up and necessary interventions. Patients remotely monitor blood pressure, pulse rate and blood sugar level via mobile health applications or wearable devices. IoMT devices wirelessly send and receive health information from the healthcare provider and from the patient to ensure health status can be observed at any one time and act appropriately in case of an emergency. Such a continuous monitoring may also enable early identification of any health complications to prevent occurrence and subsequent hospitalization. Similarly, telemedicine applications have characteristics including video consultation, tele-prescribing, and EHR which also improve the care of the patient [66]. This not only eliminates direct contacts with other people but also decreases the time and costs patients spend on travel to and from clinics.

7.7.2 Health Information Exchange Platform

Health Information Exchange is a digital framework through which patient health data is exchanged without compromising patient privacy and maintain security [67]. IoMT has a strategic role in the creation of HIE platforms to facilitate flow of medical records as well as driving the vision of a well-integrated healthcare system. Through connected, diverse technologies, IoMT-driven HIE platforms consist of different devices, healthcare apps, and data systems to make the patient records available to the healthcare providers to track and share patient details. This helps to make better decisions, decrease the probability of committing medical errors and provide best treatment to the patient. Moreover, IoMT lead to integrated HIE platforms that ensure that the data collected by different healthcare systems integrated to be compatible with other platforms. These platforms should therefore integrate analytics and ML models to give the healthcare providers proper information. Another benefit of HIE is that it reduces the healthcare costs since individual patient data prevents duplication of tests, and other treatments [68]. This integrated approach not only explains how processes can be made more effective, but it also discharges the patients to become more practical in managing health more easily through easy access to patient records. In general, IoMT driven HIE platforms play an important role in creating a more integrated, improved and a patient-centric health ecosystem.

7.7.3 AI-Assisted Diagnosis and Personalized Medicine

AI diagnosis influences devices and health data to analyse patient's data, patronize patterns with higher precision [69]. In radiology, AI is useful when the specific patterns in x-rays, CT or MRI scans and matches are matched to possible disease at early stage. This not only improves diagnostic precision but also accelerates the same, putting translational action into motion and helping the patients. The IoMT-aided personalized medicine not only employs AI to predict patient's responses to treatments depending on genotypes and phenotypes, but also to analyse patient's behaviour and environment, as well as to develop highly individualized approach

[70]. It increases the effectiveness of medical treatment, decreases the rate of errors connected with the administration of medicines and enhances the result of treatment for patients. Additionally, the use of IoMT in diagnostics, combined with AI- driven decision-making and personalized medicine, enhances patient care by enabling continuous monitoring and providing constant feedback through the internet. This results in enhancing the ability of the health care provider to make changes to the treatment plan in response to the condition of a patient in particular and thus, improving the health outcomes.

7.7.4 Wearable Heathy Technology and Monitoring Devices

Wearable has been adopted as a way of enhancing the IoMT ecosystem since it allows people to track health state without relying on the healthcare system. Smartwatches, fitness trackers, wireless sensors, and other wearable technologies have to track different degrees of thorough health indicators like heart rate, activities, sleep, and glucoses levels [71]. Connected wearables are based on IoMT and collect the real-time physiological and wellness information of the patients and deliver it to the healthcare providers for subsequent biometric analysis and intervention. Also, wearable technology makes people to learn a healthier way of living and help them to be active. There are still new features being added on both smart and wearable devices in relation to AI analytics and personalized health suggestions that are helping to monitoring health and make people more involved [72]. Such applications help provide large-scale health investigations, using new knowledge on population's health state and contributing to design necessary preventions and treatments. Moreover, with the advancement in the technology, wearables are increasingly compatible with other technology devices in the healthcare system, helping in sharing information between various trends hence enhancing the connectivity of health technology.

7.7.5 Mobile Health Applications and Patient Engagement Tools

Patient engagement tools and mHealth applications have become essential components of the IoMT framework, providing a means of healthcare providers to communicate with patients and manage health needs effectively [73]. IoMT helps clients to track medications, have scheduled appointments, keep track of the required healthcare related information, which can use to increase their compliance with prescribed doctor's recommendations. It also provides advantages of having the services of telemedicine extended in the form of mobile apps available for patients so that they can relate to the healthcare professionals and have an individual consultation. Moreover, IoMT-driven mHealth applications act as a tool for real-time alerts and reminders for medication administration to enhance compliance and thereby patients' health status [74]. The application of IoMT will allow for the prioritization of patient-centric care in the healthcare sector and guarantee better results for both the patients as well as the physician. Also, the popularity of the concept of providing remote access to health information also benefits all patients with mobility restricted issues; everyone have access to enhanced healthcare development.

7.7.6 Healthcare Infrastructure Planning and Facility Management

IoMT has expanded into a broader segment of planning and running healthcare facilities bringing benefits to the healthcare institutions both in terms of efficiency and the quality of performing their functions. Smart devices and IoT sensors allow to track the condition in the facility, including temperature, humidity, and the air quality so that patients and employees feel comfortable and healthy. Facility management systems based on IoMT minimizes the operational costs while increasing the effectiveness of allocated resources [75] in tasks that might otherwise need manual assigning and tracking, including inventory management, energy consumption monitoring, and equipment maintenance. Moreover, IoMT improves the health outcome of patients by connecting to EHRs to enable appropriateness of real-time information, improve the decision-making process of health practitioners. Moreover, through the use of IoT, predictive analysis makes it possible to identify failure or even notice that equipment needs to be maintained to avoid a break in services. The significance of IoMT in enhancing a connected and adaptive healthcare system remains vital as it helps in accelerating positive changes and growth within this industry [76]. IoMT technology is also vital in the assessment of calamity impact and as part of planning for disaster responses. Through objects, components, connected devices, and sensors, it is possible to observe the real-time status of resource availability, at times of emergencies.

Today's society is adorned with medical things that have made a significant impact that has modernized the healthcare sector throughout the world. The use of IoMT in applications such as telemedicine, Health Information Exchange Platforms, AI in diagnosis, Wearable Health Technology, Mobile Health Application and in healthcare infrastructure development have changed the era of health care delivery and has shifted focus more towards the patient.

7.8 Open Research Challenges and Research Directions

The rapid evolution of digital technologies has significantly influenced the transformation of healthcare systems toward smarter, more connected, and patient-centric model. Despite remarkable progress, several challenges remain that prevents the full implementation of intelligent healthcare infrastructures. Figure 7.14 illustrates the open research challenges and potential directions in developing smart and connected healthcare systems.

7.8.1 Data Privacy and Security

Healthcare is critical to data security due to the reliance on data collected from various sources, such as wearable devices, smartphones, and electronic health records. This creates a significant risk of data theft or cyberattack. To address these challenges, further research is needed to develop enhanced security mechanism,

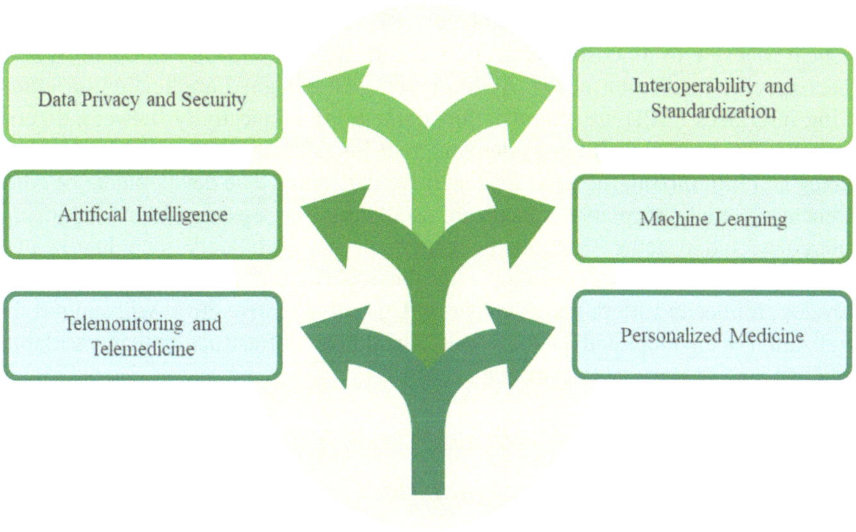

Fig. 7.14 Open research challenges and future directions in smart and connected healthcare systems

including encryption methods, access controls, and authentication process to protect patient data [77]. Encryption plays a crucial role in ensuring that sensitive information is transferred and stored securely, accessible only to those with the proper authorization. Additionally, it is essential to implements security policies that restrict access to patient data, granting permissions only to authorized personnel, and regularly reviewing access logs to detect any unauthorized activity. In addition to technological measures, organizations should conduct ongoing training sessions on data privacy and cybersecurity for employees to minimize human errors [78]. By focusing on both technical and human factors, healthcare systems can better safeguard patient information and reduce the risk associated with data breaches and cyberattacks [79].

7.8.2 Interoperability and Standardization

Health information exchange is important for achieving interoperability and standardization, as it significantly enhances the efficiency and effectiveness of healthcare systems [80]. Proper data integration across various healthcare systems and devices enables the secure and seamless sharing of patient information, improving coordination and continuity of healthcare [81]. Open-source solution also plays a critical role due to its cost-effectiveness and flexibility, allowing customization to meet specific healthcare needs. Additionally, well-designed application programming interfaces (API) are essential for maintaining connectivity between diverse systems. These APIs facilitate systematic, real-time data exchange, reducing redundancy and minimizing the need for repetitive data entry. The development of comprehensive and adaptable healthcare standards is essential for supporting interconnected systems. Continued engagement of stakeholders, including healthcare providers, technology developers, policymakers, and patients is necessary to develop, refine, and adopt these standards. This collaborative effort will support the realization of an integrated, patient-centred healthcare infrastructure that is scalable, efficient, and responsive to evolving needs [82].

7.8.3 Artificial Intelligence and Machine Learning

Artificial intelligence and machine learning are highly effective in diagnostics, as it process large volume of data much faster than humans [83]. This ability allows for early detection of diseases such as tumours, cancer, and cardiovascular conditions, leading to on-time and more effective treatment. AI also plays important role in areas like genomics, where it helps in developing treatment plans based on a patient's genetic makeup and lifestyle, thereby improving the quality of healthcare. In addition to clinical applications, AI reduce the amount of paperwork and other routine tasks for healthcare professionals and administrators providing more time to focus on patient care. The integration of both the technical and health professions will ensure full exploitation of the potential of AI and ML to revolutionize the health systems across the globe [84, 85].

7.8.4 Telemonitoring and Telemedicine

Telemonitoring and telemedicine are essential components of modern healthcare, which require stronger regulations to ensure the secure transfer of patient information through digital channels [86]. This brings for the implementation of robust data protection policies to safeguard patient privacy. Technical integration is equally important to enable seamless data exchange between the various systems and technologies used in healthcare management [87]. Such integration enhances the interconnectivity of healthcare systems, improving efficiency and coordination. It is also vital to train healthcare practitioners to use telemedicine tools effectively and confidently. At the same time, efforts should be made to educate patient on how to access and navigate these tools to ensure successful engagement with remote healthcare services. To support the effective development to telehealth, governments should establish clear guidelines and policies that not only promote the safe use of these technologies but also prioritize equitable access, especially for populations in rural areas.

7.8.5 Personalized Medicine

Personalized medicine, also known as precision medicine, uses omics analysis including genomics and biotechnology to improve understanding of the factors that influence human health [88]. A key focus is identifying biomarkers that have predictive value for specific diseases, as well as exploring the relationships between genotypes and phenotypes to understand how genetic variations affect disease susceptibility and treatment response. By integrating these methods, therapies are more effectively designed, preventing traditional approaches. Personalized medicine also uses genetic information to improve disease prediction and monitor progression, leading to more accurate, individualized care and better health outcomes for patients [89].

7.9 Summary

The transition of healthcare systems from traditional to smart and connected systems is promoting the use of technology to improve the wellbeing of patients. Health care, driven by the Internet of Medical Things and smart and connected health care powered by technological advancements, plays a vital role in improving citizens' health. IoMT devices like the remote health monitoring, and wearable technologies assists patients round the clock. The development of AI diagnostic aids and personalized medicine initiate social changes in the treatment for healthcare 5.0. Mobile health applications and patient engagement tools integrate into the idea of democratizing healthcare and empowering individuals to make proper decisions in health-related issues. The integration of smart and connected healthcare establishes a new version for intelligent healthcare systems. The future of healthcare management is likely to be shaped by addressing current research challenges through unified and advanced approached that enhance access, utilization, efficiency, and patient-centred care.

References

1. M. Wazid, A.K. Das, N. Mohd, Y. Park, Healthcare 5.0 security framework: Applications, issues and future research directions. IEEE Access **10**, 129429–129442 (2022). https://doi.org/10.1109/ACCESS.2022.3228505
2. S. Rashid, A. Nemati, Human-centered IoT-based health monitoring in the healthcare 5.0 era: Literature descriptive analysis and future research guidelines. Discov. Internet Things **4**(1), 26 (2024). https://doi.org/10.1007/s43926-024-00082-5
3. P. Mishra, G. Singh, Smart healthcare in sustainable smart cities, in *Sustainable Smart Cities*, (Springer, Cham, 2023), pp. 195–219. https://doi.org/10.1007/978-3-031-33354-5_9
4. G. Aceto, V. Persico, A. Pescapé, The role of information and communication technologies in healthcare: Taxonomies, perspectives, and challenges. J. Netw. Comput. Appl. **107**, 125–154 (2018). https://doi.org/10.1016/J.JNCA.2018.02.008
5. A.N. Navaz, M.A. Serhani, H.T. El Kassabi, N. Al-Qirim, H. Ismail, Trends, technologies, and key challenges in smart and connected healthcare. IEEE Access **9**, 74044–74067 (2021). https://doi.org/10.1109/ACCESS.2021.3079217
6. Y. Xie et al., Integration of artificial intelligence, Blockchain, and wearable Technology for Chronic Disease Management: A new paradigm in smart healthcare. Curr. Med. Sci. **41**(6), 1123–1133 (2021). https://doi.org/10.1007/s11596-021-2485-0
7. B. Rothman, J.C. Leonard, M.M. Vigoda, Future of electronic health records: Implications for decision support. Mount Sinai J. Med. J. Transl. Personal. Med. **79**(6), 757–768 (2012). https://doi.org/10.1002/msj.21351
8. B. Stanberry, Telemedicine: Barriers and opportunities in the 21st century. J. Intern. Med. **247**(6), 615–628 (2000). https://doi.org/10.1046/j.1365-2796.2000.00699.x
9. M.I. Pramanik, R.Y.K. Lau, H. Demirkan, M.A.K. Azad, Smart health: Big data enabled health paradigm within smart cities. Expert Syst. Appl. **87**, 370–383 (2017). https://doi.org/10.1016/J.ESWA.2017.06.027
10. A. Garg, V.V. Venkataramani, A. Karthikeyan, U.D. Priyakumar, Modern AI/ML methods for healthcare: Opportunities and challenges, in *Distributed Computing and Intelligent*

Technology. ICDCIT 2022, Lecture Notes in Computer Science, vol. 13145, (Springer, Cham, 2022), pp. 3–25. https://doi.org/10.1007/978-3-030-94876-4_1
11. S. Tian, W. Yang, J.M. Le Grange, P. Wang, W. Huang, Z. Ye, Smart healthcare: Making medical care more intelligent. Glob. Heal. J. **3**(3), 62–65 (2019). https://doi.org/10.1016/J.GLOHJ.2019.07.001
12. M.D. Ruiz Morilla, M. Sans, A. Casasa, N. Giménez, Implementing technology in healthcare: Insights from physicians. BMC Med. Inform. Decis. Mak. **17**(1), 92 (2017). https://doi.org/10.1186/s12911-017-0489-2
13. D.C. Nguyen et al., Federated learning for smart healthcare: A survey. ACM Comput. Surv. **55**(3), 1–37 (2023). https://doi.org/10.1145/3501296
14. K. Kumar, K. Chaudhury, S.L. Tripathi, Future of machine learning (ML) and deep learning (DL) in healthcare monitoring system, in *Machine Learning Algorithms for Signal and Image Processing*, (Wiley, 2022), pp. 293–313. https://doi.org/10.1002/9781119861850.ch17
15. N. Noorbakhsh-Sabet, R. Zand, Y. Zhang, V. Abedi, Artificial intelligence transforms the future of health care. Am. J. Med. **132**(7), 795–801 (2019). https://doi.org/10.1016/J.AMJMED.2019.01.017
16. E. Kim, S.M. Rubinstein, K.T. Nead, A.P. Wojcieszynski, P.E. Gabriel, J.L. Warner, The evolving use of electronic health records (EHR) for research. Semin. Radiat. Oncol. **29**(4), 354–361 (2019). https://doi.org/10.1016/J.SEMRADONC.2019.05.010
17. T. Haidegger, S. Speidel, D. Stoyanov, R.M. Satava, Robot-assisted minimally invasive surgery—Surgical robotics in the data age. Proc. IEEE **110**(7), 835–846 (2022). https://doi.org/10.1109/JPROC.2022.3180350
18. R.P. Araldi et al., Medical applications of clustered regularly interspaced short palindromic repeats (CRISPR/Cas) tool: A comprehensive overview. Gene **745**, 144636 (2020). https://doi.org/10.1016/J.GENE.2020.144636
19. B. Munos et al., Mobile health: The power of wearables, sensors, and apps to transform clinical trials. Ann. N. Y. Acad. Sci. **1375**(1), 3–18 (2016). https://doi.org/10.1111/nyas.13117
20. I.H. Sarker, Machine learning: Algorithms, real-world applications and research directions. SN Comput. Sci. **2**(3), 160 (2021). https://doi.org/10.1007/s42979-021-00592-x
21. A. Fuller, Z. Fan, C. Day, C. Barlow, Digital twin: Enabling technologies, challenges and open research. IEEE Access **8**, 108952–108971 (2020). https://doi.org/10.1109/ACCESS.2020.2998358
22. B. Farahani, F. Firouzi, V. Chang, M. Badaroglu, N. Constant, K. Mankodiya, Towards fog-driven IoT eHealth: Promises and challenges of IoT in medicine and healthcare. Futur. Gener. Comput. Syst. **78**, 659–676 (2018). https://doi.org/10.1016/j.future.2017.04.036
23. A.D. Dwivedi, G. Srivastava, S. Dhar, R. Singh, A decentralized privacy-preserving healthcare Blockchain for IoT. Sensors **19**(2), 326 (2019). https://doi.org/10.3390/s19020326
24. G.G. Dagher, J. Mohler, M. Milojkovic, P.B. Marella, Ancile: Privacy-preserving framework for access control and interoperability of electronic health records using blockchain technology. Sustain. Cities Soc. **39**, 283–297 (2018). https://doi.org/10.1016/j.scs.2018.02.014
25. J. Xie et al., A survey of Blockchain technology applied to smart cities: Research issues and challenges. IEEE Commun. Surv. Tutor. **21**(3), 2794–2830 (2019). https://doi.org/10.1109/COMST.2019.2899617
26. Y. Wu, K. Zhang, Y. Zhang, Digital twin networks: A survey. IEEE Internet Things J. **8**(18), 13789–13804 (2021). https://doi.org/10.1109/JIOT.2021.3079510
27. J. Shi et al., Smart textile-integrated microelectronic systems for wearable applications. Adv. Mater. **32**(5) (2020). https://doi.org/10.1002/adma.201901958
28. N. Du, H. Dai, R. Trivedi, U. Upadhyay, M. Gomez-Rodriguez, L. Song, Recurrent marked temporal point processes, in *Proceedings of the 22nd ACM SIGKDD International Conference on Knowledge Discovery and Data Mining*, (ACM, New York, 2016), pp. 1555–1564. https://doi.org/10.1145/2939672.2939875

29. S. Borchardt, G.B. Vignola, D. Buscaglia, M. Maroni, *Mapping EU Policies with the 2030 Agenda and SDGs-Fostering Policy Coherence through text-based SDG Mapping* (2022). https://doi.org/10.2760/110687
30. G. Prieto-Avalos, N.A. Cruz-Ramos, G. Alor-Hernández, J.L. Sánchez-Cervantes, L. Rodríguez-Mazahua, L.R. Guarneros-Nolasco, Wearable devices for physical monitoring of heart: A review. Biosensors (Basel) **12**(5), 292 (2022). https://doi.org/10.3390/bios12050292
31. P. Mishra, G. Singh, Internet of medical things healthcare for sustainable smart cities: Current status and future prospects. Appl. Sci. **13**(15), 8869 (2023). https://doi.org/10.3390/app13158869
32. M. Haghjoo, Cardiac implantable electronic devices, in *Practical Cardiology: Principles and Approaches*, (Elsevier, 2022), pp. 279–288. https://doi.org/10.1016/B978-0-323-80915-3.00002-8
33. W. Gao, C. Yu, Wearable and implantable devices for healthcare. Adv. Healthc. Mater. **10**(17) (2021). https://doi.org/10.1002/adhm.202101548
34. P. Peng et al., Emerging ELISA derived technologies for in vitro diagnostics. TrAC Trends Anal. Chem. **152**, 116605 (2022). https://doi.org/10.1016/J.TRAC.2022.116605
35. R. An et al., Emerging point-of-care technologies for anemia detection. Lab Chip **21**(10), 1843–1865 (2021). https://doi.org/10.1039/D0LC01235A
36. J. Lorkowski, O. Kolaszyńska, M. Pokorski, Artificial intelligence and precision medicine: A perspective, in *Integrative Clinical Research. Advances in Experimental Medicine and Biology*, vol. 1375, (Springer, Cham, 2021), pp. 1–11. https://doi.org/10.1007/5584_2021_652
37. J. Jagarapu, R.C. Savani, A brief history of telemedicine and the evolution of teleneonatology. Semin. Perinatol. **45**(5), 151416 (2021). https://doi.org/10.1016/J.SEMPERI.2021.151416
38. V. Sharma, M. Feldman, R. Sharma, Telehealth Technologies in Diabetes Self-management and Education. J. Diabetes Sci. Technol. **18**(1), 148–158 (2024). https://doi.org/10.1177/19322968221093078
39. T. Kim et al., Comparison of X-ray computed tomography and magnetic resonance imaging to detect pest-infested fruits: A pilot study. Nucl. Eng. Technol. **54**(2), 514–522 (2022). https://doi.org/10.1016/J.NET.2021.07.015
40. J. Urlings, G. de Jong, T. Maal, D. Henssen, Views on augmented reality, virtual reality, and 3D printing in modern medicine and education: A qualitative exploration of expert opinion. J. Digit. Imaging **36**(4), 1930–1939 (2023). https://doi.org/10.1007/s10278-023-00833-w
41. A. Gyrard et al., Knowledge engineering framework for IoT robotics applied to smart healthcare and emotional Well-being. Int. J. Soc. Robot. **15**(3), 445–472 (2023). https://doi.org/10.1007/s12369-021-00821-6
42. A. Garcia-Gonzalez, R.Q. Fuentes-Aguilar, I. Salgado, I. Chairez, A review on the application of autonomous and intelligent robotic devices in medical rehabilitation. J. Braz. Soc. Mech. Sci. Eng. **44**(9), 393 (2022). https://doi.org/10.1007/s40430-022-03692-8
43. T.A.J. Schoonderwoerd, W. Jorritsma, M.A. Neerincx, K. van den Bosch, Human-centered XAI: Developing design patterns for explanations of clinical decision support systems. Int. J. Hum. Comput. Stud. **154**, 102684 (2021). https://doi.org/10.1016/J.IJHCS.2021.102684
44. E.A. Adeniyi, R.O. Ogundokun, J.B. Awotunde, IoMT-based wearable body sensors network healthcare monitoring system, in *IoT in Healthcare and Ambient Assisted Living. Studies in Computational Intelligence*, vol. 933, (Springer, Singapore, 2021), pp. 103–121. https://doi.org/10.1007/978-981-15-9897-5_6
45. A.H. Mohd Aman, W.H. Hassan, S. Sameen, Z.S. Attarbashi, M. Alizadeh, L.A. Latiff, IoMT amid COVID-19 pandemic: Application, architecture, technology, and security. J. Netw. Comput. Appl. **174**, 102886 (2021). https://doi.org/10.1016/J.JNCA.2020.102886
46. N. Garg, M. Wazid, J. Singh, D. P. Singh, and A. K. Das, "Security in IoMT -driven smart healthcare: A comprehensive review and open challenges," Secur. Priv., vol. 5, no. 5, 2022, https://doi.org/10.1002/spy2.235

47. P. Mishra, P. Thakur, G. Singh, Sustainable smart city to society 5.0: State-of-the-art and research challenges. SAIEE Afr. Res. J. **113**(4), 152–164 (2022). https://doi.org/10.23919/SAIEE.2022.9945865
48. E. Minvielle, A. Fourcade, T. Ricketts, M. Waelli, Current developments in delivering customized care: A scoping review. BMC Health Serv. Res. **21**(1), 575 (2021). https://doi.org/10.1186/s12913-021-06576-0
49. C. Wu, Y.M. Tang, W.T. Kuo, H.T. Yip, K.Y. Chau, Healthcare 5.0: A secure and distributed network for system informatics in medical surgery. Int. J. Med. Inform. **186**, 105415 (2024). https://doi.org/10.1016/J.IJMEDINF.2024.105415
50. E. Hickmann, P. Richter, H. Schlieter, All together now—Patient engagement, patient empowerment, and associated terms in personal healthcare. BMC Health Serv. Res. **22**(1), 1116 (2022). https://doi.org/10.1186/s12913-022-08501-5
51. H.R. Chi, M. de Fátima Domingues, H. Zhu, C. Li, K. Kojima, A. Radwan, Healthcare 5.0: In the perspective of consumer internet-of-things-based fog/cloud computing. IEEE Trans. Consum. Electron. **69**(4), 745–755 (2023). https://doi.org/10.1109/TCE.2023.3293993
52. K.R.M. Blenman, R. Hull, C. Maimela, T. Molefi, R. Khanyile, Z. Dlamini, Society 5.0 healthcare: Ethics, legal rights, human rights, safety and security, in *Society 5.0 and Next Generation Healthcare*, (Springer Nature Switzerland, Cham, 2023), pp. 267–291. https://doi.org/10.1007/978-3-031-36461-7_12
53. S.M. Karunarathne, N. Saxena, M.K. Khan, Security and privacy in IoT smart healthcare. IEEE Internet Comput. **25**(4), 37–48 (2021). https://doi.org/10.1109/MIC.2021.3051675
54. A. Rehman, K. Haseeb, S.M. Fati, J. Lloret, L. Peñalver, Reliable bidirectional data transfer approach for the internet of secured medical things using ZigBee wireless network. Appl. Sci. **11**(21), 9947 (2021). https://doi.org/10.3390/app11219947
55. N.S. Chilamkurthy, O.J. Pandey, A. Ghosh, L.R. Cenkeramaddi, H.-N. Dai, Low-power wide-area networks: A broad overview of its different aspects. IEEE Access **10**, 81926–81959 (2022). https://doi.org/10.1109/ACCESS.2022.3196182
56. G.M. Bianco et al., UHF RFID and NFC point-of-care—Architecture, security, and implementation. IEEE J. Radio Freq. Ident. **7**, 301–309 (2023). https://doi.org/10.1109/JRFID.2023.3268422
57. O. Ebere et al., NFC tag-based mHealth patient healthcare tracking system, in *2022 3rd International Conference on Next Generation Computing Applications (NextComp)*, (IEEE, 2022), pp. 1–6. https://doi.org/10.1109/NextComp55567.2022.9932185
58. A. Moglia et al., 5G in healthcare: From COVID-19 to future challenges. IEEE J. Biomed. Health Inform. **26**(8), 4187–4196 (2022). https://doi.org/10.1109/JBHI.2022.3181205
59. E.M. Ar-Reyouchi, K. Ghoumid, D. Ar-Reyouchi, S. Rattal, R. Yahiaoui, O. Elmazria, Protocol wireless medical sensor networks in IoT for the efficiency of healthcare. IEEE Internet Things J. **9**(13), 10693–10704 (2022). https://doi.org/10.1109/JIOT.2021.3125886
60. S. Caputo, G. Borghini, S. Jayousi, A. Rashid, L. Mucchi, Visible light communications for healthcare applications: Opportunities and challenges, in *2023 IEEE 17th International Symposium on Medical Information and Communication Technology (ISMICT)*, (IEEE, 2023), pp. 1–6. https://doi.org/10.1109/ISMICT58261.2023.10152290
61. M. Jaušovec, B. Gabrovec, Architectural evaluation of healthcare facilities: A comprehensive review and implications for building design. Buildings **13**(12), 2926 (2023). https://doi.org/10.3390/buildings13122926
62. Z. Mohammadzadeh, H.R. Saeidnia, A. Lotfata, M. Hassanzadeh, N. Ghiasi, Smart city healthcare delivery innovations: A systematic review of essential technologies and indicators for developing nations. BMC Health Serv. Res. **23**(1), 1180 (2023). https://doi.org/10.1186/s12913-023-10200-8
63. J. Berniak-Woźny, M. Rataj, Towards green and sustainable healthcare: A literature review and research agenda for green leadership in the healthcare sector. Int. J. Environ. Res. Public Health **20**(2), 908 (2023). https://doi.org/10.3390/ijerph20020908

64. P. Mishra, G. Singh, *Sustainable Smart Cities* (Springer, Cham, 2023). https://doi.org/10.1007/978-3-031-33354-5
65. M. Stoltzfus, A. Kaur, A. Chawla, V. Gupta, F.N.U. Anamika, R. Jain, The role of telemedicine in healthcare: An overview and update. Egypt J. Intern. Med. **35**(1), 49 (2023). https://doi.org/10.1186/s43162-023-00234-z
66. A.R. Bhalotra, Teleconsultation for anaesthesiologists - update and limitations. Curr. Med. Res. Pract. **12**(5), 238–243 (2022). https://doi.org/10.4103/cmrp.cmrp_68_22
67. B.E. Dixon, Introduction to health information exchange, in *Health Information Exchange: Navigating and Managing a Network of Health Information Systems*, (Academic Press, 2023), pp. 3–20. https://doi.org/10.1016/B978-0-323-90802-3.00013-7
68. E. Chakravartty et al., Clinical decision support leveraging health information exchange improves concordance with patient's resuscitation orders and end-of-life wishes. Appl. Clin. Inform. (2025). https://doi.org/10.1055/a-2591-9040
69. X. Wang, Y. Wang, Analysis of trust factors for AI-assisted diagnosis in intelligent healthcare: Personalized management strategies in chronic disease management. Expert Syst. Appl. **255**, 124499 (2024). https://doi.org/10.1016/J.ESWA.2024.124499
70. F. Muheidat, L.A. Tawalbeh, AIoMT artificial intelligence (AI) and Internet of Medical Things (IoMT): applications, challenges, and future trends, in *Computational intelligence for medical internet of things (MIoT) applications: Machine intelligence applications for IoT in healthcare*, (2023), pp. 33–54. https://doi.org/10.1016/B978-0-323-99421-7.00013-1
71. R.K. Kher, D.M. Patel, A comprehensive review on wearable health monitoring systems. Open Biomed. Eng. J. **15**(1), 213–225 (2021). https://doi.org/10.2174/1874120702115010213
72. D.E. Id, A. Djurovic, J. Lark, The future of personalized healthcare: ai-driven wearables for real-time health monitoring and predictive analytics article info. Curr. Rès. Health Sci. **2**(2), 10–14 (2024). https://doi.org/10.58613/crhs223
73. A.S. Albahri et al., Development of IoT-based mhealth framework for various cases of heart disease patients. Health Technol (Berl) **11**(5), 1013–1033 (2021). https://doi.org/10.1007/s12553-021-00579-x
74. I. Al Khatib, A. Shamayleh, M. Ndiaye, Healthcare and the internet of medical things: Applications, trends, key challenges, and proposed resolutions. Informatics **11**(3), 47 (2024). https://doi.org/10.3390/informatics11030047
75. J.K.C. Chen, H.-H. Ho, Asset resource optimization solution for smart hospital facilities and energy management through an interpretive structural model. Buildings **13**(12), 3064 (2023). https://doi.org/10.3390/buildings13123064
76. T.V. Bezerra, G. Callou, F. Airton, E. Tavares, A modeling-based approach for performance and availability assessment of IoMT systems. Electronics (Basel) **14**(6), 1157 (2025). https://doi.org/10.3390/electronics14061157
77. A. Vyas, S. Abimannan, R. Hwang, Sensitive Healthcare Data: Privacy and Security Issues and Proposed Solutions, in *Emerging Technologies for Healthcare*, (Wiley, 2021), pp. 93–127. https://doi.org/10.1002/9781119792345.ch4
78. K.Y. Yigzaw et al., Health data security and privacy: Challenges and solutions for the future, in *Roadmap to Successful Digital Health Ecosystems: A Global Perspective*, (Academic Press, 2022), pp. 335–362. https://doi.org/10.1016/B978-0-12-823413-6.00014-8
79. A.Y. Balogun, Strengthening compliance with data privacy regulations in U.S. healthcare cybersecurity. Asian J. Res. Comput. Sci. **18**(1), 154–173 (2025). https://doi.org/10.9734/ajrcos/2025/v18i1555
80. M. Gaynor, F. Yu, C.H. Andrus, S. Bradner, J. Rawn, A general framework for interoperability with applications to healthcare. Health Policy Technol. **3**(1), 3–12 (2014). https://doi.org/10.1016/J.HLPT.2013.09.004
81. C.J. Bradley, L. Penberthy, K.J. Devers, D.J. Holden, Health services research and data linkages: Issues, methods, and directions for the future. Health Serv. Res. **45**(5), 1468–1488 (2010). https://doi.org/10.1111/j.1475-6773.2010.01142.x

82. A. Aragues et al., Trends and challenges of the emerging technologies toward interoperability and standardization in e-health communications. IEEE Commun. Mag. **49**(11), 182–188 (2011). https://doi.org/10.1109/MCOM.2011.6069727
83. A. Bhardwaj, Promise and provisos of artificial intelligence and machine learning in healthcare. J. Healthc. Leadersh. **14**, 113–118 (2022). https://doi.org/10.2147/JHL.S369498
84. A. Nayyar, L. Gadhavi, N. Zaman, Machine learning in healthcare: Review, opportunities and challenges, in *Machine Learning and the Internet of Medical Things in Healthcare*, (Academic Press, 2021), pp. 23–45. https://doi.org/10.1016/B978-0-12-821229-5.00011-2
85. Y.Y.M. Aung, D.C.S. Wong, D.S.W. Ting, The promise of artificial intelligence: A review of the opportunities and challenges of artificial intelligence in healthcare. Br. Med. Bull. **139**(1), 4–15 (2021). https://doi.org/10.1093/bmb/ldab016
86. S.F.A. Alsgaer, X. Tao, J. Zhang, H. Wang, Z. Guo, Telemedicine and telemonitoring in healthcare, in *Brain and Health Informatics. BHI 2013*, Lecture Notes in Computer Science, vol. 8211, (Springer, Cham, 2013), pp. 201–209. https://doi.org/10.1007/978-3-319-02753-1_20
87. P. Jennett, M. Watanabe, Healthcare and telemedicine: Ongoing and evolving challenges. Dis. Manag. Health Out. **14**(Suppl 1), 9–13 (2006). https://doi.org/10.2165/00115677-200614001-00004
88. D.E. Pritchard, F. Moeckel, M.S. Villa, L.T. Housman, C.A. McCarty, H.L. McLeod, Strategies for integrating personalized medicine into healthcare practice. Per. Med. **14**(2), 141–152 (2017). https://doi.org/10.2217/pme-2016-0064
89. S.S.-J. Lee, Personalized medicine and pharmacogenomics: Ethical and social challenges. Per. Med. **2**(1), 29–35 (2005). https://doi.org/10.1517/17410541.2.1.29

Chapter 8
Smart Mobility Systems in Sustainable Smart Cities

Contents

P. Mishra, G. Singh, *Sustainable Smart Cities 2.0*,
https://doi.org/10.1007/978-3-032-01102-2_8

8.1 Introduction

Smart mobility systems is an essential element of smart city development as it mitigates transport infrastructure with minimal negative influence to the environment [1]. These advanced systems provide information, communication as well as automation technologies designed to yield intelligent, effective, and environmentally sustainable means of transport in urban areas. Smart cities use real-time data, big data analysis, and internet of things (IoT) to enhance traffic patterns [2]. The application of smart mobility systems results in improving the quality of urban living and traffic flow in cities all over the world [3]. These advantages are attainable only if cities invest in and integrate public transport, cycling, pedestrian infrastructure,

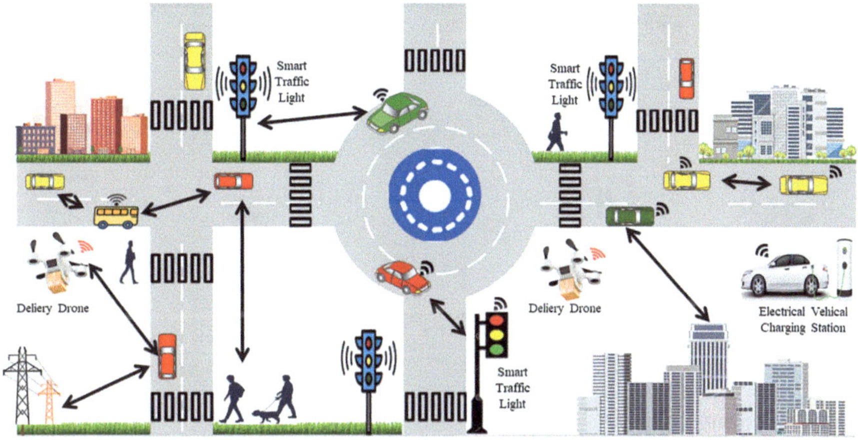

Fig. 8.1 Overview of smart mobility systems in sustainable smart city

intelligent transport system (ITS), and green mobility. Figure 8.1 presents an overview of smart mobility systems in a sustainable smart city and depicts integrated essential elements such as smart traffic lights, electric vehicle (EVs), charging stations, delivery drones, and pedestrian-friendly infrastructure for efficient urban transportation.

8.1.1 Origins and Development of Smart Mobility

The progressive evolution of transportation systems moves from traditional modes to the era of electric and autonomous mobility. The transformation phase begins with horse carts and early automobiles, followed by mass motorization and digital integration using GPS and traffic cameras. The revolution phase introduces artificial intelligence (AI) and IoT technologies that enable real-time data and ride sharing applications, leading to a future featuring EVs, self-driving cars, and AI-powered urban mobility as illustrated in Fig. 8.2. Smart mobility is the optimal use of transportation resources in order to modify and develop transport services while optimizing the transport systems and environmental impacts [4]. It has evolved from the greater integration of Information and Communication Technology (ICT) in every sector including transport [5]. Smart mobility has been recognized as the evolution of ITS, which was started in the late twentieth century with the help of global positioning systems (GPS), mobile communication devices, and new-generation sensors.

The concept of smart cities emerged in half of the twentieth century when the global population rate increased and urbanization became inevitable for cities that have to address rising traffic congestion and its effect on air pollution [6]. The advancement of smarter mobility continued through the emergence of ITS as it

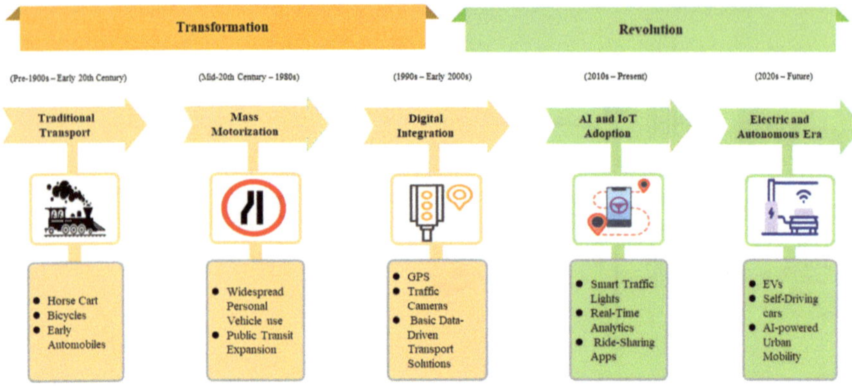

Fig. 8.2 Evolution of smart mobility

refers to the number of technologies and solutions that focused on optimizing and making transport safer and more environmentally friendly [7]. ITS solutions involve traffic management system, advanced public transportation systems, vehicle to infrastructure (V2I) communication systems, etc. Several factors such as rise in the mobility needs of people in cities, negative impacts of transport to the environment, and poor quality of life of citizens in urban areas has initiated the development of smart mobility. As such, the approaches to establishing smart mobility systems in smart cities have been developed based on the principle of sustainability, efficiency and safety. In smart cities, sustainable smart mobility refers to the sustainable solutions that can be employed in vehicles, roads and other element of smart mobility so as to decrease the levels of emissions from transport. This includes walking, cycling, use of public transport, and electric or hybrid cars. Realization of smart mobility systems is crucial in green smart cities, which requires collaboration between urban planners, transport authorities, technology experts, and the private sector. This integration is intended to promote an integrated transport planning and management for all the users and modes of transport. Sustainable smart city for smart mobility aims to make a shift towards new generation transport, which has less and minimum negative effects on the environment and enhances the quality of life of citizens [8].

8.1.2 Smart Mobility Significance for Urban Growth

Smart cities will remain as the main focus of urban development and expansion in the recent past years. Smart mobility is the most critical in defining the future of sustainable cities and territories. Advanced smart mobility solutions involve the use of advanced ICTs including IoT, big data, and AI in enabling the intelligent transport in the city [9]. The application of these technologies supports the core principle of smart cities by enabling effective management of transport networks, preventing traffic congestion, reducing greenhouse gas emission, and ensuring security and convenience. A key benefit of smart mobility system is the ability to promote sustainable urban development. Implementing these systems reduces the need for constructing new roads, parking lots, or public transport line, which positively impacts the environmental footprint of urbanization. In addition, smart mobility technologies help reduce greenhouse gas emissions through electric and hybrid vehicles, car sharing programs, and bike sharing services, making cities more environmentally friendly [10]. Beyond sustainability, smart mobility systems improve quality of life for city residents. Optimizing traffic flow with smart technologies increases efficiency by reducing travel time and enhances safety by decreasing the risk of accidents, making urban transport more effective for everyone.

8.1.3 Transportation System: Recent Advances in Smart Mobility

Transportation is one of the most important components in the smart cities as it influences mobility within the cities. Smart mobility systems in sustainable smart cities are characterized by an integration of advanced technology and infrastructure in the developmental process. These systems apply current data, prognosis, and IoT to improve transport services, customer satisfaction, and interact with the environment cohesively. Figure 8.3 illustrates the recent advancement in transportation systems for smart city.

8.1.3.1 Integrated Mobility Management

An important innovation in transport systems is the interconnected concept of transport that brings together public, private, and shared transportation means. Smart city concepts make use of big data and AI analytics to manage traffic and minimize the traffic jam [11]. It utilizes real-time data collected from sensors, cameras, and portable devices to analyse traffic flow and to forecast the demand for public transport and taxi services. This information is used to dynamically adjust service cost roots and availability to improve resource efficiency and enhance convenience for citizens.

8.1.3.2 Electric and Autonomous Vehicles

Smart cities prefer using electric vehicles because of their impacts on the environment. They are so widespread availability of effective charging technology and favourable government regulations. In addition, smart cities are also developing fast

Fig. 8.3 Recent advances in smart mobility systems

in the use of self-driving cars [12]. Self-driving cars could revolutionize how citizens live, they should not cause traffic accidents, resolve traffic jam, improve mobility of the disabled, etc. These are actually in a category of other systems such as smart traffic systems, connected infrastructures, and cooperative ITS.

8.1.3.3 Smart Infrastructure

In smart cities, infrastructure is developed alongside appropriate transportation systems to support sustainable mobility [13]. These include overhead pedestrian bridges, dedicated lanes for buses/bicycles, specialized roads and underground charging points for EVs. Additionally, smart cities employ wired and wireless sensors to monitor infrastructure conditions schedule maintenance and enhance safety.

8.1.3.4 Last-Mile Delivery

Last-mile delivery is essential in smart city transportation for smooth delivery without traffic [14]. Intelligent cities employ AI and machine learning (ML) techniques in delivery van routing and schedules to reduce the emission of CO_2. Another solution that the smart cities are looking at for solving the problems of urban and last-mile delivery include drones.

8.1.3.5 Citizen Engagement and Smart Mobility Apps

Smart cities allow the engagement of citizens in an active manner in the planning on transport systems, parking spaces and air quality of cities [15]. Apps as an aspect of smart mobility enable citizens to make better choices about the transport they take and be more environmentally friendly when it comes to mobility within cities.

8.2 Current Research Trends of Smart Mobility in Smart Cities

This study investigates the trends, contributions, and impact of research in smart cities with smart mobility from 2015 to 2025 through a bibliometric analysis of Scopus data. The Bibliometrix R package along with Origin software were utilized to analyse and depict the data. This approach ensures that researchers are presented with the most current research developments in the field enabling them to stay informed on emerging trends and advancements. In March 2025, a bibliometric analysis was conducted utilizing the Scopus core collection database to thoroughly investigate the concept and understanding of smart cities. Ultimately, 4473 articles were exported and saved in BibTex format before being included and examined in the bibliometric study. The primary information, syntax, and specific criteria used in screening for this study are shown in Table 8.1.

8.2.1 Annual Scientific Production and Top Ten Most Published Journals/Cited Paper

Figure 8.4 shows the annual publication trends from 2015 to 2025, highlighting the growth in research output. The early years shows a steady increase in publications, rising from 2015 (142 doc) to 2021 (605 doc), indicating a growing interest in the field. There is a slight decrease in the number of documents in 2022 (523), followed by a small increase in 2023 (543). There is a very slight decrement in

Table 8.1 Summary of data characteristics and main information of the bibliometric study

Description	Specific Standard Requirements/Results
Research database	Scopus
Search query	("smart") AND ("cities") AND ("smart") AND ("transportation")
Time span	2015–2025
Document type	Articles, conference paper, book chapter, etc.
Language	English
Data export format	BibTex
Documents	4473
Annual growth rate %	−3.07
Document average age	4.53
Average citations per doc	14.61
Keywords plus (ID)	14,944
Author's keywords (DE)	10,391
Authors	12,250
Authors of single-authored docs	421
Single-authored docs	604
Co-authors per doc	3.54
International co-authorships %	22.92

2024 (541). In 2025 (104), the number continues to rise and may continue increasing until the end of the year. Figure 8.5 presents the highest number of articles in sustainable development goals in the smart city research domain published in "Sustainability (Switzerland)" (130 docs), "Lecture Notes in Networks and Systems" (88 docs), "Lecture notes in Computer Science (Including subseries lecture notes in Artificial Intelligence and Lecture notes in Bioinformatics)" (78 docs), and "Communications in Computer and Information Science" (76 docs), which indicates a higher output of articles compared to the other journals. In addition, some other journals, such as the "Transportation Research Procedia" (74 docs), "Sustainable Cities and Society" (54 docs), "ACM International Conference Proceeding Series" (53 docs), "Advances in Intelligent Systems and Computing" (53 docs), "AIP Conference Proceedings" (45 docs), and "IOP Conference Series: Earth and Environmental Science" (44 docs), are in this domain. Table 8.2 displays the top ten most cited papers from the study which presents both the ranking and authors with their published works. This section presents information about authors, article titles and their corresponding publications in journals with documentation for publication years and citation frequency. The ranking focuses on engineering field smart cities articles to demonstrate their total citation (TC) counts and annual citation rates and normalized total citation (NTC) data which collectively signify their impact in the field.

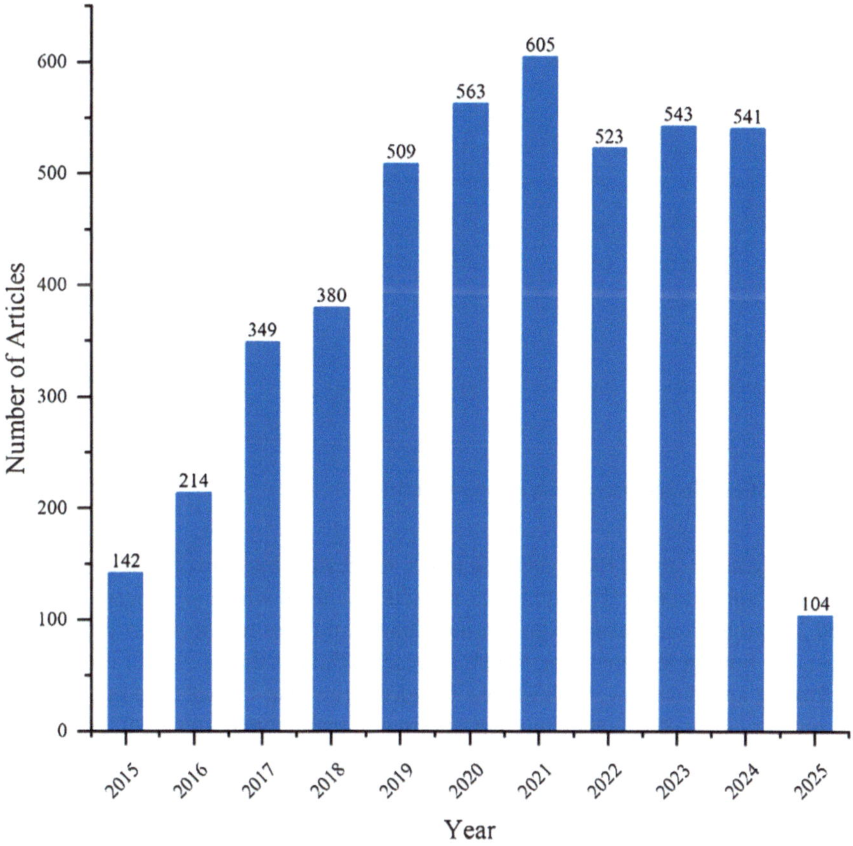

Fig. 8.4 Yearly trend of publication number

8.2.2 *Sankey Diagram (Three-Field Plot and Thematic Map)*

Figure 8.6 Sankey diagram illustrates the three-field plot thematic progression to visualize the flow and the distribution in scientific research. In this case, keywords plus, abstract, and title are added, acting as indicators to highlight the most influential research papers in the subject. These keywords indicate vital concepts and themes that stand out in academic discussions of the smart city domain. The diagram's keywords are sourced from earlier published studies and were then thematically organized

Figure 8.7 categorizes various smart city themes based on their urban development goals and relevant degrees. In the top right quadrant (motor themes), which represents themes that are highly developed and highly relevant, terms like "smart mobility", "sustainability", and "transport". In quadrant left (niche themes) there are represents "public transport", "smart card data", and "public transportation".

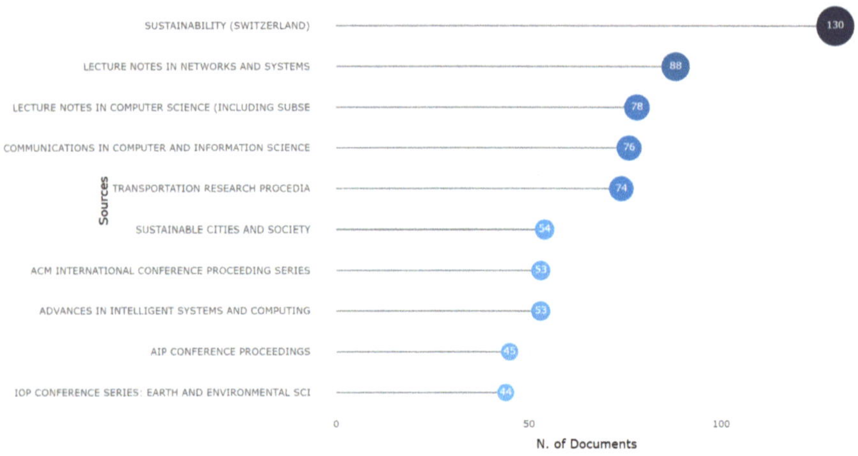

Fig. 8.5 Most relevant sources with number of articles

This indicates a strong focus on these areas, suggesting they are central to smart city research and practice. This quadrant reflects the areas where significant advances and investments are made, highlighting smart mobilities role in smart city. In contrast, the bottom left quadrant, which represents themes that are either emerging or declining and have lower relevance, includes terms like "vanet". Additionally, the terms like "smart city", and "internet of things" are the foundational information for further analysis and serve as a base for more advanced analyses. These themes might be considered foundational but must be emphasized more in smart cities. This visual categorization provides valuable insights for educators, curriculum developers, and researchers, helping them understand which areas are gaining traction and which might need more attention or innovation.

8.2.3 Co-occurrence Keyword Trends and Network Visualization

Figure 8.8 depicts the most prominent themes and keywords related to research in conceptual change. The most significant and most central terms, such as "smart city," "urban transportation", "internet of things", "intelligent systems", and "public transport" indicate the core focus of the research on how researchers grasp and alter their conceptual knowledge, particularly within smart city contexts. An illustration in Fig. 8.9 displays word occurrence frequencies from the sustainable development goals in smart city research network. Research focuses mainly on "smart city" as this term appears 1599 times throughout the literature. Research on smart mobility goals in smart cities indicates the importance of "urban transportation" based on its 663 words occurrences. The "internet of things" (528) and "intelligent systems"

Table 8.2 Most global cited documents smart transportation in smart cities

Rank	Authors name	Paper Title	Journal's Name	TC	TC/ Year	NTC
1	Sharifi and Garmsir (2020) [16]	The COVID-19 pandemic: Impacts on cities and major lessons for urban planning, design, and management	Science of the Total Environment	772	128.67	35.50
2	Ismagilova et al. (2019) [17]	Smart cities: Advances in research—An information systems perspective	International Journal of Information Management	652	93.14	35.44
3	Yang et al. (2017) [18]	Big data and cloud computing: Innovation opportunities and challenges	International Journal of Digital Earth	620	68.89	32.11
4	Li et al. (2019) [19]	Blockchain in the built environment and construction industry: A systematic review, conceptual models and practical use cases	Automation in Construction	480	68.57	26.09
5	Moustafa et al. (2019) [20]	An ensemble intrusion detection technique based on proposed statistical flow features for protecting network traffic of internet of things	IEEE Internet of Things Journal	420	60.00	22.83
6	Singh et al. (2020) [21]	Convergence of blockchain and artificial intelligence in IoT network for the sustainable smart city	Sustainable Cities and Society	410	68.33	18.85
7	Calvillo et al. (2016) [22]	Energy management and planning in smart cities	Renewable and Sustainable Energy Reviews	388	38.80	13.99
8	Javed et al. (2022) [23]	Future smart cities: Requirements, emerging technologies, applications, challenges, and future aspects	Cities	352	88.00	30.23
9	Huovila et al. (2019) [24]	Comparative analysis of standardized indicators for smart sustainable cities: What indicators and standards to use and when?	Cities	352	50.29	19.14
10	Sharma et al. (2017) [25]	Block-VN: A distributed blockchain based vehicular network architecture in smart city	Journal of Information Processing Systems	337	37.44	17.46

(483) combinations show the growth and sustainability in cities. The sustainability challenges in development are shown through terms like "public transport" (449) along with "traffic congestion" (339). The documents place emphasis on data-driven planning through combined mentions of "sustainable development" (318) and "urban transport" (283). Additionally, "intelligent vehicle highway systems" (273)

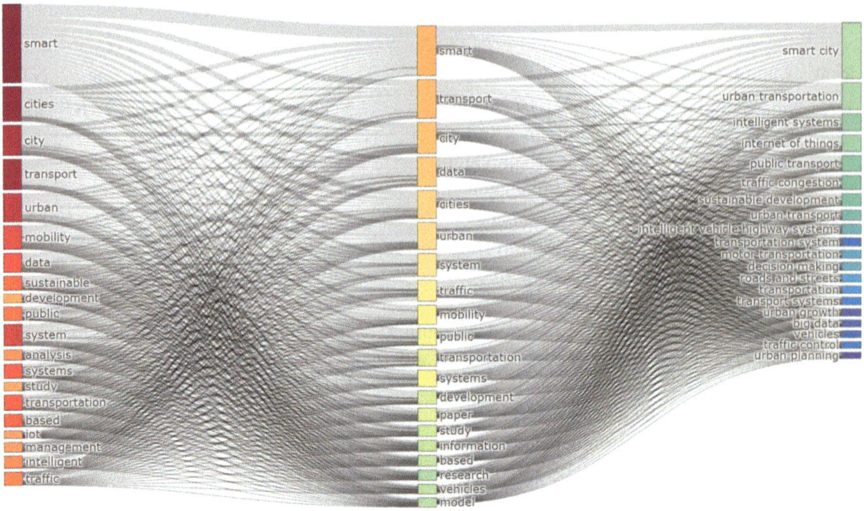

Fig. 8.6 Three field plot (thematic development) using sankey diagram

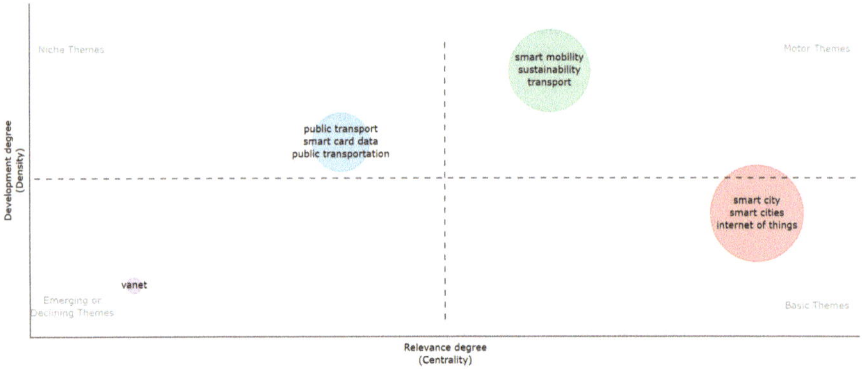

Fig. 8.7 Thematic map of smart mobility with smart city research domain

and "decision making" (217) showcase the important role of sustainability and smart technologies in urban environments. The data shows that smart mobility and smart city research revolves primarily around several major subjects.

8.2.4 *Countries' Scientific Production*

The scientific publication frequencies of various countries in the world are represented in Fig. 8.10. Scientific contributions from India stand at 1356 which demonstrates their powerful role in research development. The growing scientific power of

Fig. 8.8 Co-occurrence keyword trends in smart city domain with smart mobility

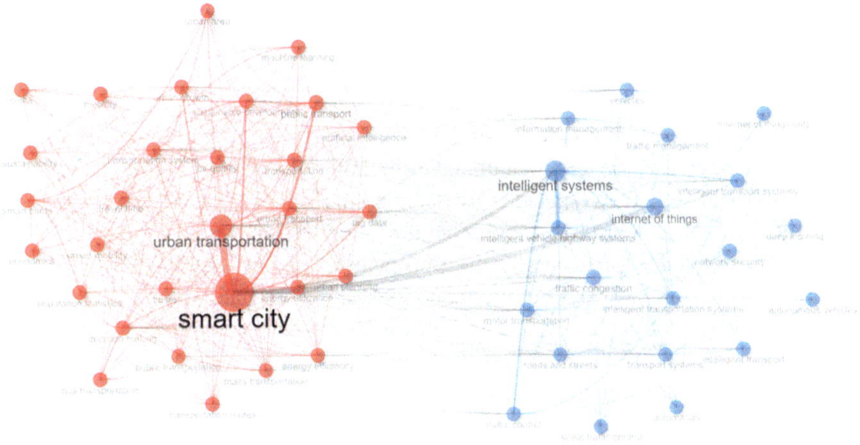

Fig. 8.9 Co-occurrence network of words in smart city research domain with smart mobility

China displays itself through its 904 scientific publications which rank second after India. The research hub of the Italy produces 529 scientific papers among its total contributions. Numerous publications emerge from UK and USA together with Spain given their research output comprising 488, 392, and 329 papers, respectively. Academic and industrial research sectors of Australia and Poland produce approximately 308 and 251 scholarly works between them. Germany and Portugal have smaller publication numbers but their contributions amount to totals of 247 and 197 research items, respectively.

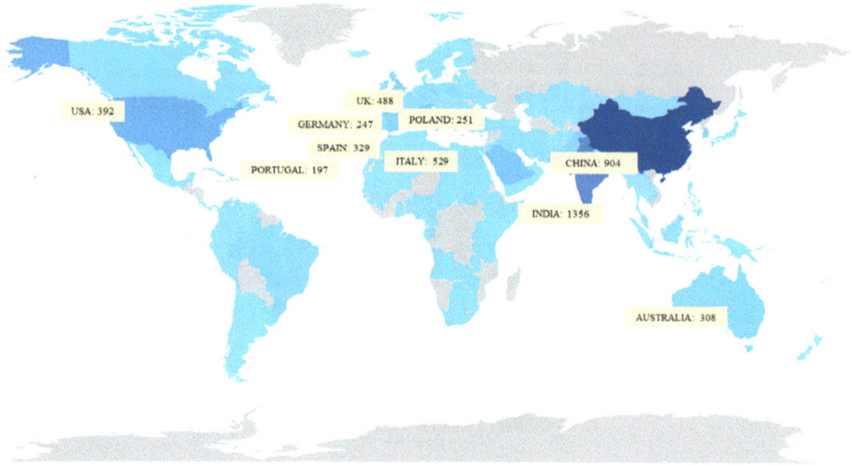

Fig. 8.10 Top ten countries' scientific production of various countries based on frequency

8.2.5 Mapping Smart Mobility with Smart City Research to Sustainable Development Goals (SDGs)

SDGs, their targets (bubble series), and the percentage (bar graph) of corresponding keywords detected in the text (calculated as the ratio of keywords in one Goal to the total number of keywords detected) are depicted in Fig. 8.11. For SDG mapping, the study utilized the SDG Mapper tool from the "Knowledge Base for the Sustainable Development Goals," provided by the European Commission [26]. This tool uses bubble charts to depict SDG interconnections and relevance among bibliometric records. The size of each bubble correlates with the proportion of associated keywords found, which is calculated as the number of goal-specific keywords divided by the total keyword count. As such, SDG 11, marked by 112 keyword occurrences. This is followed by SDG 7, with 13 occurrences, and SDG 9, with 7 occurrences, underlining their relative prominence in the research database. Within SDG 11, there are several key subtargets that the research address: Subtarget 11.2 is shown as a distinct bubble within the larger SDG 11 circle, indicating focused research on mobility in smart city. Subtarget 11.3, which aims in settlement and management, is another bubble within the SDG 11 area. This highlights the efforts in research to improving the cities. Subtarget 11.6 focuses on reducing the adverse per capita environmental impact of cities in smart cities solutions. Figure 8.11 shows that the size of the SDG 7 bubble implies that advancements in smart city are being focussed on affordable and clean energy that prioritize sustainability. Similarly, SDG 9 (Industry, innovation and infrastructure) is also closely linked to SDG 11 and aims to enhance scientific research and affordable access to internet. The detailed description is depicted in Table 8.3.

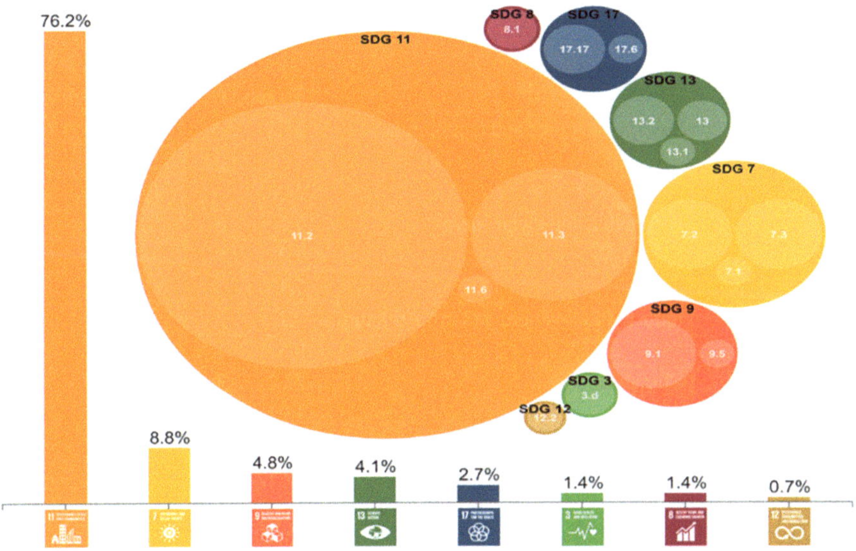

Fig. 8.11 Mapping mobility based smart city research to relevant SDGs and targets

8.3 Smart Mobility Infrastructure

Smart mobility infrastructure is the key to the modern, robust system of transportation. It adopts modern technologies to enhance the delivery of transport services as well as the safety of the services. Smart Infrastructures include smart traffic systems, intelligent collection, and distribution of traffic data and information exchange between the vehicle-traffic environment. Smart technology integrated with green transport thus forms a sustainable, smart, and efficient transport system for a smart city [27].

8.3.1 Smart Mobility Infrastructure Requirements

The current transport systems depend on advanced infrastructure so as to make convenient, efficient, and green forms of transport. With the development of smart cities, great emphasis is set on the development of infrastructure for utilizing new transportation technologies. Following are the smart mobility infrastructure requirements as depicted in Fig. 8.12.

8.3.1.1 Communication and Connectivity

Smart mobility infrastructure establishes the integration of different interconnected systems for connections in transport system. With the help of the smart devices and IoT sensors collects data on traffic levels, transport schedule, and road conditions in

Table 8.3 SDGs ranking, targets, and relevance in mobility-based smart city research domain with SDGs

SDGs (By Rank)	Description	Targets	No. of Occurrences		Relevance%
SDG 11	Sustainable cities and communities	11.2	89	112	76.2
		11.3	22		
		11.6	1		
SDG 7	Affordable and clean energy	7.2	6	13	8.8
		7.3	6		
		7.1	1		
SDG 9	Industry, innovation, and infrastructure	9.1	6	7	4.8
		9.5	1		
SDG 13	Climate action	13.2	3	6	4.1
		13	2		
		13.1	1		
SDG 17	Partnerships for the goals	17.17	3	4	2.7
		17.6	1		
SDG 3	Good health and well-being	3.d	2	2	1.4
SDG 8	Decent work and economic growth	8.1	2	2	1.4
SDG 12	Responsible consumption and production	12.2	1	1	0.7

Fig. 8.12 Requirements for building smart mobility infrastructure

real-time. Such information is used to monitor and control traffic conditions, as well as to schedule and regulate public transport. Moreover, with Vehicle-to-Everything (V2X) communication systems, it is possible to combine vehicles and infrastructures' capabilities and enhance safety, avoid congestion, and deliver better driving experiences [28].

8.3.1.2 Intelligent Transportation Systems

Intelligent transportation systems are required for platforming and optimizing various modes and services of transportation [29]. Advanced traffic management systems, public transport schedule and location, and the traveller's information system are part of the ITS. These systems are open source as this allows for the integration of the services provided by the transport sector for efficient management.

8.3.1.3 Energy and Sustainable Infrastructure

Smart mobility infrastructures must create importance on energy efficiency and sustainability due to an increased international focus on environmental issues. Transportation for citizens and goods benefits from design of transportation infrastructure that encourages use of electric and hybrid cars, trucks, buses, trolleys, and cycles, as well as other transportation means [30]. In addition, services like the installation of charging points for EVs, bicycle lanes, and walkways make it more useful and valuable.

8.3.1.4 Multimodal Integration

For the development of smart mobility infrastructure, it important to consider all users and all modes of transport must be connected through seamless connections [31]. Accessibility should be provided for parking space, public transport interchange areas, cycle stands, and pedestrian area.

8.3.1.5 Scalability and Flexibility

The evolving nature of transportation technology and shifting mobility needs demands that smart mobility infrastructure be designed to accommodate future developments. Infrastructure must remain flexible and broad enough to allow the integration of emerging technologies and transport systems [32]. This is achievable through the adoption of standardized approaches, modular system designs, and the use of open-source platforms where feasible.

8.3.1.6 Data Management and Analytics

In smart mobility infrastructures, the collection and analysis of data from multiple sources is essential for successful implementation [2]. Information collected through sensors and IoT devices should be properly captured and stored using reliable data systems. These data support decisions help improve the efficiency and effectiveness of transportation services.

8.3.1.7 Security and Privacy

Smart mobility infrastructure depends on the continuous exchange of information through interconnected digital systems such as sensors, communication networks, and real-time monitoring. This reliance on data collection and connectivity makes security and privacy key priorities. Enhancing infrastructure security helps prevent hacking and unauthorized access. Protecting user information involves data protection measures to ensure privacy and trust [33].

8.3.2 Intelligent Transportation Systems

Intelligent transportation systems are defined as the integration of technologies and service into the transportation system environment, vehicles, and infrastructure to enhance the transport system. Smart mobility infrastructure is one of the parts of ITS which is data and network based, aligned and automated to enhance the transport system. It is achieved through supply of operative information to the users, better management of traffic flow and the services to the travellers and the communication between the users and the management of the traffic systems. Smart infrastructure includes highways, roads, interchanges, and other transportations facilities that are equipped with intelligence devices such as sensors, actuators, cameras etc. Vehicle-to-vehicle (V2V) communication, vehicle-to-infrastructure (V2I) communication, and management systems are used to facilitate the exchange of information [34]. Traffic information is real-time data collected, processed, and analysed by various data management systems derived from different sources and used for decision-making. Furthermore, smart pavements detect and manage traffic and thus minimize traffic jam and occurrence of accidents [35]. Advanced communication networks provide real-time traffic information to the traveller to change the routes and consequently reducing travel time. Real-time information and connected vehicle technologies also facilitate the better interaction of public travel services such as buses, trains, and carpooling services.

8.3.3 Automated and Connected Vehicles

Automated and connected vehicles have emerged as a key element of smart mobility systems [36]. Due to advance in technology there has been shift of model from mechanical to intelligent vehicles that are capable to operate automatically. Intelligent connected vehicles employ sensor system and wireless communication technologies, V2V, V2I, and Vehicle-to-Pedestrian (V2P) to interact in real time with other vehicles, traffic control systems, and infrastructures. However, automated vehicles have driver assistance systems with adaptive cruise control, lane keeping assist, and collision avoidance systems [37]. These systems direct control of steering, acceleration, brakes, and other functionalities of a car in order to provide safer and easier means of transport. The primary purpose is minimization of errors made by humans as they are the main causes of traffic incidents. These systems utilize various technologies like IoT, Big data, AI, and cloud computing to gather, analyse, and process real-time data from connected vehicle and other sources. Smart infrastructure comprises of intelligent roadside units, variable message signs, adaptive traffic signals, smart parking systems, and advanced highway ramp meters for interaction between vehicles and environment. The focus is on the development of different applications that address the issues for road safety, congestion, automated toll systems, and emergency response management. Consequently, automatic and networked vehicles are revolutionizing the mobility and create the foundation for ITS. It aims to design and develop a cost-effective and secure transport infrastructure to support economic growth and improvements in living standards of the population.

8.4 Wireless Technologies in Mobility System

In the mobility system, wireless technologies act as the main enabler for communication between vehicles and other components within the transportation system. The key wireless technologies that include vehicular communications, cellular mobile communications, short-range static communications, depicted in Fig. 8.13, these technologies play the pivotal roles to enhance the successful implementation of IoV applications. These technologies make possible connection and also speed in transfer of information to and from the various components of transportation network.

8.4.1 Vehicular Communications System

8.4.1.1 Dedicated Short-Range Communications

Dedicated Short-Range Communications (DSRC) is a wireless technology specifically designed for operation in the 5.9 GHz band for enabling very fast exchange of information between vehicles and road facilities. The DSRC-based

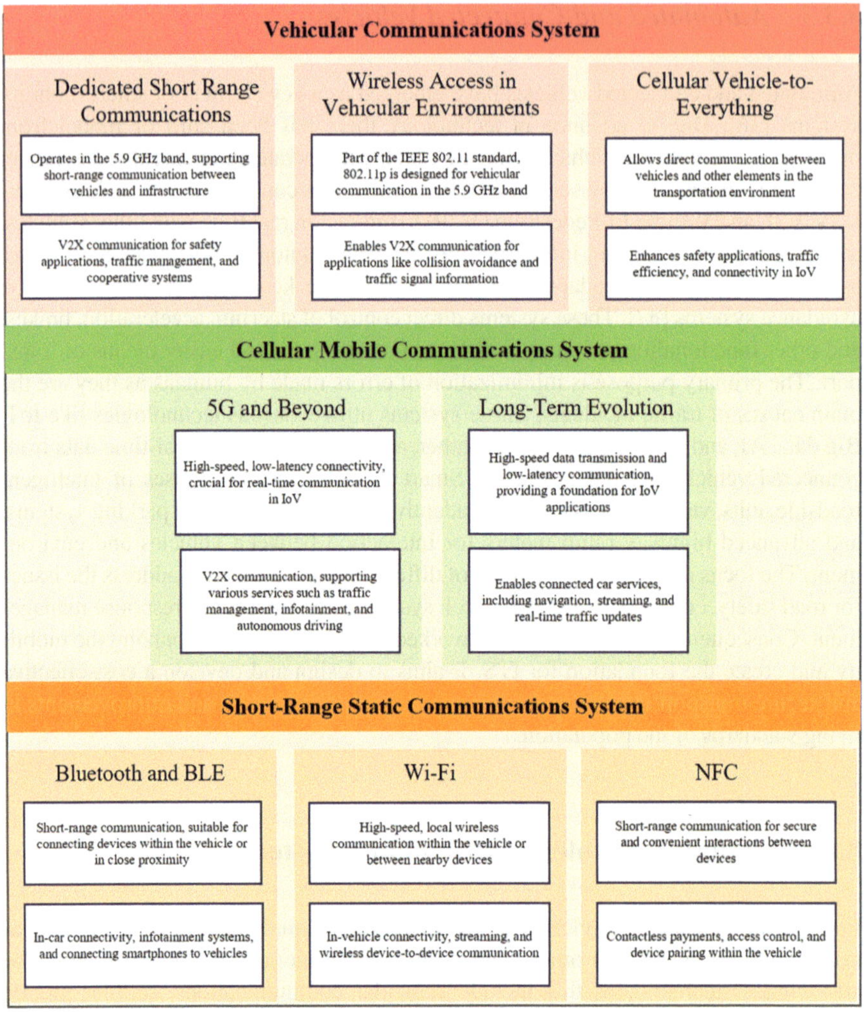

Fig. 8.13 Wireless technologies in smart mobility systems

Short-Range Vehicle Communication system works on the same 5.9 GHz band and has been designed for the V2X communication [38]. V2X technology, which mainly uses DSRC, creates a dynamic and integrated space for various applications of intelligent transportation systems and ensures the safety of vehicular communication by sharing real-time data information including location, speed and trajectory.

8.4.1.2 IEEE 802.11p (WAVE—Wireless Access in Vehicular Environments)

IEEE 802.11p is a substandard of wireless technology known as IEEE 802.11 technology that has been designed to meet requirements for transporting data in vehicular systems [39]. It works in the specific 5.9 GHz frequency spectrum for the real-time data exchange requirements of vehicles and roadway equipment. The main purpose of IEEE 802.11p is to create the foundation for V2X services, which enables communicating between vehicles and related objects. Also, 802.11p is used for V2X communication for traffic information in ITS, which in turn helps to increase the efficiency of traffic. Therefore, IEEE 802.11p provides an important prospective towards the future of connected and cooperative vehicular communication.

8.4.1.3 Cellular Vehicle-to-Everything

Cellular Vehicle-to-Everything (C-V2X) networks offers improvement in cellular communications, information exchange between vehicles as well as other components [40]. Instead, using the openness of cellular networks, cars are directly connected to other cars, road infrastructure, and different connected components. This connectivity framework plays an important role in improving vehicle safety by supporting applications like collision notification and emergency braking systems which reduces the chances of an incident. Furthermore, C-V2X is useful in traffic optimization due to the enablement of coordination of traffic management strategies. Its applicability reaches to mobility system, where it adopts a more profound, perceptive and engaged smart mobility environment. The integration of C-V2X in vehicular communication offers a safer, efficient, and more advanced transportation system for the citizens.

8.4.2 Cellular Mobile Communications System

8.4.2.1 5G and Beyond

5G and its further evolutions represents a revolutionary era in cellular mobile communications, particularly for mobility system. These technologies require availability of high speed and low latency connectivity to support the real-time requirement of mobility system applications. It is a strong foundation for V2X communication which goes beyond cellular communication to enable real-time exchange of information between vehicles and virtually everything in their environment within the transportation system. Cellular mobile communications which evolved through 5G and the generations realize the importance in the future in which vehicles are not only connected but are intelligent, consistent, and near-perfect transportation systems [41].

8.4.2.2 Long-Term Evolution

Long-Term Evolution (LTE) is an evolution from the current mobile cellular communication services, which provides high-speed and low latency [42]. This robust data transmission capability enables to create continuous connection for diverse mobility system applications. The mobility system has multiple uses for LTE for different applications, such as navigation systems that require latest information to provide the best routes. It also powers streaming services into cars thus improving entertainment in automobiles. In other words, LTE serves as a building block for the formation of the mobility system since it creates a network connection that enables automobiles and their onboard systems to function in an environment enhanced by modern communication technologies and increase both the on-road driving comfort and driving safety and performance.

8.4.3 Short-Range Static Communications System

8.4.3.1 Bluetooth and Bluetooth Low Energy

The mobility system has a large number of standards which are highly flexible and are mainly based on Bluetooth low energy (BLE) [43]. These technologies are most effective when the devices intend to have a local interaction within a car or in a limited area. Both standards of Bluetooth: traditional and low energy are also significant for the formation of in-car connectivity that allows to communicate easily with each other. Bluetooth and BLE provides the integration of smartphones with automobiles for voice callings, music, and data transfer. These flexible technologies are crucial to the establishment of a smart and engaging in-vehicle surroundings.

8.4.3.2 Wi-Fi (802.11a/b/g/n/ac)

Protocols like 802.11a/b/g/n/ac enable fast wireless transmission within a specific area, making the use of Wi-Fi. These wireless standards are flexible and useful for mobility system. Moreover, Wi-Fi standards offer effective streaming of data, voice, video etc., thereby enhancing the experience of entertainment in a car [44]. In addition to entertainment, these Wi-Fi standards also allow wireless device-to-device connections, allowing users to exchange information between their devices within the car with security.

8.4.3.3 Near Field Communication

Near Field Communication (NFC) is a primary enabler of a fully wireless and connected vehicle system under the mobility system since it helps support the fast and secure exchange of data between devices within the short range [45]. It is involved in various smart mobility system such as without contactless payment, authenticated, cost-effective, secure, and flexible, and help to realize a connected consumer experience for passengers in vehicles.

8.5 Sustainable Mobility Solutions in Smart Cities

Sustainable mobility solutions are critical elements for the creation of smart mobility systems in smart sustainable cities. It includes clean, integrated and affordable transport systems for citizens. Sustainable mobility involves the use of shared cars, buses, bicycles and EVs along with the development of infrastructure such as bike lanes and pedestrian paths. These solutions help in the removal of traffic congestion, emissions of pollutants, particularly the greenhouse gases in the city. Figure 8.14 illustrates the framework for smart and sustainable mobility solutions and highlights how environmental impact; sustainable urban transport initiative and active transportation support collectively enhance urban mobility and reduce ecological footprint.

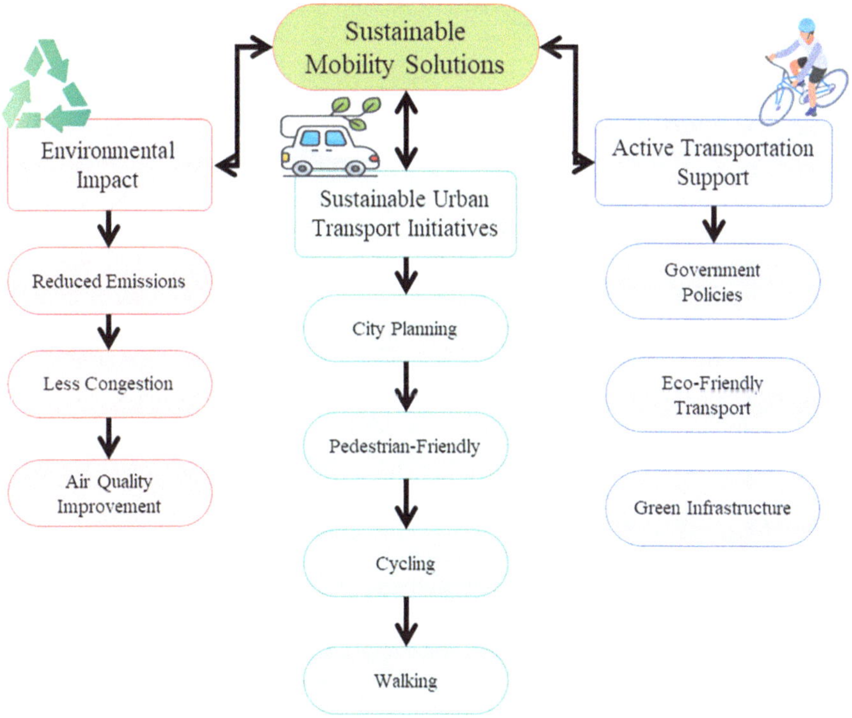

Fig. 8.14 The smart and sustainable mobility solutions

8.5.1 The Effect of Smart Mobility Solutions on the Environment

Urbanization leads to the expansion of city regions, often resulting in increased congestion and environmental pollution from vehicle emissions. This contributes to higher level of greenhouse gases which significantly impact human health. Smart mobility solutions are a key component of smart cities aimed at maximizing the efficiency of transportation infrastructure while minimizing environmental impact. Smart mobility systems are transport solutions built on the principles of interoperability, connectivity, and data-driven decision-making to improve service quality [46]. These systems help address traffic congestion by reducing the time vehicle spend in traffic, which improves fuel efficiency and lowers emissions. For example, smart traffic signal technologies use real-time traffic data to adjust signal timings, redirect vehicles to less congested roots, and provide drivers with guidance on optimal paths. Such improvements can significantly reduce travel time and distance, leading to a measurable decrease in transportation-related pollution in urban areas. The rise of shared mobility options also reduces the reliance on single occupancy vehicles, conserving fuel and easing the burden on road networks. Furthermore, integrating electric and hybrid vehicles into shared mobility services further reduce greenhouse gas emission and support the transition toward sustainable urban transport [47].

8.5.2 Support of Active Transportation

The motivation and promotion of active transport corresponds to mobility strategies for smart sustainable cities that benefit the citizen. Through the adoption of active transport modes alongside smart mobility systems, urban areas reduce the emissions of greenhouse gases achieve better air quality and promote healthy living of the people [48]. This implies that infrastructure provision, policy adoption and social marketing strategies constitute a coherent and comprehensive strategy for sustainable smart cities. These cities achieve improved quality of life, reduced personal transport costs and congestion by increasing cycling and walking. The infrastructure for cycling and walking is an important aspect of active transportation strategy in smart cities. These includes building of bicycle lanes, bike sharing and fine walkways for pedestrians. All these infrastructure enhancements also provide ease to active transport while also ensuring that people use this transport instead of private cars. However, infrastructure development alone cannot ensure the development of a sustainable smart city, supportive policies and programmes for active transportation must also be taken into consideration. This leads to the growing public trust in active transportation modes and provision of the cities that are pleasant and safe for all travellers. Promotion and awareness are the other important determinants of use of active transport. In addition, active transportation in smart cities is also reinforced by technology and data analytics. Smart sensors and GPS devices are used to track traffic flow, infrastructure facilities and movement pattern of the vehicle. Thus, promotion of active transportation within smart mobility systems remains as essential for the achievement of environmentally sustainable smart cities. The appropriate political commitment, the construction of special facilities, the adoption of favourable regulations, and a relationship with the population will allow the establishment of such conditions that would inspire and facilitate choosing rational, environmentally friendly, and active movement.

8.5.3 Initiatives for Sustainable Urban Transportation

Public transportation is a central feature of smart cities as it aims to offer residents eco-friendly and efficient transport systems that minimize environmental harm. Green and intelligent transport systems supported by ICTs along with integrated urban planning provide the foundation for developing intelligence cities [49]. These systems contribute by transforming public transport improving infrastructure and promoting sustainable transport options. An important area of focus in sustainable urban transportation is smart public transit. This approach involves the use of advanced technologies such as real-time tracking systems, smart card payment options and unified ticketing across transit modes. These innovations improve urban mobility and encourage more people to use public transport. As a result, the number of private vehicles on the roads is reduced leading to reduce traffic congestion and environmental impact. This is a crucial phase for smart cities to implement transit-oriented development, where the design of urban areas and neighbourhood on placing housings, shops, offices and other essential services near public transport facilities. This approach helps reduce dependency on private vehicles and encourages more sustainable travel behaviour in the city. It is also implementing incentives program and tax exemptions for employees who choose environmentally friendly transport modes. Alongside, infrastructure is expanding to support EVs through accessible charging facilities. Achieving sustainable urban transportation in smart cities often depends on collaborative efforts involving public and private sectors, local governments, transport authorities and technology providers [50]. These partnerships support the development of intermodal transportation systems that apply data and technology, guided by intelligent urban planning, to deliver effective and sustainable mobility solutions. These initiatives play a vital role in shaping future of smart cities and establishing sustainable transportation systems that promote healthier and more environmentally conscious urban living for future generations.

8.6 Smart Mobility Systems Real-Life Applications

Smart transportation systems are innovative sub systems of smart city that support the system and promote efficient use of resources by the use of advanced technologies, such as intelligent traffic management system, autonomous vehicles, and demand responsive transit system. The smart mobility concept is essential in sustainable development hence improving the quality of life in smart cities. Figure 8.15 illustrates the real-life applications of smart mobility systems in sustainable smart cities.

8.6.1 Infrastructure and Charging Systems for Electric Vehicles

Smart mobility systems play a crucial role in enabling the EVs charging ecosystem. By integrating IoT technologies, these systems offer real-time monitoring, management, and control of charging stations. Advanced smart mobility models use big data analytics and AI to support accurate decision-making in Evs' traffic prediction, root optimization, intelligent parking, and charging management. These intelligent services help prevent congestion at charging stations and ensure smooth and user-friendly EVs operations. Stakeholders' involvement is essential in developing an efficient and accessible electrical vehicle charging infrastructure. Public private partnerships (PPPs) are particularly valuable as they combine the expertise resources and networks of both the public and private sectors to facilitate the deployment of charging stations [51]. Local governments also play an important role by introducing regulations or incentives to encourage electric vehicles adoption and expand charging infrastructure. Utility and energy companies supports the implementation of smart charging solutions and contribute significantly to the integration of EVs

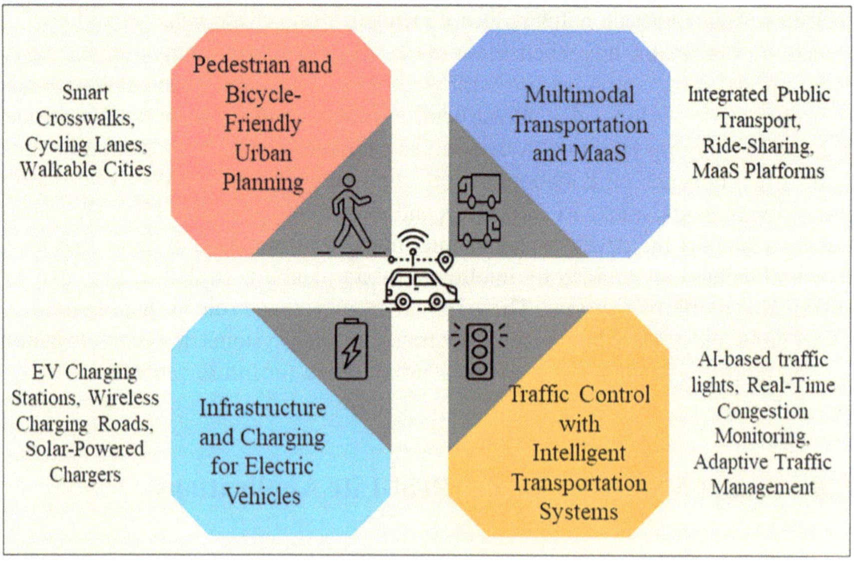

Fig. 8.15 Real-life applications of smart mobility systems

with the power grid. In addition, EVs manufacturers and charging station operators offers customized and convenient services that meet the diverse needs of users [52]. The development of an intelligent and unified approach to integrating EVs charging infrastructure is essential for advancing sustainable mobility. By considering various charger types, smart charging systems and intelligent mobility technologies cities promotes EVs use reduce environmental impact and build a long-term vision that supports transportation and energy systems with evolving societal needs.

8.6.2 Traffic Control with Intelligent Transportation Systems

Intelligent transportation systems are based on the ideas that communication systems, sensors, and real-time data processing should be used to control and coordinate transportation infrastructure. These technologies support improved traffic flow and more effective traffic management through optimized traffic signals, real-time traffic information systems, and intelligent traffic safety systems. In smart mobility systems traffic control relies on ITS components such as intelligent traffic signal control, advanced travel information systems and V2I communication. When implemented these technologies enable cities to build transportation networks that are responsive flexible and capable of adapting to fluctuations in traffic patterns. Modern travellers' information systems help computers by providing real-time updates on traffic conditions and suggesting alternative routes to avoid delays. V2I communication represents a significant advancement in ITS by enabling vehicle to interact with roadside equipment such as traffic signals, toll booth and navigation system [53]. This communication enables real-time information to be transmitted directly to vehicles. In addition, vehicle-to-vehicle communication, also known as connected vehicle technology, facilitates the exchange of key data such as speed, direction, and environmental conditions between vehicles to improve safety and reduce accidents. The application of ITS in the context of sustainable smart cities offers several benefits including reducing traffic congestion, increasing the efficiency and productivity of transportation systems and improving the road safety [29].

8.6.3 Integration of Multimodal Transportation and MaaS

Intermodal transportation refers to systems that combine various transport modes such as bus, train, car, bicycle, and walking in a coordinated way to make urban transport more efficient and accessible. This approach allows users to choose between different transportation options based on available time and financial resources. By offering a variety of connected travel options, these systems help reduce travel time pollution and traffic congestion. A key feature for smart mobility is the adoption of interconnected multimodal transportation systems along with Mobility as a Service (MaaS) solution [54]. MaaS brings together multiple forms of transport throughout digital platforms that allow users to access all services in one place. These web-based applications act as a single point of contact where users can plan their journeys, purchase tickets, and receive real-time travel information from multiple service providers.

8.6.4 Pedestrian and Bicycle-Friendly Urban Planning

Sustainable smart cities emphasize on efficient and clean means of transport that will benefit the citizens of a city. This comprises the incorporation of methods of designing cities for walking and bicycling as basic aspects of smart mobility solutions [55]. It is essential to focus more on pedestrians and cycling can generate numerous environmental, social, and economic advantages for cities. Another key advantage of pedestrian and bicycle-friendly cities is the ability to reduce road congestion. When these transport options are supported with well-developed and accessible infrastructure the number of vehicles on the road may decrease. This leads to less traffic and minimum travel times for the citizen. Designing cities to support walking and cycling is a vital aspect of building sustainable smart cities. Prioritizing these modes of transport helps create environmentally sustainable and healthy urban environments that provide efficient and reliable mobility for all [56]. As the world continues to urbanize it become essential for future smart cities to adopt locally driven strategies. These approaches should promote the integration of smart mobility systems along with sustainable and inclusive transportation options that serve the needs of the citizens.

8.7 Open Research Challenges and Research Directions

Smart mobility systems are technologies implemented in transport networks aimed at enhancing capacity, flexibility, and sustainability. However, achieving smart and sustainable mobility requires addressing numerous research challenges and exploring new opportunities. The following discussion highlights these issues and potential research directions for using smart mobility to build sustainable smart cities, as illustrated in Fig. 8.16.

8.7.1 Open Research Challenges in Smart Mobility

8.7.1.1 System Integration of Composite Systems

It is essential to develop an environment that integrates various mobility systems and technologies [57]. Smart cities include multiple transport mode such as public transit, ride sharing, micro-mobility options like e-scooters and bicycles, and self-driving cars

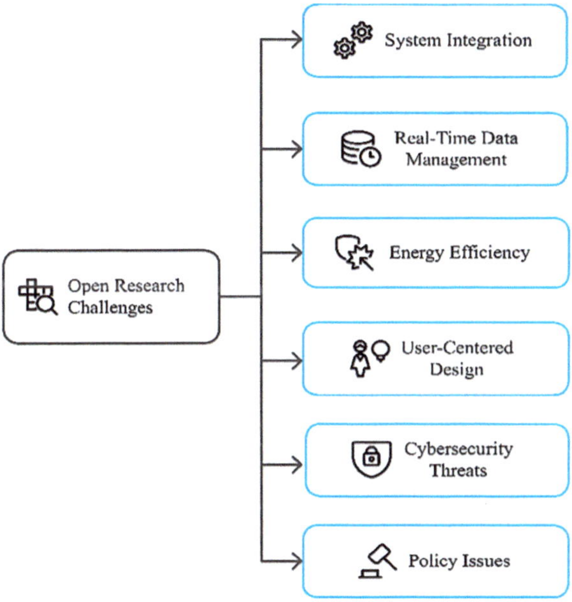

Fig. 8.16 Open research challenges in smart mobility systems

[58]. This requires data integrators, along with long-term storage, processing capabilities, and data conversion protocols. Well-equipped networks are necessary to support high volume of real data exchange. Solutions must also support enterprises in transitioning to modern architecture that connect existing systems with new technologies.

8.7.1.2 Real-Time Data Management with Big Data and Analytics

Smart mobility systems produce huge data from sensors, connected automobiles, smartphone applications, and physical infrastructures. Managing, analysing, and utilizing this data effectively constitutes several challenges. Building the necessary system to support the analysis of the high-volume data is essential [59]. Issues related to data privacy and security to a smaller extent data retention are especially important when handling sensitive user information.

8.7.1.3 Energy Efficiency and Emissions Reduction

Smart mobility systems are designed to reduce environmental impact and achieving energy efficiency and minimizing emissions remain significant challenges. Key areas of focus include managing the charging demands and energy consumption of electric vehicles along with the supporting charging infrastructure, developing intelligent traffic management systems that prevent congestion, reduce emissions and integrate the renewable energy sources into transport networks [60].

8.7.1.4 User-Centred Design and Accessibility

It is essential to design access modes that are inclusive and usable by citizens in order to promote the smart mobility services [61]. Key challenges include addressing the needs of specific user groups, Such as individuals with disabilities and those with limited economic means. In terms of user experience and user face design, creating intuitive and user-friendly interfaces for mobile applications and other digital platforms is important. Another potential challenge is achieving affordability and mobility solutions without compromising the financial sustainability of service providers.

8.7.1.5 Cybersecurity and Privacy Threats

The information systems that support mobility networks are becoming increasingly interconnected which also makes them more vulnerable to cyber threats [62]. Key challenges in this area include protecting transportation infrastructure from cyber-attacks, ensuring secure communication between devices and systems and balancing data sharing with user privacy through the use of advanced encryption techniques.

8.7.1.6 Policy and Governance Issues

There are always policy and regulatory concerns that affect the implementation of smart mobility systems. Several emerging technologies, Such as self-driving cars often encounter legal and regulatory challenges across different regions. Addressing these issues requires coordinated policy efforts across jurisdictions to develop unified solutions to mobility-related issues [63].

8.7.2 Open Research Directions in Smart Mobility

There is a need to explore new approaches and models within the existing literature, as it offers promising research directions to address the identified challenges. These potential research directions are discussed below and also illustrated in Fig. 8.17.

8.7.2.1 Interoperable Platform Creation

Research should focus on developing architectures that support the integration of smart mobility systems. As the emphasis on intelligent transport continues to grow, it is important to design platforms that enable smooth interaction among various mobility services. Key areas of focus include creating open APIs and data flow standards to support flexible system integration, using blockchain technology to ensure

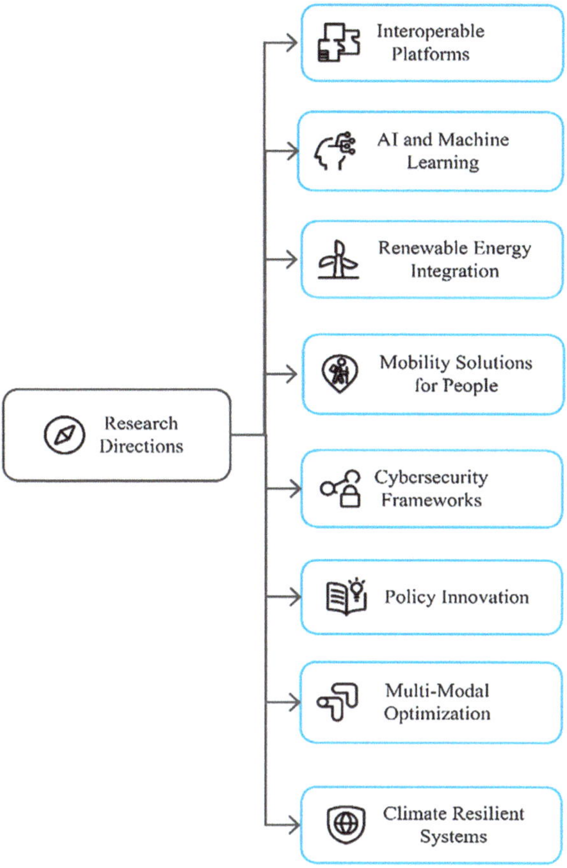

Fig. 8.17 Open research directions in smart mobility systems

reliable and transparent data exchange and applying design patterns that utilize services to build modular architectures [64]. These architectures should be structured to allow easy upgradation.

8.7.2.2 Artificial Intelligence and Machine Learning

Artificial intelligence and machine learning prove to be very effective for improving smart mobility systems [65]. Research directions in this domain include constructing models for prediction of traffic conditions and traffic demand. Computing the shortest path to the destination with least amount of time to be consumed. Adaptable learning systems play a significant role in improving decision-making for self-driving cars operating in complex urban environments.

8.7.2.3 Electric Power System and Renewable Energy Integration

Research into energy efficient technologies and the integration of renewable energy is essential for advancing sustainable smart mobility [66]. In the area of EVs, efforts include improving battery technologies to extend driving range and reduce charging time. Another important direction is the development of smart grids that support sustainable transport by incorporating renewal energy sources [67]. Additionally, policies should be developed to promote energy efficient urban infrastructures, such as the use of solar power systems in transportation networks and related buildings.

8.7.2.4 Mobility Solutions for People

Future research must prioritize user experience and inclusivity in smart mobility systems. Key focus areas include incorporating user-centred design approaches to actively involve diverse user groups in the design process, exploring pricing strategies that make smart mobility solutions accessible to price sensitive populations, and developing mobility solutions that are responsive to specific social context and demographic characteristics [68].

8.7.2.5 Building Up on the Cybersecurity Frameworks

Addressing cybersecurity risks in smart mobility networks requires advanced research in several keys area [62]. These include designing intrusion detection systems that align with the specific characteristics of smart mobility environments, exploring the use of AI for real-time threat detection and response, and establishing a comprehensive framework of best practices for secure software development within the transportation sector.

8.7.2.6 Policy Innovation and Governance Models

There is a need to develop new models of governance to address policy barriers in urban mobility systems [63]. Research directions include creating a regularity framework that adapt to ongoing technological changes, examining the effectiveness of stimulating policy implementation through digital replicas known as digital twins, and encouraging international collaboration on mobility standards and practices to support consistency and interoperability.

8.7.2.7 Integration and Multi-Modal Optimization

Integrated and efficient mobility-on-demand (MODAL) systems are essential for ensuring smooth urban transportation. Future research should focus on several keys areas: implementing a unified payment and ticketing system across various modes of transport, optimizing the allocation and coordination of connection points to minimize delays, and utilizing micro-mobility to support short-distance travel and enhance connectivity with public transportation networks [69].

8.7.2.8 Climate Resilient Mobility Systems

Urban mobility systems are required to be adaptable to natural occurrences such as climate change. The solution is to build the infrastructure that will be resilient to floods, heat waves, and other disasters. Adapting existing traffic management systems to incorporate real-time weather conditions and evaluating the integration of insurance solutions with mobility systems are essential for effectively addressing climate-related risk [70].

8.8 Summary

The mobility system is an important development within the vehicle interaction environment and has gained attention as the part of the broader IoT. It is a key technology for enabling communication between vehicles, between vehicles and people and between vehicles and infrastructure. This area is expected to play a major role in solving traffic-related problems, improving road safety, and addressing other transportation challenges. The main focus of the new mobility system is to enhance security, improve vehicle performance and support the development of vehicle-based communication networks. To develop effective solutions, it is important for engineers and developer to understand the systems architecture, the different network model it uses and the challenges it may face. Furthermore, the development of an intelligent transport system is essential for the success of smart cities. However, there are several challenges that need to be addressed. These include system integration, energy consumption, user concerns, data security, and policy issues. The use of artificial intelligence, renewal energy, and inclusive design practices helps in creating mobility systems that are efficient, environmentally friendly, and accessible to all.

References

1. L. Butler, T. Yigitcanlar, A. Paz, Smart urban mobility innovations: A comprehensive review and evaluation. IEEE Access **8**, 196034–196049 (2020). https://doi.org/10.1109/ACCESS.2020.3034596
2. S. Paiva, M.A. Ahad, G. Tripathi, N. Feroz, G. Casalino, Enabling technologies for urban smart mobility: Recent trends, opportunities and challenges. Sensors **21**(6), 1–45 (2021). https://doi.org/10.3390/S21062143
3. P. Ribeiro, G. Dias, P. Pereira, Transport systems and mobility for smart cities. Appl. Syst. Innov. **4**(3), 61 (2021). https://doi.org/10.3390/asi4030061
4. J.B. Greenblatt, S. Shaheen, Automated vehicles, on-demand mobility, and environmental impacts. Curr. Sustain. Renew. Energy Rep. **2**(3), 74–81 (2015). https://doi.org/10.1007/S40518-015-0038-5
5. M. Savastano, M.C. Suciu, I. Gorelova, G.A. Stativă, How smart is mobility in smart cities? An analysis of citizens' value perceptions through ICT applications. Cities **132**, 104071 (2023). https://doi.org/10.1016/J.CITIES.2022.104071
6. P. Mishra, P. Thakur, G. Singh, Sustainable smart city to society 5.0: State-of-the-art and research challenges. SAIEE Afr. Res. J. **113**(4), 152–164 (2022). https://doi.org/10.23919/SAIEE.2022.9945865
7. Z. Mahrez, E. Sabir, E. Badidi, W. Saad, M. Sadik, Smart urban mobility: When mobility systems meet smart data. IEEE Trans. Intell. Transp. Syst. **23**(7), 6222–6239 (2022). https://doi.org/10.1109/TITS.2021.3084907
8. P. Mishra, G. Singh, Internet of vehicles for sustainable smart cities, in *Sustainable Smart Cities*, (Springer, Cham, 2023), pp. 175–193. https://doi.org/10.1007/978-3-031-33354-5_8
9. P. Mishra, G. Singh, Enabling technologies for sustainable smart city, in *Sustainable Smart Cities*, (Springer, Cham, 2023), pp. 59–73. https://doi.org/10.1007/978-3-031-33354-5_3
10. S. Goumiri, S. Yahiaoui, S. Djahel, Smart mobility in smart cities: Emerging challenges, recent advances and future directions. J. Intell. Transp. Syst., 81–117 (2023)
11. M.M. Rathore, A. Paul, W.H. Hong, H.C. Seo, I. Awan, S. Saeed, Exploiting IoT and big data analytics: Defining smart Digital City using real-time urban data. Sustain. Cities Soc. **40**, 600–610 (2018). https://doi.org/10.1016/J.SCS.2017.12.022
12. C. Badue et al., Self-driving cars: A survey. Expert Syst. Appl. **165**, 113816 (2021). https://doi.org/10.1016/J.ESWA.2020.113816
13. G. Dispenza, V. Antonucci, F. Sergi, G. Napoli, L. Andaloro, Development of a multi-purpose infrastructure for sustainable mobility. A case study in a smart cities application. Ener. Proc. **143**, 39–46 (2017). https://doi.org/10.1016/J.EGYPRO.2017.12.645
14. L. Butler, T. Yigitcanlar, A. Paz, W. Areed, How can smart mobility bridge the first/last mile gap? Empirical evidence on public attitudes from Australia. J. Transp. Geogr. **104**, 103452 (2022). https://doi.org/10.1016/J.JTRANGEO.2022.103452
15. D.L. Angarita Lozano, S.E. Diaz Marquez, M.E. Morales Puentes, Sustainable and smart mobility evaluation since citizen participation in responsive cities. Transport. Res. Proc. **58**, 519–526 (2021). https://doi.org/10.1016/J.TRPRO.2021.11.069
16. A. Sharifi, A.R. Khavarian-Garmsir, The COVID-19 pandemic: Impacts on cities and major lessons for urban planning, design, and management. Sci. Total Environ. **749**, 142391 (2020). https://doi.org/10.1016/j.scitotenv.2020.142391
17. E. Ismagilova, L. Hughes, Y.K. Dwivedi, K.R. Raman, Smart cities: Advances in research—An information systems perspective. Int. J. Inf. Manag. **47**, 88–100 (2019). https://doi.org/10.1016/j.ijinfomgt.2019.01.004
18. C. Yang, Q. Huang, Z. Li, K. Liu, F. Hu, Big data and cloud computing: Innovation opportunities and challenges. Int. J. Digit. Earth **10**(1), 13–53 (2017). https://doi.org/10.1080/17538947.2016.1239771

19. J. Li, D. Greenwood, M. Kassem, Blockchain in the built environment and construction industry: A systematic review, conceptual models and practical use cases. Autom. Constr. **102**, 288–307 (2019). https://doi.org/10.1016/j.autcon.2019.02.005
20. N. Moustafa, B. Turnbull, K.-K.R. Choo, An ensemble intrusion detection technique based on proposed statistical flow features for protecting network traffic of internet of things. IEEE Internet Things J. **6**(3), 4815–4830 (2019). https://doi.org/10.1109/JIOT.2018.2871719
21. S. Singh, P.K. Sharma, B. Yoon, M. Shojafar, G.H. Cho, I.-H. Ra, Convergence of blockchain and artificial intelligence in IoT network for the sustainable smart city. Sustain. Cities Soc. **63**, 102364 (2020). https://doi.org/10.1016/j.scs.2020.102364
22. C.F. Calvillo, A. Sánchez-Miralles, J. Villar, Energy management and planning in smart cities. Renew. Sust. Energ. Rev. **55**, 273–287 (2016). https://doi.org/10.1016/j.rser.2015.10.133
23. A.R. Javed et al., Future smart cities: Requirements, emerging technologies, applications, challenges, and future aspects. Cities **129**, 103794 (2022). https://doi.org/10.1016/j.cities.2022.103794
24. A. Huovila, P. Bosch, M. Airaksinen, Comparative analysis of standardized indicators for smart sustainable cities: What indicators and standards to use and when? Cities **89**, 141–153 (2019). https://doi.org/10.1016/j.cities.2019.01.029
25. P.K. Sharma, S.Y. Moon, J.H. Park, Block-VN: A distributed blockchain based vehicular network architecture in smart city. J. Inform. Process. Syst. (2017). https://doi.org/10.3745/JIPS.03.0065
26. S. Borchardt, G.B. Vignola, D. Buscaglia, M. Maroni, *Mapping EU Policies with the 2030 Agenda and SDGs-Fostering Policy Coherence through text-based SDG Mapping*. https://doi.org/10.2760/110687
27. M.M. Abdel Wahed Ahmed, N. Abd El Monem, Sustainable and green transportation for better quality of life case study greater Cairo – Egypt. HBRC J. **16**(1), 17–37 (2020). https://doi.org/10.1080/16874048.2020.1719340
28. T. Aldhanhani, A. Abraham, W. Hamidouche, M. Shaaban, Future trends in smart green IoV: Vehicle-to-everything in the era of electric vehicles. IEEE Open J. Vehic. Technol. **5**, 278–297 (2024). https://doi.org/10.1109/OJVT.2024.3358893
29. R. Mangiaracina, A. Perego, G. Salvadori, A. Tumino, A comprehensive view of intelligent transport systems for urban smart mobility. Int. J. Log. Res. Appl. **20**(1), 39–52 (2017). https://doi.org/10.1080/13675567.2016.1241220
30. A. García-Olivares, J. Solé, O. Osychenko, Transportation in a 100% renewable energy system. Energy Convers. Manag. **158**, 266–285 (2018). https://doi.org/10.1016/J.ENCONMAN.2017.12.053
31. P.P. Kumar, M. Parida, M. Swami, Performance evaluation of multimodal transportation systems. Proc. Soc. Behav. Sci. **104**, 795–804 (2013). https://doi.org/10.1016/J.SBSPRO.2013.11.174
32. G. Yannis, A. Chaziris, Transport system and infrastructure. Transport. Res. Proc. **60**, 6–11 (2022). https://doi.org/10.1016/J.TRPRO.2021.12.002
33. E. Zavvos, E.H. Gerding, V. Yazdanpanah, C. Maple, S. Stein, M.C. Schraefel, Privacy and trust in the internet of vehicles. IEEE Trans. Intell. Transp. Syst. **23**(8), 10126–10141 (2022). https://doi.org/10.1109/TITS.2021.3121125
34. B. Bučko, M. Michálek, K. Papierniková, K. Zábovská, Smart mobility and aspects of vehicle-to-infrastructure: A data viewpoint. Appl. Sci. **11**(22), 10514 (2021). https://doi.org/10.3390/app112210514
35. X. Shi, More than smart pavements: Connected infrastructure paves the way for enhanced winter safety and mobility on highways. J. Infrastruct. Preserv. Resil. **1**(1), 13 (2020). https://doi.org/10.1186/s43065-020-00014-x
36. H. Farah, S.M.J.G. Erkens, T. Alkim, B. van Arem, Infrastructure for automated and connected driving: state of the art and future research directions, in *Road Vehicle Automation 4*, Lecture Notes in Mobility, (Springer, Cham, 2018), pp. 187–197. https://doi.org/10.1007/978-3-319-60934-8_16

37. A. Vahidi, A. Eskandarian, Research advances in intelligent collision avoidance and adaptive cruise control. IEEE Trans. Intell. Transp. Syst. **4**(3), 143–153 (2003). https://doi.org/10.1109/TITS.2003.821292
38. L. Cheng, B. Henty, D. Stancil, F. Bai, P. Mudalige, Mobile vehicle-to-vehicle narrow-Band Channel measurement and characterization of the 5.9 GHz dedicated short range communication (DSRC) frequency band. IEEE J. Sel. Areas Commun. **25**(8), 1501–1516 (2007). https://doi.org/10.1109/JSAC.2007.071002
39. F. Arena, G. Pau, A. Severino, A review on IEEE 802.11p for intelligent transportation systems. J. Sens. Actuat. Netw. **9**(2), 22 (2020). https://doi.org/10.3390/jsan9020022
40. L. Bai, Z. Huang, Y. Ge, R. Yu, L. Wang, X. Cheng, Cellular vehicle-to-everything (C-V2X) testing: From theory to practice. IEEE Netw., 1–1 (2025). https://doi.org/10.1109/MNET.2025.3534196
41. L. Guevara, F. Auat Cheein, The role of 5G technologies: Challenges in smart cities and intelligent transportation systems. Sustainability **12**(16), 6469 (2020). https://doi.org/10.3390/su12166469
42. W. Li, X. Ma, J. Wu, K.S. Trivedi, X.-L. Huang, Q. Liu, Analytical model and performance evaluation of long-term evolution for vehicle safety services. IEEE Trans. Veh. Technol. **66**(3), 1926–1939 (2017). https://doi.org/10.1109/TVT.2016.2580571
43. J.F. Ensworth, M.S. Reynolds, BLE-backscatter: Ultralow-power IoT nodes compatible with Bluetooth 4.0 low energy (BLE) smartphones and tablets. IEEE Trans. Microw Theory Tech. **65**(9), 3360–3368 (2017). https://doi.org/10.1109/TMTT.2017.2687866
44. D. Dzemydienė, Wireless communication network infrastructure analysis for smart service system functioning, in *Development of Smart Context-Aware Services for Cargo Transportation*, International Series in Operations Research & Management Science, vol. 330, (Springer, Cham, 2022), pp. 29–60. https://doi.org/10.1007/978-3-031-07199-7_3
45. C. Dai, S. Bao, S. Chen, S. Garg, G. Kaddoum, M.M. Hassan, Precision-adaptive task offloading and resource allocation for efficient positioning and sensing in near-field IoV systems. IEEE Internet Things J. **12**, 1–1 (2025). https://doi.org/10.1109/JIOT.2025.3557431
46. R.G. Mugion, M. Toni, H. Raharjo, L. Di Pietro, S.P. Sebathu, Does the service quality of urban public transport enhance sustainable mobility? J. Clean. Prod. **174**, 1566–1587 (2018). https://doi.org/10.1016/j.jclepro.2017.11.052
47. C. Wu, X. Tang, X. Xu, System design, analysis, and control of an intelligent vehicle for transportation in greenhouse. Agriculture **13**(5), 1020 (2023). https://doi.org/10.3390/agriculture13051020
48. J. Zawieska, J. Pieriegud, Smart city as a tool for sustainable mobility and transport decarbonisation. Transp. Policy (Oxf) **63**, 39–50 (2018). https://doi.org/10.1016/J.TRANPOL.2017.11.004
49. K. Ahmad, H. Khujamatov, A. Lazarev, N. Usmanova, M. Alduailij, M. Alduailij, Internet of things-aided intelligent transport systems in smart cities: Challenges, opportunities, and future. Wirel. Commun. Mob. Comput. **2023**, 1–28 (2023). https://doi.org/10.1155/2023/7989079
50. M. Visan, S.L. Negrea, F. Mone, Towards intelligent public transport systems in smart cities; collaborative decisions to be made. Proc. Comput. Sci. **199**, 1221–1228 (2022). https://doi.org/10.1016/J.PROCS.2022.01.155
51. L. Zhang, Z. Zhao, Z. Kan, Private-sector partner selection for public-private partnership projects of electric vehicle charging infrastructure. Energy Sci. Eng. **7**(5), 1469–1484 (2019). https://doi.org/10.1002/ese3.367
52. M. Shafiei, A. Ghasemi-Marzbali, Fast-charging station for electric vehicles, challenges and issues: A comprehensive review. J. Ener. Stor. **49**, 104136 (2022). https://doi.org/10.1016/J.EST.2022.104136
53. C. Creß, Z. Bing, A.C. Knoll, Intelligent transportation systems using roadside infrastructure: A literature survey. IEEE Trans. Intell. Transp. Syst. **25**(7), 6309–6327 (2024). https://doi.org/10.1109/TITS.2023.3343434
54. G. Smith, D.A. Hensher, Towards a framework for mobility-as-a-service policies. Transp. Policy (Oxf) **89**, 54–65 (2020). https://doi.org/10.1016/J.TRANPOL.2020.02.004

55. S. Handy, Making US cities pedestrian- and bicycle-friendly, in *Transportation, Land Use, and Environmental Planning*, (Elsevier, 2020), pp. 169–187. https://doi.org/10.1016/B978-0-12-815167-9.00009-8
56. C. Zhao, T.A. Carstensen, T.A.S. Nielsen, A.S. Olafsson, Bicycle-friendly infrastructure planning in Beijing and Copenhagen - between adapting design solutions and learning local planning cultures. J. Transp. Geogr. **68**, 149–159 (2018). https://doi.org/10.1016/J.JTRANGEO.2018.03.003
57. T. Alam, R. Gupta, N. Nasurudeen Ahamed, A. Ullah, A. Almaghthwi, Smart mobility adoption in sustainable smart cities to establish a growing ecosystem: Challenges and opportunities. MRS Ener. Sustain. **11**(2), 304–316 (2024). https://doi.org/10.1557/s43581-024-00092-4
58. L. Rayle, D. Dai, N. Chan, R. Cervero, S. Shaheen, Just a better taxi? A survey-based comparison of taxis, transit, and ridesourcing services in San Francisco. Transp. Policy (Oxf) **45**, 168–178 (2016). https://doi.org/10.1016/j.tranpol.2015.10.004
59. B.N. Silva et al., Urban planning and Smart City decision management empowered by real-time data processing using big data analytics. Sensors **18**(9), 2994 (2018). https://doi.org/10.3390/s18092994
60. Q. Guo, Y. Wang, X. Dong, Effects of smart city construction on energy saving and CO_2 emission reduction: Evidence from China. Appl. Energy **313**, 118879 (2022). https://doi.org/10.1016/j.apenergy.2022.118879
61. C. Schöne, S. Spundflasch, H. Krömker, User centred design of a knowledge repository to support gender smart mobility, in *HCI in Mobility, Transport, and Automotive Systems. HCII 2019*, Lecture Notes in Computer Science, vol. 11596, (Springer, Cham, 2019), pp. 413–429. https://doi.org/10.1007/978-3-030-22666-4_30
62. R. Al Mallah, D. Lopez, B. Farooq, Cyber-security risk assessment framework for blockchains in smart mobility. IEEE Open J. Intell. Transp. Syst. **2**, 294–311 (2021). https://doi.org/10.1109/OJITS.2021.3106863
63. B. Anthony Jnr, Sustainable mobility governance in smart cities for urban policy development—A scoping review and conceptual model. Smart Sustain. Built Environ. **14**(3), 649–671 (2025). https://doi.org/10.1108/SASBE-05-2023-0109
64. F. Orecchini, A. Santiangeli, F. Zuccari, A. Pieroni, T. Suppa, Blockchain technology in smart city: a new opportunity for smart environment and smart mobility, in *Intelligent Computing & Optimization. ICO 2018. Advances in Intelligent Systems and Computing*, vol. 866, (Springer, Cham, 2019), pp. 346–354. https://doi.org/10.1007/978-3-030-00979-3_36
65. P. Mishra, G. Singh, *Sustainable Smart Cities* (Springer, Cham, 2023). https://doi.org/10.1007/978-3-031-33354-5
66. A. Brown, J. Gonder, B. Repac, An analysis of possible energy impacts of automated vehicles. Lect. Note. Mobil., 137–153 (2014). https://doi.org/10.1007/978-3-319-05990-7_13
67. Y. Chen, A. Ardila-Gomez, G. Frame, Achieving energy savings by intelligent transportation systems investments in the context of smart cities. Transp. Res. D. Transp. Environ. **54**, 381–396 (2017). https://doi.org/10.1016/J.TRD.2017.06.008
68. M. Joshi, A. Vaidya, M. Deshmukh, Sustainable transport solutions for the concept of smart city, in *Sustainable Energy and Transportation. Energy, Environment, and Sustainability*, (Springer, Singapore, 2018), pp. 21–42. https://doi.org/10.1007/978-981-10-7509-4_3
69. R.L. Abduljabbar, S. Liyanage, H. Dia, The role of micro-mobility in shaping sustainable cities: A systematic literature review. Transp. Res. D. Transp. Environ. **92**, 102734 (2021). https://doi.org/10.1016/J.TRD.2021.102734
70. K.C. Dey, A. Mishra, M. Chowdhury, Potential of intelligent transportation systems in mitigating adverse weather impacts on road mobility: A review. IEEE Trans. Intell. Transp. Syst. **16**(3), 1107–1119 (2015). https://doi.org/10.1109/TITS.2014.2371455

Chapter 9
Smart Education with Immersive Technology

Contents

P. Mishra, G. Singh, *Sustainable Smart Cities 2.0*,
https://doi.org/10.1007/978-3-032-01102-2_9

9.1 Introduction

Education in the current society is under dynamic development due to the recent emergence of immersive technology that has led to smart education [1]. As the world's cities transition towards being smart, sustainable, efficient, the concept of smart learning within these sustainable smart cities must also be considered. Intelligent learning with the use of technology in education not only ensures that the students are capable to respond to challenges in a world that is continually moving toward the digital environment but also cultivates uniqueness, logic and analysis, and skills to solve issues. Virtual Reality (VR), Augmented Reality (AR), and Mixed Reality (MR) make learning effective and informative, since they mimic the real environment in an artificial way. It will help students develop the ability to learn in a more intuitive, targeted, and effective way. For smart cities that acknowledge environmental conservation as an important goal, smart education promotes educational processes with an emphasis on a safe environment. This enables the students to develop a right perception towards environment and become responsible members of the global society [2]. In addition, smart education using AR and VR technology enhances inter- and intra-personal relationships among students within smart education city. Networks and tools such as Learning Management System (LMS) and online materials are useful for the learners and also share this knowledge, regardless of distance [3]. Therefore, smart education together with VR forms the basis of learning in smart sustainable cities. Smart city solutions for learning aim to position countries toward a sustainable future based on innovation, environmental responsibility, and the use of information technology (IT) solutions.

9.1.1 Overview of the Evolution and Importance of Smart Education

Smart education is based on the idea of building educational institutions and educational processes that help the learner/reader to learn [4]. In recent centuries, many technological progresses help bring various evolutions in education, such as the radio and television for distance education in the early twentieth century and use of computers and the internet in the late twentieth century. Due to the skills of critical thinking, problem-solving, and creativity, smart education helps people to create, solve, and support the United Nations Sustainable Development Goals [5]. Smart education is the combination of different disciplines and internship, and it aims at encouraging students with competencies to understand the global problems faced around the world today and discover the solutions for sustainable development. In addition, smart education plays an important role in the implementation of the ideas of lifelong learning, as well as the development of new skills, which is a priority in the framework of the formation of a smart city and the preparation of its residents for possible changes [6]. In a smart city where, technology is thus ubiquitous in the society, the capacity to learn and be updated is the key focus. Smart learning helps a person to pursue relevant education through the use of technological solutions in learning, take courses online, and share the acquired materials with those in the similar field worldwide. The success of smart education in creating sustainable smart cities in a function of governments, educators, businesses and citizens coming up with practical ideas [7]. Fundamentally, governments need to invest into development and application of smart education technologies and methods. Teachers have no other option but to adapt to change and step into the digital world in order to produce engaging teaching and learning content. Employers need to embrace any form of training and development for workers to understand the impact of smart education in business. Last but not the least, citizens need to understand the importance of learning as well as the potential benefits it possesses in the form of smart education in order to live better and fulfilling lives as well as develop nation economically. Therefore, smart education is disrupting every aspect of learners and communities leading to the creation of sustainable smart city ecosystems [8]. The evolution and significance of smart education is illustrated in Fig. 9.1.

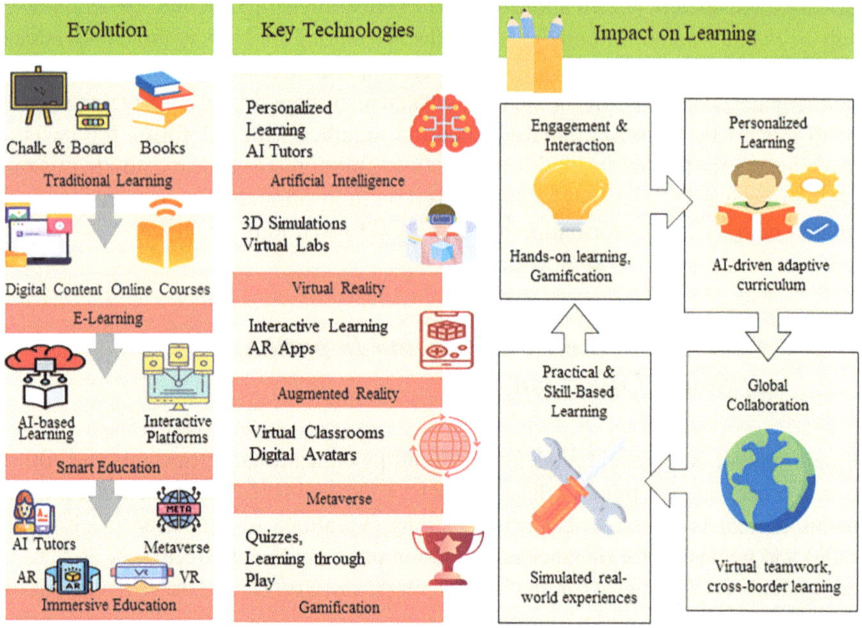

Fig. 9.1 The evolution and significance of smart education

9.1.2 The Impact of Metaverse on Learning Methodologies

The Metaverse is the evolution of the next generation VR, a 4D asymmetrical environment where the real and augmented, and augmented and virtual worlds collide. As this technology becomes more developed and available, it has a great impact on learning methodologies' organization and its connection with the concept of sustainable smart cities. First of all, the Metaverse offers an opportunity to conduct effective collaborative learning, since it allow learners from different region of the world to come to a common virtual space and practice knowledge sharing, discussions, and completion of joint tasks. Electronic learning environments helps minimize inequity in terms of learning styles, learning preferences, and learning requirements. It is easy to imagine virtual tutors and personalized teaching–learning environment in the Metaverse. These technological advancements mean that learning applications based on artificial intelligence (AI) can provide individual, real-time learning, feedback and support, and development challenge as required by individuals [9]. Also, these elements are modified, augmented, or even redesigned to new set of educational standards, curricula, practices, etc. However, in the context of sustainable smart cities, the transformative impact of the Metaverse, along with the promotion of effective learning methodologies that complement the standard progressive educational model, serves as a foundational principle for these cities. The role of the technology in smart sustainable city development is to enable the effective incorporation of technology in the improvement of human quality of life, green economic development and to minimize impact on the natural environment [10].

9.2 Current Research Trends of Smart Education in Smart Cities

This study investigates the trends, contributions, and impact of research in smart cities with smart education from 2015 to 2025 through a bibliometric analysis of Scopus data. The Bibliometrix R package along with origin software were utilized to analyse and depict the data. This approach ensures that researchers are presented with the most current research developments in the field enabling to focus on emerging trends and advancements. In March 2025, a bibliometric analysis was conducted utilizing the Scopus core collection database to thoroughly investigate the concept and understanding of smart cities. Ultimately, 2300 articles were exported and saved in BibTex format before being included and examined in the bibliometric study. The primary information, syntax, and specific criteria inclusion used in screening for this study are shown in Table 9.1.

Table 9.1 Summary of data characteristics and main information of the bibliometric study

Description	Specific Standard Requirements/Results
Research database	Scopus
Search query	("smart") AND ("cities") AND ("smart") AND ("education")
Time span	2015–2025
Document type	Articles, conference paper, book chapter, etc.
Language	English
Data export format	BibTex
Documents	2300
Annual growth rate %	5.76
Document average age	4.24
Average citations per doc	11.07
Keywords plus (ID)	9120
Author's keywords (DE)	5855
Authors	6142
Authors of single-authored docs	344
Single-authored docs	534
Co-authors per doc	3.19
International co-authorships %	20.87

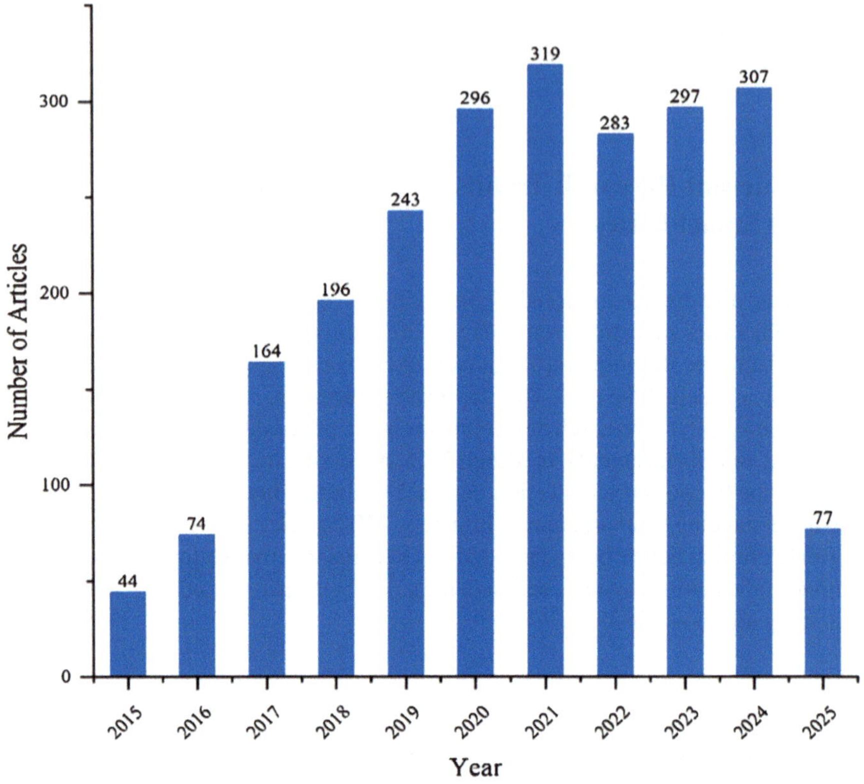

Fig. 9.2 Yearly trend of publication number in education in smart cities

9.2.1 Annual Scientific Production and Top Ten Most Published Journals/Cited Paper

Figure 9.2 shows the annual publication trends from 2015 to 2025, highlighting the growth in research output. The early years shows a steady increase in publications, rising from 2015 (44 doc) to 2021 (319 doc), indicating a growing interest in the field. There is a slight decrease in the number of documents in 2022 (283), followed by a small increase in 2023 (297) and 2024 (307). In 2025 (77), the number continues to rise and may continue increasing until the end of the year.

Figure 9.3 presents the highest number of articles in sustainable development goals in the smart city research domain published in "Advances in Intelligent Systems and Computing" (117 docs), "Lecture Notes in Networks and Systems" (69 docs), "ACM International Conference Proceeding Series" (50 docs), and "Sustainability (Switzerland)" (37 docs), which indicates a higher output of articles compared to the other journals. In addition, some other journals, such as the "Lecture notes in Computer Science (Including subseries lecture notes in Artificial Intelligence

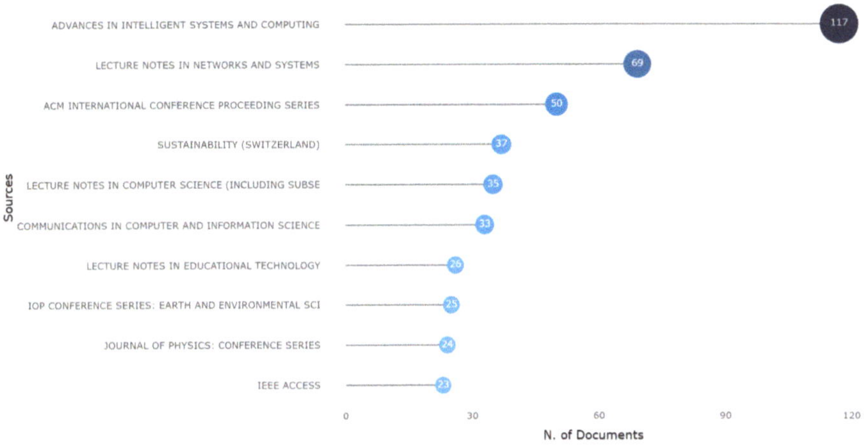

Fig. 9.3 Most relevant sources with number of articles

and Lecture notes in Bioinformatics)" (35 docs), "Communications in Computer and Information Science" (33 docs), "Lecture notes in Educational Technology" (26 docs), "IOP Conference Series: Earth and Environmental Science" (25 docs), "Journal of Physics: Conference Series" (24 docs), and "IEEE Access" (23 docs), are in this domain. Table 9.2 displays the top ten most cited papers from the study which presents both the ranking and authors with their published works. This section presents information about authors, article titles, and corresponding publications in journals with documentation for publication years and citation frequency. The ranking focuses on engineering field smart cities articles to demonstrate total citation (TC) counts and annual citation rates and normalized total citation (NTC) data which collectively signify impact in the field.

9.2.2 *Sankey Diagram (Three-Field Plot and Thematic Map)*

Figure 9.4 Sankey diagram illustrates the three-field plot thematic progression to visualize the flow and the distribution in scientific research. In this case, keywords plus, abstract, and title are added, acting as indicators to highlight the most influential research papers in the subject. These keywords indicate vital concepts and themes that stand out in academic discussions of the smart city domain. The diagram's keywords are sourced from earlier published studies and were then thematically organized by a specific researcher. Figure 9.5 categorizes various smart city themes based on their development and relevant degrees. In the top right quadrant (motor themes), which represents themes that are highly developed and highly relevant, terms like "smart city", and "education". In quadrant left (niche themes) there are represents "urbanization", "smart learning", "smart mobility", "augmented

Table 9.2 Most global cited documents of smart education in smart city research domain

Rank	Authors name	Paper Title	Journal's Name	TC	TC/ Year	NTC
1	Oztemel and Gursev (2020) [11]	Literature review of industry 4.0 and related technologies	Journal of Intelligent Manufacturing	1453	242.17	67.24
2	Nuaimi et al. (2015) [12]	Applications of big data to smart cities	Journal of Internet Services and Applications	719	65.36	24.72
3	Ahad et al. (2020) [13]	Enabling technologies and sustainable smart cities	Sustainable Cities and Society	434	72.33	20.09
4	Hariri et al. (2019) [14]	Uncertainty in big data analytics: Survey, opportunities, and challenges	Journal of Big data	431	61.57	31.67
5	Javed et al. (2022) [15]	Future smart cities: Requirements, emerging technologies, applications, challenges, and future aspects	Cities	352	88.00	35.13
6	Yigitcanlar et al. (2020) [16]	Contributions and risks of artificial intelligence (AI) in building smarter cities: Insights from a systematic review of the literature	Energies	328	54.67	15.18
7	Tanwar et al. (2020) [17]	Machine learning adoption in blockchain-based smart applications: The challenges, and a way forward	IEEE Access	303	50.50	14.02
8	Khatoun and Zeadally (2016) [18]	Smart cities: Concepts, architectures, research opportunities	Communications of the ACM	285	28.50	12.91
9	Bhushan et al. (2020) [19]	Blockchain for smart cities: A review of architectures, integration trends and future research directions	Sustainable Cities and Society	281	46.83	13.00
10	Sodiq et al. (2019) [20]	Towards modern sustainable cities: Review of sustainability principles and trends	Journal of Cleaner Production	262	37.43	19.25

reality", "virtual reality", and "AI". This indicates a strong focus on these areas, suggesting they are central to education-based smart city research and practice. This quadrant reflects the areas where significant advances and investments are made, highlighting their critical role in smart city. In contrast, the bottom left quadrant, which represents themes that are either emerging or declining and have lower relevance, includes terms like "technology", and "infrastructure". Additionally, the terms like "smart campus", "higher education", "digital transformation", "internet of things", and "artificial intelligence" are the foundational information for further analysis and serve as a base for more advanced analyses. These themes might be

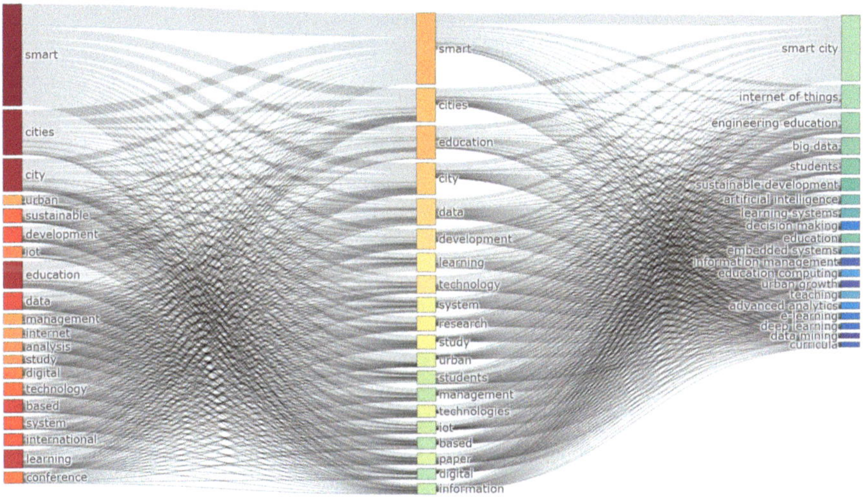

Fig. 9.4 Three field plot (thematic development) using sankey diagram

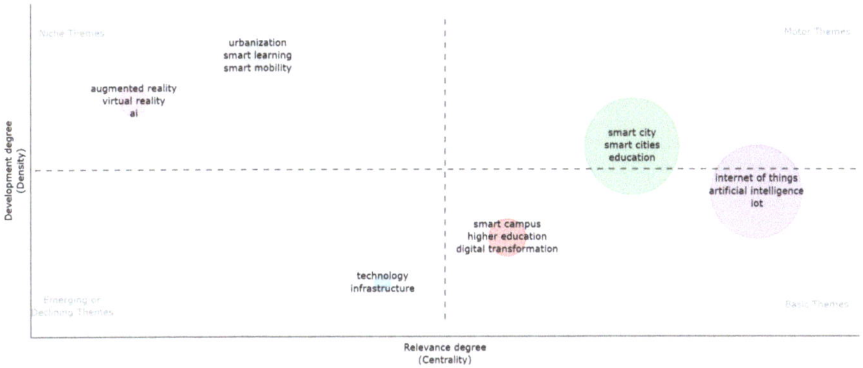

Fig. 9.5 Thematic map of smart education in smart city research domain

considered foundational but must be emphasized more in education-based smart cities. This visual categorization provides valuable insights for educators, curriculum developers, and researchers, helping them understand which areas are gaining traction and which might need more attention or innovation

9.2.3 *Co-occurrence Keyword Trends and Network Visualization*

Figure 9.6 depicts the most prominent themes and keywords related to research in conceptual change. The most significant and most central terms, such as "smart city," "internet of things", "engineering education", "big data", and "students",

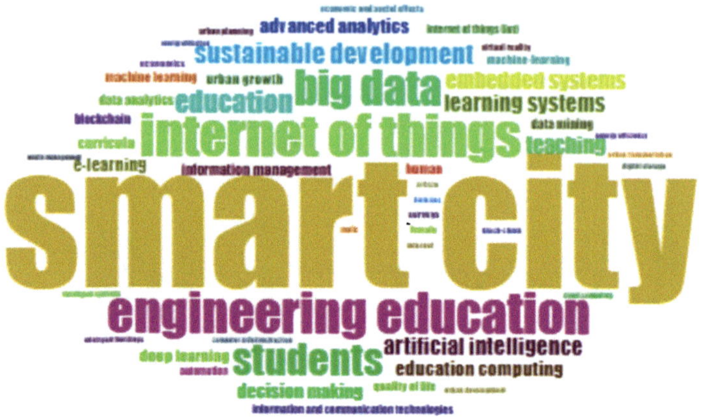

Fig. 9.6 Co-occurrence keyword trends in smart city domain and smart education

Fig. 9.7 Co-occurrence network of words in smart city research domain with smart education

indicate the core focus of the research on how researchers grasp and alter their conceptual knowledge, particularly within smart city contexts.

An illustration in Fig. 9.7 displays word occurrence frequencies from the sustainable development goals in smart city research network. Research focuses mainly on "smart city" as this term appears 865 times throughout the literature. Research on smart cities indicates the importance of "internet of things" based on its 277 word occurrences. Additionally, "engineering education" (265), "big data" (229), "students" (214), "education" (147), "sustainable development" (138), "artificial intelligence" (118), "learning systems" (117), and "embedded systems" (116) combinations gain crucial importance in emerging technology fields. The data shows that smart city research revolves primarily around several major subjects.

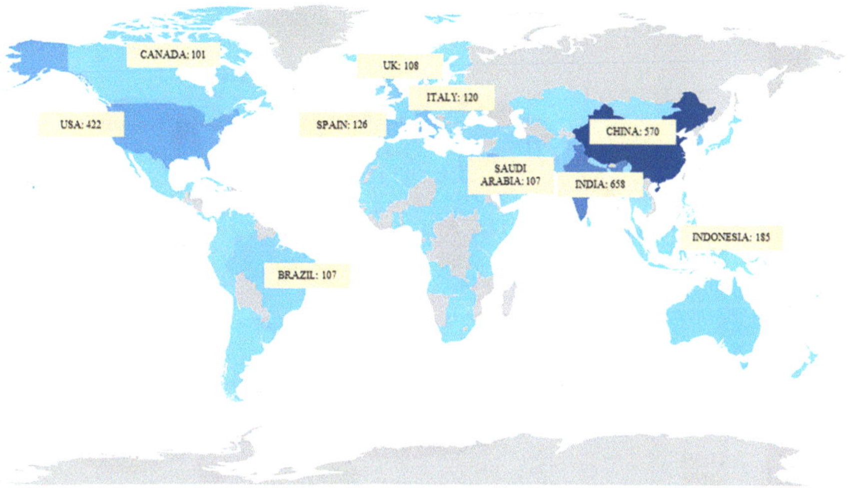

Fig. 9.8 Top ten countries' scientific production of various countries based on frequency

9.2.4 *Countries' Scientific Production*

The scientific publication frequencies of various countries in the world are represented in Fig. 9.8. Scientific contributions from India stand at 658 which demonstrates their powerful role in research development. The growing scientific power of China displays itself through its 570 scientific publications which rank second after India. The research hub of the USA produces 422 scientific papers among its total contributions. Numerous publications emerge from Indonesia and Spain together with Italy given their research output comprises 185, 126, and 120 papers, respectively. Academic and industrial research sectors of the UK and Brazil produce approximately 108 and 107 scholarly works between them. Saudi Arabia and Canada have smaller publication numbers but their contributions amount to totals of 107 and 101 research items, respectively. The measured data indicates Asian countries significantly influence the global scientific research output.

9.2.5 *Mapping Smart Education in Smart City Research to Sustainable Development Goals (SDGs)*

SDG Goals, its targets (bubble series) and the percentage (bar graph) of corresponding keywords detected in the text (calculated as the ratio of keywords in one goal to the total number of keywords detected) are depicted in Fig. 9.9. For SDG mapping, the study utilized the SDG Mapper tool from the "Knowledge Base for the Sustainable Development Goals," provided by the European Commission [21]. This

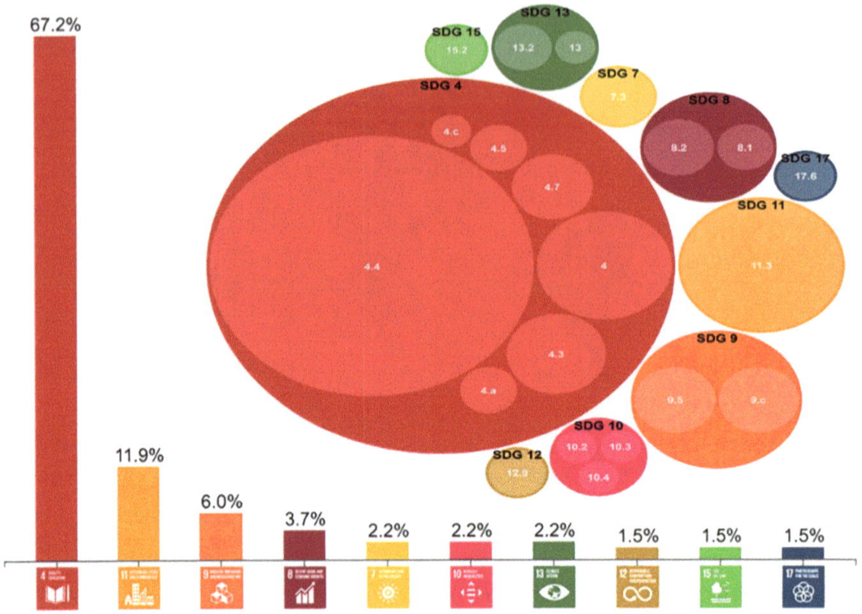

Fig. 9.9 Mapping smart education in smart city research to relevant SDGs and targets

tool uses bubble charts to depict SDG interconnections and relevance among bibliometric records. The size of each bubble correlates with the proportion of associated keywords found, which is calculated as the number of goal-specific keywords divided by the total keyword count. As such, SDG 4, marked by 90 keyword occurrences. This is followed by SDG 11, with 16 occurrences, and SDG 9, with 8 occurrences, underlining their relative prominence in the research database. Within SDG 4, there are several key subtargets that the research addresses: Subtarget 4.4 is shown as a distinct bubble within the larger SDG 4 circle, indicating focus in increasing the number of people with relevant skills for financial success. Subtarget 4.3, which aims in ensuring equal access for all women and men to affordable and quality technical, vocational, and tertiary education, including university, is another bubble within the SDG 4 area. This highlights the efforts in research to improving the cities. Subtarget 4.7 focuses on education for sustainable development and global citizenship. Subtarget 4.5 focuses on eliminating all discrimination in education. Subtarget 4.a aims to build and upgrade education facilities that are child, disability, and gender-sensitive, ensuring safe, non-violent, inclusive, and effective learning environments for all. Subtarget SDG 4.c aims to substantially increase the supply of qualified teachers by 2030. Figure 9.9 shows that the size of the SDG 11 bubble implies that advancements in smart city are being focussed on upgrading the cities and prioritize sustainability. Similarly, SDG 9 (industry, innovation, and infrastructure) is also closely linked to SDG 11 and aims to enhance scientific research and affordable access to internet. The detailed description is depicted in Table 9.3.

Table 9.3 SDGs ranking, targets and relevance in education based smart city research domain with SDGs

SDGs (By Rank)	Description	Targets	No. of Occurrence		Relevance%
SDG 4	Quality education	4.4	64	90	67.2
		4	11		
		4.3	6		
		4.7	4		
		4.5	2		
		4.a	2		
		4.c	1		
SDG 11	Sustainable cities and communities	11.3	16	16	11.9
SDG 9	Industry, innovation and infrastructure	9.5	4	8	6.0
		9.c	4		
SDG 8	Decent work and economic growth	8.2	3	5	3.7
		8.1	2		
SDG 7	Affordable and Clean Energy	7.3	3	3	2.2
SDG 10	Reduced inequalities	10.2	1	3	2.2
		10.3	1		
		10.4	1		
SDG 13	Climate action	13.2	2	3	2.2
		13	1		
SDG 12	Responsible consumption and production	12.8	2	2	1.5
SDG 15	Life on land	15.2	2	2	1.5
SDG 17	Partnerships for the goals	17.6	2	2	1.5

9.3 E-Learning Platforms and Virtual Classrooms

Education has an important function in the development of a society's knowledge-based economy in the era of smart cities. As the process of urbanization goes faster, the need for quality education rises too. This informs why e-learning platforms as well as virtual classroom are crucial important in providing education in a more unique, dynamic, and more sustainable ways [22]. E-learning is an internet-based application that provides resources for learning and teaching on any device, at any time, and from any location [23]. These platforms use multimedia, web conferencing and mobile to integrate various technologies and deliver comprehensive and customized kinds of learning. Since e-learning platforms make education accessible to a large number of people, they contribute to the development of sustainable smart city by promoting the use of technology while reducing CO_2 emissions from student transportation and paper usage in traditional classrooms. Online classes that have the same structure as the normal class, where the students interact with the teachers and the students also in real time, are referred to as virtual classrooms [24]. These virtual classrooms enable the students to learn more solve problems independently and share knowledge in an efficient manner. It provides a sense of classroom interaction where students can discuss with doubts, exchange documents, do group assignments, and attend a webinar, all in one virtual classroom. Figure 9.10 illustrates the various aspects of digital learning.

9.3.1 The Rise of e-Learning Platforms: Features and Benefits

9.3.1.1 Features of e-Learning Platforms

E-learning platforms are now becoming an important part of sustainable smart cities. These platforms come with enhanced features and benefits that are needed for instructors and students [25]. E-learning platforms are popular in the education sector as they provide several important characteristics that make learning easy. Some of these features include the following:

9.3.1.1.1 Multimedia-Rich Content

Online courses are diversified and individualized with teachers utilizing various media types such as the text, audio, video, and animations. This makes learning easier and more effective through utilizing multimedia to suit different learning modes of the learners. Text is more descriptive and sourcing it is easier as it enables one to easily quote or paraphrase, while in cases of audio and videos, learners are exposed to lectures, interviews, and demonstrations. It also includes elements such

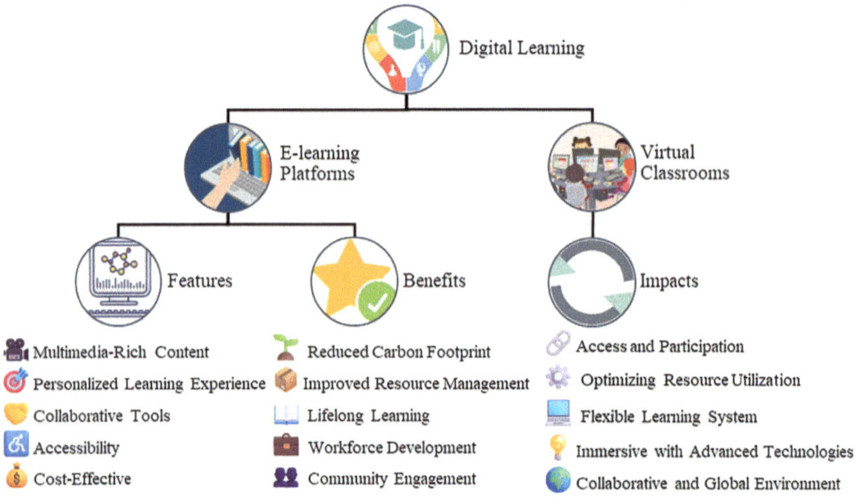

Fig. 9.10 Aspects of digital learning in smart education

as the use of questionnaires and discussion boards, which increase engagement and information mastery. The use of different media types will therefore enhance the diversified learning ability of the classroom where this course will benefit both the visually and the audibly inclined learner [26].

9.3.1.1.2 Personalized Learning Experience

By using AI and analytics technology e-learning platforms monitors and review the learner's progress and demonstrate the areas to target for the improvements [27]. Learner information is utilized to modify, delivery and provide the appropriate tools and services to achieve desired learning outcomes. These make learning interesting and interactive as it provides the learner with possibilities to practice in real-life situations making it easier to deal with the concepts in practice. The AI-powered analytics monitors social learning interactions in these environments and provides teachers with information on tool usage and outcomes. The integration of the technology of AI, analytics, and immersions brings change to the conventional learning methodologies making them efficient, engaging, and effective.

9.3.1.1.3 Collaborative Tools

E-learning is an important interactive tool such as discussion boards, real-time chat, and video conferencing, which help enable communication of students with students and among the students and tutor [28]. Other than conventional teaching methods that are characterized by knowledge delivery, e-learning adopted incorporates collaborative tools and other learning models that create productive learning. These tools overcome geographical barriers that may bar interaction between students coming from different backgrounds.

9.3.1.1.4 Accessibility

Online courses help learners to access education easily by enabling them to learn from any place and at any time [29]. This is especially useful for those that cannot attend physical class in learning institutions due to other numerous responsibilities or lack of easily accessible education centres in community or work. Online courses are also available with a varying list of subjects and focus areas, which gives students freedom to choose a subject they are interested in and would like to make a career out of. It is also flexible regarding the speed since learners can take their time depending on the difficulties they encounter in learning progress. In addition, interaction occurs with the participation of students where class activities like forum, webinars and peer reviews are conducted online which also helps to build a form of group since everyone is not in the same physical location.

9.3.1.1.5 Cost-Effective

Since e-learning generally does not require physical infrastructure, it also promotes economic efficiency, as it cost less than traditional learning facilities and removes the need for transportation expenses. It makes it possible that a large number of students be used by the same set of learning materials at the same time at low incremental cost. This scalability minimizes the need of hiring more instructors or rental or establishment of more instructional facilities to accommodate many learners. Using digital platforms, it is easier to reuse resources such as videos and course materials, which can be developed or modified at a much lower cost compared to rewriting textbooks or creating new physical materials [30].

9.3.1.2 Benefits of e-Learning Platforms

In the context of sustainable smart cities, e-learning platforms offer a multitude of benefits, such as the following:

9.3.1.2.1 Reduced Carbon Footprint

E-learning platforms have contributed to the sustainability of education by reducing fuel consumption for transportation, as well as minimizing the need for buildings and related infrastructures [31]. The shift from traditional classrooms to virtual learning offers several benefits, including eliminating the energy required for powering educational facilities such as lighting, heating, and air conditioning. This helps in reducing greenhouse gas emissions and reduction of energy consumption, thus making energy consumption more efficient [32]. Moreover, its application safeguards the environment by cutting down deforestation and emission of carbon related to production of paper for educational materials. It reduces the extent of polluting emissions in construction-related activities, for instance, emissions arising from building materials. Additionally, the e-learning platform only accommodates the use of digital text and other teaching–learning material which is eco-friendly. The e-learning platforms for expanding the information and communication (ICT)-based education will provide an important contribution in the direction towards sustainable eco-smart city development [33].

9.3.1.2.2 Improved Resource Management

Inefficient use of resources is eliminated because e-learning platforms are one of the defining features of sustainable smart cities [34]. Since e-learning eliminates the necessity of infrastructure like physical classrooms and libraries, it limits the use of products including timber, paper, and metals. This also helps reduce deforestation and mining, thereby preserving valuable natural resources and maintaining environmental quality. Moreover, information in computerized forms can be revised and disseminated to other students without changing the physical copies of the textbooks, which contributes to reducing paper usage and carbon emissions from the process of manufacturing and distribution of books. Using electronic gadgets and online platforms also creates the need for efficiency in energy utilization which has a positive impact on the current energy consumption. In addition, the decreased mobility between school and other institutions leads to a decreased demand for fuel, and therefore lowering the emission of greenhouse gases in urban areas.

9.3.1.2.3 Lifelong Learning

Continuous learning is important for citizens of smart environment because it enables people to acquire the necessary knowledge for utilizing innovations and improving living conditions in modern environment [35]. E-learning solutions provide a means through which residents continue learning and take any course or program of interest that transform lives or careers. Furthermore, learning for a lifetime via e-learning has the benefits of flexibility hence promote among learners more resiliency to face the various challenges arising from the current lifestyle in urban contexts.

9.3.1.2.4 Workforce Development

E-learning platforms can greatly contribute to building a modern-type capable workforce by providing needed courses and certification programs in areas such as sustainability and green technology [36]. It is easier and convenient to update the knowledge of the employees via the e-learning platforms employed and create packages that helps the workers to be conversant with the latest in sustainability measures and smart features. Accessibility to education on these platforms brings convenience of learning to underprivileged groups in the society and employees restricted to workplace learning.

9.3.1.2.5 Community Engagement

People of different culture and geographical backgrounds form part of e-learning therefore it becomes easy for them build a sense of community through learning. However, mobile learning platforms of sustainable smart cities help the communities promote innovation, inclusion, and access [37]. People of all ages are also granted educational materials irrespective of the country they live in and get an education. Furthermore, e-learning involves interaction sessions such as forums and group discussions where learners post their ideas, incidents, view, or feelings, which boosts learner interaction. The element of liberal learning promotes community interaction and leads to optimal achievement of local challenges hence enhancing the enhancement of sustainable urban developments. Additionally, continuous learning through e-learning is beneficial to the communities since it makes them continually learn, contributing to a culture of change. This interaction and co-learning may help build up united and sustainable communities that are indispensable to making smart cities.

9.3.2 Virtual Classrooms: Bridging the Gap Between Traditional and Digital Education

Education has grown through the ages since the traditional way of conducting classes in the classroom. Due to development in ICTs, virtual classrooms have become the most popular model of education delivery giving students a better experience. The virtual classroom plays a significant role in connecting traditional teaching and learning with new technologies to improve access, efficiency, and sustainability of education. Through bridging the gap between the IT-enablement of education, especially within conventional classrooms, virtual classrooms positively impact the smart cities vision as depicted in Fig. 9.11.

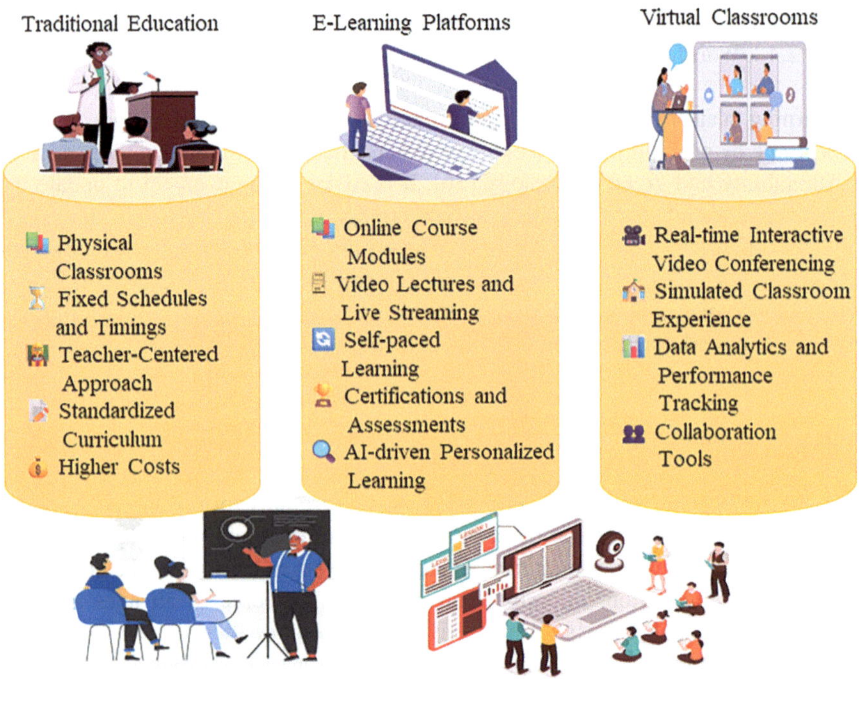

Fig. 9.11 Comparison between traditional, and digital education

9.3.2.1 Improving the Access and Participation

Adaptive classrooms eliminate the spatial and architectural constraints of schools and provide learning to learners in distant or inaccessibility areas [38]. This is essential because it ensures that the learners from all levels of societies are provided with quality education, which is a foundational requirement in establishment of smart cities. Flexible tools and platforms designed for individual students with disabilities enhance the development of more inclusive educational systems.

9.3.2.2 Optimizing Resource Utilization

Traditional education is highly dependent on classroom, books and papers, transport, and mobility. Virtual classrooms reduce these needs by providing e-books, online libraries, and virtual laboratory as the means of learning [39]. It minimizes the physical impacts left by the education systems in that they bring about decreased use of paper, energy, and carbon emissions arising from a number of transports in regard to the fact that smart cities have incorporated sustainability.

9.3.2.3 A Flexible Learning System and Continuing Education

The flexibility of virtual classrooms is considered one of the biggest advantages of e-learning environment. The adaptability of the concept of smart city ecosystems, the duration of the learning process, opportunities for students to gain knowledge at their own pace, and enrolling in a large number of courses facilitates lifelong learning. Virtual classrooms are also useful in offering upskilling and reskilling activities to produce a skilled workforce sufficient to drive smart city technologies [40].

9.3.2.4 Immersive with Advanced Technologies

Virtual classrooms use modern technologies which includes AI, AR, and VR to make learning efficient [41]. Cognitive-based smart systems deliver differentiated learning solutions and AR and VR give tangible learning experience in the form of simulated environment. Such innovations enable the students to close the gap between what they learned in the class and what they are likely to experience out in the industry.

9.3.2.5 Building a Collaborative and Global Environment

The current technology added the social aspect to learning by creating collaboration between students and educators across the globe and across different cultures. Such interconnections help to encourage diversity and prepare students for future life in the globalized city.

9.4 Digital Skills Training for Urban Workforce Development

Workforce development in urban areas is increasingly dependent on digital skills training, while smart education using immersive technology plays a vital role. The training is a means of preparing the workers for an advanced technological working environment. It is an integrated technology that utilizes VR and AR to create more realistic and engaging learning experiences. These methods prove to be more effective and stimulating compared to traditional classroom teaching techniques. Therefore, combining smart education and immersive technology with digital skill training significantly enhances urban work force development, leading to more effective outcome.

9.4.1 Identifying Key Digital Skills for the Modern Workforce

In the current dynamically changing environment, the important activities and skills that a modern employee should master are becoming important. With smart education being supported by immersive technology in schools and universities, there arises the need to prepare learners with relevant digital skills [42]. First and foremost, key competencies for the modern workstation are represented by the capacity of using digital technology to engage in productive work. Computer competencies include essential computer skills, web search, multi-media and information communication, and assessment of information. Second, critical thinking skills and problem-solving skills still apply because the world is more digitized and information is more frequent. These skills are essential for the employee to acquire to excel and make incredible decisions in a fast-growing digital world. Digital fluency has become essential in today's professional landscape, enabling individuals to navigate and utilize various digital tools and platforms effectively. Digital literacy enables individuals to adapt efficiently to the evolving information technology and perform tasks within digital context. Additionally, skills in data analysis and data visualization have become essential due to the growing need of big data across various industries. Smart education incorporating virtual technologies help students with these critical skills and enhancing employability. Furthermore, digital citizenship and security awareness are fundamental competencies for the modern workforce. As work and daily life become more digitized, it is essential for individuals to adopt practices that protect personal and organizational data while exercising caution regarding online presence. Thus, prioritizing the skills like, digital literacy, critical thinking, collaboration, fluency, and data analysis along with digital citizenship will help the schools and university to prepare the student for career growth in this digital world. These are some recommended digital skills for the modern workers, use and a brief explanation as follows in Table 9.4.

Table 9.4 Digital skills for the modern workspace

Digital Skill	Purpose	Description
Digital literacy	Interacting with technology and exploring technology tools and applications	The competence of an individual in managing digital technology well, in order to meet certain needs, using digital resources in a way that is efficient and to the intended purpose
Critical thinking and problem solving	Accurately assess situations, problems and make well-informed decisions	Activities that involves the analysis of data, recognition of problems and determination of the best course of action to take based on thinking through processes
Collaboration and communication	Ability to work with people in their group and hold meetings	The ability to communicate effectively and cooperate with people face-to-face and online with the use of such methods as video calls and various platforms
Data analysis and visualization	Identify and process big data	This is the manner in which students are to become aware on how to behave responsibly with the use of the social media, how to protect individual as well as organizational data as well as learn what actually constitutes a threat and how it can be prevented
Digital citizenship and security awareness	The abilities of the person in using digital systems and tools for conquering the task with ease and efficiency	The degree of efficiency of the individual in the use of distinct social networks, cloud services, content management systems, and other working tools

9.4.2 Strategies and Programs for Enhancing Digital Skills in Urban Areas

The world right now becomes a digital world were having internet connectivity to explore opportunities within the community. Hence, there exist immense possibilities of improving the digital skills through smart education in urban areas with the introduction of reality-based technology. Solutions to increase people's informativeness in virtual school with innovative technology to promote digital competence in urban environment are as follows:

9.4.2.1 Integration of Immersive Technology in the Curriculum

Immersive technology should therefore be included in schools, colleges, and training institutions, and its use should start at the elementary levels [43]. This includes employing virtual replicas, animated and related multimedia products, and AR/VR-based exercises to convey numerous topics, including, science, history, and geography. In this manner, educators may make learning less monotonous thus promoting effective development of appropriate and constructive digital literacy skills among learners.

9.4.2.2 Development of Specialized Courses

It has been found that urban areas benefit from the development of more focused and intensive courses in immersive technologies. These courses may be designed to serve a broad audience including students, professionals, and individuals at various stages of their careers. For example, local institutions can provide courses such as credit hours for developing VR, designing AR games, and developing MR applications. Most of such courses can be offered at local institutions or through online learning platforms. These courses will assist people in gaining increased levels of competency in deployed immersive technology tools and enhance their knowledge about the current topic [44].

9.4.2.3 Collaboration Between Educational Institutions and Technology Companies

Colleges can work with the IT industry to create new effective programs [45]. Such programs should aimed towards the improvement of multiple digital competencies based on VR application in industries including medical, engineering, architecture, and entertainment industries. In this case, both parties will be able to tap into each other's knowledge base, resources, and technical skills, with the outcome being the generation of high quality education programs and a pool of well-trained human resource.

9.4.2.4 Infrastructure Development for Immersive Technology

Areas must establish the necessary infrastructure to support immersive technology-based learning [46]. This includes providing dedicated facilities for its operation, ensuring the availability of VR/AR headsets and determining the appropriate level of Internet connectivity needed to maintain effective usage. All stakeholders, including government bodies through ministries of education, schools and technology firms, promote an environment that supports the growth of digital skills.

9.4.2.5 Training and Upskilling Programs for Educators

The teachers have a very important function at the successful application of immersion technologies in the learning process [47]. This staff must be sufficiently trained and reskilled to enable them to apply these tools within teaching practices. Training programs may focus at different levels of learning including instructional methods, content, and problem solving. In this way, the use of the immersive technology in the context of smart education can be advanced, and educators can be given necessary skills for this. The use of immersive technology in urban education means has the potential of improving digital skills characteristics. With proper formulation of a strategic program as well as partnership, cities should be ready to take the lead and provide the society with skilled employees to face the challenges in the future.

9.5 Role of Technologies for Inclusive Learning Education in Smart Cities: 5G/6G in Higher Education

In smart cities, 5G/6G technologies are used to support inclusive learning education in higher education since technologies provides student with equal opportunity to quality education. Technology tools and social media tools support students' interaction and enhance dissemination of knowledge and experience to create inclusive and interactive learning communities. These technologies should be fully adopted in smart cities so that quality and inclusive higher education can be embraced for entire populations.

9.5.1 Fifth Generation (5G)

5G is the new generation of mobile networks which offer much higher speed, capacity, and response. With data transfer rate of up to 20 Gbps and very little delay, it would mean there could be so much opportunities of the combination of so many technologies in different fields, among them being education. Building on the advancements achieved by 4G in communication and connectivity, 5G further enhances institutions capabilities to expand virtual learning, improve access to resources and strengthen global connections. When it comes to this to the idea of applying it in the sphere of education, it becomes even more critical to have technologies like 5G inclusive in learning. The features of 5G technology in education systems are illustrated in Fig. 9.12. Some of the key areas where these advancements have a significant impact includes:

Smart cities are vital to future development and hold a responsibility for higher education institutions to prepare professionals for the growing technological market. Therefore, the adoption of 5G technology is necessary to deliver high quality application that meets current demands. The following highlights key ways in which 5G impacts higher education in smart cities:

9.5.1.1 Enhanced Learning Experiences

Through 5G, students are able to utilize VR- and AR-based learning modules as effective learning from anywhere across the globe [48]. This not only helps to improve the ability to memorize information but also filling the gap between the abstract knowledge learnt in class and the "real-life" application.

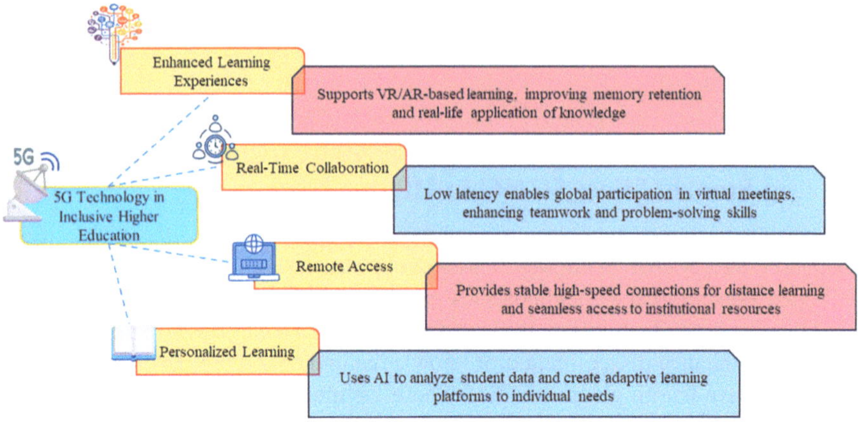

Fig. 9.12 Features of 5G technology in education system

9.5.1.2 Real-Time Collaboration

The low latency characteristics of 5G technology facilitate simultaneous participation of students and faculty in a single meeting, irrespective of geographical locations [49]. Inclusion of such integration can enhance the communication and coordination between team members and eventually will improve creativity, problem solving, and effective teamwork.

9.5.1.3 Remote Access

5G's high-speed and very stable connections allow students to use institutional resources and access digital-learning resources smoothly [50]. Moreover, distance learning platform supported by this technology become more advanced, thereby making education more accessible and convenient for students unable to attend physical classes due to factors such as geographic distance.

9.5.1.4 Personalized Learning

5G technology enables the collection of data on students' works in activities, facilitating a more individualized approach to education [50]. Effective implementation of these capabilities leads to the development of adaptable learning platforms that serve as instructional tools, enhancing educational outcomes by adjusting to the diverse learning styles of students.

9.5.2 *Sixth Generation (6G)*

6G is the sixth-generation wireless technology that is expected to deliver excellent internet speed, very low latency, and high connection quality. These enhancements to the technical features will dramatically change the nature of learning processes and the ways the students perceive the surrounding reality. In general, the concept of smart cities with the use of 6G technology introduce many incredible solutions such as remote learning, education for different types and needs, or even education based on experience. This not only improves the general education that students get but it also provides a new ground to both teachers and school principals to broaden up new ways of teaching methodologies and administrative management. The features of 6G technology in education systems are illustrated in Fig. 9.13. Some of the key areas where these advancements have a significant impact include the following:

9.5.2.1 Personalized Learning

6G technology enhances personalized learning by enabling real-time data processing and connectivity [51]. This supports adaptive educational system that adjust content and activities to meet individual student needs. The high speed and low latency of 6G also allows the integration of immersive technologies like VR and AR, creating interactive and engaging learning experiences. Additionally, continuous monitoring of progress enables timely support and improving learning effectiveness.

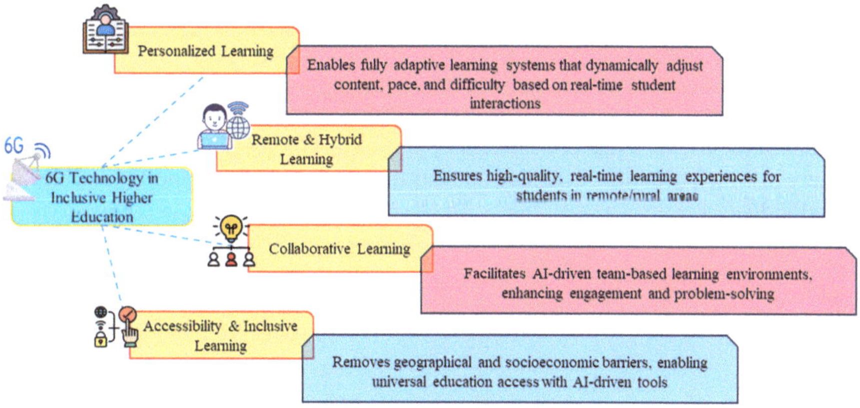

Fig. 9.13 Features of 6G technology in education system

9.5.2.2 Remote and Hybrid Learning

The low latency and high-speed connectivity offered by 6G technology make remote and hybrid learning more accessible and effective [52]. Students in rural and remote areas will get high-quality real-time learning experiences, while one-on-one interactions and online activities.

9.5.2.3 Collaborative Learning

The integration of 6G technology within smart cities enables collaboration among students in both physical and remote learning [53]. VR and AR applications supported by 6G infrastructure facilitate interactive learning environment where students engage in problem-solving and teamwork. This approach enhances understanding and strengthens learning outcomes beyond conventional lecture-based study,

9.5.2.4 Accessibility and Inclusive Learning

Some of the barriers to education that can be solved by implementing the 6G technology are the geographical barriers affecting access to education, the socioeconomic barriers that prevent individual from getting quality education and barriers arising from physical disability. Students with disabilities read educational materials using text-to-speech or speech-to-text technologies, while multilingual students receive translations in real-time. On the development of the professions of education, educators, and administrators have to accompany the constant upgrading of the skills that will help to utilize 6G technologies [54]. Moreover, the exploitation of 6G technologies in smart city is indispensable in determining the future of learning in higher education institutions particularly with regard to inclusiveness. As these technologies become available it is essential that the barriers to their adoption be addressed to provide equal educational opportunities for students. It helps to build a society in which everyone is valued and has the knowledge and opportunity relevant to individual.

9.6 Gamification and Immersive Learning Experiences

The concepts of game as well as simulations may be regarded as the transition towards participative and fun-oriented learning approaches that could potentially engage and stimulate the learner's ability to memorize. The learner engagement is the use of liberal environments that engage the learners in simulation or virtual learning to increase interest and learning achievement. While using elements of points, badges, and leaderboards, gamification engages the learners; and simulations give practice in a real-life environment. The significance of gamification, VR, and AR in immersive learning is illustrated in Fig. 9.14.

9.6.1 Understanding Gamification and Its Application in Education

Gamification refers to the application of game design elements and principles in non-game context, particularly in education to increase motivation and engagement among students. By incorporating features such as point scoring, Leaderboards and challenges, gamification encourages active participation and encourage competitive and collaborative learning environment. This approach supports the development of

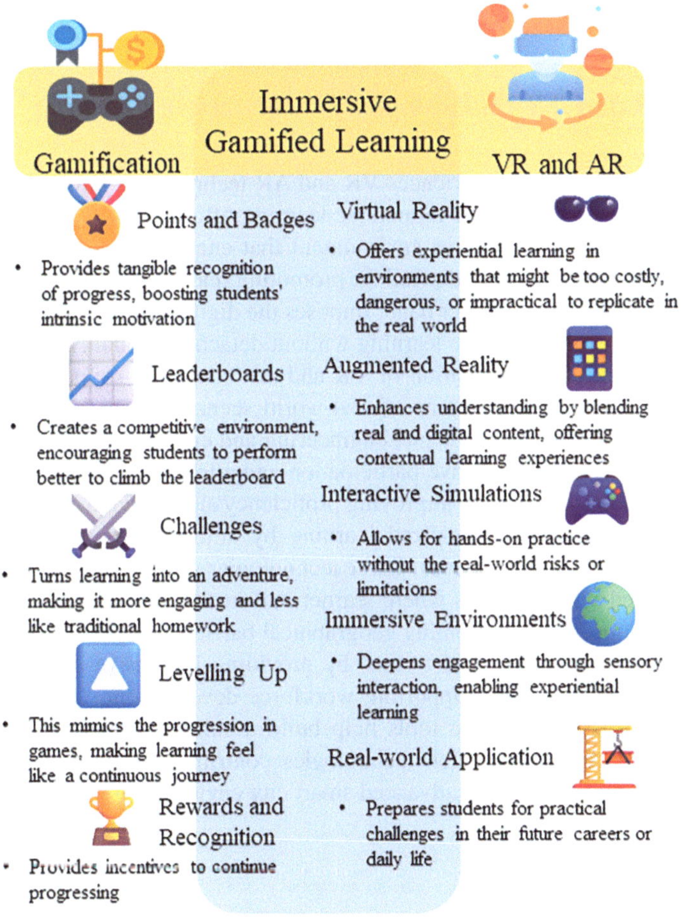

Fig. 9.14 Overview of immersive gamified learning

problem-solving skills and enhances learner interaction with educational content. In educational environment, gamification promotes sustained attention and improves information by making learning activities more dynamic and goal oriented [55]. It also provides feedback, allowing students to assess their progress and adjust strategies accordingly. Furthermore, gamification accommodates diverse learning style by integrating various interactive elements. The strategic implementation of gamification aligns with pedagogical objectives and help educators design effective instructional experiences. As a result, gamification plays an important role in transforming traditional education methods and contributes to more effective knowledge acquisition and skill development. Gamification contributes to building smart cities by encouraging a skilled and motivated workforce essential for urban development. It supports learning and innovation, enabling citizens to acquire relevant digital skills needed for smart infrastructure, sustainable practices and technology driven services. This approach promotes active civic engagement and prepares communities for future challenges and smart city environments.

9.6.2 VR and AR Technologies for Immersive Learning

To enhance educational experiences VR and AR technologies have become transformative tools in the field of immersive learning offering innovative approaches. VR provides a fully immersive environment that enables learner to engage with content in a three-dimensional printing, promoting understanding of complex concept [56] [57]. AR, on the other hand, imposes the digital information on to the real world, allowing for interactive learning without detaching students from the physical surroundings. The application of VR and AR education supports experiential learning by enabling simulations of real-world scenarios, which are particularly valuable in fields such as medicine, engineering and environmental science. These are technologies facilitate active participation and allow students to practice skills safely and repeatedly, thereby improving proficiency and confidence. Moreover, VR and AR contribute to personalized learning by adjusting content to individual learner needs and learning speeds. These technologies also encourage collaboration through shared virtual spaces where learner and instructors interact regardless of physical location, thus overcoming geographical barriers. VR and AR technologies enhance education within smart cities by providing immersive and personalized learning experiences [58]. Supporting workforce development, collaboration and improving accessibility, these tools help build a skilled population equipped to address urban challenges. Such technologies contribute to creating sustainable, inclusive and technologically advanced smart city environments.

9.7 Open Educational Resources and Digital Libraries

Open Educational Resources (OER) and digital libraries are extensively reshaping learning processes in smart cities. New generation technologies like VR and AR facilitate learning by giving the students advanced teaching material. Software-based digital libraries combined with immersion technology afford a new perspective in accessing educational resources. When implemented, these technologies enable students to retrieve and utilize instructional materials in a customized manner. Therefore, the future of education for smart city in respect of use of OER, digital libraries and immersive technology is enlightening. The framework for open educational resources and digital libraries in smart city is illustrated in Fig. 9.15.

9.7.1 Role of OER in Democratizing Education and Advancing Digital Libraries

The implementation of OER offers possibilities of developing the principles of social inclusiveness in education and supporting the value-adding strategies of digital libraries to improve the quality of education throughout urban development [59]. It is widely used in order to share free educational resources and high-quality content accessible for everyone. To provide learning, skills development, and improvement of residents of smart cities, local government and education institutions should partner to establish digital libraries containing OER. Not only does it promote the culture of learning but it also encourages people to take full responsibility of education and profession. When immersive technologies are integrated into digital libraries, learners get simulation, virtual tours, and interactive lessons to support multiple-learning styles and modalities. For example, VR and AR could be used to take students to a historical site, to do virtual lab work or even do problem solving activities. This approach in learning not only makes it possible, but also interesting and productive for everyone who is interested in acquiring knowledge. The learning management system design of this platform allows students, to learn about different subjects like science, history and art through different learning environments that are virtual. The integration of OER and immersive technology complement each other in revolutionizing the education and building smarter and more open cities [60]. Therefore, OER and IT have a potential to transform education in smart cities by sharing resources and providing learners with the meaningful and engaging educational experiences.

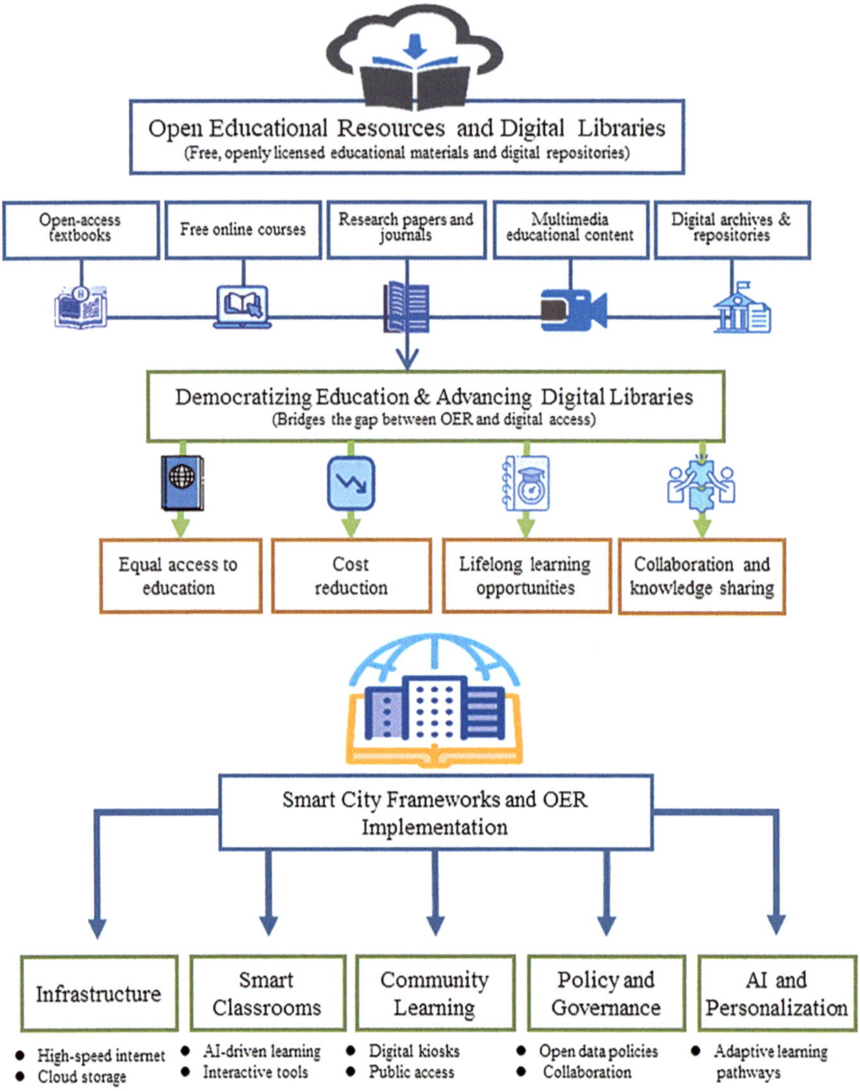

Fig. 9.15 Open educational resources and digital libraries framework for smart city

9.7.2 *Strategies for Implementing Digital Libraries and OER in Smart City Frameworks*

Digital libraries are internet-based databases of a wide range of digital sources including e-books, journals, audiobooks, and other multimedia. The use of technologies, like VR, AR, and MR enable the user to interact with the digital content in a better way [61]. OER is defined as teaching, learning, and research material, that freely available for use, sharing, adaptation, and redistribution and is available online under an open license. To effectively implement digital libraries and OER in smart city frameworks, local authorities must develop a comprehensive strategy. Broad Essential Broadband is important for increasing and developing the broadband infrastructure necessary for fast connectivity and ensuring that an effective number of users are capable of accessing digital libraries and OER. Also, the formation of protected cloud storage and management enables institutions to collect, sort, and keep digital library collections effectively [62]. Collections of high-quality digital library resources and OER also involves a coordinated effort involving local government, schools and other relevant organizations. This will help in the production of high quality and relevant contents for all users to contribute. Accessibility enables digitized library and OER interfaces for everyone and all levels of digital literacy. Moreover, because the information will be delivered with features for users with disabilities, more people will be able to access the knowledge product. Raising public awareness about navigating digital libraries and OER and accessing these tools effectively is essential. It is equally important to provide clear guidance on how to obtain support from technical teams to resolve difficulties, inconsistent and equitable access to digital learning resources [63]. Therefore, the integration of digital libraries and OER in smart city with ICT immersive technology, is likely to transform the citizens' use of information, learning systems and cultural content.

9.8 Open Research Challenges and Research Directions

Smart education with the use of immersed technology is still an area of research that has been used to conceptualize the smart cities. The process of implementing a variety of technologies in ways that help students and instructors to have efficient and effective education. Technologies include artificial intelligence and machine learning as well as the other techniques like Virtual and augmented reality, gamification and so on. In reference to smart city, use of immersive technology in education is capable of developing an efficient, effective and easily accessible education system. This section will review the current and potential issues in the area of smart education with Immersive technology in sustainable smart cities and its classification is illustrated in Fig. 9.16. These technologies are used to augment the educational system and identify the challenges that prevent the expansion of those technologies.

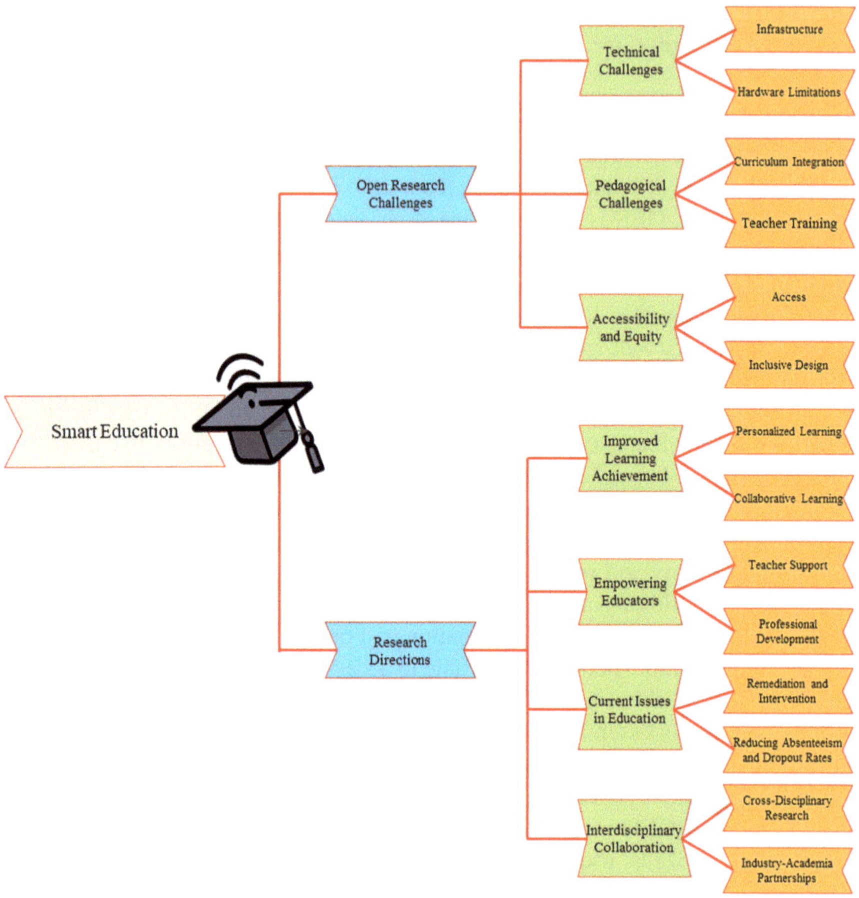

Fig. 9.16 Open research challenges and research directions in smart education

9.8.1 Open Research Challenges in Smart Education

9.8.1.1 Technical Challenges

9.8.1.1.1 Infrastructure

A key issue, which smart cities face regarding the integration of immersive technology is connectivity and more so the availability of high bandwidth. This is particularly important when it comes to streaming high quality content for good quality of material and for efficient delivery of knowledge [64]. Further studies are required to ascertain which strategies to boost the efficiency scale for extension of internet connectivity in cities.

9.8.1.1.2 Hardware Limitations

Available hardware, such as VR headsets and AR glasses, is often expensive and uncomfortable to wear [65]. Researchers need to design cheaper and readily accessible devices that can be easily incorporated in to the system of learning. Further research is also required to design cheap, small-sized and energy friendly identification hardware that are portable and implementable in different environments.

9.8.1.2 Pedagogical Challenges

9.8.1.2.1 Curriculum Integration

Implementing innovative technology in the current learning system may be difficult because it means customizing them [66]. This availability also emphasizes the necessity of studies on the most effective approaches to incorporating immersive technology into student learning and formulating theories of how these technologies support instructions [67].

9.8.1.2.2 Teacher Training

Information technology plays a significant role in teacher training, necessitating relevant and continuous professional development to ensure effective integration. Future research is required to identify the most suitable professional development programs, tools, and implementation strategies considering the potential resistance among the educators to adopt new pedagogical approaches [68].

9.8.1.3 Accessibility and Equity

9.8.1.3.1 Access

Technology is not only equally accessible to all students, leading to increased disparities in learning outcomes [66]. Further research is necessary to explore the feasibility of integrating VR classroom environments and to develop methods for identifying and supporting students.

9.8.1.3.2 Inclusive Design

Some difficulties which the students with disabilities encounters when using the immersive technology includes. There is not enough research in making sure these technologies will be implemented to be beneficial to the disabled and creating learning environments for the disabled.

9.8.2 *Research Directions in Smart Education*

9.8.2.1 Improved Learning Achievement

9.8.2.1.1 Personalized Learning

AR can be incorporated in a classroom as a tool to enhance individualized learning targeted at each student [69]. Further research is required on how to apply the algorithms and AI systems to make content and delivery methods responsive to student performance and feedback responsive.

9.8.2.1.2 Collaborative Learning

Immersive technologies facilitate distance learning as student have the ability to physically create school environments which promote the communication and group formation necessary for the group projects [66]. Further, study should be conducted regarding best practices in designing collaborative learning environments and motivate students' involvement.

9.8.2.2 Empowering Educators

9.8.2.2.1 Teacher Support

AR and other forms of information technology helps the educational process and give teachers an opportunity to monitor learning outcomes, give immediate feedback or differentiate the content [66]. Research should extend to the production of cells and examination in relation to how they should be incorporated into the education system.

9.8.2.2.2 Professional Development

Future research should examine how to effectively teach educators in the use of technology as well as how to facilitate professional learning for educators in the context of digital media. It involves establishment of online tools, live meetings and web-based courses or presentations and professional learning networks [70].

9.8.2.3 Current Issues in Education

9.8.2.3.1 Remediation and Intervention

Immersive technology is effectively utilized in teaching and providing targeted interventions for students who are failing behind or require additional practice in specific areas. Future studies should explore which applications of immersive technologies optimized remediation and intervention, whether specific forms of immersive technologies are better suited for particular interventions, and how given approaches are capable into the scope of schooling.

9.8.2.3.2 Reducing Absenteeism and Dropout Rates

Educational technologies can be used to bring students back into classes and even keep them from dropping out in the first place. Research should explore ways in which these technologies can be deployed to increase and encourage student participation and increase student participation as well as support students' needs to provide a more encouraging learning atmosphere for each student [71].

9.8.2.4 Interdisciplinary Collaboration

9.8.2.4.1 Cross-Disciplinary Research

The research in the area of immersive technology should be conducted under the disciplines including education, psychology, computer science, and urban planning [72]. This is essential for enhancing ongoing efforts and simplifying procedures in smart city education, making it more efficient and less challenging.

9.8.2.4.2 Industry–Academia Partnerships

Knowledge and ideas are transformed into practical solutions, making University–Industry collaborations in a vital bridge between academic research and real-world application [73]. Scholars should explore opportunities to collaborate with IT firms, educational institutions, and government agencies to develop innovative and practical immersive technologies that address the complex challenges faced by smart cities.

9.9 Summary

The integration of immersive technologies is revolutionizing smart education by enabling more interactive, personalized, and accessible learning experiences across the diverse environments. Advancements such as e-learning platforms, virtual classrooms, and the integration of the metaverse have made learning more interactive and accessible. Digital skills training supports urban workforce, while technologies like 5G/6G enhance connectivity and higher education. Gamification and immersive tool like AR and VR increase student engagement, and OER helps make education more inclusive. Implementing digital libraries within smart city frameworks promotes wider access to quality content. Emphasis on innovation and inclusivity remains key to addressing the evolving demands of modern education.

References

1. A.D. Samala et al., Emerging technologies for global education: A comprehensive exploration of trends, innovations, challenges, and future horizons. SN Comput. Sci. **5**(8), 1175 (2024). https://doi.org/10.1007/s42979-024-03538-1
2. P. Mishra, P. Thakur, G. Singh, Sustainable smart city to society 5.0: state-of-the-art and research challenges. SAIEE Afr. Res. J. **113**(4), 152–164 (2022). https://doi.org/10.23919/SAIEE.2022.9945865
3. L.M. Nkomo, B.K. Daniel, R.J. Butson, Synthesis of student engagement with digital technologies: A systematic review of the literature. Int. J. Educ. Technol. High. Educ. **18**(1), 34 (2021). https://doi.org/10.1186/s41239-021-00270-1
4. H. Singh, S.J. Miah, Smart education literature: A theoretical analysis. Educ. Inf. Technol. (Dordr) **25**(4), 3299–3328 (2020). https://doi.org/10.1007/s10639-020-10116-4
5. P. Mishra, G. Singh, Sustainable smart city to society 5.0, in *Sustainable Smart Cities*, (Springer, Cham, 2023), pp. 39–57. https://doi.org/10.1007/978-3-031-33354-5_2
6. K.A. Demir, Smart education framework. Smart Learn. Environ. **8**(1), 29 (2021). https://doi.org/10.1186/s40561-021-00170-x
7. P.I. Mazza, Education & smart cities: The role of the goals of agenda 2030 for sustainable development of smart cities. Int. J. Innov. Stud. Sociol. Human. **6**(2), 24–31 (2021)
8. P. Mishra, G. Singh, Introduction: Importance of sustainable smart city, in *Sustainable Smart Cities*, (Springer, Cham, 2023), pp. 1–37. https://doi.org/10.1007/978-3-031-33354-5_1
9. P. Mishra, G. Singh, Artificial intelligence for sustainable smart cities, in *Sustainable Smart Cities*, (Springer, Cham, 2023), pp. 119–142. https://doi.org/10.1007/978-3-031-33354-5_6
10. P. Mishra, P. Thakur, G. Singh, Enabling technologies for IoT based Smart City, in *2021 Sixth International Conference on Image Information Processing (ICIIP)*, (IEEE, 2021), pp. 99–104. https://doi.org/10.1109/ICIIP53038.2021.9702673
11. E. Oztemel, S. Gursev, Literature review of industry 4.0 and related technologies. J. Intell. Manuf. **31**(1), 127–182 (2020). https://doi.org/10.1007/s10845-018-1433-8
12. E. Al Nuaimi, H. Al Neyadi, N. Mohamed, J. Al-Jaroodi, Applications of big data to smart cities. J. Internet Serv. Appl. **6**(1), 25 (2015). https://doi.org/10.1186/s13174-015-0041-5
13. M.A. Ahad, S. Paiva, G. Tripathi, N. Feroz, Enabling technologies and sustainable smart cities. Sustain. Cities Soc. **61**, 102301 (2020). https://doi.org/10.1016/j.scs.2020.102301
14. R.H. Hariri, E.M. Fredericks, K.M. Bowers, Uncertainty in big data analytics: Survey, opportunities, and challenges. J. Big Data **6**(1), 44 (2019). https://doi.org/10.1186/s40537-019-0206-3
15. A.R. Javed et al., Future smart cities: Requirements, emerging technologies, applications, challenges, and future aspects. Cities **129**, 103794 (2022). https://doi.org/10.1016/j.cities.2022.103794
16. T. Yigitcanlar, K. Desouza, L. Butler, F. Roozkhosh, Contributions and risks of artificial intelligence (AI) in building smarter cities: Insights from a systematic review of the literature. Energies (Basel) **13**(6), 1473 (2020). https://doi.org/10.3390/en13061473
17. S. Tanwar, Q. Bhatia, P. Patel, A. Kumari, P.K. Singh, W.-C. Hong, Machine learning adoption in Blockchain-based smart applications: The challenges, and a way forward. IEEE Access **8**, 474–488 (2020). https://doi.org/10.1109/ACCESS.2019.2961372
18. R. Khatoun, S. Zeadally, Smart cities. Commun. ACM **59**(8), 46–57 (2016). https://doi.org/10.1145/2858789
19. B. Bhushan, A. Khamparia, K.M. Sagayam, S.K. Sharma, M.A. Ahad, N.C. Debnath, Blockchain for smart cities: A review of architectures, integration trends and future research directions. Sustain. Cities Soc. **61**, 102360 (2020). https://doi.org/10.1016/j.scs.2020.102360
20. A. Sodiq et al., Towards modern sustainable cities: Review of sustainability principles and trends. J. Clean. Prod. **227**, 972–1001 (2019). https://doi.org/10.1016/j.jclepro.2019.04.106
21. S. Borchardt, G.B. Vignola, D. Buscaglia, M. Maroni, *Mapping EU Policies with the 2030 Agenda and SDGs-Fostering Policy Coherence through text-based SDG Mapping*. https://doi.org/10.2760/110687

22. U. Matthew, J. Kazaure, N. Okafor, Contemporary development in E-learning education, cloud computing technology & Internet of Things. EAI Endors. Trans. Cloud Syst., 169173 (2018). https://doi.org/10.4108/eai.31-3-2021.169173
23. S. Luckyardi, L. Rahman, Application of e-learning system in the world of education. Jurnal Teknologi Informasi dan Pendidikan **14**(1), 47–52 (2021). https://doi.org/10.24036/tip.v14i1.327
24. M. Manegre, K.A. Sabiri, Online language learning using virtual classrooms: An analysis of teacher perceptions. Comput. Assist. Lang. Learn. **35**(5–6), 973–988 (2022). https://doi.org/10.1080/09588221.2020.1770290
25. S. Sharma, S. Garg, S.K. Mittal, Role of e-trainings in building smart cities. Proc. Comput. Sci. **111**, 24–30 (2017). https://doi.org/10.1016/J.PROCS.2017.06.005
26. J.L. Ross, R.A. Schulz, Using the world wide web to accommodate diverse learning styles. Coll. Teach. **47**(4), 123–129 (1999). https://doi.org/10.1080/87567559909595801
27. I. Gligorea, M. Cioca, R. Oancea, A.-T. Gorski, H. Gorski, P. Tudorache, Adaptive learning using artificial intelligence in e-learning: A literature review. Educ. Sci. (Basel) **13**(12), 1216 (2023). https://doi.org/10.3390/educsci13121216
28. M. Liu, D. Yu, Towards intelligent E-learning systems. Educ. Inf. Technol. (Dordr) **28**(7), 7845–7876 (2023). https://doi.org/10.1007/s10639-022-11479-6
29. A. Selvaraj, V. Radhin, N. Ka, N. Benson, A.J. Mathew, Effect of pandemic based online education on teaching and learning system. Int. J. Educ. Dev. **85**, 102444 (2021). https://doi.org/10.1016/J.IJEDUDEV.2021.102444
30. A. Badshah, A. Ghani, A. Daud, A. Jalal, M. Bilal, J. Crowcroft, Towards smart education through internet of things: A survey. ACM Comput. Surv. **56**(2), 1–33 (2024). https://doi.org/10.1145/3610401
31. P. Mishra, G. Singh, Internet of vehicles for sustainable smart cities, in *Sustainable Smart Cities*, (Springer, Cham, 2023), pp. 175–193. https://doi.org/10.1007/978-3-031-33354-5_8
32. P. Mishra, G. Singh, Energy management systems in sustainable smart cities based on the internet of energy: A technical review. Energies (Basel) **16**(19), 6903 (2023). https://doi.org/10.3390/en16196903
33. G. Emmanuel and A. S. Sife, "Challenges of managing information and communication technologies for education: Experiences from Sokoine National Agricultural Library," 2008. [Online]. Available: http://snalwww.ac.tz
34. P. Mishra, G. Singh, *Sustainable Smart Cities* (Springer, Cham, 2023). https://doi.org/10.1007/978-3-031-33354-5
35. R. Eynon, L. Malmberg, Lifelong learning and the internet: Who benefits most from learning online? Br. J. Educ. Technol. **52**(2), 569–583 (2021). https://doi.org/10.1111/bjet.13041
36. S. Akselrod et al., Building an interdisciplinary workforce for prevention and control of non-communicable diseases: The role of e-learning. BMJ **381**, e071071 (2023). https://doi.org/10.1136/bmj-2022-071071
37. M. Abduljawad, A. Ahmad, *Multicultural Education an Analysis of Mobile Learning (M-Learning) in Education*. https://doi.org/10.5281/zenodo.7665894
38. Y.S. Rizvi, A. Nabi, Transformation of learning from real to virtual: An exploratory-descriptive analysis of issues and challenges. J. Res. Innov. Teach. Learn. **14**(1), 5–17 (2021). https://doi.org/10.1108/JRIT-10-2020-0052
39. T. Tate, M. Warschauer, Equity in online learning. Educ. Psychol. **57**(3), 192–206 (2022). https://doi.org/10.1080/00461520.2022.2062597
40. P. Mishra, G. Singh, Enabling technologies for sustainable smart city, in *Sustainable Smart Cities*, (Springer, Cham, 2023), pp. 59–73. https://doi.org/10.1007/978-3-031-33354-5_3
41. F. Nurhasanah, A.S.C. Nugraheni, B. Kusharjanta, R. Ardiansyah, S. Widono, K. Saddhono, A design of virtual classrooms through AI, ML and DL to improve the level of learning, in *2024 4th International Conference on Advance Computing and Innovative Technologies in Engineering (ICACITE)*, (IEEE, 2024), pp. 940–945. https://doi.org/10.1109/ICACITE60783.2024.10616641

42. C. Lazou, A. Tsinakos, Critical immersive-triggered literacy as a key component for inclusive digital education. Educ. Sci. (Basel) **13**(7), 696 (2023). https://doi.org/10.3390/educsci13070696
43. J. Mogas, R. Palau, M. Fuentes, G. Cebrián, Smart schools on the way: How school principals from Catalonia approach the future of education within the fourth industrial revolution. Learn. Environ. Res. **25**(3), 875–893 (2022). https://doi.org/10.1007/s10984-021-09398-3
44. C. Seufert, S. Oberdörfer, A. Roth, S. Grafe, J.L. Lugrin, M.E. Latoschik, Classroom management competency enhancement for student teachers using a fully immersive virtual classroom. Comput. Educ. **179**, 104410 (2022). https://doi.org/10.1016/J.COMPEDU.2021.104410
45. J. Southworth et al., Developing a model for AI across the curriculum: Transforming the higher education landscape via innovation in AI literacy. Comput. Educ. Artif. Intell. **4**, 100127 (2023). https://doi.org/10.1016/J.CAEAI.2023.100127
46. C.H. Wu, Y.M. Tang, Y.P. Tsang, K.Y. Chau, Immersive learning design for technology education: A soft systems methodology. Front. Psychol. **12** (2021). https://doi.org/10.3389/fpsyg.2021.745295
47. F. Tang, Understanding the role of digital immersive technology in educating the students of english language: Does it promote critical thinking and self-directed learning for achieving sustainability in education with the help of teamwork? BMC Psychol. **12**(1), 144 (2024). https://doi.org/10.1186/s40359-024-01636-6
48. B. Kizilkaya, G. Zhao, Y.A. Sambo, L. Li, M.A. Imran, 5G-enabled education 4.0: Enabling technologies, challenges, and solutions. IEEE Access **9**, 166962–166969 (2021). https://doi.org/10.1109/ACCESS.2021.3136361
49. J. Rong, Innovative research on intelligent classroom teaching mode in the '5G' era. Mob. Inf. Syst. **2022**, 1–6 (2022). https://doi.org/10.1155/2022/9297314
50. B. Vamsi Krishna, K. Suresh Kumar, S. Gildhiyal, M. Srinivas, W.K. Ibrahim, M. Bader Alazzam, An empirical study on application of 5G smart system in education, in *2023 3rd International Conference on Advance Computing and Innovative Technologies in Engineering (ICACITE)*, (IEEE, 2023), pp. 637–642. https://doi.org/10.1109/ICACITE57410.2023.10182680
51. P. Mishra, G. Singh, 6G-IoT framework for sustainable smart city: Vision and challenges, in *Sustainable Smart Cities*, (Springer, Cham, 2023), pp. 97–117. https://doi.org/10.1007/978-3-031-33354-5_5
52. Y. Liu, M.R. Razman, S.Z. Syed Zakaria, L.K. Ern, A. Hussain, V. Chamola, Utilizing ubiquitous learning to foster sustainable development in rural areas: Insights from 6G technology. Comput. Human Behav. **161**, 108418 (2024). https://doi.org/10.1016/J.CHB.2024.108418
53. R.R. Reddy, M.V.K. Reddy, R. Uyyala, C.M. Kumar, S. Nimmala, A.K. Tyagi, Educational implications of 6G technologies, in *Human-Centric Integration of 6G-Enabled Technologies for Modern Society*, (Academic Press, 2025), pp. 283–300. https://doi.org/10.1016/B978-0-443-27434-3.00019-2
54. Y. Zhang, M.F.A. Bin Azizui, C. Yan, The application of 6G and augmented reality technology in education and training. Wirel. Pers. Commun. (2024). https://doi.org/10.1007/s11277-024-11038-x
55. M.A. Kuhail, A. ElSayary, S. Farooq, A. Alghamdi, Exploring immersive learning experiences: A survey. Informatics **9**(4), 75 (2022). https://doi.org/10.3390/informatics9040075
56. G. Lampropoulos, Kinshuk, Virtual reality and gamification in education: A systematic review. Educ. Technol. Res. Dev. **72**(3), 1691–1785 (2024). https://doi.org/10.1007/s11423-024-10351-3
57. C. Schelly, G. Anzalone, B. Wijnen, J.M. Pearce, Open-source 3-D printing technologies for education: Bringing additive manufacturing to the classroom. J. Vis. Lang. Comput. **28**, 226–237 (2015). https://doi.org/10.1016/J.JVLC.2015.01.004
58. M. Keppell, Personalised learning strategies for higher education. Int. Perspect. High. Educ. Res. **12**, 3–21 (2014). https://doi.org/10.1108/S1479-362820140000012001

59. H.M. Adil, S. Ali, M. Sultan, M. Ashiq, M. Rafiq, Open education resources' benefits and challenges in the academic world: A systematic review. Glob. Knowl. Memory Commun. **73**(3), 274–291 (2024). https://doi.org/10.1108/GKMC-02-2022-0049
60. I.S. Abeywardena, OXREF: Open XR for education framework. Int. Rev. Res. Open Distrib. Learn. **24**(3), 185–206 (2023). https://doi.org/10.19173/irrodl.v24i3.7109
61. S. Mystakidis, Metaverse. Encyclopedia **2**(1), 486–497 (2022). https://doi.org/10.3390/encyclopedia2010031
62. M. Halder, S. Prasad, Library services in e-learning environment: Its different application area. Libr. Prog. Int. **41**(1), 66–71 (2021). https://doi.org/10.5958/2320-317X.2021.00009.X
63. M.A. Camilleri, A.C. Camilleri, Digital learning resources and ubiquitous technologies in education. Technol. Knowl. Learn. **22**(1), 65–82 (2017). https://doi.org/10.1007/S10758-016-9287-7
64. I. Yaqoob, K. Salah, R. Jayaraman, M. Omar, Metaverse applications in smart cities: Enabling technologies, opportunities, challenges, and future directions. Internet of Things **23**, 100884 (2023). https://doi.org/10.1016/J.IOT.2023.100884
65. B. Kress, E. Saeedi, V. Brac-de-la-Perriere, The segmentation of the HMD market: Optics for smart glasses, smart eyewear, AR and VR headsets, in *Proceedings Volume 9202, Photonics Applications for Aviation, Aerospace, Commercial, and Harsh Environments V*, (2014), p. 92020D. https://doi.org/10.1117/12.2064351
66. A. Haleem, M. Javaid, M.A. Qadri, R. Suman, Understanding the role of digital technologies in education: A review. Sustain. Operat. Comput. **3**, 275–285 (2022). https://doi.org/10.1016/J.SUSOC.2022.05.004
67. S. Namestiuk, On the issue of teaching psychological and pedagogical disciplines at universities using immersive technologies. Futur. Edu., 33–42 (2022). https://doi.org/10.57125/fed/2022.10.11.27
68. S. Mhlongo, K. Mbatha, B. Ramatsetse, R. Dlamini, Challenges, opportunities, and prospects of adopting and using smart digital technologies in learning environments: An iterative review. Heliyon **9**(6), e16348 (2023). https://doi.org/10.1016/J.HELIYON.2023.E16348
69. V. Prain et al., Personalised learning: Lessons to be learnt. Br. Educ. Res. J. **39**(4), 654–676 (2013). https://doi.org/10.1080/01411926.2012.669747
70. N. Ndzinisa, R. Dlamini, Responsiveness vs. accessibility: Pandemic-driven shift to remote teaching and online learning. High. Educ. Res. Dev. **41**(7), 2262–2277 (2022). https://doi.org/10.1080/07294360.2021.2019199
71. E.E. Tanner-Smith, S.J. Wilson, A meta-analysis of the effects of dropout prevention programs on school absenteeism. Prev. Sci. **14**(5), 468–478 (2013). https://doi.org/10.1007/s11121-012-0330-1
72. E. Gordon, S. Schirra, J. Hollander, Immersive planning: A conceptual model for designing public participation with new technologies. Environ. Plann. B. Plann. Des. **38**(3), 505–519 (2011). https://doi.org/10.1068/b37013
73. J. Berbegal-Mirabent, D. Gil-Doménech, D.E. Ribeiro-Soriano, Fostering university-industry collaborations through university teaching. Knowl. Manag. Res. Pract. **18**(3), 263–275 (2020). https://doi.org/10.1080/14778238.2019.1638738

Chapter 10
Advanced Smart Energy Systems: The Future of the Energy Revolution

Contents

P. Mishra, G. Singh, *Sustainable Smart Cities 2.0*,
https://doi.org/10.1007/978-3-032-01102-2_10

10.1 Introduction

With increasing global energy demand, advanced smart energy systems become a critical element in creating sustainable urban infrastructures for generations to come [1]. The dominant infrastructures of the energy industry have traditionally been centralized, consisting of large-scale power stations and extensive energy transmission systems. Sustainable smart energy systems play an important role in smart cities integrating advanced technologies that secure the optimal utilization of energy supply, distribution, and consumption [2]. These systems are part of renewable energy sources that integrate solar, wind, along with advanced storage systems, smart grids, and Internet of Things (IoT)-enabled devices. It uses real-time data analytics and artificial intelligence (AI) to manage energy efficiently, reduce carbon emissions, and enhance reliability. These systems help build smart cities with a resilient, low carbon, urban environment that promotes economic growth and environmental management. However, these systems show signs of being outdated, energy degrading, and incapable of adapting to changing energy requirements. However, Advanced Smart Energy Systems (ASES) use the new technology solutions such as AI, IoT, block chain, and data analytics to establish an improved and advanced energy system [3]. It is efficient because it increases efficiency of energy generation and utilization and also actively involve the consumer in the process. These systems help the organization to save energy resources, reduce wastage, and to ensure recycling, thus helping in preserving natural resources and reducing emissions of greenhouse gases. The shift to a circular model addresses society's needs while also creating an economic win-win situation that unlocks innovations across the value chain and generates new jobs [4]. Figure 10.1 illustrates the transition from smart energy systems to a smart city energy ecosystem. The smart energy systems consist of key components such as renewable energy sources, smart grid, energy storage, and IoT devices. These systems offer key features including efficiency, sustainability, cost saving, and reduced carbon footprint. The flow of energy involves generation, storage, optimization, and distribution. In the smart city energy ecosystem, energy integrates with city elements like transportation, buildings, industries, and household. Interconnected systems such as AI-driven energy management and demand response

Fig. 10.1 The transition from smart energy systems to a smart city energy ecosystem

mechanisms enhance efficiency. Communication networks, including IoT, sensors, big data, and cloud computing, enable real-time monitoring and control for smarter urban energy management.

10.1.1 Importance of Smart Energy Systems

In the age of rapid urbanization, sustainable smart cities have risen up as the primary solution to the increasing environmental and societal problems [5]. The backbone of sustainable smart cities are smart energy systems, which include an integrated network of advanced technologies, data analytics, and renewable energy solutions that optimize energy generation, distribution, and consumption. Not only it promotes environmental management, it also helps to promote economic efficiency and improve the urban quality of life. Cities could lead the way toward an energy-efficient and sustainable future by investing and adopting advanced systems and become more sustainable by the means of implementing smart energy systems. With efficient use of these renewable energy sources including solar, wind, and geothermal energies, it reduces the dependence on fossil fuels as well as greenhouse gas emissions. These systems contribute to cleaner air, less carbon footprints, and healthier living by integrating renewable energy into the urban energy grid. A key feature of smart energy systems enables in real-time monitoring and management of energy consumption. City planners and residents assess the value of the data from smart meters, sensors, and IoT devices to provide energy usage decisions. Smart cities also require energy storage solutions, like advanced battery technologies, so as to build a reliable energy supply. These systems save some of the energy that's generated when there is low demand for it and use it when required. It enhances energy resilience and guarantees the stable and continuous power supply for the urban infrastructure. Further, smart energy systems also help to create microgrids, localized energy networks that can operate in isolation, or in conjunction with the main grid. Microgrids increase energy security and help the decentralization of energy production ending with lower risks associated with large-scale outages. Furthermore, smart energy systems are used to integrate electric vehicles (EVs) into urban ecosystems [6]. Additionally, charging stations run on renewable energy also help decrease air pollution while providing for a sustainable manner of transporting people.

10.1.2 Introduction to Smart Cities and Energy Systems

Smart cities have become an important transformative approach to the development of a city with an aim to overcome the challenges of rapid urbanization, climatic changes, and the existing scarcity of resources. The core of this idea is a fusion of technology, data, and sustainable practices to form efficient, inclusive, and eco-friendly urban environments. Energy systems are one of the principal components of a smart city and play a key role in achieving sustainability goals and improving the quality of citizens' life [7]. It utilizes latest technologies such as IoT, AI, and big data analytics to improve the urban operations and services. Smart cities are supported by connecting physical infrastructure with digital systems to help monitor physical objects in real time, efficiently allocate resources, and make better decisions. It seeks to create an urban ecosystem resilient and adaptive while meeting present and future generations' needs. Smart city development is built upon sustainability. The aim is to minimize the size of their environmental footprint through the promotion of renewable energy, energy efficiency, and sustainable solutions in mobility. The backbone of any smart city is energy systems as main enabler of urban functionalities. A smart city includes its generation, distribution, and consumption of energy in a sustainable energy system that tends to emphasize efficiency and environmental responsibility as illustrated in Fig. 10.2. Key elements of such systems include the following.

10.1.2.1 Renewable Energy Integration

Renewable energy sources like solar, wind, and hydropower are preferred as part of smart cities. Rooftop solar panels and small-scale wind turbines are the examples of distributed energy resources which enable citizens to generate more of their own energy while reducing dependence on the decreasing supply of fossil fuels.

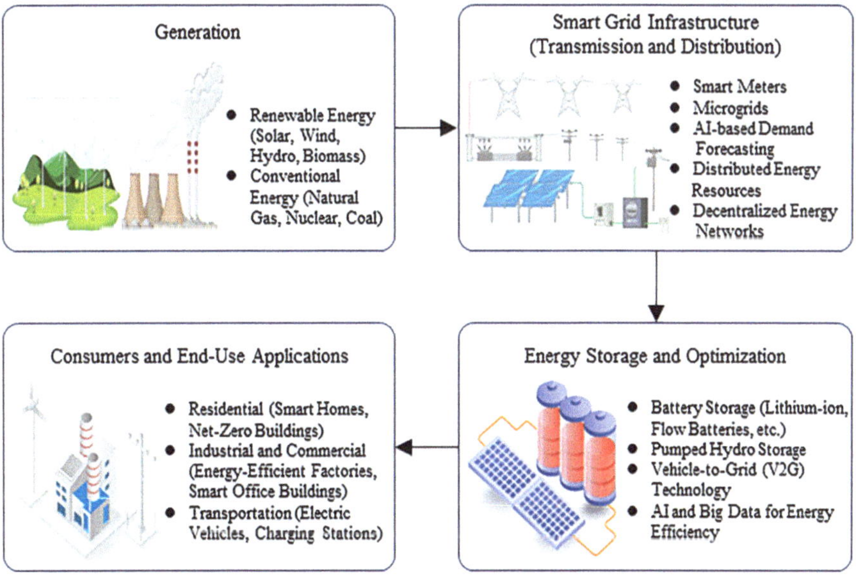

Fig. 10.2 Generation, distribution, optimization, and consumption of energy in smart city

10.1.2.2 Smart Grids

A smart grid is an intelligent electricity network that uses two-way communications to monitor and control transmission of electricity [8]. Smart grids facilitate two-way communication between energy providers and consumers, enabling energy distribution in an optimized fashion, decrease losses, and adopt renewable energy sources.

10.1.2.3 Energy Storage Solutions

Battery systems and other storage technologies play an important role in managing the variable supply of renewable energy [9]. These systems ensure that there's a consistent power supply, even when the renewable sources are unavailable by storing surplus energy during the days of low demand.

10.1.2.4 Energy Efficiency

Energy-efficient solutions are applied in multiple domains of a smart city, like buildings, transportation, and industry [10]. Reducing energy consumption and its associated emissions is achieved through smart lighting systems, energy-efficient appliances, and green building standards.

10.1.2.5 Decentralized Energy Systems

Communities and individuals generate, share, and consume energy locally through microgrids and peer-to peer energy trading platforms [11]. Decentralization of production leads to higher energy resilience and lesser transmission losses.

10.2 Current Research Trends of Smart Energy in Smart Cities

This study investigates the trends, contributions, and impact of research in smart cities with smart energy from 2015 to 2025 through a bibliometric analysis of Scopus data. The Bibliometrix R package along with origin software were utilized to analyse and depict the data. This approach ensures that researchers are presented with the most current research developments in the field enabling them to stay informed on emerging trends and advancements. In March 2025, a bibliometric analysis was conducted utilizing the Scopus core collection database to thoroughly investigate the concept and understanding of smart cities. Ultimately, 4588 articles were exported and saved in BibTex format before being included and examined in the bibliometric study. The primary information, syntax and specific criteria inclusion that used in screening for this study are shown in Table 10.1.

Table 10.1 Summary of data characteristics and main information of the bibliometric study

Description	Specific Standard Requirements/Results
Research database	Scopus
Search query	("smart") AND ("cities") AND ("smart") AND ("energy")
Time span	2015–2025
Document type	Articles, conference paper, book chapter, etc.
Language	English
Data export format	BibTex
Documents	4588
Annual growth rate %	11.03
Document average age	3.8
Average citations per doc	26.69
Keywords plus (ID)	18,714
Author's keywords (DE)	12,308
Authors	13,480
Authors of single-authored docs	313
Single-authored docs	337
Co-authors per doc	4.12
International co-authorships %	37.62

10.2.1 *Annual Scientific Production and Top Ten Most Published Journals/Cited Paper*

Figure 10.3 shows the annual publication trends from 2015 to 2025, highlighting the growth in research output. The early years shows a steady increase in publications, rising from 2015 (79 doc) to 2022 (674 doc), indicating a growing interest in the field. There is a slight decrease in the number of documents in 2023 (666), followed by a increase in 2024 (762). In 2025 (225), the number continues to rise and may continue increasing until the end of the year. Figure 10.4 presents the highest number of articles in sustainable development goals in the smart city research domain published in "IEEE ACCESS" (189 docs), "ENERGIES" (170 docs), "SUSTAINABILITY (SWITZERLAND)" (168 docs), and "SUSTAINABLE CITIES AND SOCIETY" (159 docs), which indicates a higher output of articles compared to the other journals. In addition, some other journals, such as the "IEEE INTERNET OF THINGS JOURNAL" (105 docs), "APPLIED ENERGY" (98 docs), "SENSORS" (77 docs), "SENSORS (SWITZERLAND)" (70 docs), "ENERGY" (69 docs), and "APPLIED SCIENCES (SWITZERLAND)" (66 docs), are in this domain.

Table 10.2 displays the top ten most cited paper from the study which presents both the ranking and authors with their published works. This section presents information about authors, article titles, and their corresponding publications in journals with documentation for publication years and citation frequency. The

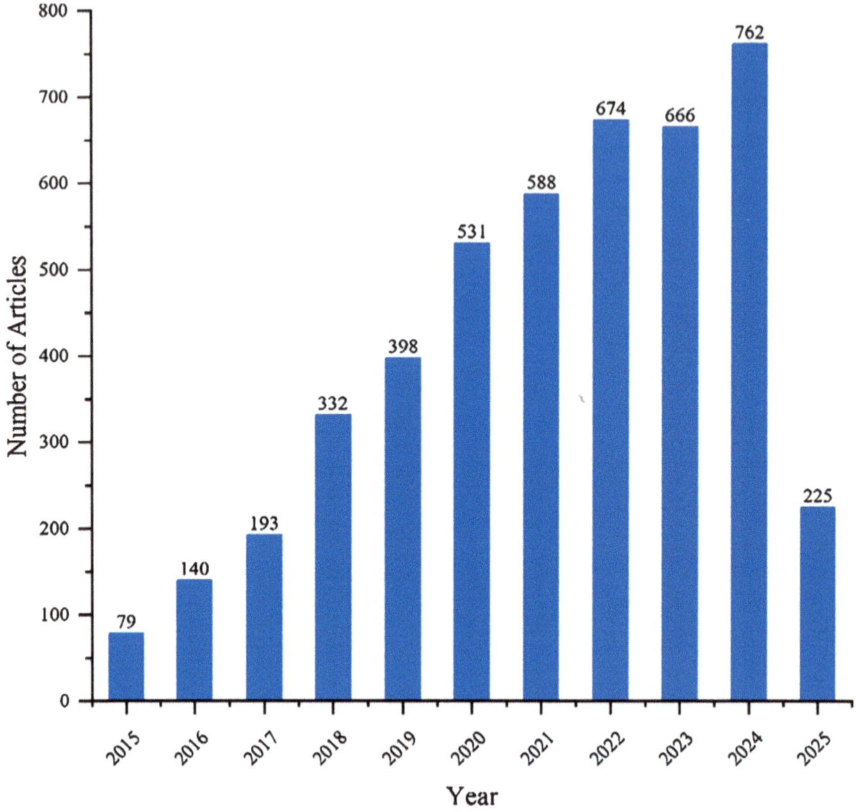

Fig. 10.3 Yearly trend of publication number for smart energy-based smart city

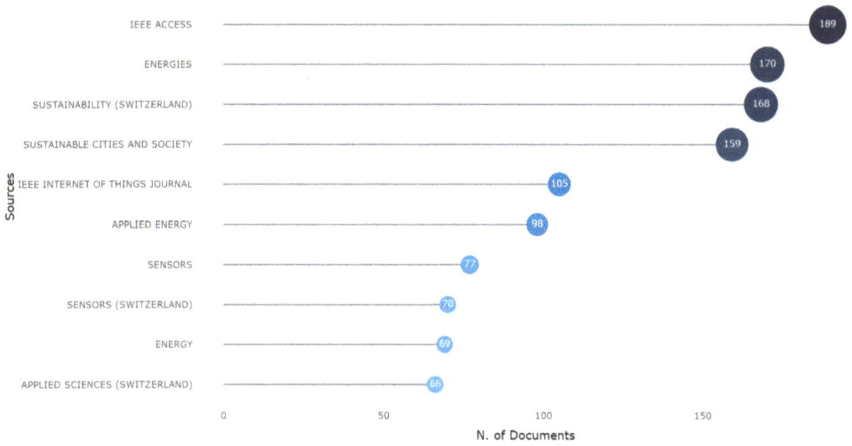

Fig. 10.4 Most relevant sources with number of articles for smart energy-based smart city

Table 10.2 Most global cited documents of smart energy-based smart city research domain

Rank	Authors name	Paper Title	Journal's Name	TC	TC/ Year	NTC
1	Mora et al. (2017) [12]	On-demand high-capacity ride-sharing via dynamic trip-vehicle assignment	Proceedings of the National Academy of Sciences	897	99.67	15.35
2	Kumar et al. (2019) [13]	Internet of things is a revolutionary approach for future technology enhancement: a review	Journal of Big Data	793	113.29	17.47
3	Nuaimi et al. (2015) [14]	Applications of big data to smart cities	Journal of Internet Services and Applications	719	65.36	13.40
4	Nižetić et al. (2020) [15]	Internet of things (IoT): Opportunities, issues and challenges towards a smart and sustainable future	Journal of Cleaner Production	679	113.17	16.64
5	Minoli et al. (2017) [16]	IoT considerations, requirements, and architectures for smart buildings—Energy optimization and next-generation building management systems	IEEE Internet of Things Journal	672	74.67	11.50
6	Rahman et al. (2018) [17]	Predicting electricity consumption for commercial and residential buildings using deep recurrent neural networks	Applied Energy	601	75.13	10.95
7	Bibri (2018) [18]	The IoT for smart sustainable cities of the future: An analytical framework for sensor-based big data applications for environmental sustainability	Sustainable Cities and Society	579	72.38	10.55
8	Zhang et al. (2017) [19]	Security and privacy in smart city applications: Challenges and solutions	IEEE Communications Magazine	508	56.44	8.69
9	Plageras et al. (2018) [20]	Efficient IoT-based sensor BIG Data collection–processing and analysis in smart buildings	Future Generation Computer Systems	491	61.38	8.95
10	Li et al. (2019) [21]	Blockchain in the built environment and construction industry: A systematic review, conceptual models and practical use cases	Automation in Construction	480	68.57	10.57

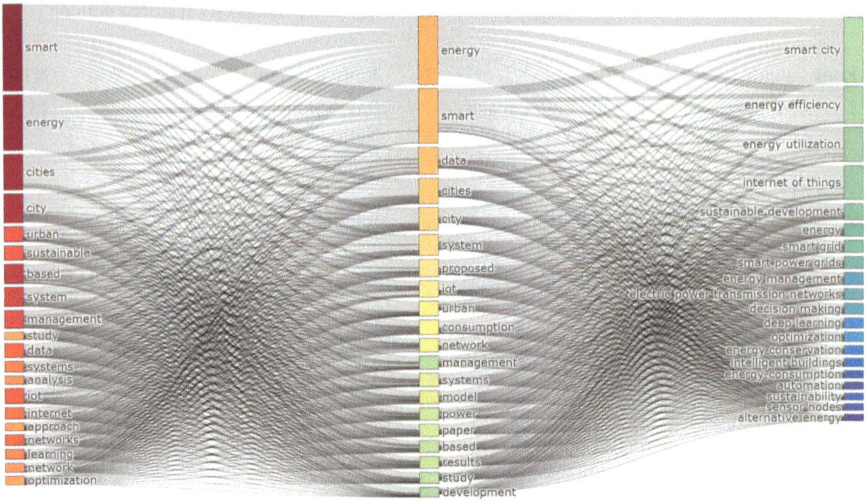

Fig. 10.5 Three field plot (thematic development) using sankey diagram

ranking focuses on engineering field smart cities articles to demonstrate their total citation (TC) counts and annual citation rates and normalized total citation (NTC) data which collectively signify their impact in the field.

10.2.2 Sankey Diagram (Three-Field Plot and Thematic Map)

In Fig. 10.5, Sankey diagram illustrates the three-field plot thematic progression to visualize the flow and the distribution in scientific research. In this case, keywords plus, abstract, and title are added, acting as indicators to highlight the most influential research papers in the subject. These keywords indicate vital concepts and themes that stand out in academic discussions of the smart city domain. The diagram's keywords are sourced from earlier published studies and were then thematically organized by a specific researcher. Figure 10.6 categorizes various smart city themes based on their energy development and relevant degrees. In the top right quadrant (motor themes), which represents themes that are highly developed and highly relevant, terms like "internet of things", and "energy efficiency". In quadrant left (niche themes) there are represents "sustainable development", "sustainable cities", and "energy transition". This indicates a strong focus on these areas, suggesting they are central to energy-based smart city research and practice. This quadrant reflects the areas where significant advances and investments are made, highlighting their critical role in smart city. In contrast, the bottom left quadrant, which represents themes that are either emerging or declining and have lower relevance, includes terms like "smart grid", "electric vehicles", and "solar energy". Additionally, the terms like "smart city" and "sustainability" are the foundational information for further analysis and serve as a base for more advanced analyses. These themes

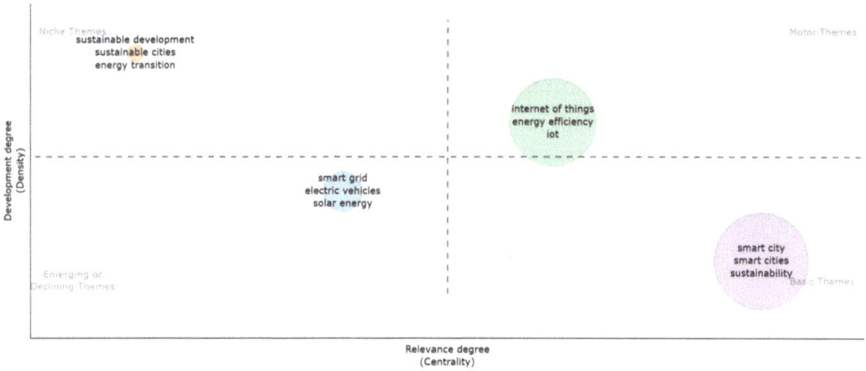

Fig. 10.6 Thematic map of smart energy-based smart city research domain

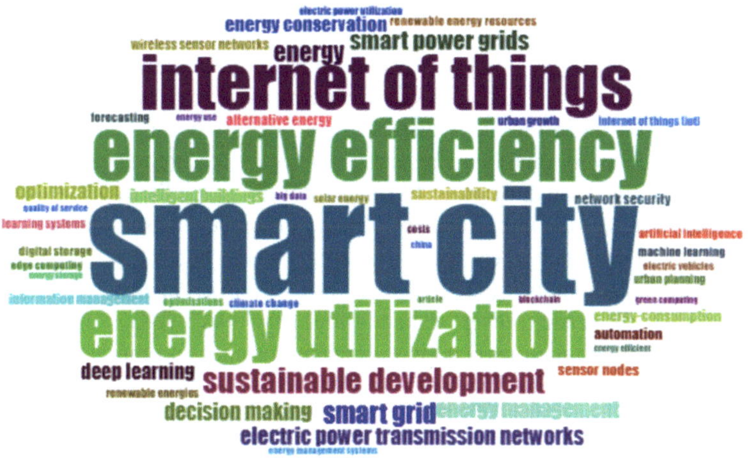

Fig. 10.7 Co-occurrence keyword trends of smart energy-based smart city research domain

might be considered foundational but must be emphasized more in energy-based smart cities. This visual categorization provides valuable insights for educators, curriculum developers, and researchers, helping them understand which areas are gaining traction and which might need more attention or innovation.

10.2.3 Co-occurrence Keyword Trends and Network Visualization

Figure 10.7 depicts the most prominent themes and keywords related to research in conceptual change. The most significant and most central terms, such as "smart city," "energy efficiency", "energy utilization", "internet of things", and

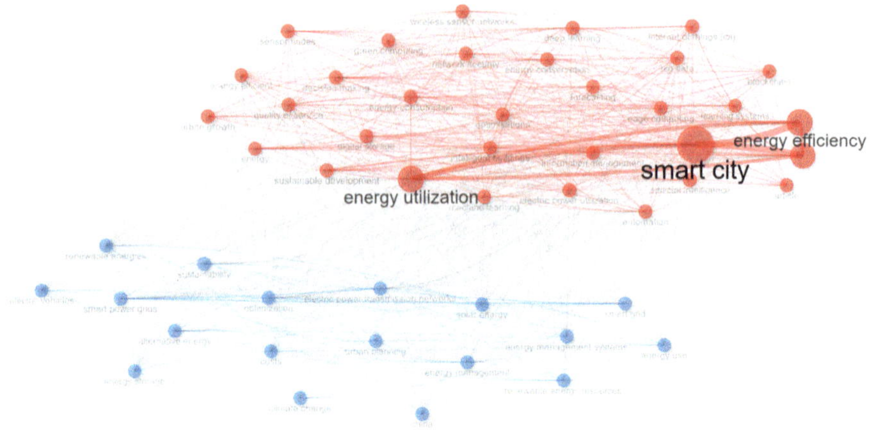

Fig. 10.8 Co-occurrence network of words in smart city research domain with smart energy

"sustainable development" indicate the core focus of the research on how researchers grasp and alter their conceptual knowledge, particularly within smart city contexts. An illustration in Fig. 10.8 displays word occurrence frequencies from the smart energy goals in smart city research network. Research focuses mainly on "smart city" as this term appears 1740 times throughout the literature. Research on smart cities indicates the importance of "energy efficiency" based on its 1024 word occurrences. Additionally, "energy utilization" (931), "internet of things" (911), "sustainable development" (428), "smart grid" (346), "energy" (325), "smart power" (311), "electric power transmission networks" (288), and "decision-making" (284) combinations occupy crucial importance in emerging technology fields. The data shows that smart city research revolves primarily around several major subjects.

10.2.4 Countries' Scientific Production

The scientific publication frequencies of various countries in the world are represented in Fig. 10.9. Scientific contributions from China stand at 2131 which demonstrates their powerful role in research development. The growing scientific power of India displays itself through its 1565 scientific publications which rank second after China. The research hub of the USA produces 780 scientific papers among its total contributions. Numerous publications emerge from Italy and Saudi Arabia together with Spain given their research output comprises 588, 567 and 486 papers respectively. Academic and industrial research sectors of the UK and South Korea produce approximately 455 and 365 scholarly works between them. Iran and Canada have smaller publication numbers but their contributions amount to totals of 240 and 235

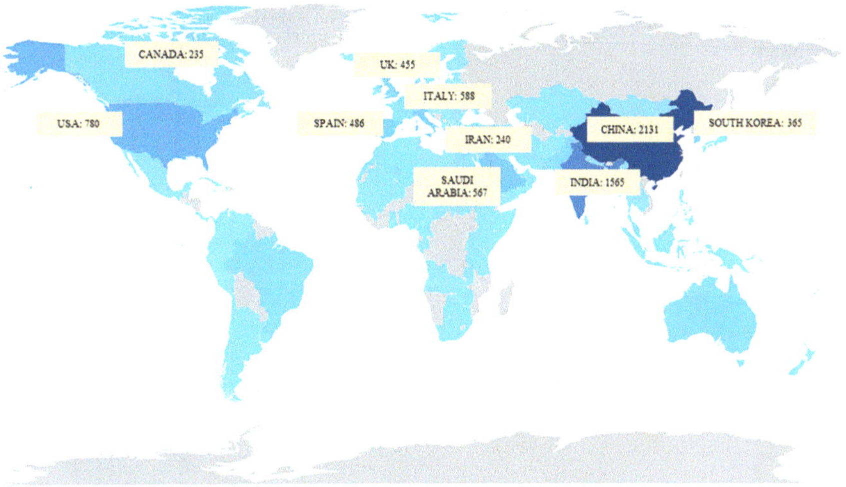

Fig. 10.9 Top ten countries' scientific production of various countries based on frequency

research items respectively. The measured data indicates Asian countries significantly influence the global scientific research output.

10.2.5 Mapping Smart City Research to Sustainable Development Goals (SDGs)

SDG Goals, its targets (bubble series) and the percentage (bar graph) of corresponding keywords detected in the text (calculated as the ratio of keywords in one goal to the total number of keywords detected) are depicted in Fig. 10.10. For SDG mapping, the study utilized the SDG Mapper tool from the "Knowledge Base for the Sustainable Development Goals," provided by the European Commission [22]. This tool uses bubble charts to depict SDG interconnections and relevance among bibliometric records. The size of each bubble correlates with the proportion of associated keywords found, which is calculated as the number of goal-specific keywords divided by the total keyword count. As such, SDG 7, marked by 220 keyword occurrences. This is followed by SDG 11, with 20 occurrences, and SDG 13, with 15 occurrences, underlining their relative prominence in the research database. Within SDG 7, there are several key subtargets that the research address: Subtarget 7.2 is shown as a distinct bubble within the larger SDG 7 circle, indicating focuses on increasing the share of renewable energy in the global energy mix. Subtarget 7.3, which aims for doubling the global rate of improvement in energy efficiency, is another bubble within the SDG 7 area. This highlights the efforts in research to improving the cities. Subtarget 7.1 aims to ensure universal access to affordable, reliable, and modern energy services by 2030. These comes under the SDG 7 which

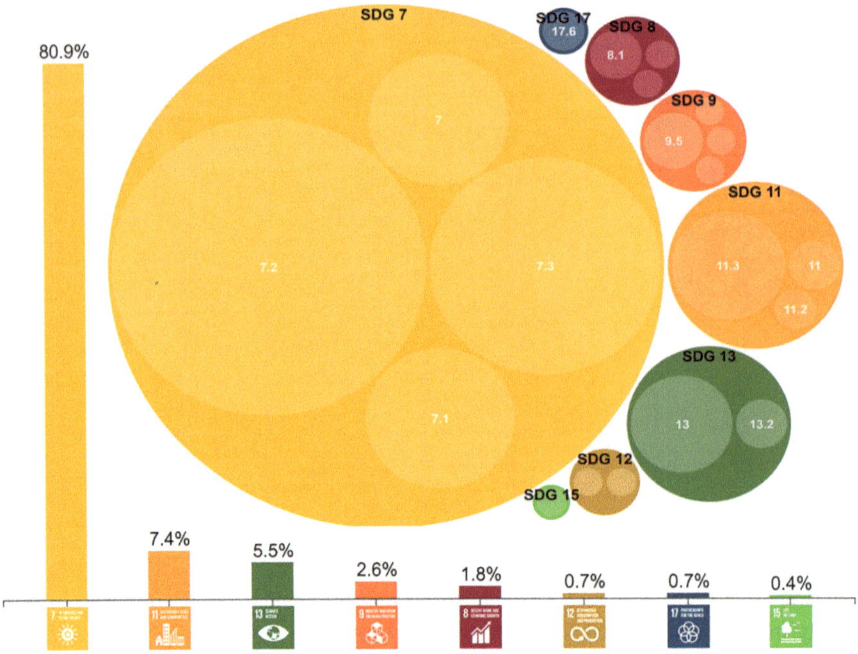

Fig. 10.10 Mapping energy-based smart city research to relevant SDGs and targets

aims for affordable and clean energy that prioritize sustainability. Another bubble with SDG 11, aims in sustainable cities with Subtarget 11.3 is shown as a distinct bubble within the larger SDG 11 circle, indicating focused research on smart city. Subtarget 11.2 focuses on providing the safe, accessible and sustainable smart transportation. Subtarget 11, which aims sustainable cities and communities, is another bubble within the SDG 11 area. This highlights the efforts in research to improving the cities. Similarly, SDG 13 (Climate action) is also closely linked to SDG 11 and suggests that research to integrate climate change measures into national policies, strategies, and planning. The other related SDGs with detailed description are depicted in Table 10.3.

10.3 Energy Requirements of Sustainable Smart Cities

Smart and sustainable cities require integrative and innovative power technologies, energy-efficient application and renewable power resources [23]. Smart cities utilize various sources such as wind, solar, and hydropower with advanced technology applications like electric vehicles, energy storage systems, and efficient buildings. Advanced digital tools including sensors and communication networks support efficient energy delivery. This approach promotes sustainability, optimizes energy use, and improves environmental quality.

Table 10.3 SDGs ranking, targets, and relevance in energy-based smart city research domain with SDGs

SDGs (By Rank)	Description	Targets	No. of Occurrence		Relevance%
SDG 7	Affordable and clean energy	7.2	113	220	80.9
		7.3	60		
		7.1	25		
		7	22		
SDG 11	Sustainable cities and communities	11.3	15	20	7.4
		11.2	3		
		11	2		
SDG 13	Climate action	13	12	15	5.5
		13.2	3		
SDG 9	Industry, innovation and infrastructure	9.5	4	7	2.6
		9.2	1		
		9.1	1		
		9	1		
SDG 8	Decent work and economic growth	8.1	3	5	1.8
		8.2	1		
		8.3	1		
SDG 12	Responsible consumption and production	12.4	1	2	0.7
		12.5	1		
SDG 17	Partnerships for the goals	17.6	2	2	0.7
SDG 15	Life on land	15.3	1	1	0.4

10.3.1 Urbanization and Increased Energy Consumption

Urbanization is the social process whereby people from rural background shift to live in cities [24]. The rapid increase in urban populations has resulted in a significant rise in energy consumption across buildings, transportation networks and industries. This growing demands for energy offer major challenges associated with urban expansion and resource usage. As more individuals relocate to urban areas huge amount of energy are required to support infrastructure and daily living needs. Nevertheless, with advancements in intelligent energy systems, smart cities are actively exploring innovative approaches to address these issues. These developments aim to promote long-term sustainability and improve energy efficiency, contributing to more environmentally responsible and efficient urban environments. To meet the rising energy requirements of citizens living in smart cities, cities are now employing efficient smart energy systems to address the issue of demand and supply of power. Smart energy systems which integrate energy storage facilities, smart grids, and renewable energy sources with energy-efficient technologies offers an intelligent energy system. Smart grids are another component in the contemporary smart energy systems since they enable efficient interconnection of renewable energy resources into the power grid. Furthermore, it enables the handling of uncertainties in supply and demand in the energy sector and make a contribution to the stabilization of the energy supply for energy supply fluctuations in the renewable energy supply. Smart grids also include better communication, associated with automation as well as control features to facilitate the management of the energy system and give the consumer real-time information on the usage of electrical energy. Smartness in city transport infrastructure like charging solutions for electric cars and bike sharing lowers energy expenses that arise from use of transport. Therefore, the process of urbanization and growth of energy consumption are considered to be among the key challenges and problems of modern society. Nevertheless, through the specific development of advanced smart energy systems in smart cities, society offers the best solution to these problems. Integrating renewable power, energy storage facilities, intelligent grid, and efficient technologies help smart cities build a sustainable, secure, and effective energy environment [25]. With more people moving to urban areas, deployment of these advanced smart energy systems will be inevitable in fulfilling the rising energy needs of cities with minimal effects on the world.

10.3.2 Need for Sustainable Energy Solutions

The past few years have brought a lot of focus on the concept of smart cities, which utilize great technological solutions and data-driven approaches to the respective cities' management in order to enhance the standards of living of people in these cities and to provide sustainable development [26]. Another important characteristic of these smart cities is integration of efficient energy solutions that would help to make the environment cleaner, energy consumption higher, and reduce greenhouse effect. These goals are complemented by modern smart energy systems since they create the necessary basis infrastructure and framework for managing energy effectively. Renewable energy technologies refer to energy technologies that can be sustainably from renewable resources in a given geographical area, energy resources which cause no harm to the environment when used for energy production. These sources of energies are crucial to be adopted in smart cities due to the impact of fossil energy such as coal and oil that emit large amount of greenhouse gases which causes climate change. Furthermore, these renewable energy sources involve low costs of operation hence making it easier for smart cities to direct resources well while directing more of funds to other sectors. The smart management of energy systems as part of critical success factors is the engine for the realization of sustainable energy systems in smart cities [27]. This integration helps cities to improve the efficiency of energy consumption and minimize the losses of energy resources of cities. This approach not only optimizes demand on the energy system but also encourages consumers to actively contribute to achieving sustainable energy objectives. Smart energy systems promote responsible energy use by integrating intelligent technologies such as energy-efficient lightning, appliances and buildings. These technologies are designed with low carbon footprints and high energy efficiency results in reduced energy consumption. Increased awareness and participation, users become more accountable in managing energy resources thereby reinforcing the broader goals of sustainability and environmental responsibility within urban infrastructure. City-based energy management may receive abundant data through sensors and smart meters, which helps estimate general energy-use trends and implement specific measures to decrease energy consumption, associated with sectors such as transport and construction. This is done by devising intelligent transportation systems that increase efficiency on the road and encourage the use of environmentally friendly transport, or simply by adding efficient insulation, windows and climate control systems in buildings [28]. Another important competency in an advanced smart energy system is the opportunity to integrate energy storage systems that enable the storage of energy produced by renewable sources and its future use in cities. The adoption of sustainable energy solutions and smart cities is essential as it contributes to reducing environmental impact, easing pressure on infrastructure, improving affordability and enhancing living standards. Modern smart energy systems provide the necessary framework for cities to integrate these technologies effectively in the energy system.

10.4 The Evolution of the Energy Revolution

Energy has been a fundamental element throughout the human history, shaping the development of civilizations. From early use of fire to burn wood and animal dung, power of water and wind, energy has played a vital role in daily life. Over centuries, human progress is linked to advances in knowledge and control of energy often referred to as energy revolution. This evolution has transformed energy systems from simple Energy 1.0 to more advanced and connected systems in Energy 5.0 reflecting ongoing innovations and improvement as illustrated in Fig. 10.11.

10.4.1 Energy 1.0: The Age of Coal and Steam

Energy 1.0, the age of coal and steam, began the industrial revolution by providing powerful and reliable energy for factories, transportation, and machinery [29]. Coal-powered steam engines transformed production and enabled large-scale industrial growth, shaping modern economies and society through increased manufacturing and expanded transportation networks.

10.4.2 Energy 2.0: The Rise of Oil and Electricity

Energy 2.0, defined by the rise of oil and electricity, shifted energy use from coal to more efficient sources [30]. This era enabled widespread industrialization, transportation advancement, and urban development. The increased use of oil and electricity

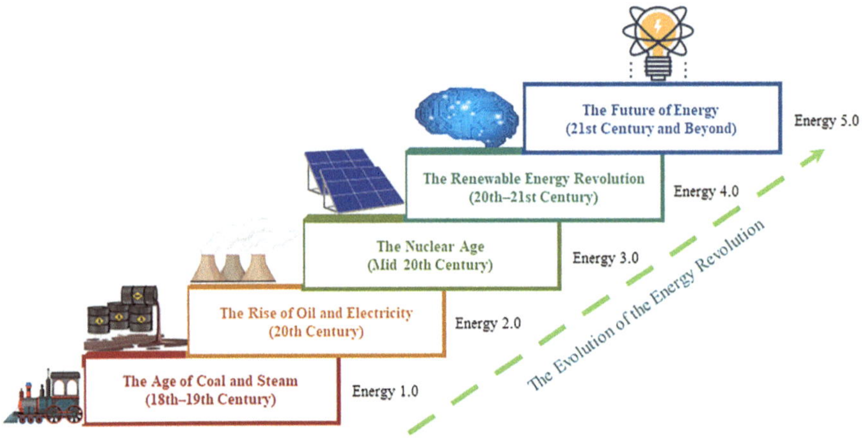

Fig. 10.11 The evolution of the energy revolution

transformed economies and societies, driving technological progress infrastructure growth.

10.4.3 Energy 3.0: The Nuclear Age

Energy 3.0, known as the nuclear age introduce nuclear power as a major energy source. This era focused on connecting nuclear fusion for large-scale electricity generation, providing a high energy output with low greenhouse emissions [30]. Despite safety concerns and challenges related to waste disposal, nuclear energy significantly contributed to meeting the rapidly increasing global demand for reliable and large-scale electricity, supporting industrial growth across various countries.

10.4.4 Energy 4.0: The Renewable Energy Revolution

Energy 4.0 represents the integration of digital technologies into energy systems to create smarter, more efficient and flexible energy management [30]. It combines advanced sensors, data analytics and IoT, AI and automation to optimize energy generation, distribution and consumption. This approach enables real-time monitoring, predictive maintenance and demand response. Energy 4.0 supports the integration of renewable energy sources and decentralizes power production allowing for more resilient and sustainable energy infrastructure. It transforms traditional energy system into intelligent network, which improves efficiency and promote environmental sustainability.

10.4.5 Energy 5.0: The Future of Energy

Energy 5.0 represents the future of energy systems, focusing on integration, sustainability, and digital transformation. It combines advanced technologies such as AI, Blockchain, IoT, and smart grid to create intelligent and decentralized energy networks. It emphasizes environmental responsibility, efficient energy use and equitable access to clean energy. It also encourages the use of emerging solutions like hydrogen fuel cells, advanced nuclear technologies, and improved energy storage [31]. This stage reflects a shift that is more responsive, adaptable, and aligned with climate goals. Energy 5.0 supports the creation of smarter, cleaner, and more resilient energy infrastructure.

10.5 Integration of Renewable Energy Sources in Smart Cities

Renewable energy source can be described as all types of energies derived from natural resources; these are resources replenished endlessly. There are five classified categories of renewable energy sources; these include solar energy, wind energy, water or hydraulic energy, geothermal energy, and biomass energy. These sources offer significant advantages over conventional fossil fuels, as they are environmentally friendly, produce low greenhouse gas emissions, and are renewable. Therefore, deployment of those smart entities in smart cities is fundamental towards the realization of smart sustainable cities. Smart cities adopt renewable energy sources and practices that make urban environments resource efficient, sustainable, and resilient. Some of the ways that smart cities integrate renewable energy source include and illustrated in Fig. 10.12:

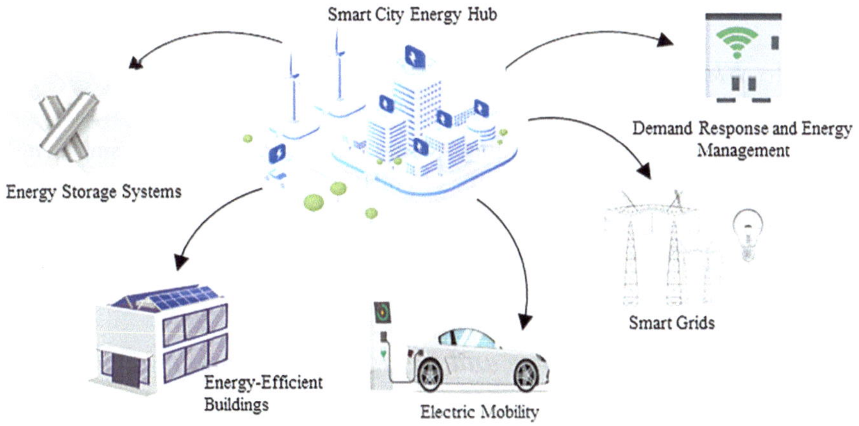

Fig. 10.12 Integration of renewable energy sources in smart cities

10.5.1 Smart Grids

Smart grid includes an enhanced electrical grid system that incorporates AI-driven solutions for monitoring, controlling, and managing energy usage [32]. It offers the easy connection of renewable energy source into the current energy system where electrical energy is transmitted from renewable resources to consumers. In addition, smart grids enhance grid reliability and contain the potential for failure of electrical power.

10.5.2 Energy Storage Systems

The fluctuating and random nature of renewable sources of energy such as solar and wind power, there are unique challenges when adopting them in cities. Battery storage allows electricity generated during period of low demand to be stored for use during times when power demand is higher and balances supply to ensure efficient energy usage [33].

10.5.3 Energy-Efficient Buildings

Planning of smart cities involves the means to use more enhanced building infrastructures which incorporate renewable energy systems for power supply. Green building elements such as the use of building materials with environmentally friendly characteristics, passive solar systems and the incorporation of renewable energy source on rooftops minimize energy utilization and costs incurred by occupants in a building [34].

10.5.4 Electric Mobility

Electric vehicles (EVs) and charging stations are now being used in smart cities. The transport sector is a major source of pollution in cities today, affordable EVs that are charged by renewable energy will cause a drastic reduction in greenhouse gas emissions [35].

10.5.5 Demand Response and Energy Management

Smart cities use complex parameters and IoT to vary, regulate, and optimize the consumption of energy in real time. Demand response programs motivate consumers to consume energy during low demand hours, whereas energy management systems assist in matching demand with renewable supply [36]. It promotes more efficient, sustainable, and secure energy systems.

10.6 Latest Trends in the Energy Sector

New trends in smart energy systems using advanced technology have been developed as the most efficient way of utilizing energies from the resources. These systems utilize green power, advanced electricity networks, and power storage technologies to reduce emissions while enhancing reliability and promoting sustainability. The components also include IoT devices, artificial intelligence and big data analytics to support improved decision-making and management of power supply. As smart city solutions progress into the future, the role of smart energy systems will still play a critical role in defining the future of smart urban energy systems. Following are the latest techniques that support smart energy systems as illustrated in Fig. 10.13.

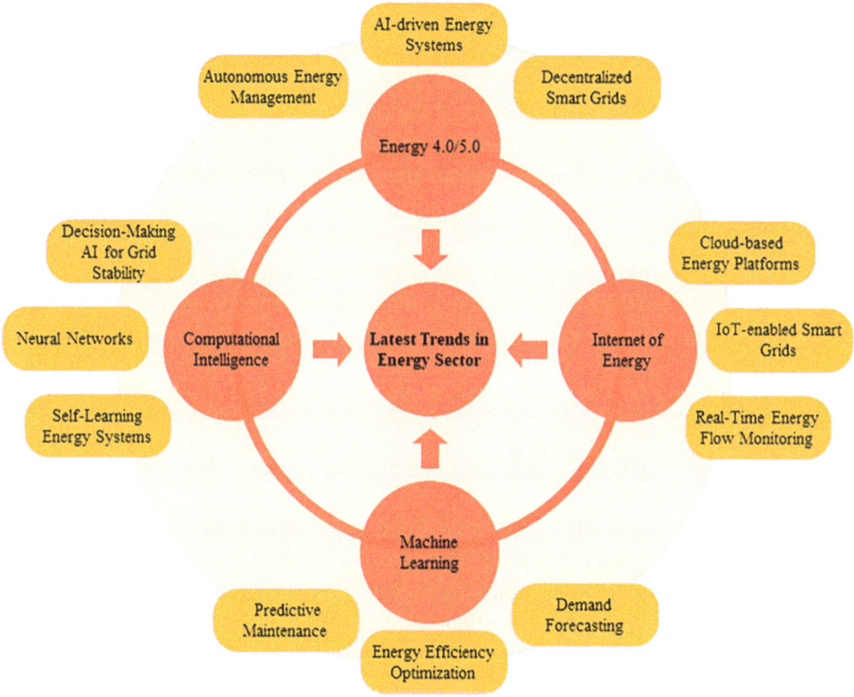

Fig. 10.13 Advanced technologies in smart energy systems in smart cities

10.6.1 Energy 4.0/5.0

The increasing rate of people moving to urban areas and the pressing need for urgent electricity in the world demands for sustainable energy solutions. Due to this demand an advanced smart energy system, Energy 4.0/5.0, is revolutionizing the smart cities. Energy 4.0/5.0 is a vision to optimistically redesign the generation, distribution, transmission, and storage of energy with enhanced efficacy to better manage the scarcer resources and the emission of greenhouse gases. Energy 4.0/5.0 refers to the use of information technology (IT) in IoT, AI, machine learning (ML), and big data analysis for efficient integration and operation of the energy systems [37]. The pillars of Energy 4.0/5.0 are smart metering infrastructure, demand management, distributed energy resources and grid enhancement. Through these components, smart cities provide for an efficient, reliable, and enduring energy system that will contribute as a key provider of energy for the community while minimizing undesirable effects on the environment. Therefore, the foundation of an Energy 4.0/5.0 system is smart metering infrastructure to monitor and track energy usage patterns in real time. Advanced metering infrastructure extends beyond traditional meters by providing consumers with specific information about energy consumption patterns. A key component of the Energy 4.0/5.0 systems is demand respond management, which involves the intelligent regulation of energy use through real-time algorithms and analytics to address variations in supply and demand [38]. This approach not only reduces the risk of power failure but also contributes to building a more reliable and efficient energy resources. Additionally, storage technologies like batteries and pumped hydro storage support stable electricity supply and efficiency. Another important component of an Energy 4.0/5.0 system is grid modernization [39]. It refers to strengthening of the existing energy infrastructure with a growing demand for power and integration of renewable energy. Consequently, Energy 4.0/5.0 is the transition to new smart energy systems that are essential for establishing smart cities. Integrated with advanced technology, Energy 4.0/5.0 systems provide the sustainable, effective, and adaptive energy supply and demand management, contributing to the construction of more robust and lowest-carbon city.

10.6.2 Internet of Energy

The Internet of Energy (IoE or IoT for energy) is an innovative model which is essential to advancing energy management system in smart cities. This concept integrates several smart energy systems including smart grid, renewable energy supplies, and energy storage systems whereby each serves as a single system linking different energy systems to improve on the effective use of energy, efficient interaction and communication. The IoE uses modern technologies, including sensors and smart meters in real-time data collection and processing. This makes it possible for cities to forecast energy requirement, supply energy in adequacy, and enhance the general management using energy. A functional block of the IoE is the smart grid. A smart grid refers to an enhanced electric network that supports the flow of electricity from producers through consumers while using communication, IT, and automation to achieve optimal performance [40]. Another key component of the IoE is the incorporation of renewable energy into the system and using power from the sun and wind. Meanwhile, smart cities effectively utilize IoT technologies in energy systems to manage and optimize the use of renewable resources [41]. These systems enhance performance based on climatic conditions and support the forecasting of power output from renewable energy resources. It also aids cities to cut down reliance with the conventional non renewable energy sources such as fossil fuel and emissions which has had a very negative impact on the environment. Energy storage systems also include the IoE environment. The integration of cost-effective storage solutions in smart city ensures a continuous supply of energy during periods when renewable energy generation is limited. This approach enhances the reliability of the energy system and supports the consistent and efficient use of clean energy sources. Battery storage systems are employed and charged during the optimal renewable energy production times and dispatched during high demand times or when production is low. These elements form the IoE that is considered to support well-developed smart energy systems of smart cities.

10.6.3 Machine Learning

Smart energy system involves the use of smart renewable energy resources, smart energy network, and complex energy storage systems for the sustainable, intelligent, and reliable energy supply. Smart energy systems with machine learning techniques work in real time to process huge amounts of information regarding energy consumption and production to improve the efficiency of the energy system [42]. Demand response management is one of the areas widely applying machine learning in smart energy systems. Machine learning helps the algorithm compute rate of consumption, weather patterns, and use data to estimate maximum consumption tendency in a given period and schedule an automatic regulation of energy consumption. This ability to predict accurately enables the development of successful demand response programs in cities, decrease in energy importation costs and greenhouse gas output. Furthermore, machine learning improves the effects of system integration of renewable energy resources including the solar and wind energy into the smart grid [43]. These energy sources are unpredictable because they depend on weather; therefore, it is difficult to predict how much energy is going to be produced. Machine learning algorithms enable cities to better control energy usage and integrate renewable energy, as well as make energy in cities more productive and sustainable. The promising technological advancements in the future of smart cities and their even smarter energy systems are certain to incorporate machine learning taking an even more central position enabling the advancements of a cleaner, greener and sustainable environment of urban living.

10.6.4 Computational Intelligence

Artificial intelligence branch which designs and implements intelligent systems and algorithms to solve problems is called computational intelligence. It is now an indispensable part of modern smart energy systems in smart cities as it contributes to the effective and efficient management and improvement of the energy distribution and consumption systems [44]. This is made possible by the complex algorithms that calculates and model different energy systems, assess the effectiveness of the systems, and forecast the future demand and supply patterns. A smart city is supported by a complex of energy systems, such as smart grids, distributed renewable energy generation, energy storage, and energy management systems. Effective smart energy systems for many hierarchical levels of power management are developed and enhanced with help of computational intelligence methods such as genetic algorithms, artificial neural networks, fuzzy systems, and swarm optimization algorithms [45]. Such systems and sources management can be achieved efficiently and safely while reducing wastage, CO_2 emissions in the city. Artificial neural network is applied in analysing behaviour of smart grids given the stochastic nature of renewable energy resources. These neural networks are trained on the basis of historical data and the current states of the system to forecast short-term and long-term load demands. Fuzzy systems help to the regulation of electricity use in constructing through controlling the central heating, ventilation and air conditioning systems. Energy distributing networks can also be optimized using swarm optimization algorithms, leading to reduction in energy losses and achieving higher reliability in the system.

10.7 Smart Energy Technologies Used in Smart Cities

Smart cities have developed intelligent smart energy types that enable efficient use of energy as well as the minimization of wastage. It includes smart grids, digital twin, smart metering, energy storage, and energy-efficient technology systems as illustrated in Fig. 10.14. These technologies function as an innovative platform for smart energy management within new urban environment in smart cities.

10.7.1 Smart Metering Technology

Smart metering technology is the modern technology associated with the energy management and efficiency in smart cities. These smart meters are employed to determine the usage of electricity, water, or gas at home. In smart cities, this technology serves as the basis of ultra-smart energy systems, enhancing the aspects of performance, stability and eco-friendliness [46]. It includes sensors and data

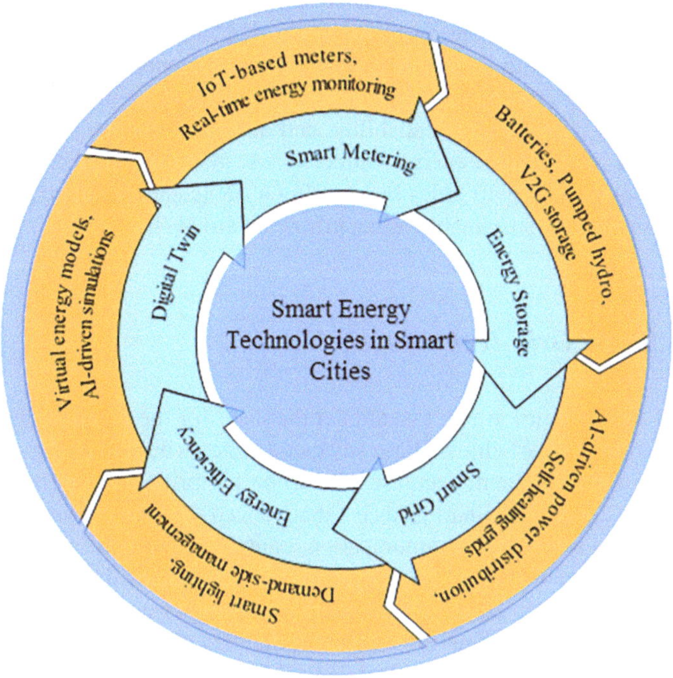

Fig. 10.14 Smart energy technologies in smart cities

analytics instruments, through which energy use are monitored remotely, automatically, and constantly. In addition to accurately recording daily power consumption, smart meters are designed to predict energy usage patterns and support in integrating renewable energy into the power grid. As a component of advanced smart energy systems, smart metering enhances energy utility, availability, security, and sustainability. It also plays a significant role in promoting energy conservation among consumers. Through the adoption of smart meters, cities develop intelligent and sustainable energy systems that provide benefits for both residents and the environment.

10.7.2 Energy Storage Technologies

The energy storage systems are critical in optimizing smart cities and effective in utilization of renewable energy [2]. Energy storage technologies used in smart cities are classified into four major categories: electrochemical storage, mechanical storage, chemical storage, and heat storage or thermal storage. Electrochemical storage including batteries and fuel cells plays the central role in smart energy systems by enabling the storage and transformation of electrical energy. Chemical storage methods such as hydrogen and methane store energy in the chemical form produce from for renewable sources during period of low demand. Hydrogen storage focuses on storing heat energy to be supplied when required supporting consistent energy delivery. The integration of various storage types leads to improved use of renewable energy, strengthen energy reliability, and addresses challenges related to the variable nature of renewable energy sources. These systems contribute significantly to the development of smart cities by ensuring a stable energy supply, energy efficiency and promoting sustainable urban infrastructure [47].

10.7.3 Smart Grid Technologies

Smart grid technologies are an essential component for the creation and effective implementation of smart cities. These advanced systems built to optimize the function of energy management, decrease the carbon footprint, and optimize the grid. Therefore, the concept of smart grids is far advanced than the conventional supply systems as it has integrated communication, control and computing in the generation, supply and use of energy. These systems use data obtained from smart meters, sensors, or any IoT device to control energy consumption, identify and mitigate power loss situations, and determine efficient delivery of energy in real-time. Such an advanced degree of monitoring and control is essential to ensure that energy is only sent where it is needed and to an acceptable extent. Such incorporation is a key factor to achieving the goal for the green and sustainable future since such integration helps to avoid excessive utilization of non-sustainable energy resources and minimize negative impact of climatic change [48]. The progressive transformation of the energy sector in the direction of an eco-friendly infrastructures supported by smart Grids, with the capacity of proposing a stable and responsive environment for the integration of renewable energy. Therefore, smart grid technologies are the key elements of the smart cities, enabling decreasing energy loss, enhancing energy use and laying the foundation of a more efficient and functional energy future [49].

10.7.4 Energy Efficiency Technologies

Energy efficiency technologies play a vital role in smart cities by using advanced energy systems that improve both energy usage and efficiency. These technologies focus on the use of energy resources, reduction of energy losses and minimization of greenhouse gas emission, supporting the development of environmentally sustainable and urban areas [31]. For example, LED lightning has contributed significantly to reduce electricity consumption and producing maintenance cost in urban infrastructure. Smart lightning systems use control and sensors to adjust light intensity, temperature, and distribution based on specific locations and time of day. Another important aspect in energy management is real-time monitoring, control, and optimization of energy consumption in buildings, transportation, and industry activities. These technologies are essential to the advancement of smart energy systems enabling the connection of multiple energy resources and the efficient management of energy use. It contributes significantly to the creation of sustainable cities that are environmentally responsible and helps to meet future energy challenges.

10.7.5 Digital Twin

A digital twin is a virtual representation of a physical object, process or system updated with real-time data and simulation algorithms to reflect actual conditions. In smart cities, digital twins support a range of application and are useful in managing complex smart energy systems [50]. By the use of data analytics, cities optimize the management of renewal energy resources, energy storage and smart grids. City officials gain the ability to understand and predict the behaviour of various energy components leading to better decision-making and improved efficiency. Continuous monitoring allows detection inefficiencies, anticipation of potential failures, maintenance, and contributing to energy savings. It also supports energy trading by enabling cities to manage and distribute energy generated from renewable sources or stored-in energy systems. In addition, digital models assist in addressing climate change by promoting energy-efficient practices and reducing greenhouse gas emissions [51]. Based on the energy usage trends, digital twins contribute to the development of sustainable technologies and play an important role in shaping future urban energy management systems.

10.8 Open Research Challenges and Research Directions

The use of smart technologies in infrastructure supports urban growth as urbanization continues to rise globally. Smart energy systems play a vital role by enhancing energy efficiency, ensuring reliable supply and minimizing environmental impact. This section discusses current challenges and feature research opportunities in smart energy systems for smart cities as illustrated in Fig. 10.15.

10.8.1 Open Research Challenges in Smart Energy Systems

10.8.1.1 Integration of Renewable Energy Sources

The fluctuating and unpredictable nature of renewable energy sources, such as solar and wind power, requires both storage solutions and efficient grid systems to ensure a stable electricity supply. Batteries, flywheels, and pumped hydro storage systems are

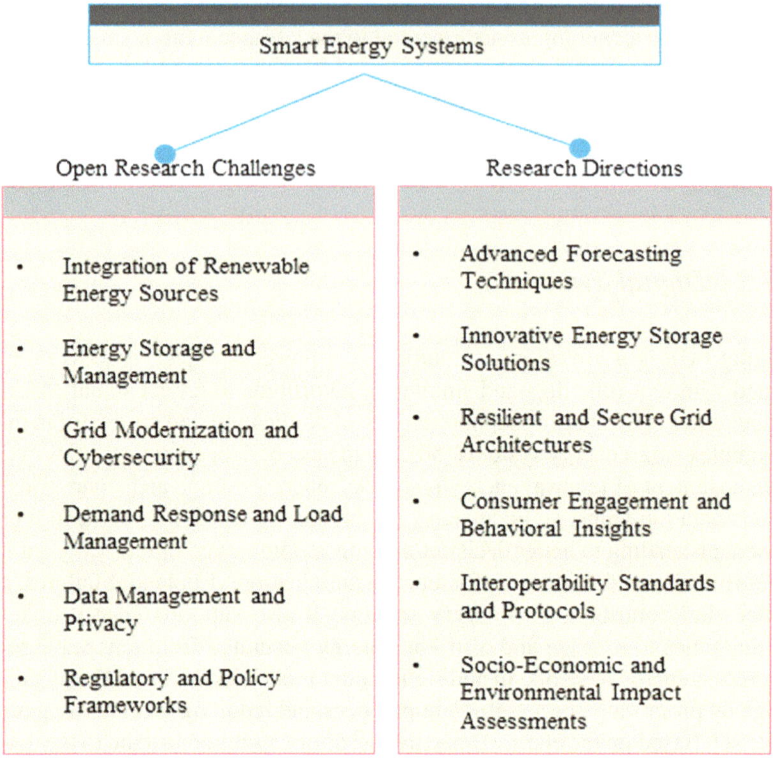

Fig. 10.15 Open research challenges and research directions in smart energy systems

good energy storage technologies that need to improve on efficiency, cycle durability as well as a reduction in the cost [52]. There is a need to ensure that such technologies facilitate real-time and demand response and control of the supply and demand [53]. To overcome these challenges, it is essential to conduct multi-disciplinary research and partnerships between academia, industry, and government to address issues of sustainable and resilient energy systems towards smart city.

10.8.1.2 Energy Storage and Management

Energy storage is critical challenge that must be addressed to ensure demand is met according to supply availability leading to the development of efficient and cost-effective storage solutions [54]. It is important to explore alternative options such as flow batteries, compressed air energy storage, and thermal storage. Smart energy management systems are also necessary to manage diverse energy sources and storage methods for optimal balance between supply and demand.

10.8.1.3 Grid Modernization and Cybersecurity

Renewable energy grid integration and cybersecurity are two of the most significant research areas related to further enhancement and utilization of smart energy system in smart cities. The integration of the advanced grid communication requires the conversion of the current systems to more intelligent systems that supports two-way communication and streaming of data in real time [55]. The modernization of existing smart grids must address cybersecurity concerns as these grids are increasingly vulnerable to cyber attack due to dependence on information technology networks that control and monitor their operations.

10.8.1.4 Demand Response and Load Management

Demand response refers to the process of reducing power demand during peak time by encouraging consumers to adopt energy-efficient practices. Developing accurate algorithms to predict and control energy demand in real time is still a significant challenge. Addressing these challenges requires interdisciplinary collaboration, innovative and appropriate policymaking, and the development of adaptable models. These efforts will support the creation of future smart cities that are both sustainable and resilient in energy management [56].

10.8.1.5 Data Management and Privacy

Large amounts of automatically collected data for smart energy systems must be effectively managed. The issues of data privacy and security are important when data sharing is needed for system optimization. Further research is required to identify an effective method for managing these concerns in a balanced way [57]. As smart city systems become more complex, data management and the protection of citizens' privacy are becoming critical component of smart city design.

10.8.1.6 Regulatory and Policy Frameworks

The research challenges of advanced smart energy system integration in smart cities include regulatory and policy aspects. However, many of these frameworks must be changed and developed to deal with the complexity of highly connected and diversely organized energy systems, new technologies, and different forms of organization [58]. In order to overcome these issues, policymakers, academics, and business entities should come up with policy support measures that shape a favourable environment for smart energy systems.

10.8.2 Research Directions in Smart Energy Systems

10.8.2.1 Advanced Forecasting Techniques

The intelligent and collective management of various energy systems is vital when it comes to planning future smart energy systems of smart cities. With advancement in technology and increased interconnectivity of several cities, the need for appropriate energy management systems is essential. The focus should be on solutions using advanced technology, such as ML, AI, and big data in order to increase the prediction of energy and its reliability [42]. Through proper utilization of ML and AI techniques these models are utilized in renewable energy sources into the grid.

10.8.2.2 Innovative Energy Storage Solutions

The advancement in smart city technology is key for the efficiency and effectiveness in the utilization of the energy storage technologies. The advanced research to smart energy system is vital when it comes to developing new solutions of energy storage which are vital for the future life in cities. Research is being conducted on advanced materials such as graphene and silicon and anode to improve battery performance and reduce charging time. These materials offer high energy density and faster charging compared to conventional technology. It is highly suitable for applications in modern energy storage systems and electric mobility solutions. These research directions therefore highlight the need for low-cost, efficient, and sustainable energy storage technologies that could be necessary for efficient incorporation of renewable energy into smart cities [59].

10.8.2.3 Resilient and Secure Grid Architectures

Developing a reliable grid as architecture is essential to manage and overcome disruptions such as cyber threats. In advanced smart energy management in smart cities, resilient and secure grid represents a promising area of research. As cyber threats continue to grow, enhancing cyber security becomes critical to ensure protection against risks. Additionally, grid architecture must support emerging technologies like EVs and IoT devices. This includes using blockchain platforms for secure energy transactions along with intelligent systems for threat detection in EVs [60]. Through predictive analysis and real-time data processing, smart city planning can ensure dependable and secure energy systems.

10.8.2.4 Consumer Engagement and Behavioural Insights

The research on consumer engagement and behaviour is essential for advancing smart energy systems in urban areas [61]. Understanding energy usage patterns helps improve resource efficiency. Applying behavioural science in cities supports the development of customized energy services that address the needs of the population.

10.8.2.5 Interoperability Standards and Protocols

Interoperability standard and protocol are important research areas for the growth of smart energy systems in smart cities. As cities become increasingly digital, integration of smart systems and devices is essential. Research focuses on interoperable smart grids, renewable energy, and IoT-based appliances using common models and standards. Advancing adaptive protocols with AI and ML is essential to ensure sustainable energy systems in growing smart cities [62].

10.8.2.6 Socio-Economic and Environmental Impact Assessments

Socio-Economic and Environmental Impact Assessment (SEEIA) is one of the most important directions of research in the creation of progressive smart energy systems in smart cities. The use of smart technologies to improve energy efficiency and urban buildings has become increasingly important. These assessments consider the social impact of such technologies identifying positive outcomes such as job creation, cost savings, and improved quality of life and negative effects including displacement from housing and unequal access to technological advancements. Future research should aim to develop advanced methods for SEEIAs using big data analytics and IoT Technology to enable real-time monitoring and control of smart energy systems [63]. To extend the benefits of SEEIAs, smart cities should implement sustainable energy systems that promote the development of socially and economically integrated communities.

10.9 Summary

The rapid growth of global cities and rising energy consumption have intensified the need for efficient energy management giving rise to smart cities and smart energy systems. Key innovations include renewable energy integration, IoE, smart metering, energy storage systems, and smart grids. These advancements along the evolution towards Energy 4.0/5.0 and computational intelligence such as machine learning, have significantly enhanced energy production and distribution through automation and optimization. Digital twin technology provides a virtual representation of a city's energy system that offers further support for decision-making and operational efficiency in smart cities. However, there are several critical research challenges including cyber security for energy infrastructure, integrating renewable energy into existing grids, managing increased energy demand, and minimizing greenhouse gas emissions. Additionally, the development of improved energy storage, smarter grid systems, and advanced energy management techniques is essential for enhancing sustainability.

References

1. M. Shahidehpour, Z. Li, M. Ganji, Smart cities for a sustainable urbanization: Illuminating the need for establishing smart urban infrastructures. IEEE Electrif. Mag. **6**(2), 16–33 (2018). https://doi.org/10.1109/MELE.2018.2816840
2. P. Pandiyan, S. Saravanan, K. Usha, R. Kannadasan, M.H. Alsharif, M.K. Kim, Technological advancements toward smart energy management in smart cities. Energ. Rep. **10**, 648–677 (2023). https://doi.org/10.1016/J.EGYR.2023.07.021
3. H. Lund, J.Z. Thellufsen, P.A. Østergaard, P. Sorknæs, I.R. Skov, B.V. Mathiesen, EnergyPLAN—Advanced analysis of smart energy systems. Smart Energ. **1**, 100007 (2021). https://doi.org/10.1016/J.SEGY.2021.100007
4. P. Mishra, P. Thakur, G. Singh, Sustainable smart city to society 5.0: State-of-the-art and research challenges. SAIEE Afr. Res. J. **113**(4), 152–164 (2022). https://doi.org/10.23919/SAIEE.2022.9945865
5. P. Mishra, G. Singh, *Sustainable Smart Cities* (Springer, Cham, 2023). https://doi.org/10.1007/978-3-031-33354-5
6. P. Mishra, G. Singh, Internet of vehicles for sustainable smart cities, in *Sustainable Smart Cities*, (Springer, Cham, 2023), pp. 175–193. https://doi.org/10.1007/978-3-031-33354-5_8
7. S.J. Wang, P. Moriarty, Energy savings from smart cities: A critical analysis. Energy Procedia **158**, 3271–3276 (2019). https://doi.org/10.1016/j.egypro.2019.01.985
8. C. Huang et al., Smart meter pinging and reading through AMI two-way communication networks to monitor grid edge devices and DERs. IEEE Trans. Smart Grid **13**(5), 4144–4153 (2022). https://doi.org/10.1109/TSG.2021.3133952
9. M. Rezaeimozafar, R.F.D. Monaghan, E. Barrett, M. Duffy, A review of behind-the-meter energy storage systems in smart grids. Renew. Sust. Energ. Rev. **164**, 112573 (2022). https://doi.org/10.1016/J.RSER.2022.112573
10. C. Wang, J. Gu, O. Sanjuán Martínez, R. González Crespo, Economic and environmental impacts of energy efficiency over smart cities and regulatory measures using a smart technological solution. Sustain. Energ. Technol. Assess **47**, 101422 (2021). https://doi.org/10.1016/J.SETA.2021.101422
11. A. Nadeem, A survey on peer-to-peer energy trading for local communities: Challenges, applications, and enabling technologies. Front Comput Sci **4** (2022). https://doi.org/10.3389/fcomp.2022.1008504
12. J. Alonso-Mora, S. Samaranayake, A. Wallar, E. Frazzoli, D. Rus, On-demand high-capacity ride-sharing via dynamic trip-vehicle assignment. Proc. Natl. Acad. Sci. **114**(3), 462–467 (2017). https://doi.org/10.1073/pnas.1611675114
13. S. Kumar, P. Tiwari, M. Zymbler, Internet of things is a revolutionary approach for future technology enhancement: A review. J. Big Data **6**(1), 111 (2019). https://doi.org/10.1186/s40537-019-0268-2
14. E. Al Nuaimi, H. Al Neyadi, N. Mohamed, J. Al-Jaroodi, Applications of big data to smart cities. J. Internet Serv. Appl. **6**(1), 25 (2015). https://doi.org/10.1186/s13174-015-0041-5
15. S. Nižetić, P. Šolić, D. López-de-Ipiña González-de-Artaza, L. Patrono, Internet of things (IoT): Opportunities, issues and challenges towards a smart and sustainable future. J. Clean. Prod. **274**, 122877 (2020). https://doi.org/10.1016/j.jclepro.2020.122877
16. D. Minoli, K. Sohraby, B. Occhiogrosso, IoT considerations, requirements, and architectures for smart buildings—Energy optimization and next-generation building management systems. IEEE Internet Things J. **4**(1), 269–283 (2017). https://doi.org/10.1109/JIOT.2017.2647881
17. A. Rahman, V. Srikumar, A.D. Smith, Predicting electricity consumption for commercial and residential buildings using deep recurrent neural networks. Appl. Energy **212**, 372–385 (2018). https://doi.org/10.1016/j.apenergy.2017.12.051
18. S.E. Bibri, The IoT for smart sustainable cities of the future: An analytical framework for sensor-based big data applications for environmental sustainability. Sustain. Cities Soc. **38**, 230–253 (2018). https://doi.org/10.1016/j.scs.2017.12.034

19. K. Zhang, J. Ni, K. Yang, X. Liang, J. Ren, X.S. Shen, Security and privacy in smart city applications: Challenges and solutions. IEEE Commun. Mag. **55**(1), 122–129 (2017). https://doi.org/10.1109/MCOM.2017.1600267CM
20. A.P. Plageras, K.E. Psannis, C. Stergiou, H. Wang, B.B. Gupta, Efficient IoT-based sensor BIG data collection–processing and analysis in smart buildings. Futur. Gener. Comput. Syst. **82**, 349–357 (2018). https://doi.org/10.1016/j.future.2017.09.082
21. J. Li, D. Greenwood, M. Kassem, Blockchain in the built environment and construction industry: A systematic review, conceptual models and practical use cases. Autom. Constr. **102**, 288–307 (2019). https://doi.org/10.1016/j.autcon.2019.02.005
22. S. Borchardt, G.B. Vignola, D. Buscaglia, M. Maroni, *Mapping EU Policies with the 2030 Agenda and SDGs-Fostering Policy Coherence Through Text-Based SDG Mapping*. https://doi.org/10.2760/110687
23. P. Mishra, P. Thakur, G. Singh, Enabling technologies for IoT based smart city, in *2021 Sixth International Conference on Image Information Processing (ICIIP)*, (IEEE, 2021), pp. 99–104. https://doi.org/10.1109/ICIIP53038.2021.9702673
24. J. Hussain et al., Social welfare maximization with efficient energy management of community microgrid considering customer behavioral response using MDCLPIS. Energ. Build. **315**, 114277 (2024). https://doi.org/10.1016/J.ENBUILD.2024.114277
25. P. Mishra, G. Singh, Enabling technologies for sustainable smart city, in *Sustainable Smart Cities*, (Springer, Cham, 2023), pp. 59–73. https://doi.org/10.1007/978-3-031-33354-5_3
26. D. Szpilko, X. Fernando, E. Nica, K. Budna, A. Rzepka, G. Lăzăroiu, Energy in smart cities: Technological trends and prospects. Energies (Basel) **17**(24) (2024). https://doi.org/10.3390/EN17246439
27. M.G.M. Almihat, M.T.E. Kahn, K. Aboalez, A.M. Almaktoof, Energy and sustainable development in smart cities: An overview. Smart Cities **5**(4), 1389–1408 (2022). https://doi.org/10.3390/smartcities5040071
28. B. Gajdzik et al., Encouraging residents to save energy by using smart transportation: Incorporating the propensity to save energy into the UTAUT model. Energies (Basel) **17**(21) (2024). https://doi.org/10.3390/EN17215341
29. J. Bailey, The steam age—Evolution of steam engines and the 1st steam locomotive, in *Inventive Geniuses Who Changed the World*, (Springer, Cham, 2022), pp. 23–36. https://doi.org/10.1007/978-3-030-81381-9_3
30. P. Mishra, G. Singh, Energy management systems in sustainable smart cities based on the internet of energy: A technical review. Energies (Basel) **16**(19), 6903 (2023). https://doi.org/10.3390/en16196903
31. Y. Zhou, J. Liu, Advances in emerging digital technologies for energy efficiency and energy integration in smart cities. Energ. Build. **315**, 114289 (2024). https://doi.org/10.1016/J.ENBUILD.2024.114289
32. M.Q. Raza, A. Khosravi, A review on artificial intelligence based load demand forecasting techniques for smart grid and buildings. Renew. Sust. Energ. Rev. **50**, 1352–1372 (2015). https://doi.org/10.1016/j.rser.2015.04.065
33. J. Mitali, S. Dhinakaran, A.A. Mohamad, Energy storage systems: A review. Energ. Stor. Sav. **1**(3), 166–216 (2022). https://doi.org/10.1016/J.ENSS.2022.07.002
34. K.M. Al-Obaidi, M. Hossain, N.A.M. Alduais, H.S. Al-Duais, H. Omrany, A. Ghaffarianhoseini, A review of using IoT for energy efficient buildings and cities: A built environment perspective. Energies (Basel) **15**(16) (2022). https://doi.org/10.3390/EN15165991
35. M. Hable, C. Schwaegerl, L. Tao, A. Ettinger, R. Köberle, E.P. Meyer, Requirements on electrical power infrastructure by electric vehicles, in *2010 Emobility - Electrical Power Train, EEPT 2010*, vol. 2010. https://doi.org/10.1109/EMOBILITY.2010.5668076
36. S. Karnouskos, Demand side management via prosumer interactions in a smart city energy marketplace, in *IEEE PES Innovative Smart Grid Technologies Conference Europe*, (2011). https://doi.org/10.1109/ISGTEUROPE.2011.6162818

37. K. Zhou, C. Fu, S. Yang, Big data driven smart energy management: From big data to big insights. Renew. Sust. Energ. Rev. **56**, 215–225 (2016). https://doi.org/10.1016/J.RSER.2015.11.050
38. H.T. Nguyen, A.H. Ba-Alawi, N. Somu, R. Chang-Silva, C.K. Yoo, Smart multi-stage energy-emission nexus framework for sustainable supply-demand management and CO2 mitigation. Renew. Sust. Energ. Rev. **214**, 115551 (2025). https://doi.org/10.1016/J.RSER.2025.115551
39. J. Rice, N. Martin, B. Rice, *Renewables and Energy Storage Enabled Smart Grid Modernisation: A Directed Graph Analysis* (2024). https://doi.org/10.2139/ssrn.4975971
40. G. Dileep, A survey on smart grid technologies and applications. Renew. Energ. **146**, 2589–2625 (2020). https://doi.org/10.1016/J.RENENE.2019.08.092
41. P. Mishra, G. Singh, Internet of things for sustainable smart city, in *Sustainable Smart Cities*, (Springer, Cham, 2023), pp. 75–96. https://doi.org/10.1007/978-3-031-33354-5_4
42. T. Ahmad, R. Madonski, D. Zhang, C. Huang, A. Mujeeb, Data-driven probabilistic machine learning in sustainable smart energy/smart energy systems: Key developments, challenges, and future research opportunities in the context of smart grid paradigm. Renew. Sust. Energ. Rev. **160**, 112128 (2022). https://doi.org/10.1016/J.RSER.2022.112128
43. P. Kiran, K.R.M. Vijaya Chandrakala, S. Balamurugan, T.N.P. Nambiar, M. Rahmani-Andebili, A new agent-based machine learning strategic electricity market modelling approach towards efficient smart grid operation. Power Syst., 1–29 (2022). https://doi.org/10.1007/978-3-030-94522-0_1
44. Y. Wang, J. Fu, Computational intelligence in smart cities and smart energy systems, in *Cutting Edge Applications of Computational Intelligence Tools and Techniques. Studies in Computational Intelligence*, vol. 1118, (Springer, Cham, 2023), pp. 305–325. https://doi.org/10.1007/978-3-031-44127-1_14
45. M.A. Abdullah, T. Al-Hadhrami, C.W. Tan, A.H. Yatim, Towards green energy for smart cities: Particle swarm optimization based MPPT approach. IEEE Access **6**, 58427–58438 (2018). https://doi.org/10.1109/ACCESS.2018.2874525
46. T.S. Choi, K.R. Ko, S.C. Park, Y.S. Jang, Y.T. Yoon, S.K. Im, Analysis of energy savings using smart metering system and IHD (In-Home Display), in *Transmission and Distribution Conference and Exposition: Asia and Pacific, T and D Asia 2009*, vol. 2009. https://doi.org/10.1109/TD-ASIA.2009.5356956
47. D.A. Elalfy, E. Gouda, M.F. Kotb, V. Bureš, B.E. Sedhom, Comprehensive review of energy storage systems technologies, objectives, challenges, and future trends. Energ. Strat. Rev. **54**, 101482 (2024). https://doi.org/10.1016/J.ESR.2024.101482
48. X. Zhang, G. Manogaran, B.A. Muthu, IoT enabled integrated system for green energy into smart cities. Sustain. Energ. Technol. Assess. **46**, 101208 (2021). https://doi.org/10.1016/J.SETA.2021.101208
49. M. Farmanbar, K. Parham, O. Arild, C. Rong, A widespread review of smart grids towards smart cities. Energies (Basel) **12**(23) (2019). https://doi.org/10.3390/EN12234484
50. Y. Wang, X. Kang, Z. Chen, A survey of Digital Twin techniques in smart manufacturing and management of energy applications. Green Energ. Intell. Transport. **1**(2) (2022). https://doi.org/10.1016/j.geits.2022.100014
51. Y. Zhou, Low-carbon urban–rural modern energy systems with energy resilience under climate change and extreme events in China—A state-of-the-art review. Energ. Buildings **321**, 114661 (2024). https://doi.org/10.1016/J.ENBUILD.2024.114661
52. A. Aghmadi, O.A. Mohammed, Energy storage systems: Technologies and high-power applications. Batteries **10**(4), 141 (2024). https://doi.org/10.3390/batteries10040141
53. S. Choudhury, Flywheel energy storage systems: A critical review on technologies, applications, and future prospects. Int. Trans. Electr. Energ. Syst. **31**(9) (2021). https://doi.org/10.1002/2050-7038.13024
54. H. Golpîra, S. Bahramara, Internet-of-things-based optimal smart city energy management considering shiftable loads and energy storage. J. Clean. Prod. **264**, 121620 (2020). https://doi.org/10.1016/j.jclepro.2020.121620

55. J. Yan, M. Menghwar, E. Asghar, M. Kumar Panjwani, Y. Liu, Real-time energy management for a smart-community microgrid with battery swapping and renewables. Appl. Energ. **238**, 180–194 (2019). https://doi.org/10.1016/j.apenergy.2018.12.078
56. C.F. Calvillo, A. Sánchez-Miralles, J. Villar, Energy management and planning in smart cities. Renew. Sust. Energ. Rev. **55**, 273–287 (2016). https://doi.org/10.1016/j.rser.2015.10.133
57. Q. Yang, H. Wang, Privacy-preserving transactive energy management for IoT-aided smart homes via Blockchain. IEEE Internet Things J. **8**(14), 11463–11475 (2021). https://doi.org/10.1109/JIOT.2021.3051323
58. P. Mishra, G. Singh, 6G-IoT framework for sustainable smart city: Vision and challenges, in *Sustainable Smart Cities*, (Springer, Cham, 2023), pp. 97–117. https://doi.org/10.1007/978-3-031-33354-5_5
59. D. Kong, M. Miyatake, Energy management of superconducting magnetic energy storage applied to urban rail transit for regenerative energy recovery, in *23rd International Conference on Electrical Machines and Systems, ICEMS 2020*, (2020), pp. 2073–2077. https://doi.org/10.23919/ICEMS50442.2020.9290891
60. N. Lasla, M. Al-Ammari, M. Abdallah, M. Younis, Blockchain based trading platform for electric vehicle charging in smart cities. IEEE Open J. Intell. Transport. Syst. **1**(1), 80–92 (2020). https://doi.org/10.1109/OJITS.2020.3004870
61. P. Mishra, G. Singh, Introduction: Importance of sustainable smart city, in *Sustainable Smart Cities*, (Springer, Cham, 2023), pp. 1–37. https://doi.org/10.1007/978-3-031-33354-5_1
62. Y. Zhou, S. Zheng, Machine-learning based hybrid demand-side controller for high-rise office buildings with high energy flexibilities. Appl. Energ. **262**, 114416 (2020). https://doi.org/10.1016/J.APENERGY.2019.114416
63. M. Babar, A.S. Khattak, M.A. Jan, M.U. Tariq, Energy aware smart city management system using data analytics and internet of things. Sustain. Energ. Technol. Assess. **44**, 100992 (2021). https://doi.org/10.1016/j.seta.2021.100992

Chapter 11
Smart Governance in Sustainable Smart Cities

Contents

P. Mishra, G. Singh, *Sustainable Smart Cities 2.0*,
https://doi.org/10.1007/978-3-032-01102-2_11

11.1 Introduction

Smart governance helps sustainable smart cities grow by uniting participants and making decisions more effectively through open processes [1]. Smart governance uses advanced technology platforms to improve how cities are planned while making resources and citizen services more effective [2]. City leaders utilize smart governance methods to overcome major urban issues and build communities that work for citizens. Under this strategy, citizens take an active role in making decisions about public services. Through smart governance, cities become strong and adaptable meeting citizens' needs by supporting economic development, protecting nature, and improving residents' quality of life [3]. Figure 11.1 represents a framework of smart governance in sustainable smart cities. It shows the main governance principles including transparency, citizen engagement, digital efficiency, accountability, and interoperability. These principles guide the development of key components such as artificial intelligence (AI) decision systems, digital identity services, public dashboard, and online consultation platforms. The smart government connects governance principles with practical applications. The framework also includes enabling technologies like cloud computing, blockchain, mobile apps, AI, and standardized systems [4]. These technologies support the delivery of services and improve coordination between agencies.

11.1.1 Traditional Governance Models Versus Smart Government

Conventional management systems operate through hierarchical infrastructure, which is slow to respond and lack the flexibility to effectively address the complex challenges of modern cities. Figure 11.2 represents a comparison between conventional and smart governance approaches. Traditional governance relies on paper-based processes, hierarchical decisions and limited citizen participation. In contrast, smart governance uses digital tools like blockchain, AI, and data-driven decision-making process [5]. It promotes faster policy implementation,

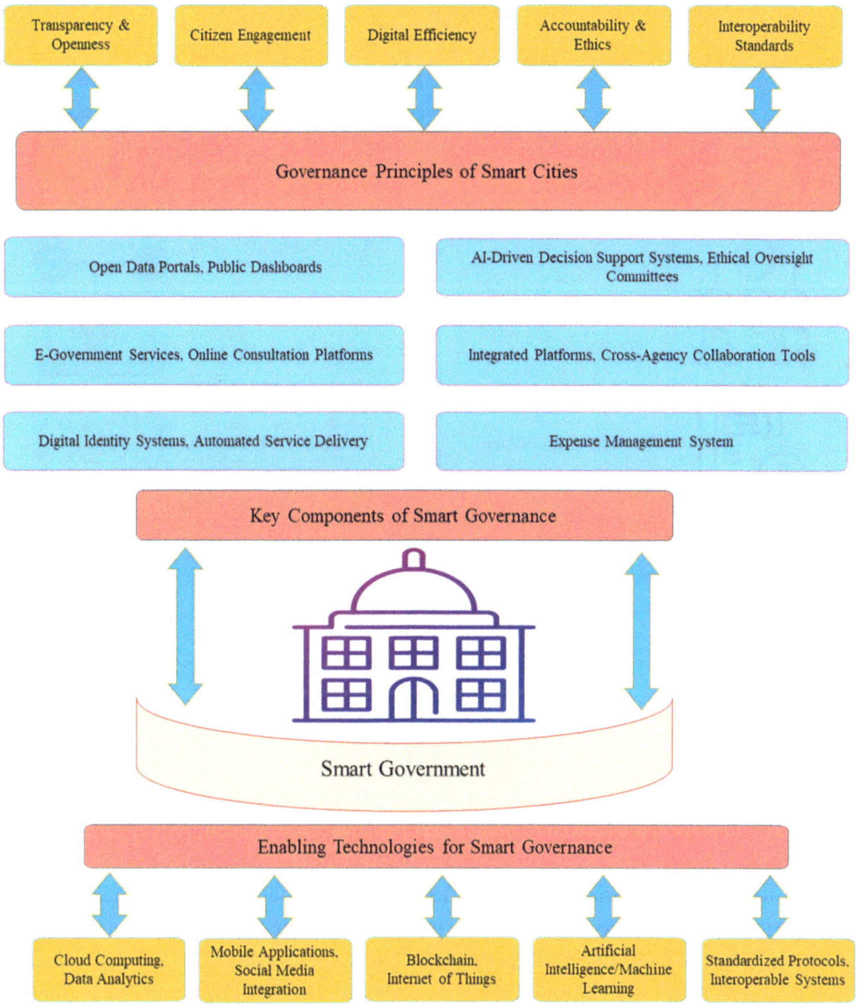

Fig. 11.1 Framework of smart governance in sustainable smart cities

transparency and active citizen engagement through e-government platforms. Traditional governmental systems follow a directive management style where agencies set up solutions and provide services independently without seeking substantial participation from residents [6]. Traditional administrative systems helped smaller communities but cannot handle fast-evolving urban problems. Smart governance drives smart government by using technology-based decentralized participatory processes which strengthen public decision-making through enhanced efficiency transparency and responsiveness [7]. Through big data analytics, AI and Internet of Things (IoT) technologies smart governance collects and analyses

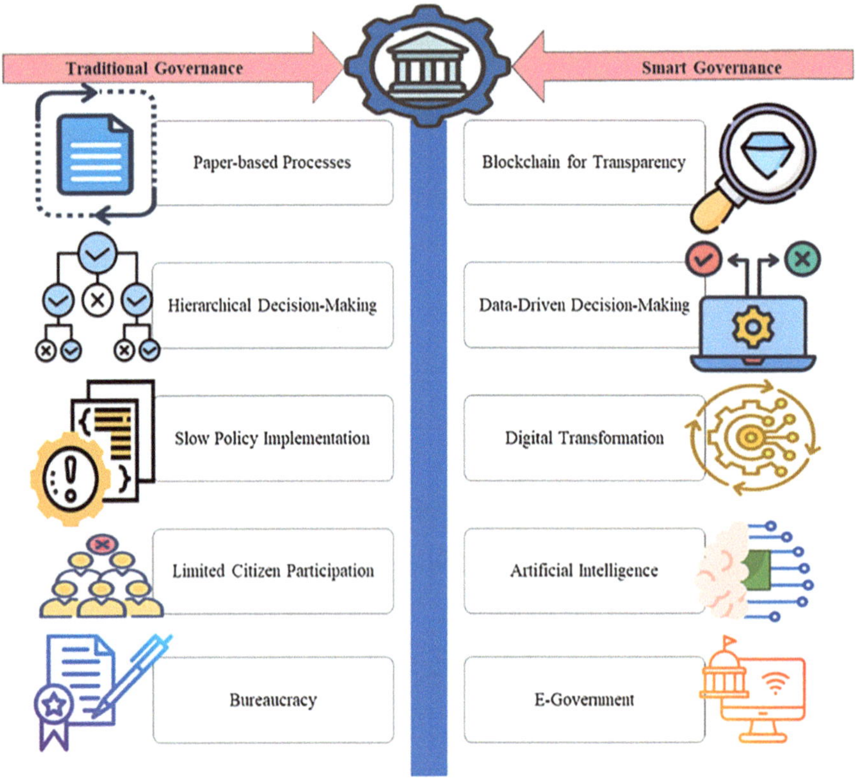

Fig. 11.2 Comparison between convention and smart governance system

data to make better decisions in real-time. This approach empowers government authorities to anticipate and address urban challenges in advance, rather than reaching after issues have emerged. Smart governance supports the implementation of intelligent traffic management systems, predictive waste management solutions and energy optimization mechanisms, contributing to the creation of more sustainable cities [8]. Smart governance makes citizen involvement a key part of its operations. Instead of keeping citizens away from the policy decisions like traditional governance does smart governance includes them through electronic government platforms and digital applications. Public engagement through digital platforms builds people's trust and gives citizens a sense of security and responsibility. Smart governance changes the way sustainable smart cities operate for the better. The modern governance model improves daily operations while building strong city foundations through sustainable development and equal access to the citizens. Traditional models prevent cities from keeping with changing world but smart governance offers urban areas to succeed and build communities that serve each and every citizen.

11.1.2 Evolution and Importance of Governance in Smart Cities

The evolution of urban leadership models highlights the need for governance systems that integrate modern technologies with flexible and inclusive decision-making processes [9]. Traditional governance infrastructure, focused on centralized control and directive operations have become inadequate for addressing the growing complexity of urban challenges. Rapid urbanization, technological advancement and environmental concerns have made conventional urban management models less effective. Smart city governance is built on cooperation between public agencies, private sector and local communities [10]. By collecting real-time data through IoT devices and analysing it using AI and big data tools, cities can better understand needs and implement on-time solutions. This includes improving traffic, increasing energy efficiency, and enhancing disaster management. Smart governance promotes transparency and public involvement in decision-making, building trust and supporting long-term development goals. In addition, it contributes to global sustainability through the adoption of renewable energy, effective waste management, and sustainable infrastructure [11]. It enables cities to remain responsive to new challenges, including environmental shifts and urban congestion. By using live data and open communication, smart governance helps create cities that meet current needs and future demands.

11.1.3 Information and Communication Technologies in Smart Governance

The integration of information and communication technologies (ICT) in urban governance creates smart governance systems that improve connectivity, transparency, and inclusiveness. It enables efficient government operations by connecting digital tools with data networks allowing decision-making and effective management of urban challenges. Smart governance uses real-time data collection through the IoT, cloud computing and big data analytics to optimize resource use and provide services that meet citizens need [12]. Digital monitoring technologies such as environmental sensors offer important information about urban activities helping authorities to develop appropriate policies. Furthermore, ICT increases citizen engagement through digital platforms including web portals and social media application, enabling residents to access services and actively participates in public decision-making [13]. This communication promotes transparency, accountability, and public trust which are essential for creating including urban environments. Additionally, ICT supports cooperation among government agencies, organizations, and educational institutions facilitating coordinated efforts towards sustainable development goals. Intelligent urban planning based on ICT improves the disaster preparation and municipal service delivery. Ultimately, smart governance combines traditional administrative methods with technological advancement to develop adaptable, sustainable, and equitable urban communities.

11.2 Current Research Trends of Smart Governance in Smart Cities

This study investigates the trends, contributions, and impact of research in smart cities with smart governance from 2015 to 2025 through a bibliometric analysis of Scopus data. The Bibliometrix R package along with origin software were utilized to analyse and depict the data. This approach ensures that researchers are presented with the most current research developments in the field enabling them to stay informed on emerging trends and advancements. In March 2025, a bibliometric analysis was conducted utilizing the Scopus core collection database to thoroughly investigate the concept and understanding of smart cities. Ultimately, 3548 articles were exported and saved in BibTex format before being included and examined in the bibliometric study. The primary information, syntax, and specific criteria used in screening for this study are shown in Table 11.1.

Table 11.1 Summary of data characteristics and main information of the bibliometric study

Description	Specific Standard Requirements/Results
Research database	Scopus
Search query	("smart") AND ("cities") AND ("smart") AND ("governance")
Time span	2015–2025
Document type	Articles, conference paper, book chapter, etc.
Language	English
Data export format	BibTex
Documents	3548
Annual growth rate %	2.94
Document average age	4.12
Average citations per doc	18.21
Keywords plus (ID)	8710
Author's keywords (DE)	7624
Authors	8005
Authors of single-authored docs	622
Single-authored docs	819
Co-authors per doc	2.98
International co-authorships %	25.03

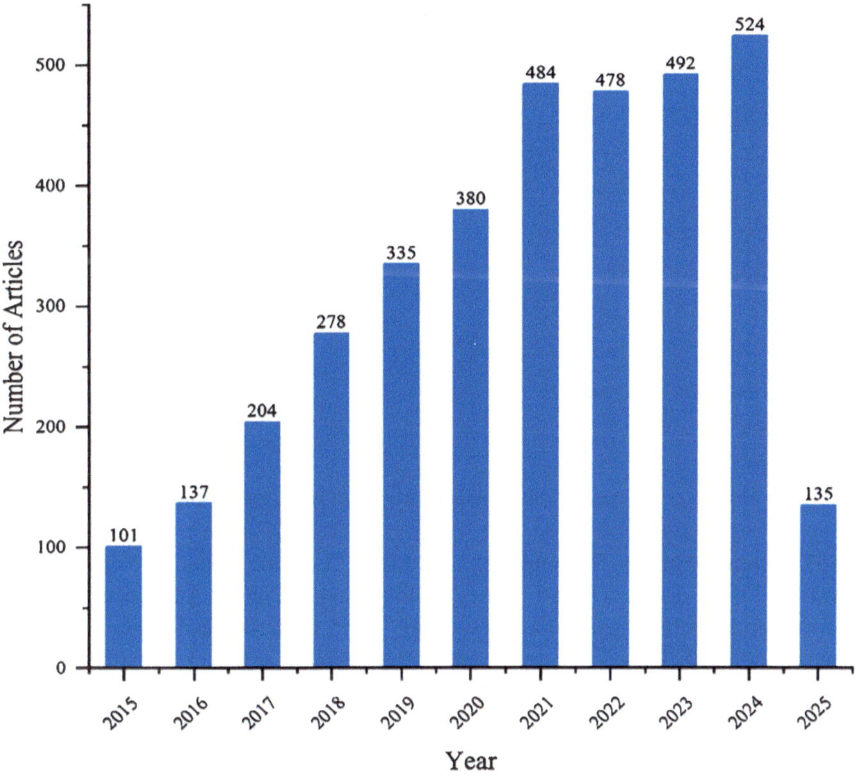

Fig. 11.3 Yearly trend of publication number of smart governance in smart cities

11.2.1 Annual Scientific Production and Top Ten Most Published Journals/Cited Paper

Figure 11.3 shows the annual publication trends from 2015 to 2025, highlighting the growth in research output. The early years shows a steady increase in publications, rising from 2015 (101 doc) to 2021 (484 doc), indicating a growing interest in the field. There is a slight decrease in the number of documents in 2022 (478), followed by an increase in 2023 (492) and 2024 (524). In 2025 (135), the number continues to rise and may continue increasing until the end of the year. Figure 11.4 presents the highest number of articles in sustainable development goals in the smart city research domain published in "Sustainability (Switzerland)" (130 docs), "ACM International Conference Proceeding Series" (104 docs), "Lecture Notes in Computer Science (Including Subseries Lecture Notes in Artificial Intelligence and Lecture Notes in Bioinformatics)" (76 docs), and "Cities" (72 docs), which indicates a higher output of articles compared to the other journals. In addition, some other journals, such as the "Sustainable Cities and Society" (57 docs), "Advances in

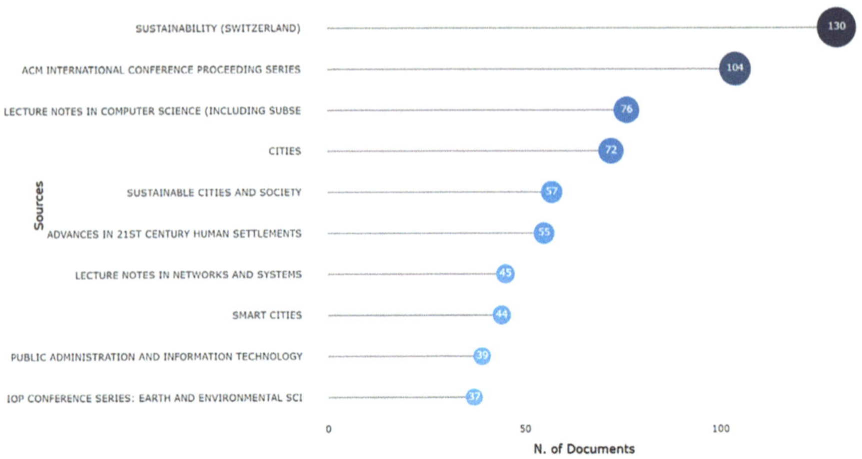

Fig. 11.4 Most relevant sources with number of articles of smart governance in smart cities

21ST Century Human Settlements" (55 docs), "Lecture Notes in Networks and Systems" (45 docs), "Smart Cities" (44 docs), "Public Administration and Information Technology" (39 docs), and "IOP Conference Series: Earth and Environmental Science" (37 docs), are in this domain. Table 11.2 displays the top ten most cited papers from the study which presents both the ranking and authors with their published works. This section presents information about authors, article titles and their corresponding publications in journals with documentation for publication years and citation frequency. The ranking focuses on engineering field smart cities articles to demonstrate the total citation (TC) counts and annual citation rates and normalized total citation (NTC) data which collectively signify their impact in the field.

11.2.2 Sankey Diagram (Three-Field Plot and Thematic Map)

Figure 11.5 Sankey diagram illustrates the three-field plot thematic progression to visualize the flow and the distribution in scientific research. In this case, keywords plus, abstract, and title are added, acting as indicators to highlight the most influential research papers in the subject. These keywords indicate vital concepts and themes that stand out in academic discussions of the smart city domain. The diagram's keywords are sourced from earlier published studies and were then thematically organized by a specific researcher.

Figure 11.6 categorizes various smart city themes based on their energy development and relevant degrees. In the top right quadrant (motor themes), which represents themes that are highly developed and highly relevant, terms like "governance", "technology", "infrastructure", "urban governance", "urban planning", and

Table 11.2 Most global cited documents of smart governance in smart cities

Rank	Authors name	Paper Title	Journal's Name	TC	TC/ Year	NTC
1	Silva et al. (2018) [14]	Towards sustainable smart cities: A review of trends, architectures, components, and open challenges in smart cities	Sustainable Cities and Society	1291	161.38	34.76
2	Meijer and Bolívar (2016) [15]	Governing the smart city: a review of the literature on smart urban governance	International Review of Administrative Sciences	1046	104.60	22.09
3	Sharifi and Garmsir (2020) [16, 17]	The COVID-19 pandemic: Impacts on cities and major lessons for urban planning, design, and management	Science of the Total Environment	772	128.67	28.58
4	Allam and Dhunny (2019) [17]	On big data, artificial intelligence and smart cities	Cities	770	110.00	30.97
5	Hollands (2015) [18]	Critical interventions into the corporate smart city	Cambridge Journal of Regions, Economy and Society	692	62.91	19.20
6	Shelton et al. (2015) [19, 20]	The 'actually existing smart city'	Cambridge Journal of Regions, Economy and Society	628	57.09	17.42
7	Ahad et al. (2020)	Enabling technologies and sustainable smart cities	Sustainable Cities and Society	434	72.33	16.06
8	Yigitcanlar et al. (2018) [21]	Understanding 'smart cities': Intertwining development drivers with desired outcomes in a multidimensional framework	Cities	421	52.63	11.34
9	Arasteh et al. (2016) [22]	Iot-based smart cities: A survey	IEEE 16th International Conference on Environment and Electrical Engineering	418	41.80	8.83
10	Ejaz et al. (2017) [23]	Efficient energy management for the internet of things in smart cities	IEEE Communications Magazine	417	46.33	15.39

"sustainable development". In quadrant left (niche themes) there are represents "smart mobility", "smart economy", and "smart environment". This indicates a strong focus on these areas, suggesting they are central to governance-based smart city research and practice. This quadrant reflects the areas where significant advances and investments are made, highlighting their critical role in smart city. In contrast, the bottom left quadrant, which represents themes that are either emerging

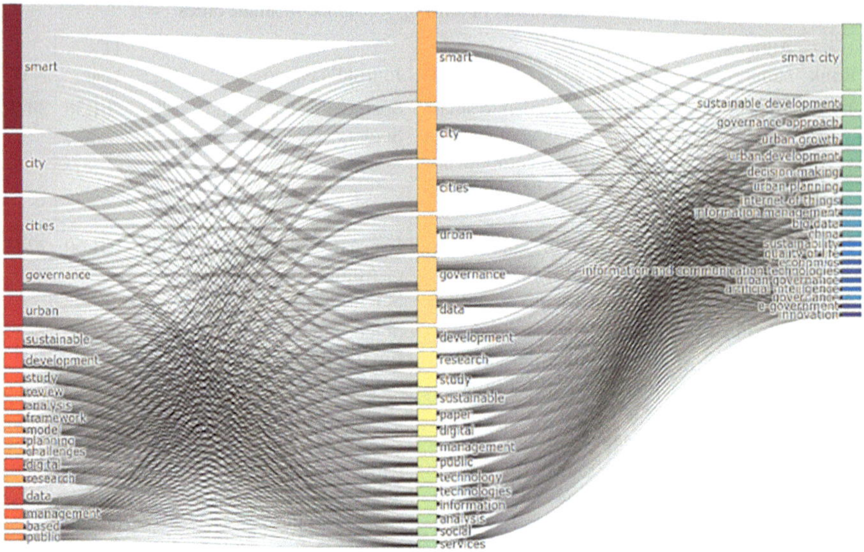

Fig. 11.5 Three field plot (thematic development) using sankey diagram

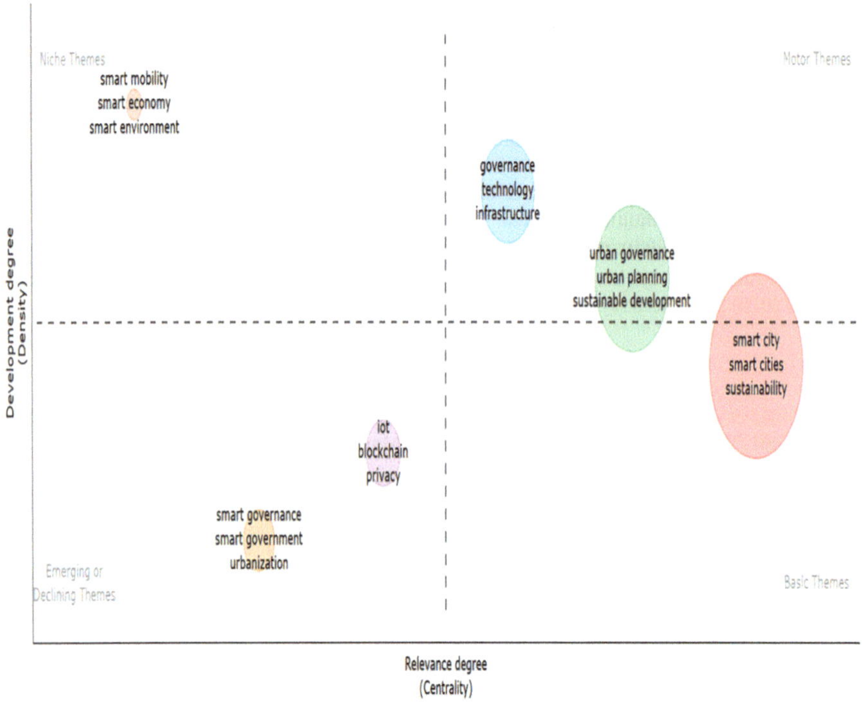

Fig. 11.6 Thematic map of smart governance in smart cities

or declining and have lower relevance, includes terms like “iot”, “blockchain”, “privacy”, “smart governance”, “smart government”, and “urbanization”. Additionally, the terms like “smart city”, and “sustainability” are the foundational information for further analysis and serve as a base for more advanced analyses. These themes might be considered foundational but must be emphasized more in energy-based smart cities. This visual categorization provides valuable insights for educators, curriculum developers, and researchers, helping them understand which areas are gaining traction and which might need more attention or innovation.

11.2.3 Co-occurrence Keyword Trends and Network Visualization

Figure 11.7 depicts the most prominent themes and keywords related to research in conceptual change. The most significant and most central terms, such as “smart city,” “sustainable development”, “governance approach”, “decision-making”, and “urban development” indicate the core focus of the research on how researchers grasp and alter the conceptual knowledge, particularly within smart city contexts.

An illustration in Fig. 11.8 displays word occurrence frequencies from the sustainable development goals in smart city research network. Research focuses mainly on “smart city” as this term appears 1472 times throughout the literature. Research on sustainable development goals in smart cities indicates the importance of “sustainable development” based on its 327 word occurrences. The “governance approach” (316) and “decision-making” (241) combinations show the growth and sustainability in cities. Additionally, “urban development” (231), “urban planning” (227), “urban growth” (220), “internet of things” (195), and “information management” (121) combinations occupy crucial importance in emerging technology fields. The data shows that smart city research revolves primarily around several major subjects.

Fig. 11.7 Co-occurrence keyword trends in smart city domain and smart governance

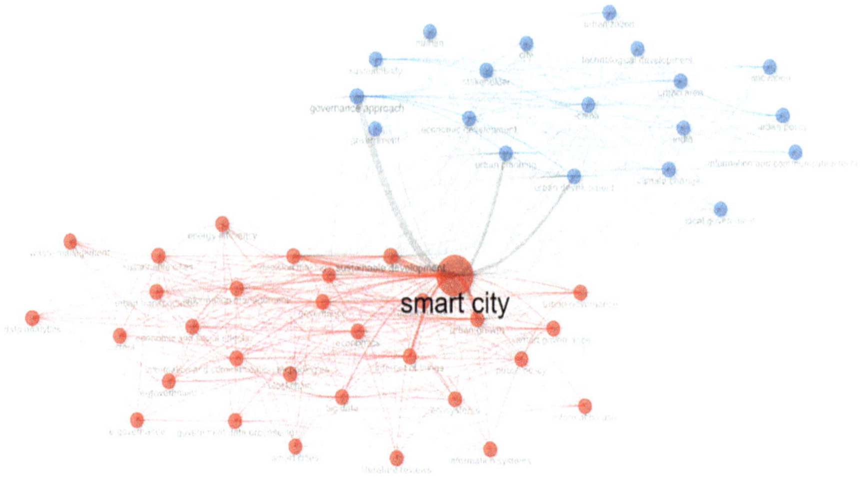

Fig. 11.8 Co-occurrence network of words in smart governance in smart cities research domain

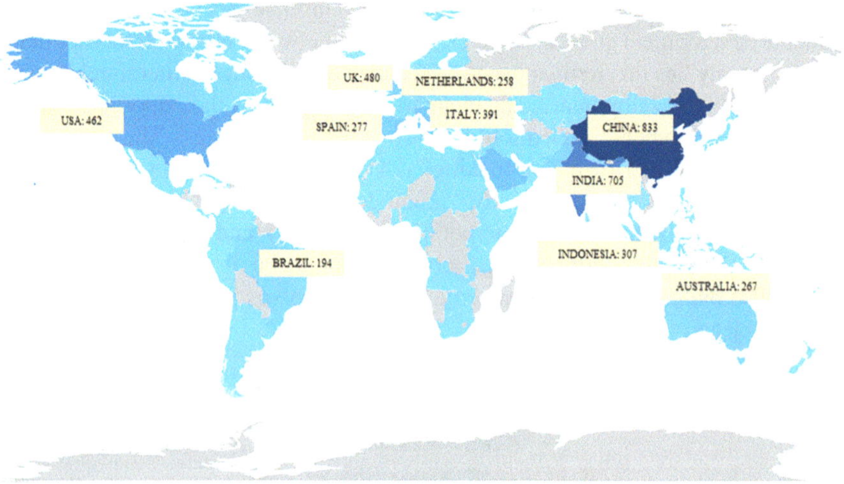

Fig. 11.9 Top ten countries' scientific production of various countries based on frequency

11.2.4 Countries' Scientific Production

The scientific publication frequencies of various countries in the world are represented in Fig. 11.9. Scientific contributions from China stand at 833 which demonstrates their powerful role in research development. The growing scientific power of India displays itself through its 705 scientific publications which rank second after

China. The research hub of the UK produces 480 scientific papers among its total contributions. Numerous publications emerge from USA and Italy together with Indonesia given their research output comprises 462, 391 and 307 papers respectively. Academic and industrial research sectors of the Spain and Australia produce approximately 277 and 267 scholarly works between them. Netherlands and Brazil have smaller publication numbers but the contributions amount to totals of 258 and 194 research items respectively.

11.2.5 Mapping Smart City Research to Sustainable Development Goals (SDGs)

SDGs, their targets (bubble series), and the percentage (bar graph) of corresponding keywords detected in the text (calculated as the ratio of keywords in one Goal to the total number of keywords detected) are depicted in Fig. 11.10. For SDG mapping, the study utilized the SDG Mapper tool from the "Knowledge Base for the Sustainable Development Goals," provided by the European Commission [24]. This tool uses bubble charts to depict SDG interconnections and relevance among bibliometric records. The size of each bubble correlates with the proportion of associated keywords found, which is calculated as the number of goal-specific

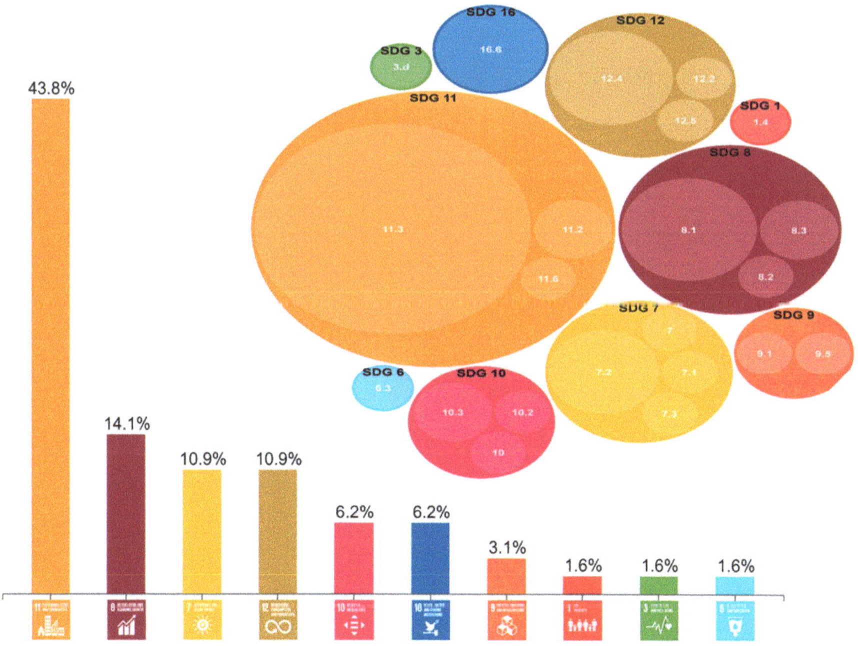

Fig. 11.10 Mapping smart governance in smart cities research to relevant SDGs and targets

keywords divided by the total keyword count. As such, SDG 11, marked by 28 keyword occurrences. This is followed by SDG 8, with 9 occurrences, and SDG 7 with 7 occurrences, underlining their relative prominence in the research database. Within SDG 11, there are several key subtargets that the research address: Subtarget 11.3 is shown as a distinct bubble within the larger SDG 11 circle, indicating focused research on smart city. Subtarget 11.2 focuses on providing the safe, accessible, and sustainable smart transportation. Subtarget 11.6 focuses on reducing the adverse per capita environmental impact of cities in smart cities solutions, is another bubble within the SDG 11 area. This highlights the efforts in research to improving the cities. Similarly, SDG 8 (Decent work and economic growth) is also closely linked to SDG 11 and suggests that research should explore how to achieve the economic growth, productive employment and decent work for all women and men. Figure 11.10 shows that the size of the SDG 7 bubble implies that advancements in smart city are being focussed on affordable and clean energy that prioritize sustainability. The detailed description of other relevant SDGs is depicted in Table 11.3.

Table 11.3 SDGs ranking, targets and relevance in governance based smart city research domain with SDGs

SDGs (By Rank)	Description	Targets	No. of Occurrence		Relevance%
SDG 11	Sustainable cities and communities	11.3	25	28	43.8
		11.2	2		
		11.6	1		
SDG 8	Decent work and economic growth	8.1	6	9	14.1
		8.3	2		
		8.2	1		
SDG 7	Affordable and clean energy	7.2	4	7	10.9
		7.3	1		
		7.1	1		
		7	1		
SDG 12	Responsible consumption and production	12.4	5	7	10.9
		12.5	1		
		12.2	1		
SDG 10	Reduced inequalities	10.3	2	4	6.2
		10.2	1		
		10	1		
SDG 16	Peace, justice, and strong institutions	16.6	4	4	6.2
SDG 9	Industry, innovation and infrastructure	9.5	1	2	3.1
		9.1	1		
SDG 1	No poverty	1.4	1	1	1.6
SDG 3	Good health and Well-being	3.d	1	1	1.6
SDG 6	Clean water and sanitation	6.3	1	1	1.6

11.3 Principles of Smart Government

Sustainable smart cities build smart government systems that promote effective administration with inclusive solutions and creative methods. People trust smart governments when open data access and transparent decision-making are ensured. Smart cities offer transparency to the citizens which helps to develop policy decisions and services at all levels of governance. Real-time data helps public administrators make better decisions while sustainability helps to protect the environment and resources appropriately. When systems work together stakeholders deliver better services to customers [25]. When cities are resilient and adaptive they faces transition issues from global warming and fast-growing neighbourhoods. Governments that follow these essential principles create better and greener environments that serve all citizens more effectively and improve the wellbeing. The major principles are illustrated in Fig. 11.11.

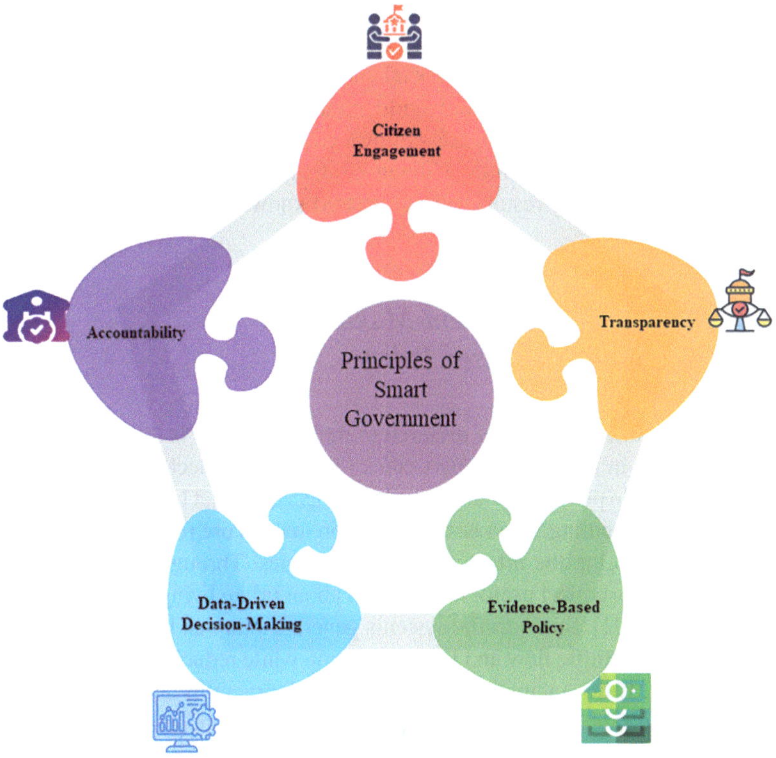

Fig. 11.11 Principles of smart government

11.3.1 Transparency, Accountability, and Citizen Engagement

Smart government values transparency, accountability, and citizen engagement along with responsible leadership and community participation to establish successful smart governance in smart sustainable urban areas [26]. These three elements develop public confidence while creating better service opportunities for all residents and helps government to work towards sustainable urban living [27]. Transparency to government information and policies defines transparent smart governance operations. Governments in smart cities employ digital platforms to make their urban operations budget details and initiatives available online for everyone in real time. Open data portals give people direct visibility into how their government handles public works and protects the environment [28]. The use of digital records and automatic procedures helps smart systems stop dishonest acts. Through digital performance measures and audits smart governance tracks how well public services and city projects work. Government trust grows when leaders accept both good and bad results through accountability. The blockchain technology helps governments maintain accountability through unchangeable records of their operational tasks and decision-making processes [29]. This system makes sure resources support the goals of sustainability and social fairness. Smart governance needs citizen participation to help decisions and build resident investment in the community. Digital budget participation programs allows citizens decide how to use city funds based on local needs and values. Citizen contributions build city inclusivity and improve problem-solving results through shared knowledge from the experts.

11.3.2 Data-Driven Decision-Making and Evidence-Based Policy

Using evidence-based policy and research to make decisions is the essential basis of smart governance in sustainable smart cities [30]. Through technology and analytics these principles help public officials make better decisions that benefit the environment and people while handling urban issues. Decision makers use real-time data collection and analysis to develop the urban plans and strategies. Through IoT sensors and AI systems smart cities collect useful information about urban living from traffic, energy, or waste disposal [31]. Smart traffic systems generate data that helps cities adjust signal patterns to ease traffic flow and lower pollution while reducing cost and protecting the environment. Cities maintains sustainable development by monitoring water and energy use in real-time to use resources more effectively. Data helps governments take preventive action instead of waiting to respond to problems that develop. Data-driven decisions help organizations achieve better results by eliminating unnecessary load and utilize resources right on time. Decision-makers use data findings to create public policies through evidence-based policy. Measurable results help governments plan the best strategies to reach their city's long-term goals while delivering focused and powerful outcomes. Data about air quality and public health helps officials design better rules for clean energy systems and eco-friendly transit options. When governments use evidence-based methods, it monitors policy results over time and then make necessary improvements. Government transparency and responsibility improve when these principles take effect. When smart governments base decisions on data it gathers public confidence and make policies easy to understand. People become active partners in government when they can access open urban data and help solve community problems. Cities governed by real-time data make decisions that serve residents by constantly adapting to their requirements. By using data these methods help governments solve challenges and promote a sustainable urban future. When smart governance combines these principles, it leads sustainable smart cities toward better performance and higher resilience while merging technology policy with citizen needs [32].

11.4 Smart Government Components

Analysing smart government components is based on the use of infrastructure, IT tools, information, and analysis to improve the delivery of services. IoT and connectivity fundamentals are advanced and respond to the needs of information exchange in real-time. Emerging technologies like AI and blockchain make decision-making and processes much more efficient services for the citizens. Both data and analysis contribute to supporting policy decisions while promoting accountability. Ethical principles in governance, particularly regarding data privacy and security policies are essential for building trust and ensuring responsible management. All these components create a strong environment that supports smart governance and develop sustainable solutions to the society demands [33]. It supports the development of a smart city focused on operational efficiency and environmental balance while serving the people's needs. The various essential components such as infrastructure, technology, data and analytics, governance and policies, and collaboration and partnerships are illustrated in Fig. 11.12.

11.4.1 Infrastructure

Government infrastructure planning needs to create solutions that withstand and adjust while following the smart practices [34]. The integration of sustainable energy systems with environmentally smart architecture helps cities meet their growth requirements without damaging the natural environment. The success of infrastructure projects depends on public–private partnerships (PPP) which fund development and maintain sustainable operation [35]. A smart city depends on its infrastructure base to provide smooth service delivery and bring technology into city systems. Smart infrastructure covers digital and physical tools that help deliver necessities including transport, energy, water, waste handling, and public protection. Three primary smart infrastructure components are integrating IoT, cloud computing, and sustainable urban architecture, which support utility services.

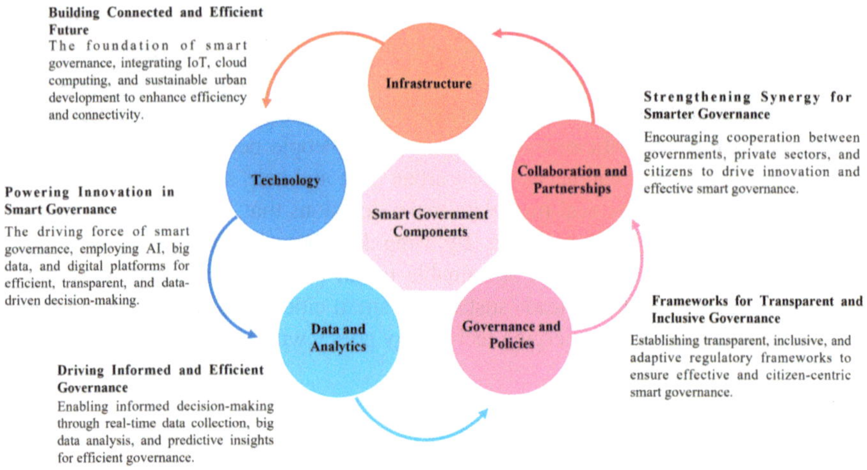

Fig. 11.12 Components of smart governance

11.4.2 Technology

Technology powers the smart initiatives of city governments in smart cities. Modern technologies including IoT, AI, blockchain, big data, and cloud systems help government departments offer better services and connect better with citizens [36]. IoT devices collect real-time data to help governments and make better decisions about city administration. Through data analysis AI systems discover useful patterns and project the next move so that governments can respond effectively. Smart systems monitor public safety while AI monitor's unusual activities captured on security cameras. Blockchain technology helps government operations run more transparently and securely especially for public procurement, personal identification systems and urban record management [37]. The use of cloud computing allows government to achieve effective data processing and scaling abilities for handling its growing urban data sets at affordable rates. Modern public service platforms including apps and digital toolkits help people to better connect with government services. These platforms allow residents to communicate with officials about issues while checking public information and joining decision-making which builds better community relations and public confidence in the government.

11.4.3 Data and Analytics

A smart government depends on data which needs analytic tools to produce important awareness in smart cities [38]. Real-time data management helps urban areas solve problems and make life better for smart city residents. A smart government uses big data analytics to study and forecast the activities in cities. Government agencies use data from environmental sensors, social media, and mobile devices to control air quality levels while forecasting traffic movements and understanding the public activities. Urban planners use geographic information system (GIS) systems to map areas and monitor location-based data. Advanced analytic systems allow government agencies to detect problems ahead of time for prevention. The use of predictive models helps to explore regions that may experience floods or infrastructure problems so that action can be taken beforehand. Data visualization tools support better decisions by showing complex information in simple visual formats that help decision makers and citizens to analyse important patterns. Data analytics offers significant benefits when robust data governance frameworks are established [39]. By releasing anonymized datasets for use by public and private entities, governments stimulate innovation while maintaining transparency and accountability in processes.

11.4.4 Governance and Policies

Governance and policies serve as basic building blocks for smart government initiatives to work efficiently. An organized system of management helps smart city projects reach their targets by making the best use of resources. Smart governance makes decisions by including all stakeholders in an active process [40]. E-governance platforms give citizens the ability to participate in government actions and rate policy decisions directly. This method increases open access to information and develops rules that match the community requirements. The smart city infrastructure depends on strong data protection rules plus cybersecurity and technology standards [41]. Government organizations need to create data rules that secure public information and encourage data sharing among all departments. Effective cybersecurity methods help defend important smart city systems from digital attacks and maintain safety. Better public services result when government departments use common platforms to exchange information and communicate directly.

11.4.5 Collaboration and Partnerships

Successful smart city projects depend on teamwork, collaboration, and partnership between different organizations [42]. Effective smart governments build partnerships between themselves, companies, schools, NGOs, and citizens. Smart city projects depend on public–private partnerships because these partnerships use private sector resources to deliver better solutions. Working with technology firms helps cities place IoT devices while renewable energy partners make eco-friendly power systems develop faster. Schools and universities help smart city progress through their research work plus technology development and by training professionals. NGOs help smart city projects include everyone and make sure the plans serve communities with fewer resources. Public involvement strengthens people's commitment to smart city projects. To empower citizens, governments need to organize budgeting discussions and encourage participation in technology innovation events and collaborative design sessions. International organizations like the Smart Cities Council and the UN's SDGs help countries share valuable knowledge through their platforms [43] [44].

11.5 Benefits of Smart Government in Smart Cities

Smart city government operations provide numerous benefits that enhance urban living conditions by advancing sustainable practices, economic growth, and improving community safety. The implementation of smart governance in these cities results in efficient public sector performance alongside an improved quality of life for residents. These environmentally sustainable approaches support innovative economic development initiatives while strengthening safety and security measures. The benefits of smart governance in smart cities are illustrated in Fig. 11.13.

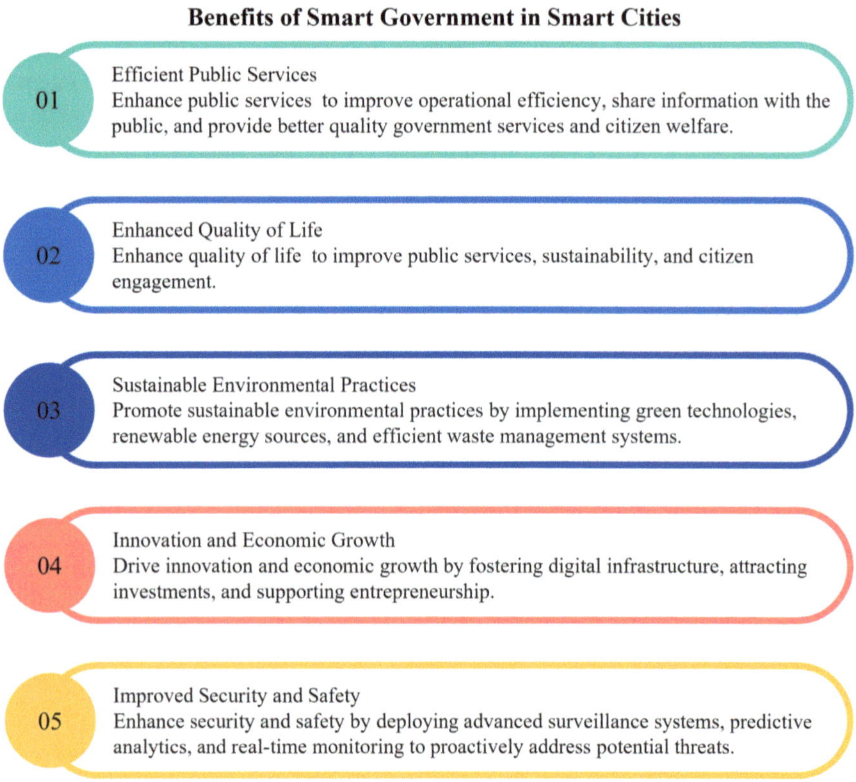

Fig. 11.13 Benefits of smart governance in sustainable smart cities

11.5.1 Efficient Public Services

Smart government strategies with regard to smart cities are changing the practice of delivering public services and improving the quality of life in cities [45]. This change is mainly driven by the use of IoT, AI, and big data such that cities become more flexible and meet the needs of its inhabitants. Through employing of these technologies smart governments enhance the efficiency of the administrative procedures and also the level of the services to be delivered. Big data and electronic sensors enable real-time monitoring of public utilities, enhancing operational efficiency. This approach also leads to cost reduction by minimizing the expenses associated with resource maintenance and replacement. Also, intelligent traffic control systems which incorporate technology involve the use of data in solving traffic issues hence decreasing traffic jams. In addition, smart governments promote the level of engagement and share the information with the public. Citizens conveniently access government services and feedback through website interfaces and mobile apps, which significantly enhances governance accountability and citizen's trust in the government [46]. Moreover, AI application in smart government improves the making of decisions. Since AI can processes large volume of information, it is able to assist policy-making by identifying patterns and trends and formulating policies most suited to address urban challenges.

11.5.2 Enhanced Quality of Life

Smart government has made its contributions to smart cities more effective in improving the quality of life for people by offering more innovative, open, and responsive communities [45]. The term smart government constitutes the application of advanced technology such as, IoT, big data, and AI with an aim on improving urban services and management. This transition enables the consideration of opportunities and risks of the city and makes possible anticipatory deployment of its resources all of which enhance public services and systems. By integrating smart traffic systems, smart cities efficiently minimize traffic density and enhance the quality of inhabitants' transport, thereby making it faster and more trustworthy [47]. It not only saves time for the residents but also maintains the quality of the environment. Another positive point about the implementation of the proposed approach is the positive effect on safety and security of citizens. Smart cities utilize advanced surveillance systems and big data to respond and prevent security threats, thus increasing security in people's day to day live. Besides, smart government systems establish ways through which the citizens can be actively involved in decision-making processes.

11.5.3 Sustainable Environmental Practices

The idea of smart government forms one of the most important aspects of turning a city into a smart city and especially in encouraging the right environmental practices [48]. Innovation and big data help smart governments to make governance better and also participate the participation in environmental sustainability. Optimization of resource use is one of the main advantages that are derived from implementing the system. In addition, smart governments create smart waste management by using IoT sensors in waste bins hence monitoring the waste level and the collection routes. This not only saves fuel to be used by waste collection vehicles but also offers the way to a higher efficiency in separating and sorting recyclables for recycling. Furthermore, government offers the citizens a platform to share views and participate in environmental affairs [49]. It not only improves the work of policies but also creates a permanent culture of sustainability among residents. Moreover, smart technologies integrated in smart cities by governments are successful not only in responding to the current environmental issue but also create a long-term planning platform for sustainable urban climate requiring urban environments to remain sustainable.

11.5.4 Innovation and Economic Growth

Smart government as part of the smart city's architecture lays a fundamental foundation for innovation economics [50]. This process is driven by the central concept of smart, or intelligent technology systems, data, and connectivity, which improve governmental and municipal functionality, services, and response, creating a smart city. It is out of this perspective that a smart government provides for the efficient management of big data and more readily usable information. It is for this reason that this data-driven approach goes a long way in enhancing cultures of transparency and accountability in policy implementation; as well as, enhances policy formulation, which effectively provides for the dynamic nature of population needs prevailing within cities. In addition, smart initiatives of smart government contribute to the creation and growth of new business opportunities for investment. Smart governance simplifies transactions between authorities and citizens thereby boosting economic activity and attracting foreign direct investment. Innovation is a significant benefit, as these governments collaborate with universities, industries and non-governmental organizations to implement advanced solutions addressing urban challenges [51]. Such partnerships contribute to the development of smart infrastructure, efficient urban planning and improved public service delivery enhancing the quality of life for residents. The interaction between smart governments and cities drives innovation and economic growth, keeping urban areas aligned with technological and economic progress. Smart governance supports the development of sustainable and urban environments.

11.5.5 Improved Security and Safety

Smart government plays a vital role in enhancing security and driving the transformation of smart cities into safe urban environments. The adoption of advanced technologies such as IoT, AI, and big data analytics enables the implementation of advanced security solutions [52]. These technologies facilitate real-time surveillance and rapid response which are important in addressing emerging threats effectively. One of the key benefits of smart governance is the ability to process vast volume of data from diverse sources, including CCTV Systems, social media platforms, and citywide sensors. This data-driven approach supports on-time decision-making and efficient resource allocation. Communication technologies further enable citizens to report incidents such as disasters or criminal activities, involving a collaborative environment between the public and authorities. Smart cities also incorporate integrated applications that ensures coordination with emergency services, including police, fire, and medical department. This integrated response framework enhances efficiency and ensures incidents are managed comprehensively. Moreover, the deployment of smart technology strengthens situational awareness and encourages community involvement in public safety initiatives [53]. Governance are leading efforts to make cities safer by embedding security. This commitment reflects a proactive model focused on resilience, responsiveness, and public trust.

11.6 Open Research Challenges and Opportunities

Digital smart governance involves the integration of digital applications within the context of governance in order to support the decisions making processes, delivery of services and empowering citizens. Assertively, in relation to the sustainable smart cities, it seeks to develop urban complexes that are effective and adaptive besides being effective and sustainable. The main aspects of smart governance include e-governance, open data, collaborative decision-making, and other components of government open data regulation. The open research challenges and opportunities are illustrated in Fig. 11.14.

11.6.1 Smart Governance Research Challenges

11.6.1.1 Data Privacy and Security

Data confidentiality and protection are the primary issues emerging, which are one of the central obstacles to smart management. One of the main issues that arises with the use of big data and related technologies is privacy and protection of data

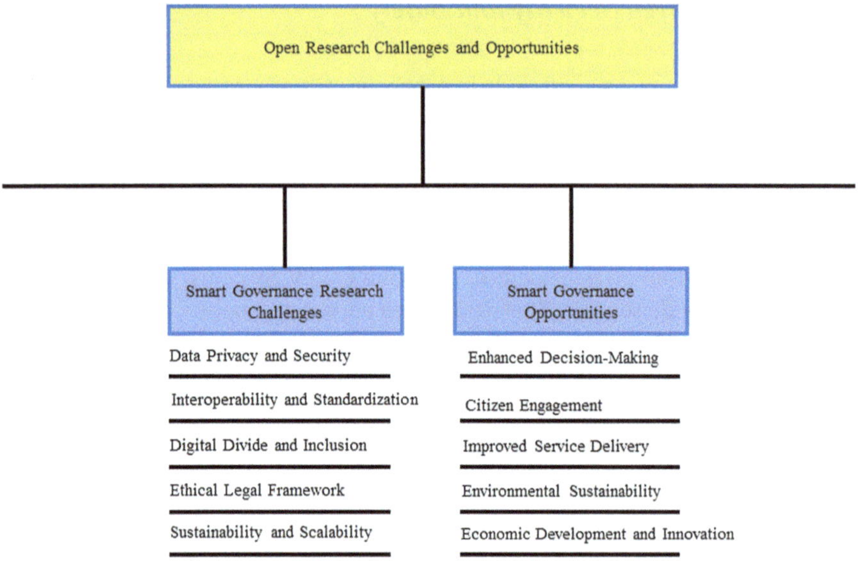

Fig. 11.14 Open research challenges and opportunities in smart governance

harvested from sources like sensors, mobile, and social platforms. In encryption and secure information sharing, it is essential to advance specialized security mechanism while also developing robust measures to prevent cybercrimes [54]. Both aspects must be strengthened to ensure the integrity and confidentiality of data in smart governance systems.

11.6.1.2 Interoperability and Standardization

One of the most important issues within and across organizations is the technical integration of various systems and technologies. Interoperability concerns arise due to the fact that smart cities tend to comprise a number of different platforms and end points which may be provided by a different range of different vendors [55]. It is therefore obvious that standardization of protocols and data formats required for the interaction between systems is critically important. Further studies should be conducted to examine general guidelines and reference models for data exchange and integration, defined as interoperability.

11.6.1.3 Digital Divide and Inclusion

Smart governance utilizes digital technology, an element that contributes to the expansion of the digital gap and increased social inequality [56]. One major problem is that all the countries' population should have the same opportunities to use digital technologies and related services irrespective of the class to which they belong. Future studies must study how to enhance the technological accessibility rates that enable beneficiaries across the digital divide.

11.6.1.4 Ethical Legal Framework

Smart governance systems raise several important yet complex issues related to performance monitoring, data ownership, intellectual property rights, and algorithmic accountability. Addressing these challenges requires the establishment of clear ethical and legal frameworks [57]. Such frameworks should ensure that innovation is guided and regulated appropriately, while simultaneously upholding individual rights and freedoms. Governments are encouraged to develop comprehensive guidelines that balance technological advancements with the protection of public interest, creating a governance environment that is both responsible and inclusive.

11.6.1.5 Sustainability and Scalability

Applying smart governance solutions on a large scale is another concern from the logistical and financial sides. These issues show that the financial sustainability of the innovation and the development of technologies that would allow optimization at large scales are key to making smart city initiatives more effective. Studies should focus on finding new financial models including Public–Private Partnership (PPP) and further the technologies that can be implemented on large scale and in various cities [58].

11.6.2 Smart Governance Opportunities

11.6.2.1 Enhanced Decision-Making

Smart governance improves the decision-making by utilizing the data obtained in the decision-making process. The application of big data techniques, ML, and AI will enable policymakers to provide strategies that seek to involve the environmental resources with efficiency and enhance service deliveries, while solving the challenges that accrue from urban development [59]. This research can concentrate on designing better analysis techniques as well as the algorithms that will display useful data required by the urban planner and other administrative personnel.

11.6.2.2 Citizen Engagement

Smart governance resolves the problem of engaging more of the populace in the decision-making process through provision of information technology platforms. E-Participation and other mobile applications and tools for using social media technology allow citizens to participate in expressing opinion and feedback to local government [60]. More work needs to be done to discover how the citizens can be engaged most effectively and participate in the decision-making process.

11.6.2.3 Improved Service Delivery

The use of information technology in the provision of governance results in effectiveness of service administration. Smart city activities help organizations to minimize the number of contracts, minimize paperwork, and offer instant services to people [61]. The possibilities of research are presented by the design of web-based and mobile solutions aimed at improving the effectiveness of public services.

11.6.2.4 Environmental Sustainability

Smart governance is essential to promote and enhance sustainability and the environmental management of the resources in cities in a manner that is contraction consistent with their ecological preconditions [62]. Cities utilize data analytics and IoT to optimize energy, improve waste management, and enhance transportation. Specifically, such research should aim at identifying efficient methods that bring to the emergence of new green and more sustainable city forms.

11.6.2.5 Economic Development and Innovation

Smart governance with the help of proper introduction may act as a powerful incentive for innovation in technologies and entrepreneurship [63]. Research can study the capability of smart cities to act as engines of growth by progressing to promoting production and entrepreneurship, employment, and investment.

11.7 Summary

The shift from traditional governance to smart government is essential for building sustainable smart cities. Conventional models characterized by central decision-making and limited public involvement are increasingly inadequate in managing the complexities of modern urbanization. Smart governance addresses these challenges by enhancing transparency, accountability, and public participation through the effective use of ICTs. It relies on data analytics and evidence-based policymaking to improve urban service delivery and governance. Key components include robust infrastructure, advanced IT systems, data technologies, and a strong governance framework, all integrated within a collaborative environment that involves innovation. This strategy not only improves public services but also supports economic growth, higher living standards, environmental sustainability, and public safety. However, challenges such as data privacy, equitable access to technology and global knowledge sharing persist. As cities involve smart government, principles will remain fundamental to achieving efficient, inclusive, and sustainable urban development.

References

1. T. Worrakittimalee, T. Pienwisetkaew, P. Naruetharadhol, The role of smart governance in ensuring the success of smart cities: a case of Thailand. Cogent Soc. Sci. **10**(1) (2024). https://doi.org/10.1080/23311886.2024.2388827
2. A. Stratigea, C.-A. Papadopoulou, M. Panagiotopoulou, Tools and Technologies for Planning the development of smart cities. J. Urban Technol. **22**(2), 43–62 (2015). https://doi.org/10.1080/10630732.2015.1018725
3. H. Kumar, M.K. Singh, M.P. Gupta, J. Madaan, Moving towards smart cities: Solutions that lead to the smart city transformation framework. Technol. Forecast Soc. Change **153**, 119281 (2020). https://doi.org/10.1016/J.TECHFORE.2018.04.024
4. P. Mishra, G. Singh, Enabling technologies for sustainable smart city, in *Sustainable Smart Cities*, (Springer, Cham, 2023), pp. 59–73. https://doi.org/10.1007/978-3-031-33354-5_3
5. E. Tan, J. Crompvoets (eds.), *The new digital era governance* (Brill | Wageningen Academic, 2022). https://doi.org/10.3920/978-90-8686-930-5
6. E. Ruijer et al., Smart governance toolbox: A systematic literature review. Smart Cities **6**(2), 878–896 (2023). https://doi.org/10.3390/smartcities6020042
7. P. Mishra, G. Singh, *Sustainable Smart Cities* (Springer, Cham, 2023). https://doi.org/10.1007/978-3-031-33354-5
8. P. Mishra, G. Singh, Introduction: Importance of sustainable smart city, in *Sustainable Smart Cities*, (Springer, Cham, 2023), pp. 1–37. https://doi.org/10.1007/978-3-031-33354-5_1
9. A. Alberti, M. Senese, *Developing Capacities for Inclusive and Innovative Urban Governance* (2020), pp. 127–152. https://doi.org/10.1007/978-981-15-2973-3_6
10. G. Viale Pereira, M.A. Cunha, T.J. Lampoltshammer, P. Parycek, M.G. Testa, Increasing collaboration and participation in smart city governance: A cross-case analysis of smart city initiatives. Inf. Technol. Dev. **23**(3), 526–553 (2017). https://doi.org/10.1080/02681102.2017.1353946
11. P. Mishra, G. Singh, Energy management systems in sustainable smart cities based on the internet of energy: A technical review. Energies (Basel) **16**(19), 6903 (2023). https://doi.org/10.3390/en16196903
12. P. Mishra, G. Singh, Internet of things for sustainable smart city, in *Sustainable Smart Cities*, (Springer, Cham, 2023), pp. 75–96. https://doi.org/10.1007/978-3-031-33354-5_4
13. D. Gagliardi, L. Schina, M.L. Sarcinella, G. Mangialardi, F. Niglia, A. Corallo, Information and communication technologies and public participation: Interactive maps and value added for citizens. Gov. Inf. Q. **34**(1), 153–166 (2017). https://doi.org/10.1016/J.GIQ.2016.09.002
14. B.N. Silva, M. Khan, K. Han, Towards sustainable smart cities: A review of trends, architectures, components, and open challenges in smart cities. Sustain. Cities Soc. **38**, 697–713 (2018). https://doi.org/10.1016/j.scs.2018.01.053
15. A. Meijer, M.P.R. Bolívar, Governing the smart city: A review of the literature on smart urban governance. Int. Rev. Adm. Sci. **82**(2), 392–408 (2016). https://doi.org/10.1177/0020852314564308
16. A. Sharifi, A.R. Khavarian-Garmsir, The COVID-19 pandemic: Impacts on cities and major lessons for urban planning, design, and management. Sci. Total Environ. **749**, 142391 (2020). https://doi.org/10.1016/j.scitotenv.2020.142391

17. Z. Allam, Z.A. Dhunny, On big data, artificial intelligence and smart cities. Cities **89**, 80–91 (2019). https://doi.org/10.1016/j.cities.2019.01.032
18. R.G. Hollands, Critical interventions into the corporate smart city. Camb. J. Reg. Econ. Soc. **8**(1), 61–77 (2015). https://doi.org/10.1093/cjres/rsu011
19. T. Shelton, M. Zook, A. Wiig, The 'actually existing smart city'. Camb. J. Reg. Econ. Soc. **8**(1), 13–25 (2015). https://doi.org/10.1093/cjres/rsu026
20. M.A. Ahad, S. Paiva, G. Tripathi, N. Feroz, Enabling technologies and sustainable smart cities. Sustain. Cities Soc. **61**, 102301 (2020). https://doi.org/10.1016/j.scs.2020.102301
21. T. Yigitcanlar et al., Understanding 'smart cities': Intertwining development drivers with desired outcomes in a multidimensional framework. Cities **81**, 145–160 (2018). https://doi.org/10.1016/j.cities.2018.04.003
22. H. Arasteh et al., Iot-based smart cities: A survey, in *2016 IEEE 16th International Conference on Environment and Electrical Engineering (EEEIC)*, (IEEE, 2016), pp. 1–6. https://doi.org/10.1109/EEEIC.2016.7555867
23. W. Ejaz, M. Naeem, A. Shahid, A. Anpalagan, M. Jo, Efficient energy management for the Internet of Things in smart cities. IEEE Commun. Mag. **55**(1), 84–91 (2017). https://doi.org/10.1109/MCOM.2017.1600218CM
24. S. Borchardt, G.B. Vignola, D. Buscaglia, M. Maroni, *Mapping EU Policies with the 2030 Agenda and SDGs-Fostering Policy Coherence through text-based SDG Mapping*. https://doi.org/10.2760/110687
25. J. Clement, M. Manjon, N. Crutzen, Factors for collaboration amongst smart city stakeholders: A local government perspective. Gov. Inf. Q. **39**(4) (2022). https://doi.org/10.1016/j.giq.2022.101746
26. D. Reforgiato Recupero et al., An innovative, open, interoperable citizen engagement cloud platform for smart government and users' interaction. J. Knowl. Econ. **7**(2), 388–412 (2016). https://doi.org/10.1007/s13132-016-0361-0
27. J. Bhanye, R. Shayamunda, *The Promise of Civic-Tech: Digital Technologies and Transparent, Accountable Governance* (2024), pp. 93–122. https://doi.org/10.1007/978-3-031-75079-3_5
28. G.V. Pereira, M.A. Macadar, E.M. Luciano, M.G. Testa, Delivering public value through open government data initiatives in a Smart City context. Inf. Syst. Front. **19**(2), 213–229 (2017). https://doi.org/10.1007/s10796-016-9673-7
29. L. Theodorakopoulos, A. Theodoropoulou, C. Halkiopoulos, Enhancing decentralized decision-making with big data and Blockchain technology: A comprehensive review. Appl. Sci. **14**(16), 7007 (2024). https://doi.org/10.3390/app14167007
30. S. Hwang, T. Nam, H. Ha, From evidence-based policy making to data-driven administration: Proposing the data vs. value framework. Int. Rev. Public Adm. **26**(3), 291–307 (2021). https://doi.org/10.1080/12294659.2021.1974176
31. P. Mishra, G. Singh, Artificial intelligence for sustainable smart cities, in *Sustainable Smart Cities*, (Springer, Cham, 2023), pp. 119–142. https://doi.org/10.1007/978-3-031-33354-5_6
32. V. Charles, N.P. Rana, L. Carter, Artificial intelligence for data-driven decision-making and governance in public affairs. Gov. Inf. Q. **39**(4), 101742 (2022). https://doi.org/10.1016/J.GIQ.2022.101742
33. P. Mishra, P. Thakur, G. Singh, Sustainable smart city to society 5.0: State-of-the-art and research challenges. SAIEE Afr. Res. J. **113**(4), 152–164 (2022). https://doi.org/10.23919/SAIEE.2022.9945865
34. A.H. Fakhimi, A.H. Khani, J.M. Sardroud, Smart-city infrastructure components, in *Solving Urban Infrastructure Problems Using Smart City Technologies: Handbook on Planning, Design, Development, and Regulation*, (Elsevier, 2021), pp. 17–54. https://doi.org/10.1016/B978-0-12-816816-5.00002-4
35. A.M. Mirzaee, J.M. Sardroud, Public–private-partnerships (PPP) enabled smart city funding and financing, in *Smart Cities Policies and Financing: Approaches and Solutions*, (Elsevier, 2022), pp. 117–131. https://doi.org/10.1016/B978-0-12-819130-9.00011-5

36. Z.R.M.A. Kaiser, Smart governance for smart cities and nations. J. Econ. Technol. **2**, 216–234 (2024). https://doi.org/10.1016/J.JECT.2024.07.003
37. R. Lin, L. Wang, B. Li, Y. Lu, Z. Qi, L. Xie, Organizational governance in the smart era: The implications of blockchain. Nankai Bus. Rev. Int. **14**(2), 197–229 (2023). https://doi.org/10.1108/NBRI-02-2021-0014
38. K. Soomro, M.N.M. Bhutta, Z. Khan, M.A. Tahir, Smart city big data analytics: An advanced review. WIREs Data Min. Knowl. Disc. **9**(5) (2019). https://doi.org/10.1002/widm.1319
39. M.N.I. Sarker, M. Wu, M.A. Hossin, Smart governance through bigdata: Digital transformation of public agencies, in *2018 International Conference on Artificial Intelligence and Big Data (ICAIBD)*, (IEEE, 2018), pp. 62–70. https://doi.org/10.1109/ICAIBD.2018.8396168
40. R.J. Tumpa, L. Naeni, Improving decision-making and stakeholder engagement at project governance using digital technology for sustainable infrastructure projects. Smart Sustain. Built Environ. **14**, 1292 (2025). https://doi.org/10.1108/SASBE-10-2024-0451
41. M. Vitunskaite, Y. He, T. Brandstetter, H. Janicke, Smart cities and cyber security: Are we there yet?A comparative study on the role of standards, third party risk management and security ownership. Comput. Secur. **83**, 313–331 (2019). https://doi.org/10.1016/J.COSE.2019.02.009
42. M. Ku, J.R. Gil-Garcia, J. Zhang, The emergence and evolution of cross-boundary research collaborations: An explanatory study of social dynamics in a digital government working group. Gov. Inf. Q. **33**(4), 796–806 (2016). https://doi.org/10.1016/j.giq.2016.07.005
43. R.P. Dameri, C. Benevolo, Governing smart cities. Soc. Sci. Comput. Rev. **34**(6), 693–707 (2016). https://doi.org/10.1177/0894439315611093
44. S. Blasi, A. Ganzaroli, I. De Noni, Smartening sustainable development in cities: Strengthening the theoretical linkage between smart cities and SDGs. Sustain. Cities Soc. **80**, 103793 (2022). https://doi.org/10.1016/J.SCS.2022.103793
45. J.C.F. De Guimarães, E.A. Severo, L.A. Felix Júnior, W.P.L.B. Da Costa, F.T. Salmoria, Governance and quality of life in smart cities: Towards sustainable development goals. J. Clean. Prod. **253**, 119926 (2020). https://doi.org/10.1016/J.JCLEPRO.2019.119926
46. N.V. Lopes, Smart governance: A key factor for smart cities implementation, in *2017 IEEE International Conference on Smart Grid and Smart Cities (ICSGSC)*, (IEEE, 2017), pp. 277–282. https://doi.org/10.1109/ICSGSC.2017.8038591
47. P. Mishra, G. Singh, Internet of vehicles for sustainable smart cities, in *Sustainable Smart Cities*, (Springer, Cham, 2023), pp. 175–193. https://doi.org/10.1007/978-3-031-33354-5_8
48. Z. Tomor, A. Meijer, A. Michels, S. Geertman, Smart governance for sustainable cities: Findings from a systematic literature review. J. Urban Technol. **26**(4), 3–27 (2019). https://doi.org/10.1080/10630732.2019.1651178
49. T. Janowski, E. Estevez, R. Baguma, Platform governance for sustainable development: Reshaping citizen-administration relationships in the digital age. Gov. Inf. Q. **35**(4), S1–S16 (2018). https://doi.org/10.1016/J.GIQ.2018.09.002
50. A. Visvizi, M.D. Lytras, E. Damiani, H. Mathkour, Policy making for smart cities: Innovation and social inclusive economic growth for sustainability. J. Sci. Technol. Policy Manag. **9**(2), 126–133 (2018). https://doi.org/10.1108/JSTPM-07-2018-079
51. Y. Liu, X. Yao, T. Wei, Energy efficiency gap and target setting: A study of information asymmetry between governments and industries in China. China Econ. Rev. **57**, 101341 (2019). https://doi.org/10.1016/j.chieco.2019.101341
52. H. Seo, S. Myeong, The priority of factors of building government as a platform with analytic hierarchy process analysis. Sustainability **12**(14), 5615 (2020). https://doi.org/10.3390/su12145615
53. S. Kumar, A.K. Verma, A. Mirza, *Artificial Intelligence-Driven Governance Systems: Smart Cities and Smart Governance* (2024), pp. 73–90. https://doi.org/10.1007/978-981-97-5656-8_5
54. J.R. Gil-Garcia, D.S. Sayogo, Government inter-organizational information sharing initiatives: Understanding the main determinants of success. Gov. Inf. Q. **33**(3), 572–582 (2016). https://doi.org/10.1016/J.GIQ.2016.01.006

55. T.A. Pardo, T. Nam, G.B. Burke, E-government interoperability: Interaction of policy, management, and technology dimensions. Soc. Sci. Comput. Rev. **30**(1), 7–23 (2012). https://doi.org/10.1177/0894439310392184
56. S. Myeong, Y. Kwon, H. Seo, Sustainable E-governance: The relationship among trust, digital divide, and E-government. Sustainability **6**(9), 6049–6069 (2014). https://doi.org/10.3390/su6096049
57. M. Hartswood, B. Grimpe, M. Jirotka, S. Anderson, Towards the ethical governance of smart society, in *Social Collective Intelligence*, (Springer, Cham, 2014), pp. 3–30. https://doi.org/10.1007/978-3-319-08681-1_1
58. M. Biygautane, C. Neesham, K.O. Al-Yahya, Institutional entrepreneurship and infrastructure public-private partnership (PPP): Unpacking the role of social actors in implementing PPP projects. Int. J. Proj. Manag. **37**(1), 192–219 (2019). https://doi.org/10.1016/J.IJPROMAN.2018.12.005
59. L. Anthopoulos, V. Kazantzi, Urban energy efficiency assessment models from an AI and big data perspective: Tools for policy makers. Sustain. Cities Soc. **76**, 103492 (2022). https://doi.org/10.1016/J.SCS.2021.103492
60. M. Ntalian, C. Costopoulou, S. Karetsos, Investigating the mobile side of e-participation. Int. J. Electron. Govern. **9**(3/4), 210 (2017). https://doi.org/10.1504/IJEG.2017.088215
61. A. Murray, S. Kuban, M. Josefy, J. Anderson, Contracting in the smart era: The implications of Blockchain and decentralized autonomous organizations for contracting and corporate governance. Acad. Manag. Perspect. **35**(4), 622–641 (2021). https://doi.org/10.5465/amp.2018.0066
62. I. Guandalini, Sustainability through digital transformation: A systematic literature review for research guidance. J. Bus. Res. **148**, 456–471 (2022). https://doi.org/10.1016/j.jbusres.2022.05.003
63. A. Ferraris, G. Santoro, A.C. Pellicelli, 'Openness' of public governments in smart cities: Removing the barriers for innovation and entrepreneurship. Int. Entrep. Manag. J. **16**(4), 1259–1280 (2020). https://doi.org/10.1007/s11365-020-00651-4

Chapter 12
Cybersecurity in Sustainable Smart Cities

Contents

P. Mishra, G. Singh, *Sustainable Smart Cities 2.0*,
https://doi.org/10.1007/978-3-032-01102-2_12

12.1 Introduction

Smart cities are new-generation cities that have emerged with the advancing technological improvement in the growing world. These urban systems include various systems such as transport systems, energy systems, health facilities, and public security, which depend on data streams [1]. However, with these developments the growing necessity to protect the tremendous amount of digital data of smart city from various cyber threats is crucial [2]. Thus, cybersecurity in a smart city is a fundamental and intrinsic aspect of secure, confidential, and available digital transactions within a modern city [3]. Its key infrastructures become smarter and more interconnected and this makes the effect of cyber threats ever more critical and potentially dangerous. A vulnerability related to one or more nodes in a smart city may lead to the disruption of its operations, negative economic impact, or even threaten people's lives. These threats imply the need to install advanced protective mechanisms that would safeguard the smart city from cyber-attacks. Cyber security extends beyond protecting data and resources and serves as the foundation of urban organizations and the public's trust in emerging technologies [4]. The integration of smart systems and applications in cities increases the complexity of smart city systems and makes them susceptible to several problems of cybersecurity [5]. These geographical regions have decentralized infrastructures that combine outdated hardware and software solutions, Internet of Things (IoT) devices, cloud technologies, artificial intelligence (AI) and big data [6]. The application of such technologies needs a robust security solution designed to address threats [7]. Further, the scale of information created and shared in a smart city environment also increases the probability of data breach. Another critical factor to consider in smart cities belongs to the information assurance that should be proactive as well as dynamic to offer optimal security. It also needs to continuously monitor and detect threats on a real-time basis to fast respond to cybersecurity threats. In addition, since the threats from the cyberspace are constantly evolving the system should be constantly improved and enhanced by using newly developed technologies. Another essential aspect of smart cities in addressing cybersecurity is collaboration among all stakeholders. The integrated nature of smart city systems, cybersecurity is not the responsibility of individual institutions but a shared commitment involving government agencies, private companies, technology providers, and end users. Therefore, the exchange of cyber threat information, and best practices across organization should be actively promoted to enhance the cities security. Ultimately, authorities play a crucial role in facilitating partnerships between the private sector and the government, thereby enabling the development of effective and resilient cybersecurity strategies through the combined expertise of professionals from diverse fields [8]. Awareness and education to the public also play a very important role in enhancing cybersecurity in smart cities. Cybersecurity awareness and environment in which a city manages cyber incidents effectively and recovers from potential disruptions. However, cyber security in smart cities extends beyond technical and operational considerations to include legal and regulatory dimensions. A fundamental responsibility of central and local authorities is to establish policies and regulations that govern data protection, safeguard privacy, and ensure robust cybersecurity measures. Hence, cybersecurity is an essential element in the development of smart cities as the foundation for security of digital systems that lie at the core of effective city management [9]. Figure 12.1 illustrates a conceptual framework highlighting the interconnection between cybersecurity, sustainability and smart city components. The central block outlines core smart city infrastructure areas such as IoT devices, smart grids, intelligent transportation, and AI assessed healthcare. These components rely on robust cybersecurity measures, including threat prevention, encryption and data governance to ensure secure and resilient operations. On the other side, pillars like

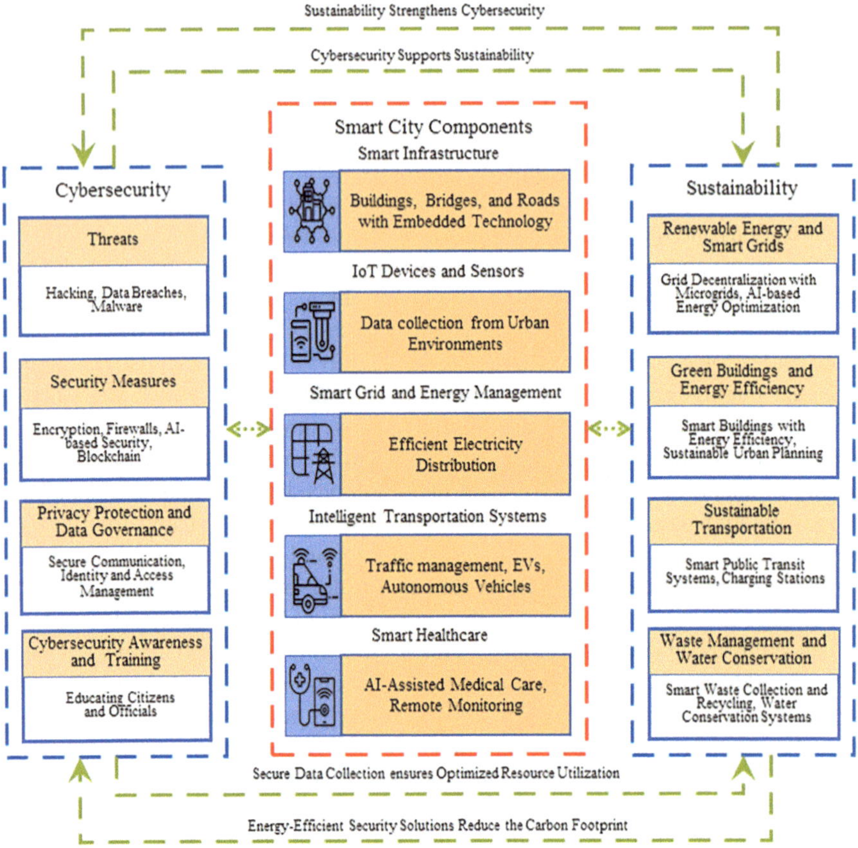

Fig. 12.1 Framework of cybersecurity integration in sustainable smart cities

renewable energy, energy-efficient buildings, and smart waste management are essential for eco-friendly smart cities [10]. This framework emphasizes the integrated role of cybersecurity in enabling secure sustainable ecosystems.

12.1.1 Cybersecurity and Sustainability in Smart Cities

In the modern world of smart urbanization, smart infrastructures and integrated systems in urban environment create a number of prospects for city development and successful implementation. Moreover, such advantages are accompanied by cybersecurity threats that mitigates the improvements. Since smart cities use data and IoT devices in decision-making such as traffic flow, energy usage, etc. Measures must ensure that sensitive information remains protected from malicious individuals, as breachers may disrupt essential city services and violates citizens' rights. At the same time, integrating sustainability into smart city is essential for long term success [11]. Proper utilization of energy, waste management and efficient transportation not only help mitigate climate change but also strengthen the resilience of smart cities against environmental challenges. Prioritizing both cybersecurity and sustainability allows smart cities to use technology that protects public safety, improves infrastructure efficiency and reduces the environmental impact of human activities [12]. This combined approach creates a balance between technological progress and environmental care, ensuring that advancements support sustainable urban development. Therefore, adopting both security and sustainability are important for smart cities to fulfil its societal potential.

12.2 Current Research Trends of Cybersecurity in Smart Cities

This study investigates the trends, contributions, and impact of research in smart cities with cybersecurity from 2016 to 2025 through a bibliometric analysis of Scopus data. The Bibliometrix R package along with origin software were utilized to analyse and depict the data. This approach ensures that researchers are presented with the most current research developments in the field enabling to focus on emerging trends and advancements. In March 2025, a bibliometric analysis was conducted utilizing the Scopus core collection database to thoroughly investigate the concept and understanding of smart cities. Ultimately, 1024 articles were exported and saved in BibTex format before being included and examined in the bibliometric study. The primary information, syntax and specific criteria used in screening for this study are shown in Table 12.1.

12.2.1 Annual Scientific Production and Top Ten Most Published Journals/Cited Paper

The annual publication trends from 2016 to 2025, highlighting the growth in research output is shown in Fig. 12.2. The early years shows a steady increase in publications, rising from 2016 (11 doc) to 2024 (267 doc), indicating a growing interest in the field. In 2025 (50), the number continues to rise and may continue increasing until the end of the year. Figure 12.3 presents the highest number of articles in cybersecurity in the smart city research domain published in "Lecture Notes in Networks and Systems" (23 docs), "IEEE ACCESS" (22 docs), "Sensors" (19 docs), and "ACM International Conference Proceeding Series" (16 docs), which indicates a higher output of articles compared to the other journals. In addition, some other journals, such as the "Communications in Computer and Information Science" (16 docs), "Lecture notes in Computer Science (Including subseries

Table 12.1 Summary of data characteristics and main information of the bibliometric study

Description	Specific Standard Requirements/Results
Research database	Scopus
Search query	("smart") AND ("cities") AND ("cybersecurity")
Time span	2016–2025
Document type	Articles
Language	English
Data export format	BibTex
Documents	1024
Annual growth rate %	18.32
Document average age	2.78
Average citations per doc	17.87
Keywords plus (ID)	4397
Author's keywords (DE)	2452
Authors	3100
Authors of single-authored docs	81
Single-authored docs	142
Co-authors per doc	3.56
International co-authorships %	32.32

lecture notes in Artificial Intelligence and Lecture notes in Bioinformatics)" (15 docs), "Sustainable Cities and Society" (15 docs), "Applied Sciences (Switzerland)" (14 docs), "IEEE Internet of Things Journal" (13 docs), and "Sustainability (Switzerland)" (11 docs), are in this domain.

Table 12.2 displays the top ten most cited papers from the study which presents both the ranking and authors with their published works. This section presents information about authors, article titles, and their corresponding publications in journals with documentation for publication years and citation frequency. The ranking focuses on engineering field smart cities articles to demonstrate their total citation (TC) counts and annual citation rates and normalized total citation (NTC) data which collectively signify their impact in the field.

12.2.2 Sankey Diagram (Three-Field Plot and Thematic Map)

Sankey diagram illustrates the three-field plot thematic progression to visualize the flow and the distribution in scientific research as depicted in Fig. 12.4. In this case, keywords plus, abstract, and title are added, acting as indicators to highlight the most influential research papers in the subject. These keywords indicate vital concepts and themes that stand out in academic discussions of the smart city domain. The diagram's keywords are sourced from earlier published studies and were then thematically organized by a specific researcher. Figure 12.5 categorizes various

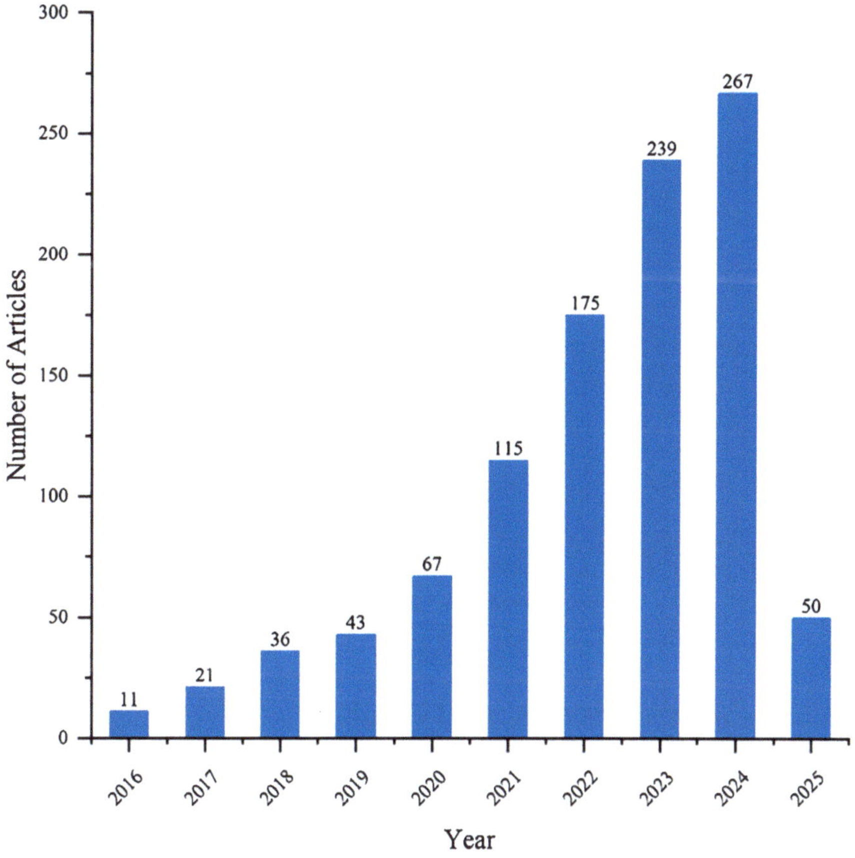

Fig. 12.2 Yearly trend of publication number of cybersecurity in smart city

smart city themes based on their energy development and relevant degrees. In the top right quadrant (motor themes), which represents themes that are highly developed and highly relevant, terms like "sustainability", "cyberattacks", "autonomous vehicles" and "big data". The left quadrant (niche themes) represents "algorithm", "optimization", "random forest", and "attack detection". This indicates a strong focus on these areas, suggesting they are central to cybersecurity-based smart city research and practice. This quadrant reflects the areas where significant advances and investments are made, highlighting their critical role in smart city. In contrast, the bottom left quadrant, which represents themes that are either emerging or declining and have lower relevance, includes terms like "communication security", "smart mobility", and "governance". Additionally, the terms like "smart city", "cyberattacks", "safety", and "cybersecurity" are the foundational information for further analysis and serve as a base for more advanced analyses. These themes might be considered foundational but must be emphasized more in energy-based smart cities. This visual categorization provides valuable insights for educators, curriculum

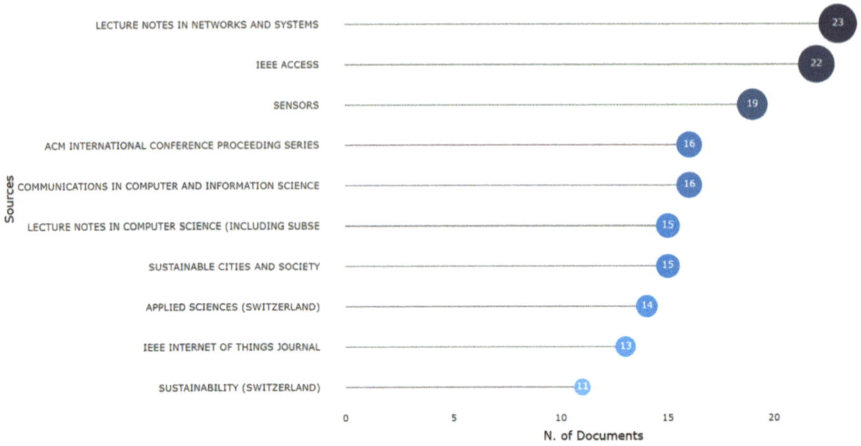

Fig. 12.3 Most relevant sources with number of articles of cybersecurity in smart city

developers, and researchers, helping identify which areas are gaining traction and which might need more attention or innovation.

12.2.3 Co-occurrence Keyword Trends and Network Visualization

Figure 12.6 depicts the most prominent themes and keywords related to research in conceptual change. The most significant and most central terms, such as "smart city," "cybersecurity", "internet of things", "network security", and "cyber-attacks", indicate the core focus of the research on how researchers grasp and alter their conceptual knowledge, particularly within smart city contexts.

An illustration in Fig. 12.7 displays word occurrence frequencies from the sustainable development goals in smart city research network. Research focuses mainly on "smart city" as this term appears 418 times throughout the literature. Research on sustainable development goals in smart cities indicates the importance of "cybersecurity" based on its 780 word occurrences. The "internet of things" (331) and "network security" (277) combinations show the growth and sustainability in cities. Additionally, "cyber-attacks" (112), "deep learning" (102), "computer crime" (93), "intrusion detection" (90), and "machine-learning" (89) combinations occupy crucial importance in emerging technology fields. The data shows that smart city research revolves primarily around several major subjects.

Table 12.2 Most global cited documents of cybersecurity in smart city

Rank	Authors name	Paper Title	Journal's Name	TC	TC/ Year	NTC
1	Sarker (2021) [13]	Machine learning: Algorithms, real-world applications and research directions	SN Computer Science	2801	560.20	58.20
2	Diro and Chilamkurti (2018) [14]	Distributed attack detection scheme using deep learning approach for Internet of Things	Future Generation Computer Systems	750	93.75	15.55
3	Minoli et al. (2017) [15]	IoT considerations, requirements, and architectures for smart buildings—Energy optimization and next-generation building management systems	IEEE Internet of Things Journal	672	74.67	12.06
4	Sarker (2022) [16]	AI-Based Modeling: Techniques, Applications and Research Issues Towards Automation, Intelligent and Smart Systems	SN Computer Science	458	114.50	25.48
5	Moustafa (2021) [17]	A new distributed architecture for evaluating AI-based security systems at the edge: Network TON_ IoT datasets	Sustainable Cities and Society	357	71.40	7.42
6	Javed et al. (2022) [18]	Future smart cities: Requirements, emerging technologies, applications, challenges, and future aspects	Cities	352	88.00	19.58
7	Kim et al. (2021) [19]	Cybersecurity for autonomous vehicles: Review of attacks and defense	Computers & Security	277	55.40	5.76
8	Vattapparamban et al. (2016) [20]	Drones for smart cities: Issues in cybersecurity, privacy, and public safety	International Wireless Communications and Mobile Computing Conference	266	26.60	4.51

(continued)

Table 12.2 (continued)

Rank	Authors name	Paper Title	Journal's Name	TC	TC/ Year	NTC
9	Alrashdi et al. (2019) [21]	Ad-iot: Anomaly detection of iot cyberattacks in smart city using machine learning	IEEE 9th Annual Computing and Communication Workshop and Conference	257	36.71	7.84
10	Yaqoob et al. (2019) [22]	Internet of things forensics: Recent advances, taxonomy, requirements, and open challenges	Future Generation Computer Systems	255	36.43	7.78

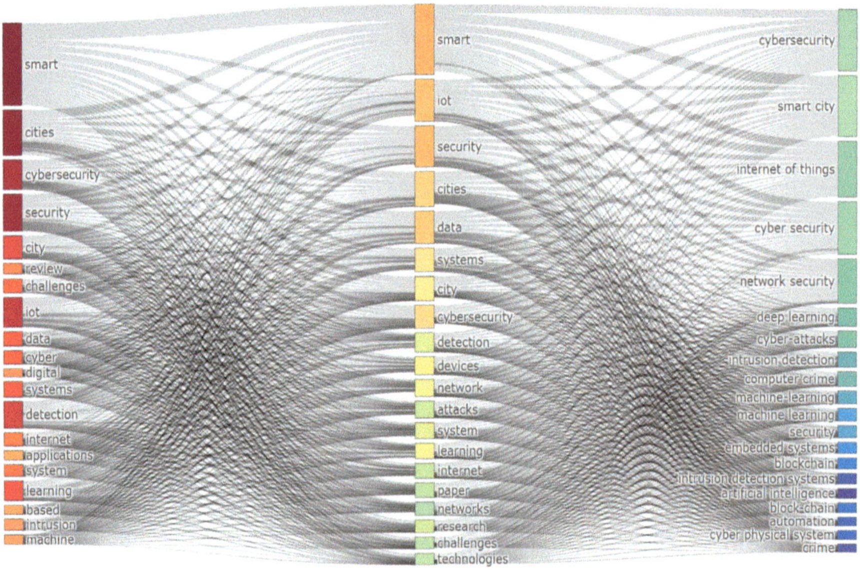

Fig. 12.4 Three-field plot (thematic development) using sankey diagram

12.2.4 Countries' Scientific Production

The scientific publication frequencies of various countries in the world are represented in Fig. 12.8. Scientific contributions from India stand at 524 which demonstrates their powerful role in research development. The growing scientific power of USA displays itself through its 268 scientific publications which rank second after India. The research hub of the Saudi Arabia produces 182 scientific papers among its total contributions. Numerous publications emerge from China and UK together with Australia given their research output comprises 140, 102, and 66 papers respectively. Academic and industrial research sectors of the South Korea and Turkey

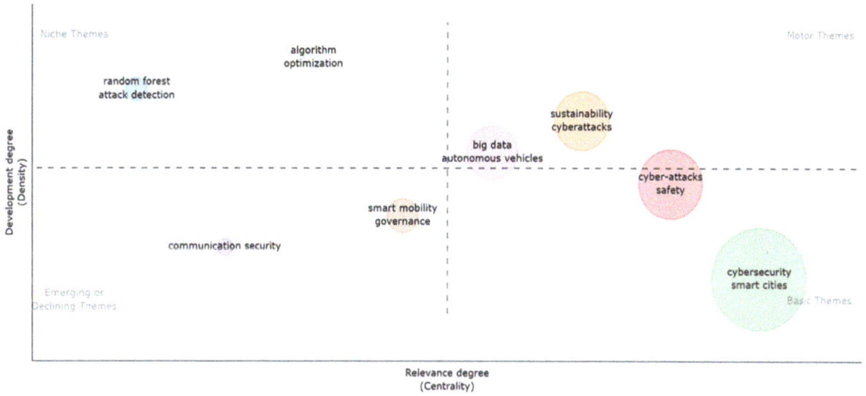

Fig. 12.5 Thematic map of cybersecurity in smart city research domain

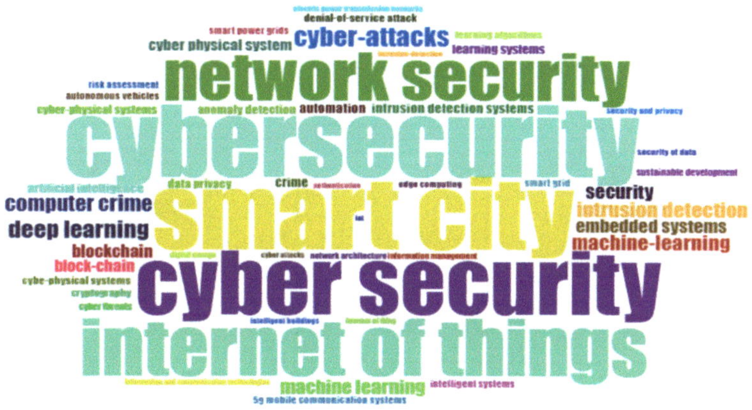

Fig. 12.6 Co-occurrence keyword trends in smart city domain with cybersecurity

produce approximately 51 and 51 scholarly works between them. Malaysia and Spain have smaller publications, with total of 50 and 48 research items, respectively. The measured data indicates Asian countries significantly influence the global scientific research output.

12.2.5 Mapping of Cybersecurity in Smart City Research to Sustainable Development Goals (SDGs)

SDGs, their targets (bubble series), and the percentage (bar graph) of corresponding keywords detected in the text (calculated as the ratio of keywords in one goal to the total number of keywords detected) are depicted in Fig. 12.9. For SDG mapping, the

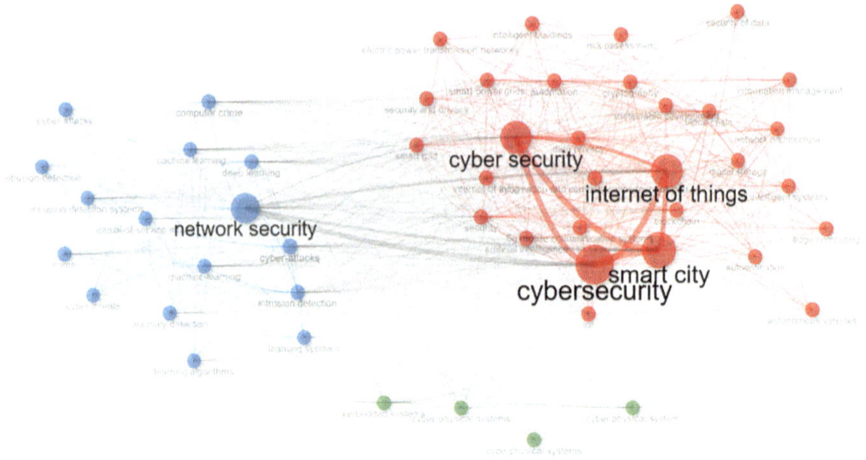

Fig. 12.7 Co-occurrence network of words in smart city research domain with cybersecurity

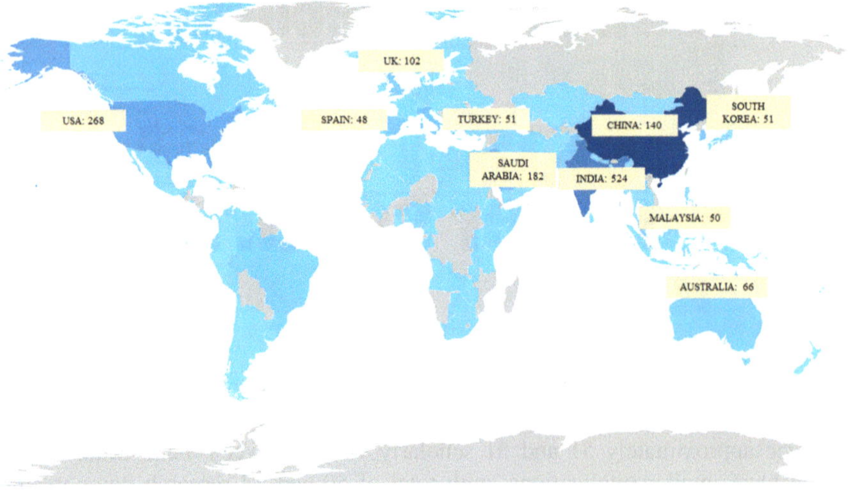

Fig. 12.8 Top ten countries' scientific production of various countries based on frequency

study utilized the SDG Mapper tool from the "Knowledge Base for the Sustainable Development Goals," provided by the European Commission [23]. This tool uses bubble charts to depict SDG interconnections and relevance among bibliometric records. The size of each bubble correlates with the proportion of associated keywords found, which is calculated as the number of goal-specific keywords divided by the total keyword count. As such, SDG 8, marked by 9 keyword occurrences. This is followed by SDG 11, with 9 occurrences, and SDG 9, with 4 occurrences, underlining their relative prominence in the research database. Within SDG 8, there are several key subtargets that the research address: Subtarget 8.1 is shown as a

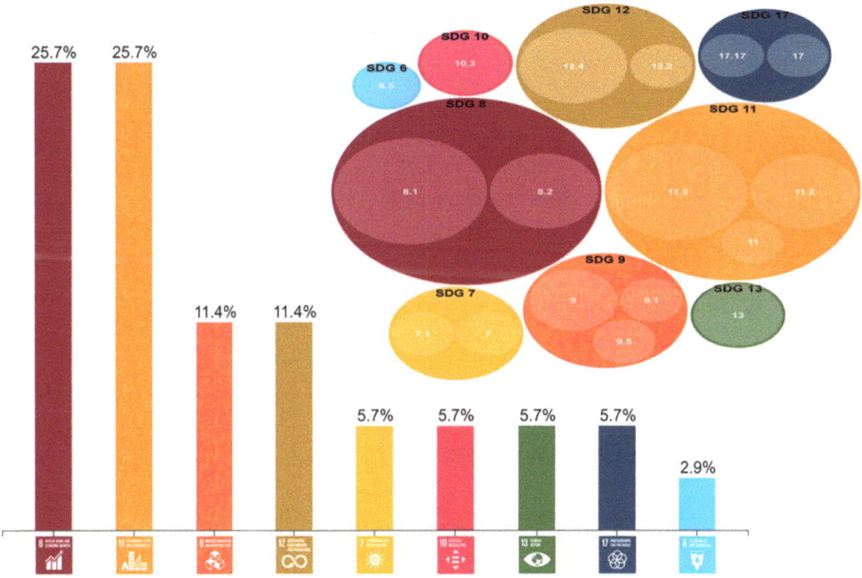

Fig. 12.9 Mapping cybersecurity in smart city research to relevant SDGs and targets

distinct bubble within the larger SDG 8 circle, indicating focused research on smart city. Subtarget 8.2, which aims technological upgradation, is another bubble within the SDG 8 area. This highlights the efforts in research and innovation to improving the cities. Another bubble with SDG 11, aims in sustainable cities with Subtarget 11.3 is shown as a distinct bubble within the larger SDG 11 circle, indicating focused research on smart city. Subtarget 11.2 focuses on providing the safe, accessible, and sustainable smart transportation. Subtarget 11, which aims sustainable cities and communities, is another bubble within the SDG 11 area. This highlights the efforts in research to improving the cities. Similarly, SDG 9 (industry, innovation, and infrastructure), which is also closely linked to SDG 11 and aims to enhance scientific research and affordable access to internet. The other related SDGs with detailed description are depicted in Table 12.3.

12.3 Cybersecurity Threats in Smart Cities

While urban cities are growing into smart city with help of advanced technologies to make them environmentally sustainable against cybersecurity threats. These integrated systems, which handle essential functions in citizens' life such as traffic signals, energy power control, and more, have become easy targets for hackers. Some of these may pose a threat to important facilities, the lives of people, and the credibility of innovation in general. With the advancement in smart infrastructure, there

Table 12.3 SDGs ranking, targets, and relevance in cybersecurity-based smart city research domain with SDGs

SDGs (By Rank)	Description	Targets	No. of Occurrence		Relevance%
SDG 8	Decent work and economic growth	8.1	6	9	25.7
		8.2	3		
SDG 11	Sustainable cities and communities	11.3	5	9	25.7
		11.2	3		
		11	1		
SDG 9	Industry, innovation, and infrastructure	9	2	4	11.4
		9.1	1		
		9.5	1		
SDG 12	Responsible consumption and production	12.4	3	4	11.4
		12.2	1		
SDG 7	Affordable and clean energy	7	1	2	5.7
		7.1	1		
SDG 10	Reduced inequalities	10.3	1	1	5.7
SDG 13	Climate action	13	2	2	5.7
SDG 17	Partnerships for the goals	17	1	2	5.7
		17.7	1		
SDG 6	Clean water and sanitation	6.5	1	1	2.9

is a need to ensure that adequate security measures involved to overcome the cybersecurity threats as illustrated in Fig. 12.10.

12.3.1 Data Breaches

A smart city employs technology such as sensors, data, and data analytics to optimize the use of limited resources and enhance the quality of life for its residents. This approach aims to improve the urban experience for inhabitants while maximizing the efficiency and effectiveness of available resources. However, the ever-increasing amount of information that is generated and passed through different platforms and networks make it easily accessible for hackers. Data breach leads to the loss of residents' data, the details of the city's critical infrastructure and operational data [24]. As the various buildings and infrastructures that comprise a smart city are affiliated to one another in a network, a breakdown in one link can affect an entire city system. However, security in smart cities is a challenge because many participants are involved. It has various governmental affiliations, companies, and third parties. Effective and secure system protection requires synchronization, comprehensive risk assessment, and continuous monitoring. Consequently, every city needs to employ a number of layers of protection from the threats such as

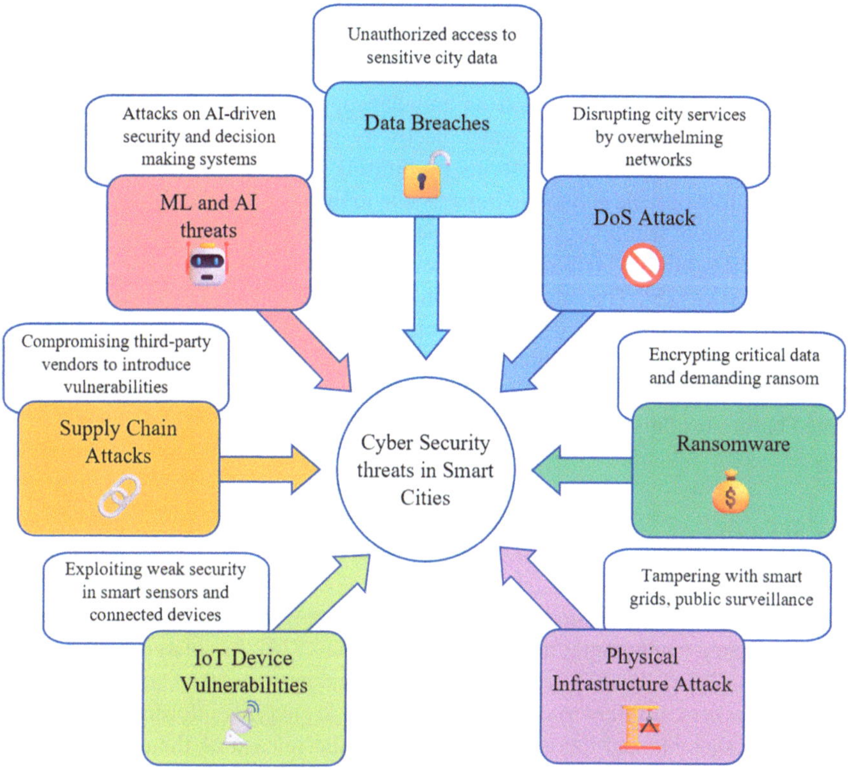

Fig. 12.10 Cybersecurity threats in smart cities

encryption technique, intrusion detection mechanisms, and security assessment to check the threats. While the advancement in smart cities, the management of risks associated with data breach is of uttermost importance. This means the city must be proactive in ensuring cybersecurity, protecting both physical and cyber systems, as well as its residents, from cybercriminal activities to maintain a safe and efficient environment.

12.3.2 Denial of Service Attack

Denial of Service (DoS) is the attack that intends to flood the network or the service with inadmissible requests, or rather send absolutely incorrect information, so that it becomes useless for other users [24]. This disruption results in serious problem in smart city where real-time information and service are imperative for operations of traffic, emergency, etc. Smart cities have many significant issues because it is composed of connected IoT devices that have different aim of processing and security. A worm attack could interrupt crucial communication systems that are used in provision of essential services such as ambulance, police, governmental/municipal organizations' operations, and water/electricity provision. DoS attack negatively impacts trust in the digital platforms by limiting use of smart technologies. This results in unfavourable economic impacts due to the costs involved in recovery. The increasing use of technology in city management and public service delivery highlights the essential need for effective cybersecurity measures against denial-of-service attacks [25]. The implementation of the advanced threat detection system, proper update of the security measures, and raising the level of cybersecurity awareness among the smart city inhabitants remain crucial steps to prevent disrupting threats. It is therefore imperative that such vulnerabilities are addressed and mitigated before becoming issues in the smart city environment.

12.3.3 Ransomware

Ransomware is a severe threat to the cybersecurity of smart cities since these are run on interrelated technologies and data-driven systems [26]. As smart cities adopt and install several IoT devices, sensors, and autonomic systems, they become easy targets for hackers. It refers to a kind of malware that encrypts information and organizational systems and demands a ransom for the release of the same. In smart cities, such an attack results in disastrous consequences affecting water and energy systems, transportation infrastructure, medical facilities, and emergency response operations. Apart from the direct impact and monetary loss, the threat of ransomware in smart city applications poses risks of loss of the public's confidence in smart city projects. Smart city systems and networks are interdependent such that an integration problem in one function may negatively impact on other functions. Protecting society from ransomware requires a robust multi-layered strategy that includes regular data backups, comprehensive training to identify cyber threats and the implementation of intelligent detection systems alongside effective response protocols. To prevent or mitigate such attacks, organizations need to take preventive measures including updating the software, ensuring access to networks and incident response plans [27]. Therefore, the management of ransomware in the smart city requires both technology solutions and policy measures as well as multi-stakeholder initiatives towards sustaining normal functioning of the city.

12.3.4 Physical Infrastructure Attack

Smart cities involve the integration of technology with the physical layout and design of urban areas aimed at improving inhabitants' quality of life through better resource utilization and reduced environmental impact [28]. However, this integration also creates vulnerability by simplifying excess for potential attackers and exposing the city to physical infrastructure threats. Such attacks may target transportation, power supply, water resources and communication networks. As IoT devices and digital networks expand to support smart city facilities these interconnected systems become increasingly susceptible to cyber physical attacks. Thus, cybersecurity is very important because reliability of data is essential in enabling effective interactions [29]. This pertains to the protection of not just the networks and data but also the development of viable risk management plans as well as security audit and assessments.

12.3.5 IoT Device Vulnerabilities

As smart cities increasingly adopt IoT to address challenges in urban living and resource management, a broad and diverse area of security threats arise [30]. Smart devices for the IoT in smart cities comprise traffic management sensors, smart street lamps, waste collectors, water and energy meters, and surveillance cameras. However, these interconnected systems are so complex that they increase the risk of cyber threats. These devices interface with wide area networks, with some intended for public access and other operating under varying protocols, resulting in heterogeneous securities throughout the network. Many IoT devices lack sufficient computing power, such as advanced processors and adequate memory to support strong security measures like robust encryption and real-time threat detection and prevention. Default credentials and unpatched firmware make these devices highly vulnerable to cyberattacks [31]. When a device is not working properly, attackers can control and use it to access other networks causing data theft or disruption of important city services. The risk increases in IoT networks due to the interconnected nature, where a single vulnerability affects the entire system. Highly advanced security programs that employ updated monitoring systems, AI threat assessments, and auto-generation of responses are needed [32]. It includes proper update, encryption, and strong authentications of the device. However, the establishment of standard security measures as well as performing routine assessment also reduce these risks. With the modern concept of smart cities growing rapidly, the overall security of such cities should remain at priority to safeguard the infrastructures of the cities for the sustainable development.

12.3.6 Supply Chain Attacks

Hackers are increasingly targeting supply chain threats which represent a relatively recent challenge in cybersecurity particularly within the context of smart cities. These attacks exploit vulnerabilities within different parts of the supply chain to gain access to more secure systems [33]. Having multiple connected systems also makes it easier for a breach. Furthermore, implementing third-party vendors and contractors increases vulnerability because they can be different organizations that do not follow similar security measures to the primary vendors. Some of the impact of such attacks includes interruption of crucial services, privacy invasion, and financial loss [34]. To counter these threats, smart cities should therefore implement security measures across most layers of the supply chain system. This involves screening of suppliers, high-level accreditation, and coordination between the government, business, and cybersecurity professionals.

12.3.7 Machine Learning and Artificial Intelligence Threats

Smart cities, which are cities in which various technologies such as machine learning (ML) and AI are used to manage urban infrastructure, it is also important to consider emerging cybersecurity issues [35]. Most of these smart systems, including traffic control, electrical power supply, water supply, and security systems, are integrated with large databases and are easy targets for hackers. Another threat of using ML and AI in smart city application is that of adversarial attacks in smart cities. Such systems are vulnerable to manipulations and may introduce false data, leading to incorrect outcomes and poor decision-making. Such manipulation might lead to disastrous results for the response capacity of such cities in cases of emergencies as well as general flow of traffic in the urban area. Also, as the focus is made on the usage of AI algorithms in decision-making, that inequality and discrimination issues are being constantly present in addition to that, there is a potential for creating intentional systemic flaws. When AI systems used for surveillance or policing are not adequately trained biased decision-making emerges causing distrust and social tension [36]. Furthermore, the development of AI models and the deployment of ML algorithms involve considerable complexity, making it difficult to detect anomalies or security breaches using conventional security mechanisms [37]. As systems are continuously upgrading, breaches may remain undetected for extended periods, becoming integrated within the system architecture. This integration greatly complicates the process of restoring and securing the system after an intrusion. In order to combat them, smart city managers and cybersecurity professionals should have in place formidable security measures, followed by the identification of potential threats, and most importantly, openness and compliance of AI applications [38]. Further, the increased creation of black-box models with little enterprise and immunity to adversarial strategies will be efficient to protect the future of urban environment.

12.4 Blockchain and Machine Learning in Prototyping of Cybersecurity in Sustainable Smart Cities

Smart cities incorporate the current development and future smart technologies to improve the life quality, sustainability, and resilience in the city. With these cities getting more developed and connected, the issue of cybersecurity becomes essential requiring new updates and methods of protecting data and infrastructures. As for the ML and blockchain, both play significant roles in cybersecurity area, especially in prototyping stage for smart cities [39]. It is vital because of its functionalities in data security and in the analysis of open ledgers for threat identification and response. However, for interdisciplinary collaboration, it becomes possible to perceive how effectively each of these technologies contributes to building strong cybersecurity in smart cities. Blockchain, which is in fact a system developed for electronic currency such as Bitcoin, now has many uses owing to its attribute of safeguarding, unchangeability, and decentralization. It solves the cybersecurity issues in smart cities by decentralization, where data are stored, and this makes difficult for the cyber attacker to get complete control of the system [40]. This is because the data is distributed over a number of nodes in a network and changes on the data would be easily accessible. This feature is especially relevant to smart cities since a huge number of IoT devices generate and send information through the networks. It is quite important that the integrity of such data has to be maintained in order to avoid false data in traffic management system, energy grids, etc. where accuracy plays a major role. In addition, through smart contracts, blockchain performs and enforce cybersecurity policies in different systems of urban regions. An advantage of blockchain is that it is open to the public which has the potential to build the trust from the stakeholders creating a record of the data, and transactions that are important for the smart city projects [41].

ML, on the other hand, brings the capacity to impose certain patterns over the huge volumes of data that might not be easily observed by human analysers. During the initial formulation of a model of security for smart city applications, ML technique may be employed to identify patterns that could be a potential security risk like unauthorized attempts at accessing [42]. With the possibility to train it over new data, the ML models are able to adjust to new threats, which a static model is unable to achieve. This capability is especially beneficial in the dynamic setting of a smart city since the characteristics of cyber threats may be significantly different from one another in such as environment. It is also important to note that the application of ML also be exploited in order to enhance the management in smart cities. If the existing data indicates ongoing or past security threats, ML algorithms can recommend the appropriate plan of action, which is faster and requires fewer resources [43]. In addition, based on the machine learning, potential risks in the city can be predicted in advance, so the security aspect can be synchronized to fix these flaws before it emerges as dangerous by the attackers [44]. Blockchain and ML are highly positively synergistic where one can overcome some of the deficiencies the other possesses [45].

While blockchain technology serves the purpose of data immutability, ML facilities the capacity of modelling the threats, analysis and response in immediate decision-making [46]. Using blockchain as a database and applications, the ML models are built with large volumes of quality data. Hence, blockchain and ML play essential roles in the prototyping of cybersecurity for sustainable smart cities as depicted in Fig. 12.11. Smart cities are characterized by many interconnected systems, hence blockchain's aims of guaranteeing data authenticity and independence as well as ML ability of detecting and responding to threats offers a strong baseline for protecting smart cities. These technologies need to be integrated in smart cities to become sustainable solutions to the future cybersecurity challenges. Incorporation of these technologies aids the smart cities in not only safeguarding infrastructures but also improving the standard of living of citizens.

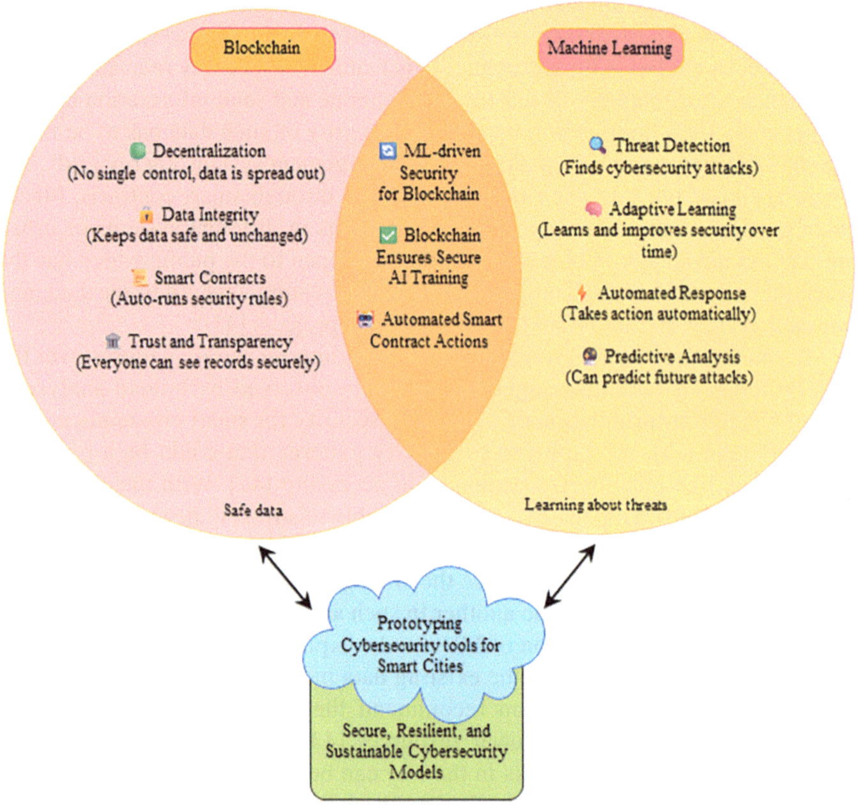

Fig. 12.11 Blockchain and machine learning in prototyping for cybersecurity in smart city

12.5 Sustainable Cybersecurity Framework for Smart Cities

Smart cities have become one of the most popular concepts for the modern world experiencing the growth of urbanization and technological progress. However, the fundamental foundations of such advanced and integrated urban infrastructure are the digitization that brings much insecurity, which makes cybersecurity a critical principle for sustainable intelligent societies. The importance of developing a sustainable cybersecurity model for these urban centers lies in its ability to reduce and potentially prevent future cyberattacks. Thus, interconnectivity of smart cities allows systems vulnerability and can be easily attacked by hackers. Smart cities are about not only an IT infrastructure of computer systems but also a network of services that should be safeguarded from the threats. For any sustainable cybersecurity strategy to be followed, it has to provide solutions to prevent, detect, respond, and recover from a cyber-attack within the urban digital environment. The sustainable cybersecurity framework for smart cities is characterized by a complex of concepts that interrelate seven categories of risk management, data protection, security by design, collaborative approach, continuous monitoring, public awareness, compliance, and resilience planning as illustrated in Fig. 12.12. Therefore, highlighting these components develops a strong protection against cyber risks that smart cities should adopt for their networks and systems' security in the future.

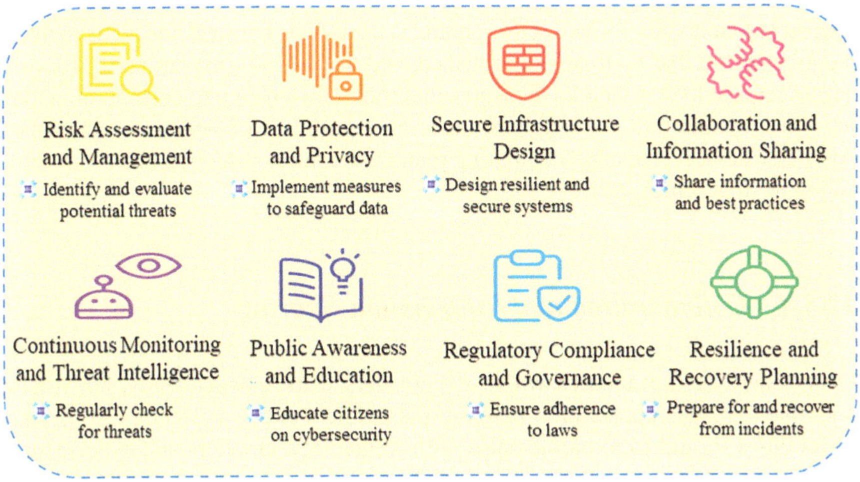

Fig. 12.12 Sustainable cybersecurity framework for smart cities

12.5.1 Risk Assessment and Management

Risk assessment is the first crucial step towards developing the program of recommended measures in any framework of cybersecurity [47]. City management in smart cities must constantly assess vulnerabilities that exist in cyber systems such as connected devices, cloud, and networks. This helps to estimate threats, predicts possible threats, and the probable effect of the threats on the entire functioning of the city. Risk management solutions should consider the usage of protective measures, regular update of threat profiles and an immediate response plan.

12.5.2 Data Protection and Privacy

Smart cities require data in almost all of its aspects ranging from traffic, energy, and health. The security of this data thus needs to be protected. To be sustainable a framework has to include data encryption, anonymization of the data, and only allow access to the data that are authorized to access [48]. Privacy policies should have regard to the general data protection regulation and other similar resolutions to ensure that the citizens' personal data should not be misused.

12.5.3 Secure Infrastructure Design

Security of smart city infrastructure should start with its construction. This includes the assumption that no trust is established, and there is no trust but verify for each access request [49]. Essential measures that should be taken into consideration during the infrastructure construction, the most significant of which is redundancy. Also, IoT devices and edge computing platform require protection due to its application in smart cities [50].

12.5.4 Collaboration and Information Sharing

Cybersecurity is not an area where stakeholders work independently, but it is an area that demands cooperation between the public and private sectors [51]. Thus, smart cities should form partnerships with technology suppliers and cybersecurity service providers, as well as cities which implemented smart technologies. The establishment of the protocols strengthen the information sharing on defence mechanisms to counter the rising threats.

12.5.5 Continuous Monitoring and Threat Intelligence

It is critical for security of systems and networks to be monitored regularly to ensure optimal security. This involves applying intelligent infrastructures to embrace advanced threat identification and recognition of cyber-attack. Threat intelligence platforms also allow to get the data in real time on the new vulnerabilities and threat which enable smart city to respond faster [52].

12.5.6 Public Awareness and Education

Establishing an effective communication system requires raising public awareness about individual responsibilities in maintaining cybersecurity. Educating citizens on common cyber threats and appropriate protective measures significantly enhances security. Equally important is the training of city employees and stakeholders to ensure all participants understand their roles and responsibilities in addressing cyber risk [53]. Education and awareness initiatives are essential components of a comprehensive cybersecurity strategies with a smart city framework.

12.5.7 Regulatory Compliance and Governance

The requirements of the legislation as well as applying strict organizational management principles are necessary in smart cities. This includes cybersecurity setting of IT policies and standards, designating liability procedures, and planning regular audits. It is recommended that the current and future governance of the frameworks should follow the national and international standards in order to ensure that there is a consistent approach of cybersecurity in the city [54].

12.5.8 Resilience and Recovery Planning

A sustainable cybersecurity should be developed to address the full cycle of a cyber breach. This includes establishing clear measures to respond effectively to any disruption of services or operations cost by a cyberattack. Evaluation and improvement of recovery plans help ensure the city recovers quickly and maintain essential services [37]. Such efforts contribute positively to the experience of residents and support the continuous delivery of critical functions within the city.

12.6 Global Standards, Policies, and Regulation for Cybersecurity in Smart Cities

Cities are gradually evolving into smart cities by adopting digital technologies to enhance infrastructure, services, and quality of life. In this transformation, the importance of strong cybersecurity remains essential. Advanced smart cities rely on interconnected networks across various sectors such as transportation, energy, water supply, and public safety, all of which are vulnerable to cyber threats. For this reason, it is important to establish clear global standards, policies, and regulations to ensure the security and resilience of smart city infrastructures [55–58] as illustrated in Fig. 12.13.

12.6.1 Global Standards for Cybersecurity in Smart Cities

12.6.1.1 International Frameworks and Guidelines

The International Organization for Standardization has published several standards related to information security management, such as, the ISO/IEC 27001 which gives the guidelines to infrastructure, implement, maintain, and review information security management systems. According to the ITU-T Y.2060 recommendation, smart cities refer to ubiquitous smart cities concerning certain key concepts and classifications of smart cities.

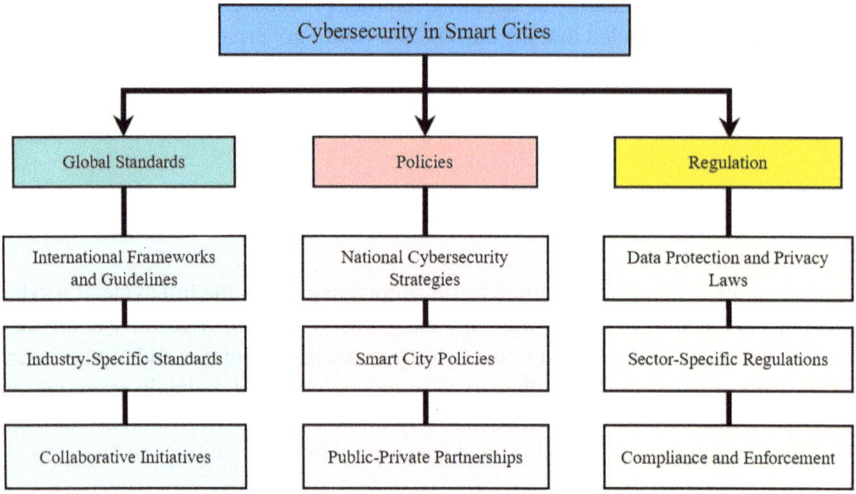

Fig. 12.13 Global standards, policies, and regulations for cybersecurity in smart cities

12.6.1.2 Industry-Specific Standards

Smart city infrastructure that incorporates IoT is safeguarded by standards that include IEC 62443 that deals with security for industrial automation and control systems. Moreover, the smart grid technologies, the IEEE 1547 standard helps to facilitate and enforce the interconnection requirements of distributed resources with the electric power systems, which involves cybersecurity and many other aspects.

12.6.1.3 Collaborative Initiatives

The global commission on the stability of cyberspace has been established to enhance the understanding of cyber stability and work on developing norms and standards in the global community. Governing body of smart city known as "Smart Cities Council" works collectively with different subgroups to ensure effective guidelines and best practices of security measures in smart city implementations.

12.6.2 Policies for Cybersecurity in Smart Cities

12.6.2.1 National Cybersecurity Strategies

Most countries have formulated national strategies to address cybersecurity in smart cities. The US' National Cyber Strategy focuses on the protection of critical infrastructure that includes the smart city components. Some of them are: "The European Union" and "'Cybersecurity of the EU and its Digital Strategy", which contain provisions to adopt the European Cybersecurity Act for the cybersecurity certification of products, services and processes.

12.6.2.2 Smart City Policies

Smart cities need to implement policies of developing the cities that incorporate cybersecurity as one of the key aspects. The cities should implement smart plan includes strategies for protection of data and infrastructure. Also, it is essential for cities to use smart technologies while providing it with appropriate security measures.

12.6.2.3 Public–Private Partnerships

Cybersecurity policies are typically developed as a public and private partnership between the state and businesses. Some of the approaches include the U.S. National Institute of Standards and Technology (NIST) Cybersecurity framework to enable the management of cybersecurity risk by both public and private enterprises.

12.6.3 Regulation for Cybersecurity in Smart Cities

12.6.3.1 Data Protection and Privacy Laws

With the General Data Protection Regulation (GDPR) enacted in the European Union passing high standards for data protection and privacy, smart city service delivery of personal data's is greatly affected. The California Consumer Privacy Act (CCPA) also offers some level of protection in the United States to regulate how smart city plans operate within the country in regard to resident's data.

12.6.3.2 Sector-Specific Regulations

The energy sector is a key component of smart cities, governed by established cybersecurity regulations such as the North American Electric Reliability Corporation (NERC), Critical Infrastructure Protection (CIP) Standards, which oversee the security of the bulk power system. Similarly, transportation systems, including connected and autonomous vehicles are subject to cybersecurity regulations under frameworks such as UNECE WP.29. These standards address the protection of vehicle-to-everything (V2X) communications and ensure the secure operation of advanced mobility systems within smart urban environments.

12.6.3.3 Compliance and Enforcement

Smart cities manage to put more efforts on regulation and compliance to ensure that smart city projects meet cybersecurity measures. The Federal Communications Commission (FCC) in the United States has decision-making power for cybersecurity regulation of broadband suppliers. In the EU context, this role is fulfilled by the European Cybersecurity Agency (ENISA) as it assists the national authorities in the enactment of the regulations pertaining to cybersecurity.

12.7 Benefits of Cybersecurity in Sustainable Smart Cities

Smart cities utilize information technologies, the IoT, and data as critical tools that help augment life in the city, manage resources, and supply services. The incorporation of smart technologies has much advantage; on the same note it has more risks. Cybersecurity has become an essential component to safeguard smart cities and ensure secure and sustainable existence [11, 59, 60]. Figure 12.14 aims to analyse the different advantages associated with the implementation of cybersecurity in relation to sustainability of smart cities.

12.7.1 Protection of Critical Infrastructure

Smart cities are developed on the basis of infrastructure that connect numerous systems, such as traffic signals, transit systems, water supply, and sanitation systems. Such facilities are often referred to critical systems and easier target for hackers. An infringement could result in serious consequences that would impact on service delivery and safety of the public. These systems are well protected from any unauthorized access or threats that may harm these systems so as to ensure they are always functional. It should be noted that this protection is vital in an effort to ensure that the existing urban structures have the capacity to function and last for long.

12.7.2 Safeguarding Citizen Data

Most smart cities are characterized by a process of collecting extensive data in order to optimize the organization of the city and delivery of services. Such data may contain personal data of citizens such as their mobility, preferences or usage of city services. Appropriate measures also ensure that this data is procured, stored, and managed securely to reduce instances of privacy infringement and identity theft on the citizens.

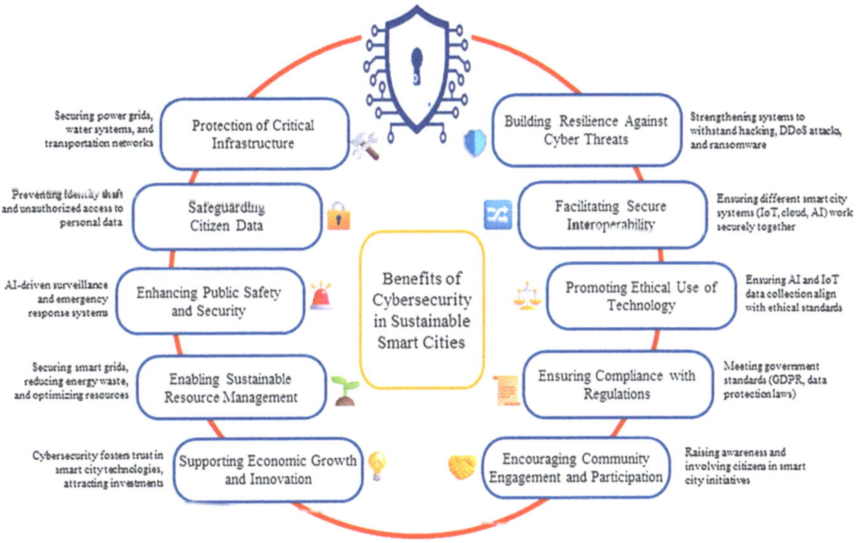

Fig. 12.14 Benefits of cybersecurity in sustainable smart cities

12.7.3 Enhancing Public Safety and Security

The security of the cyberspace gains a significant role in protecting the public interest in smart cities. Coordinated vigilance, efficient assistance, and public safety notifications require the use of various digital technologies. Security precautions do not allow the misuse of such systems with an aim of not functioning appropriately during emergencies. Thus, cities promote the security of these systems from cyber threats, which increases the ability to respond to emergencies, coordinate assistance measures efficiently, and maintain order to protect lives.

12.7.4 Enabling Sustainable Resource Management

These include a smart grid, an intelligent transportation system, and IoT in smart environmental management. Information security hence plays a critical role in protecting the information that is used to manage these resources in order to avoid disruptions through an attack. Secure data management ensures that right decisions that adopt the principles of sustainability such as energy conservation, emissions and wastes are well addressed by cities.

12.7.5 Supporting Economic Growth and Innovation

Cybersecurity helps to ensure that the necessary conditions for business and new start-ups to develop innovative technologies successfully in smart cities. Through protecting the information of intellectual property and other important business information, cybersecurity promotes the funding of new elements in the technology and services. Moreover, cybersecurity as an industry is already growing and the implementations that are being developed are considered as more business propositions and as economic stimuli for the city.

12.7.6 Building Resilience Against Cyber Threats

It is necessary for smart cities to be capable of adaptive mechanisms to respond to various cyber threats. The adoption of good cybersecurity frameworks assists the cities to resist and even overcome from cyber-attacks. Such resilience is key in enhancing the confidence of the public and to ensure that these systems in urban areas are held ready to operate in anticipation of an attack. Ensuring that cities have a culture with concern for cybersecurity and maintain proper precautions will enhance their ability to confront the threats of cyber threats.

12.7.7 Facilitating Secure Interoperability

Smart cities require the engagement of everybody and anybody that owns any stake in the smart city project; it can be government department, a private firm, or an individual. Cybersecurity provides access between disparate systems to work and exchange information over the networks that exists in an organization without being breached. Technology and innovations make it possible for smart cities to facilitate safe sharing of information, therefore improving efficiency in promoting various services.

12.7.8 Promoting Ethical Use of Technology

As the concept of smart cities continues to develop, proper use of these technologies is not possible without concerning ethical issues. Security on cyberspace is the aspect that ensures that new technologies are adopted efficiently and not be misused. This approach helps to safeguard citizen's privacy, trust and accountability with proper utilization of smart technologies.

12.7.9 Ensuring Compliance with Regulations

Governments are coming up with policies to protect data privacy and security. These regulations are met through cybersecurity, which acts as a means of ensuring that data is secure while following the necessary legal requirements on privacy. Legal noncompliance serves a purpose of saving the city from legal consequences and improves the functioning of the organizations. Cities also have the corresponding intent to protect the citizen and create a safe digital environment that complies with the regulation.

12.7.10 Encouraging Community Engagement and Participation

Cybersecurity enables citizens to make effective contribution toward smart city platforms. Citizens are also likely to engage in digital platforms or services that assure them that their data is safe. This involvement is extremely useful to understand their experience and concern and shape more effective urban environment and policies.

12.8 Open Research Challenges and Research Directions

As smart cities move toward sustainable urban development by using technology to improve the quality of life, the involvement of smart technologies such as IoT, AI, cloud computing, and big data analytics is creating smarter cities. However, these systems are highly interconnected and any such integration poses a new set of cybersecurity risks or threats that need to be addressed in order to not only protect the infrastructure used in smart cities but its privacy too. Figure 12.15 aims at depicting the challenges and further research opportunities of cybersecurity in the context of sustainable smart cities.

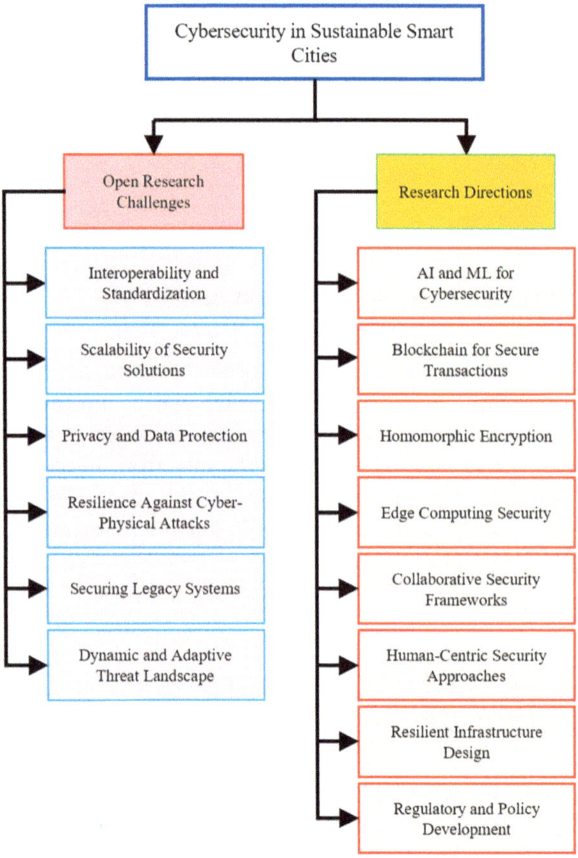

Fig. 12.15 Open research challenges and research direction of cybersecurity in sustainable smart cities

12.8.1 *Open Research Challenges*

12.8.1.1 Interoperability and Standardization

Determination of interoperability among multiple smart city systems and devices is one of the biggest challenges. This implies there are no clear architectures when it comes to such systems which are key determinants for smart cities. It will be necessary for measures in blockchain technology to be applied to the incompatibilities observed in the systems and also for common frameworks of the exchange of data to be developed for the various domains [61].

12.8.1.2 Scalability of Security Solutions

As the development of smart cities is growing, the connected devices and more data produced are rising. This might create possible vulnerabilities in current cybersecurity solutions because it is not adequately scalable enough to accommodate for the growth. The desire to create large-scale schemes of secure IoT networks in today's world means that these issues need to be taken into account [39].

12.8.1.3 Privacy and Data Protection

Smart cities generate a tremendous amount of data, which may contain personal data. It is imperative that its privacy and data protection is implemented to the maximum capacity due to several risks that deal with the breach of private data. There must be efforts aimed at promoting privacy enhancing technologies that include advanced encryption and anonymization [62].

12.8.1.4 Resilience Against Cyber-Physical Attacks

Smart city involves connecting the physical and cyber worlds and it may cause cyber-attack physically [63]. The key research directions should further focus on proactively securing such systems by identifying reliable methods for creating resilient cyber-physical systems and enforcing the identification of threats.

12.8.1.5 Securing Legacy Systems

Smart cities utilize original systems that were not initially designed to protect from the cybersecurity threats [64]. Implementing security to these systems is quite difficult, but vital, if done after the initial development of the system. Future work should be devoted to exploring the ways of protecting legacy infrastructures while aligning with the advancement by integrating them with new technologies.

12.8.1.6 Dynamic and Adaptive Threat Landscape

Information security is a very dynamic domain where innovative practices are always used by hackers to compromise a specific computer system. These threats are dynamic in nature and hence need the development of contemporary strategies that may be used as a roadmap to address such threats. It should therefore be for research to devise AI-based security mechanisms that evolve with the increasing threats in a dynamic manner [65].

12.8.2 Research Directions

12.8.2.1 Artificial Intelligence and Machine Learning for Cybersecurity

Artificial Intelligence and Machine Learning technologies are applied to improve threat detection and handling the threats [66]. More research efforts should channel and focus on training, deploying and tuning models for the purpose of performing massive computations in order to discover the symptoms of cyber threats. Also, the use of analytical models allows the usage of big data and AI for countermeasures and potential threats' prevention, which would increase the protective state of smart cities.

12.8.2.2 Blockchain for Secure Transactions

The system has the feature of decentralization and the ability to perform secure and efficient transactions and exchange of data [67]. Research in the future should concentrate on the approaches which use blockchain to enhance smart city systems security and assurance. Blockchain can be applied to IoT devices to enhance the protection of data and trace its history of transaction.

12.8.2.3 Homomorphic Encryption

Homomorphic encryption is a technique that permits computations on encrypted data without decryption because computing on encrypted data maintains its confidentiality while analysing the results [68]. Further research should examine the use of homomorphic encryption within the smart city to safeguard the data collected and processed while still maintaining usability.

12.8.2.4 Edge Computing Security

Edge computing performs the analysis of data closer to that data source but it creates new threats as many computations are carried out on numerous distributed devices [69]. The research should address the issues that are related to securing the edge computing environment and include secure data aggregation, secure authentication and access control.

12.8.2.5 Collaborative Security Frameworks

Smart cities involve a number of components in the development of smart environment that includes the government, private sectors, and the citizen. It is therefore imperative that enhanced security through collaborative frameworks by sharing information and working together protects from cybercrimes [70]. Further research needs to be conducted with regards to the governance models and communication channels to facilitate cooperation and enhance confidence between stakeholders.

12.8.2.6 Human-Centric Security Approaches

It is true that human beings have escalated the situation as far as cybersecurity is concerned. The focus should therefore be directed toward understanding user behaviour and developing security solutions that are both effective and intuitive, allowing users to engage with them easily and without complexity [52]. This includes creating mechanisms through which these threats are reported in a precise and concise manner as well as ensuring that the users are well aware of solutions that could prevent acts of cyber criminals.

12.8.2.7 Resilient Infrastructure Design

The construction of the resilient physical infrastructure should implement to lower the risks of cyberattacks [52]. Future research should examine the important issue of architectural redundancy and diversity that addresses continuity for the important services during cyber-attacks.

12.8.2.8 Regulatory and Policy Development

There is a need for supportive policies and regulations in the field of cybersecurity of smart cities. Legal research should be dedicated to analysing remaining loopholes in the current legislation regarding cybersecurity and creating the policy that would make it more beneficial to develop new technologies but at the same time would protect the society from cyber threats [43]. Also, part of this includes working with other states to deal with cyberattacks that have no borders.

12.9 Summary

Cybersecurity is fundamental to the development of smart cities that are safe, efficient, and sustainable. Blockchain technology, when combined with machine learning, offers innovative approaches to establishing a robust cybersecurity framework. These technologies not only facilitate secure and decentralized data transactions but also enable predictive threats detection, thereby strengthening the security of smart city systems. A sustainable cybersecurity framework should be integrated, comprehensive, and adaptive, enabling cities to evolve while safeguarding infrastructure in alignment with sustainable development principles. The establishment of global standards, policies, and regulations is essential to create a coherent international cybersecurity framework. Implementing secure systems within smart city environments offers numerous benefits, including increased public trust, operational efficiency, and enhanced quality of life for residents. However, challenges such as privacy concerns, shortage of cybersecurity expertise, and the complexity of coordinating multiple stakeholders internationally continue to pose serious issues. Research is necessary to address these challenges, develop effective strategies, and align cybersecurity efforts with the key objectives of sustainable and urban development. Through these efforts, cities will maximize the advantages of digital transformation while ensuring the protection of citizens and vital infrastructure from evolving cyber threats.

References

1. P. Mishra, G. Singh, Introduction: Importance of sustainable smart city, in *Sustainable Smart Cities*, (Springer, Cham, 2023), pp. 1–37. https://doi.org/10.1007/978-3-031-33354-5_1
2. A. Mishra, A.V. Jha, B. Appasani, A.K. Ray, D.K. Gupta, A.N. Ghazali, Emerging technologies and design aspects of next generation cyber physical system with a smart city application perspective. Int. J. Syst. Assur. Eng. Manag. **14**(S3), 699–721 (2023). https://doi.org/10.1007/s13198-021-01523-y
3. A. Alibasic, R. Al Junaibi, Z. Aung, W.L. Woon, M.A. Omar, Cybersecurity for smart cities: A brief review, in *Data Analytics for Renewable Energy Integration. DARE 2016*, Lecture Notes in Computer Science, vol. 10097, (Springer, Cham, 2017), pp. 22–30. https://doi.org/10.1007/978-3-319-50947-1_3
4. A.C.M. Nafrees, A.M.A. Sujah, C. Mansoor, Smart cities: Emerging technologies and potential solutions to the cyber security threads, in *2021 5th International Conference on Electrical, Electronics, Communication, Computer Technologies and Optimization Techniques (ICEECCOT)*, (IEEE, 2021), pp. 220–228. https://doi.org/10.1109/ICEECCOT52851.2021.9707994
5. P. Mishra, G. Singh, *Sustainable Smart Cities* (Springer International Publishing, Cham, 2023). https://doi.org/10.1007/978-3-031-33354-5
6. P. Mishra, G. Singh, Internet of things for sustainable Smart City, in *Sustainable Smart Cities*, (Springer International Publishing, Cham, 2023), pp. 75–96. https://doi.org/10.1007/978-3-031-33354-5_4

7. R.O. Andrade, S.G. Yoo, L. Tello-Oquendo, I. Ortiz-Garces, A comprehensive study of the IoT cybersecurity in smart cities. IEEE Access **8**, 228922–228941 (2020). https://doi.org/10.1109/ACCESS.2020.3046442
8. J. ThankGod, Public-private partnerships in strengthening cybersecurity for international trade: Examining the role of collaborative efforts between governments and private sector entities in crafting and enforcing robust cybersecurity measures for global trade. SSRN Electron. J. (2024). https://doi.org/10.2139/ssrn.4858776
9. F. Almeida, Prospects of cybersecurity in smart cities. Fut. Inter. **15**(9), 285 (2023). https://doi.org/10.3390/fi15090285
10. P. Mishra, G. Singh, Energy management systems in sustainable smart cities based on the Internet of Energy: A technical review. Energies (Basel) **16**(19), 6903 (2023). https://doi.org/10.3390/en16196903
11. R.O. Andrade, S.G. Yoo, L. Tello-Oquendo, I. Ortiz-Garcés, Cybersecurity, sustainability, and resilience capabilities of a smart city, in *Smart Cities and the UN SDGs*, (Elsevier, 2021), pp. 181–193. https://doi.org/10.1016/B978-0-323-85151-0.00012-9
12. A. Sulich, M. Rutkowska, A. Krawczyk-Jezierska, J. Jezierski, T. Zema, Cybersecurity and sustainable development. Proc. Comput. Sci. **192**, 20–28 (2021). https://doi.org/10.1016/J.PROCS.2021.08.003
13. I.H. Sarker, Machine learning: Algorithms, real-world applications and research directions. SN Comput. Sci. **2**(3), 160 (2021). https://doi.org/10.1007/s42979-021-00592-x
14. A.A. Diro, N. Chilamkurti, Distributed attack detection scheme using deep learning approach for internet of things. Futur. Gener. Comput. Syst. **82**, 761–768 (2018). https://doi.org/10.1016/j.future.2017.08.043
15. D. Minoli, K. Sohraby, B. Occhiogrosso, IoT considerations, requirements, and architectures for smart buildings—Energy optimization and next-generation building management systems. IEEE Internet Things J. **4**(1), 269–283 (2017). https://doi.org/10.1109/JIOT.2017.2647881
16. I.H. Sarker, AI-based modeling: Techniques, applications and research issues towards automation, intelligent and smart systems. SN Comput Sci **3**(2), 158 (2022). https://doi.org/10.1007/s42979-022-01043-x
17. N. Moustafa, A new distributed architecture for evaluating AI-based security systems at the edge: Network TON_IoT datasets. Sustain. Cities Soc. **72**, 102994 (2021). https://doi.org/10.1016/j.scs.2021.102994
18. A.R. Javed et al., Future smart cities: Requirements, emerging technologies, applications, challenges, and future aspects. Cities **129**, 103794 (2022). https://doi.org/10.1016/j.cities.2022.103794
19. K. Kim, J.S. Kim, S. Jeong, J.-H. Park, H.K. Kim, Cybersecurity for autonomous vehicles: Review of attacks and defense. Comput. Secur. **103**, 102150 (2021). https://doi.org/10.1016/j.cose.2020.102150
20. E. Vattapparamban, I. Guvenc, A.I. Yurekli, K. Akkaya, S. Uluagac, Drones for smart cities: Issues in cybersecurity, privacy, and public safety, in *2016 International Wireless Communications and Mobile Computing Conference (IWCMC)*, (IEEE, 2016), pp. 216–221. https://doi.org/10.1109/IWCMC.2016.7577060
21. I. Alrashdi, A. Alqazzaz, E. Aloufi, R. Alharthi, M. Zohdy, H. Ming, AD-IoT: Anomaly detection of IoT cyberattacks in Smart City using machine learning, in *2019 IEEE 9th Annual Computing and Communication Workshop and Conference (CCWC)*, (IEEE, 2019), pp. 0305–0310. https://doi.org/10.1109/CCWC.2019.8666450
22. I. Yaqoob, I.A.T. Hashem, A. Ahmed, S.M.A. Kazmi, C.S. Hong, Internet of things forensics: Recent advances, taxonomy, requirements, and open challenges. Futur. Gener. Comput. Syst. **92**, 265–275 (2019). https://doi.org/10.1016/j.future.2018.09.058
23. S. Borchardt, G.B. Vignola, D. Buscaglia, M. Maroni, *Mapping EU Policies with the 2030 Agenda and SDGs-Fostering Policy Coherence through text-based SDG Mapping*. https://doi.org/10.2760/110687

24. B. Hamid, N. Jhanjhi, M. Humayun, A. Khan, A. Alsayat, Cyber security issues and challenges for smart cities: A survey, in *2019 13th International Conference on Mathematics, Actuarial Science, Computer Science and Statistics (MACS)*, (IEEE, 2019), pp. 1–7. https://doi.org/10.1109/MACS48846.2019.9024768
25. A. Huseinovic, S. Mrdovic, K. Bicakci, S. Uludag, A survey of denial-of-service attacks and solutions in the smart grid. IEEE Access **8**, 177447–177470 (2020). https://doi.org/10.1109/ACCESS.2020.3026923
26. M. Benmalek, Ransomware on cyber-physical systems: Taxonomies, case studies, security gaps, and open challenges. IoT Cyber Phys. Syst. **4**, 186–202 (2024). https://doi.org/10.1016/J.IOTCPS.2023.12.001
27. A. Tandon, A. Nayyar, A comprehensive survey on ransomware attack: A growing havoc cyberthreat. Adv. Intell. Syst. Comput. **839**, 403–420 (2019). https://doi.org/10.1007/978-981-13-1274-8_31
28. M. Lehto, Cyber-attacks against critical infrastructure, in *Cyber Security. Computational Methods in Applied Sciences*, vol. 56, (Springer, Cham, 2022), pp. 3–42. https://doi.org/10.1007/978-3-030-91293-2_1
29. C.-W. Ten, G. Manimaran, C.-C. Liu, Cybersecurity for critical infrastructures: Attack and defense modeling. IEEE Trans. Syst. Man Cybern. Syst. Hum. **40**(4), 853–865 (2010). https://doi.org/10.1109/TSMCA.2010.2048028
30. M. Abomhara, G.M. Kien, Cyber security and the Internet of Things: Vulnerabilities, threats, intruders and attacks. J. Cyber Sec. Mob. **4**(1), 65–88 (2015). https://doi.org/10.13052/jcsm2245-1439.414
31. A.N. Lone, S. Mustajab, M. Alam, A comprehensive study on cybersecurity challenges and opportunities in the IoT world. Secur. Priv. **6**(6) (2023). https://doi.org/10.1002/spy2.318
32. M.T. Masud, M. Keshk, N. Moustafa, I. Linkov, D.K. Emge, Explainable artificial intelligence for resilient security applications in the Internet of Things. IEEE Open J. Commun. Soc. **6**, 2877–2906 (2025). https://doi.org/10.1109/OJCOMS.2024.3413790
33. T. Sobb, B. Turnbull, N. Moustafa, Supply chain 4.0: A survey of cyber security challenges, solutions and future directions. Electronics (Basel) **9**(11), 1864 (2020). https://doi.org/10.3390/electronics9111864
34. B. Hammi, S. Zeadally, J. Nebhen, Security threats, countermeasures, and challenges of digital supply chains. ACM Comput. Surv. **55**(14s), 1–40 (2023). https://doi.org/10.1145/3588999
35. S. Ahmed, M.F. Hossain, M.S. Kaiser, M.B.T. Noor, M. Mahmud, C. Chakraborty, Artificial intelligence and machine learning for ensuring security in smart cities, in *Data-Driven Mining, Learning and Analytics for Secured Smart Cities*, Advanced Sciences and Technologies for Security Applications, (Springer, Cham, 2021), pp. 23–47. https://doi.org/10.1007/978-3-030-72139-8_2
36. P. Mishra, P. Thakur, G. Singh, Sustainable smart city to society 5.0: State-of-the-art and research challenges. SAIEE Afr. Res. J. **113**(4), 152–164 (2022). https://doi.org/10.23919/SAIEE.2022.9945865
37. C. Ma, Smart city and cyber-security; technologies used, leading challenges and future recommendations. Energy Rep. **7**, 7999–8012 (2021). https://doi.org/10.1016/J.EGYR.2021.08.124
38. P.P. Groumpos, Artificial intelligence: Issues, challenges, opportunities and threats, in *Creativity in Intelligent Technologies and Data Science. CIT&DS 2019. Communications in Computer and Information Science*, vol. 1083, (Springer, Cham, 2019), pp. 19–33. https://doi.org/10.1007/978-3-030-29743-5_2
39. S.S. Sefati, R. Craciunescu, B. Arasteh, S. Halunga, O. Fratu, I. Tal, Cybersecurity in a scalable smart city framework using Blockchain and federated learning for Internet of Things (IoT). Smart Cities **7**(5), 2802–2841 (2024). https://doi.org/10.3390/smartcities7050109
40. M. Onifade, J.A. Adebisi, T. Zvarivadza, Recent advances in blockchain technology: Prospects, applications and constraints in the minerals industry. Int. J. Min. Reclam. Environ. **38**(7), 497–533 (2024). https://doi.org/10.1080/17480930.2024.2319453

41. M. Ben Farah et al., A survey on blockchain technology in the maritime industry: Challenges and future perspectives. Futur. Gener. Comput. Syst. **157**, 618–637 (2024). https://doi.org/10.1016/J.FUTURE.2024.03.046
42. K. Ahmad, M. Maabreh, M. Ghaly, K. Khan, J. Qadir, A. Al-Fuqaha, Developing future human-centered smart cities: Critical analysis of smart city security, data management, and ethical challenges. Comput. Sci. Rev. **43**, 100452 (2022). https://doi.org/10.1016/J.COSREV.2021.100452
43. M. Houichi, F. Jaidi, A. Bouhoula, Cyber security within smart cities: a comprehensive study and a novel intrusion detection-based approach. Comput. Mater. Cont. **81**(1), 393–441 (2024). https://doi.org/10.32604/cmc.2024.054007
44. A. Gharaibeh et al., *Smart Cities: A Survey on Data Management, Security, and Enabling Technologies* (Institute of Electrical and Electronics Engineers Inc., 2017). https://doi.org/10.1109/COMST.2017.2736886
45. K.C. Nosina, T. Swarna Latha, Introduction to cybersecurity with AI, ML, and blockchain, in *Next-Generation Cybersecurity. Blockchain Technologies*, (Springer, Singapore, 2024), pp. 1–21. https://doi.org/10.1007/978-981-97-1249-6_1
46. H.M. Moyeenudin, G. Bindu, R. Anandan, Blockchain Networks for Cybersecurity Using Machine-Learning Algorithms, in *Integrating Blockchain and Artificial Intelligence for Industry 4.0 Innovations*, EAI/Springer Innovations in Communication and Computing, (Springer, Cham, 2024), pp. 233–242. https://doi.org/10.1007/978-3-031-35751-0_16
47. K. AL-Dosari, N. Fetais, Risk-management framework and information-security Systems for Small and Medium Enterprises (SMEs): A meta-analysis approach. Electronics (Basel) **12**(17), 3629 (2023). https://doi.org/10.3390/electronics12173629
48. E. Weippl, S. Schrittwieser, Introduction to security and privacy, in *Introduction to Digital Humanism*, (Springer Nature Switzerland, Cham, 2024), pp. 397–414. https://doi.org/10.1007/978-3-031-45304-5_26
49. I. Leroy, I. Zolotaryova, S. Semenov, Impact of critical infrastructure cyber security on the sustainable development of smart cities: Insights from internal specialists and external information security auditors. Sustainability **17**(3), 1188 (2025). https://doi.org/10.3390/su17031188
50. Y. Jararweh, S. Otoum, I. Al Ridhawi, Trustworthy and sustainable smart city services at the edge. Sustain. Cities Soc. **62**, 102394 (2020). https://doi.org/10.1016/J.SCS.2020.102394
51. M. Kianpour, S.J. Kowalski, H. Øverby, Advancing the concept of cybersecurity as a public good. Simul. Model. Pract. Theory **116**, 102493 (2022). https://doi.org/10.1016/J.SIMPAT.2022.102493
52. W.S. Admass, Y.Y. Munaye, A.A. Diro, Cyber security: State of the art, challenges and future directions. Cyber Sec. Appl. **2**, 100031 (2024). https://doi.org/10.1016/J.CSA.2023.100031
53. N. Panteli, B.R. Nthubu, K. Mersinas, Being responsible in cybersecurity: A multi-layered perspective. Inf. Syst. Front. (2025) https://doi.org/10.1007/s10796-025-10588-0
54. S.T. Hossain, T. Yigitcanlar, K. Nguyen, Y. Xu, Cybersecurity in local governments: A systematic review and framework of key challenges. Urban Govern. **5**(1), 1–19 (2025). https://doi.org/10.1016/J.UGJ.2024.12.010
55. J.S. Oliha, P.W. Biu, O.C. Obi, Securing the smart city: A review of cybersecurity challenges and strategies. Open Access Res. J. Multidiscip. Stud. **7**(1), 94–101 (2024). https://doi.org/10.53022/oarjms.2024.7.1.0013
56. A. Chaudhuri, S. Bozkus Kahyaoglu, Cybersecurity assurance in Smart cities: A risk management perspective. EDPACS **67**(4), 1–22 (2023). https://doi.org/10.1080/07366981.2023.2165293
57. E. Estevez, K. Cenci, P. Fillottrani, T. Janowski, Review of international standards and policy guidelines for smart sustainable cities, in *Smart Cities and Smart Governance. Public Administration and Information Technology*, vol. 37, (Springer, Cham, 2021), pp. 69–99. https://doi.org/10.1007/978-3-030-61033-3_4
58. M. Vitunskaite, Y. He, T. Brandstetter, H. Janicke, Smart cities and cyber security: Are we there yet?A comparative study on the role of standards, third party risk management

and security ownership. Comput. Secur. **83**, 313–331 (2019). https://doi.org/10.1016/J.COSE.2019.02.009
59. N. Mohamed, J. Al-Jaroodi, I. Jawhar, Opportunities and challenges of data-driven cybersecurity for smart cities, in *2020 IEEE Systems Security Symposium (SSS)*, (IEEE, 2020), pp. 1–7. https://doi.org/10.1109/SSS47320.2020.9174388
60. K. Achuthan, S. Sankaran, S. Roy, R. Raman, Integrating sustainability into cybersecurity: Insights from machine learning based topic modeling. Discov. Sustain. **6**(1), 44 (2025). https://doi.org/10.1007/s43621-024-00754-w
61. S. Biswas et al., Interoperability benefits and challenges in smart city services: Blockchain as a solution. Electronics (Basel) **12**(4), 1036 (2023). https://doi.org/10.3390/electronics12041036
62. X. Ye, Y. Yan, J. Li, B. Jiang, Privacy and personal data risk governance for generative artificial intelligence: A Chinese perspective. Telecomm. Policy **48**(10), 102851 (2024). https://doi.org/10.1016/j.telpol.2024.102851
63. A. Clim, A. Toma, R.D. Zota, R. Constantinescu, The need for cybersecurity in industrial revolution and smart cities. Sensors (Basel) **23** (2023). https://doi.org/10.3390/s23010120
64. R. Al Sharif, S. Pokharel, Smart city dimensions and associated risks: Review of literature. Sustain. Cities Soc. **77**, 103542 (2022). https://doi.org/10.1016/J.SCS.2021.103542
65. R. Kaur, D. Gabrijelčič, T. Klobučar, Artificial intelligence for cybersecurity: Literature review and future research directions. Inform. Fusion **97**, 101804 (2023). https://doi.org/10.1016/J.INFFUS.2023.101804
66. N. Mohamed, Artificial intelligence and machine learning in cybersecurity: A deep dive into state-of-the-art techniques and future paradigms. Knowl. Inf. Syst. (2025). https://doi.org/10.1007/s10115-025-02429-y
67. T. Kukman, S. Gričar, Blockchain for quality: Advancing security, efficiency, and transparency in financial systems. FinTech **4**(1), 7 (2025). https://doi.org/10.3390/fintech4010007
68. L. Jiang, Z. Xia, X. Sun, Review on privacy-preserving data comparison protocols in cloud computing. Adv. Comput. **120**, 81–119 (2021). https://doi.org/10.1016/bs.adcom.2020.09.002
69. A.M. Sheikh, M.R. Islam, M.H. Habaebi, S.A. Zabidi, A.R. Bin Najeeb, A. Kabbani, A survey on edge computing (EC) security challenges: Classification, threats, and mitigation strategies. Fut. Inter. **17**(4), 175 (2025). https://doi.org/10.3390/fi17040175
70. M.R. Belgaum, Z. Alansari, R. Jain, J. Alshaer, A framework for evaluation of cyber security challenges in smart cities, in *Smart Cities Symposium 2018*, (Institution of Engineering and Technology, 2018), p. 4. https://doi.org/10.1049/cp.2018.1372

Index

P. Mishra, G. Singh, *Sustainable Smart Cities 2.0*,
https://doi.org/10.1007/978-3-032-01102-2

T

U

V

MIX
Papier aus verantwortungsvollen Quellen
Paper from responsible sources
FSC® C105338

If you have any concerns about our products,
you can contact us on
ProductSafety@springernature.com

In case Publisher is established outside the EU,
the EU authorized representative is:
Springer Nature Customer Service Center GmbH
Europaplatz 3, 69115 Heidelberg, Germany

Printed by Libri Plureos GmbH
in Hamburg, Germany